Perspectives on Molecular Materials—A Tribute to Professor Peter Day

Perspectives on Molecular Materials—A Tribute to Professor Peter Day

Editors

Lee Martin
Scott Turner
John Wallis
Hiroki Akutsu
Carlos J. Gómez García

Basel • Beijing • Wuhan • Barcelona • Belgrade • Novi Sad • Cluj • Manchester

Editors

Lee Martin	Scott Turner	John Wallis
School of Science and Technology	Chemistry Department	Department of Chemistry
Nottingham Trent University	University of Surrey	Nottingham Trent University
Nottingham	Guildford	Nottingham
United Kingdom	United Kingdom	United Kingdom

Hiroki Akutsu	Carlos J. Gómez García
Department of Chemistry	Department of Inorganic Chemistry
Osaka University	University of Valencia
Osaka	Burjasot
Japan	Spain

Editorial Office
MDPI
St. Alban-Anlage 66
4052 Basel, Switzerland

This is a reprint of articles from the Special Issue published online in the open access journal *Magnetochemistry* (ISSN 2312-7481) (available at: www.mdpi.com/journal/magnetochemistry/special_issues/tribute_to_peter_day).

For citation purposes, cite each article independently as indicated on the article page online and as indicated below:

Lastname, A.A.; Lastname, B.B. Article Title. *Journal Name* **Year**, *Volume Number*, Page Range.

ISBN 978-3-0365-2771-0 (Hbk)
ISBN 978-3-0365-2770-3 (PDF)
doi.org/10.3390/books978-3-0365-2770-3

Cover image courtesy of Santiago Alvarez Reverter

© 2024 by the authors. Articles in this book are Open Access and distributed under the Creative Commons Attribution (CC BY) license. The book as a whole is distributed by MDPI under the terms and conditions of the Creative Commons Attribution-NonCommercial-NoDerivs (CC BY-NC-ND) license.

Contents

About the Editors ix

Lee Martin, Scott S. Turner, John D. Wallis, Hiroki Akutsu and Carlos J. Gómez-García
Perspectives on Molecular Materials—A Tribute to Professor Peter Day
Reprinted from: *Magnetochemistry* 2021, 7, 152, doi:10.3390/magnetochemistry7120152 1

Nesrine Benamara, Zouaoui Setifi, Chen-I Yang, Sylvain Bernès, David K. Geiger and Güneş Süheyla Kürkçüoğlu et al.
Coexistence of Spin Canting and Metamagnetism in a One-Dimensional Mn(II) Compound Bridged by Alternating Double End-to-End and Double End-On Azido Ligands and the Analog Co(II) Compound [†]
Reprinted from: *Magnetochemistry* 2021, 7, 50, doi:10.3390/magnetochemistry7040050 7

Santiago Alvarez
Peter Day's Exploration of Time and Space
Reprinted from: *Magnetochemistry* 2021, 7, 103, doi:10.3390/magnetochemistry7070103 23

Songjie Yang, Matteo Zecchini, Andrew C. Brooks, Sara J. Krivickas, Desiree M. Dalligos and Anna M. Matuszek et al.
Synthesis of New Derivatives of BEDT-TTF: Installation of Alkyl, Ethynyl, and Metal-Binding Side Chains and Formation of *Tris*(BEDT-TTF) Systems [†]
Reprinted from: *Magnetochemistry* 2021, 7, 110, doi:10.3390/magnetochemistry7080110 29

Noemi Monni, Mariangela Oggianu, Suchithra Ashoka Sahadevan and Maria Laura Mercuri
Redox Activity as a Powerful Strategy to Tune Magnetic and/or Conducting Properties in Benzoquinone-Based Metal-Organic Frameworks
Reprinted from: *Magnetochemistry* 2021, 7, 109, doi:10.3390/magnetochemistry7080109 44

Linh Trinh, Eric Rivière, Sandra Mazerat, Laure Catala and Talal Mallah
Collective Magnetic Behavior of 11 nm Photo-Switchable CsCoFe Prussian Blue Analogue Nanocrystals: Effect of Dilution and Light Intensity
Reprinted from: *Magnetochemistry* 2021, 7, 99, doi:10.3390/magnetochemistry7070099 56

Xinghui Qi, Philippe Guionneau, Enzo Lafon, Solène Perot, Brice Kauffmann and Corine Mathonière
New Photomagnetic Ionic Salts Based on $[Mo^{IV}(CN)_8]^{4-}$ and $[W^{IV}(CN)_8]^{4-}$ Anions [†]
Reprinted from: *Magnetochemistry* 2021, 7, 97, doi:10.3390/magnetochemistry7070097 68

Samia Benmansour and Carlos J. Gómez-García
The Peter Day Series of Magnetic (Super)Conductors [†]
Reprinted from: *Magnetochemistry* 2021, 7, 93, doi:10.3390/magnetochemistry7070093 81

Hiroki Akutsu, Yuta Koyama, Scott S. Turner and Yasuhiro Nakazawa
Structures and Properties of New Organic Molecule-Based Metals, $(D)_2BrC_2H_4SO_3$ [D = BEDT-TTF and BETS]
Reprinted from: *Magnetochemistry* 2021, 7, 91, doi:10.3390/magnetochemistry7070091 127

Toby James Blundell, Michael Brannan, Joey Mburu-Newman, Hiroki Akutsu, Yasuhiro Nakazawa and Shusaku Imajo et al.
First Molecular Superconductor with the Tris(Oxalato)Aluminate Anion, β''-(BEDT-TTF)$_4$(H$_3$O)Al(C$_2$O$_4$)$_3$·C$_6$H$_5$Br, and Isostructural Tris(Oxalato)Cobaltate and Tris(Oxalato)Ruthenate Radical Cation Salts
Reprinted from: *Magnetochemistry* 2021, 7, 90, doi:10.3390/magnetochemistry7070090 140

Nabil Mroweh, Alexandra Bogdan, Flavia Pop, Pascale Auban-Senzier, Nicolas Vanthuyne and Elsa B. Lopes et al.
Chiral Radical Cation Salts of Me-EDT-TTF and DM-EDT-TTF with Octahedral, Linear and Tetrahedral Monoanions [†]
Reprinted from: *Magnetochemistry* **2021**, *7*, 87, doi:10.3390/magnetochemistry7060087 152

Tatiana G. Prokhorova, Eduard B. Yagubskii, Andrey A. Bardin, Vladimir N. Zverev, Gennadiy V. Shilov and Lev I. Buravov
Influence of the Size and Shape of Halopyridines Guest Molecules G on the Crystal Structure and Conducting Properties of Molecular (Super)Conductors of $(BEDT-TTF)_4A^+[M^{3+}(C_2O_4)_3] \cdot G$ Family
Reprinted from: *Magnetochemistry* **2021**, *7*, 83, doi:10.3390/magnetochemistry7060083 168

Bin Zhang, Yan Zhang, Guangcai Chang, Zheming Wang and Daoben Zhu
Crystal-to-Crystal Transformation from $K_2[Co(C_2O_4)_2(H_2O)_2] \cdot H_2O$ to $K_2[Co(\mu-C_2O_4)(C_2O_4)]$
Reprinted from: *Magnetochemistry* **2021**, *7*, 77, doi:10.3390/magnetochemistry7060077 183

Emmelyne Cuza, Samia Benmansour, Nathalie Cosquer, Françoise Conan, Carlos J. Gómez-García and Smail Triki
Solvent-Induced Hysteresis Loop in Anionic Spin Crossover (SCO) Isomorph Complexes
Reprinted from: *Magnetochemistry* **2021**, *7*, 75, doi:10.3390/magnetochemistry7060075 189

Francis L. Pratt, Tatiana Guidi, Pascal Manuel, Christopher E. Anson, Jinkui Tang and Stephen J. Blundell et al.
Neutron Studies of a High Spin Fe_{19} Molecular Nanodisc
Reprinted from: *Magnetochemistry* **2021**, *7*, 74, doi:10.3390/magnetochemistry7060074 203

Scott S. Turner, Joanna Daniell, Hiroki Akutsu, Peter N. Horton, Simon J. Coles and Volker Schünemann
New Spin-Crossover Compounds Containing the [Ni(mnt)] Anion (mnt = Maleonitriledithiolate)
Reprinted from: *Magnetochemistry* **2021**, *7*, 72, doi:10.3390/magnetochemistry7050072 221

Stephen J. Blundell, Tom Lancaster, Peter J. Baker, Francis L. Pratt, Daisuke Shiomi and Kazunobu Sato et al.
The Internal Field in a Ferromagnetic Crystal withChiral Molecular Packing of Achiral Organic Radicals
Reprinted from: *Magnetochemistry* **2021**, *7*, 71, doi:10.3390/magnetochemistry7050071 233

Nadia Marino, María Luisa Calatayud, Marta Orts-Arroyo, Alejandro Pascual-Álvarez, Nicolás Moliner and Miguel Julve et al.
Magnetic Switching in Vapochromic Oxalato-Bridged 2D Copper(II)-Pyrazole Compounds for Biogenic Amine Sensing [†]
Reprinted from: *Magnetochemistry* **2021**, *7*, 65, doi:10.3390/magnetochemistry7050065 243

Boris Tsukerblat, Andrew Palii and Sergey Aldoshin
In Quest of Molecular Materials for Quantum Cellular Automata: Exploration of the Double Exchange in the Two-Mode Vibronic Model of a Dimeric Mixed Valence Cell [†]
Reprinted from: *Magnetochemistry* **2021**, *7*, 66, doi:10.3390/magnetochemistry7050066 259

Tadashi Sugano, Stephen J. Blundell, William Hayes and Hatsumi Mori
Magnetic and Structural Properties of Organic Radicals Based on Thienyl- and Furyl-Substituted Nitronyl Nitroxide
Reprinted from: *Magnetochemistry* **2021**, *7*, 62, doi:10.3390/magnetochemistry7050062 273

Diana Dragancea, Ghenadie Novitchi, Augustin M. Mădălan and Marius Andruh
New Cyanido-Bridged Heterometallic 3d-4f 1D Coordination Polymers: Synthesis, Crystal Structures and Magnetic Properties
Reprinted from: *Magnetochemistry* **2021**, 7, 57, doi:10.3390/magnetochemistry7050057 . . . . . . **283**

Nataliya D. Kushch, Gennady V. Shilov, Lev I. Buravov, Eduard B. Yagubskii, Vladimir N. Zverev and Enric Canadell et al.
New Radical Cation Salts Based on BDH-TTP Donor: Two Stable Molecular Metals with a Magnetic $[ReF_6]^{2-}$ Anion and a Semiconductor with a $[ReO_4]^-$ Anion
Reprinted from: *Magnetochemistry* **2021**, 7, 54, doi:10.3390/magnetochemistry7040054 . . . . . . **297**

About the Editors

Lee Martin

Dr. Lee Martin (Associate Professor in Materials Chemistry) completed his PhD in Peter Day's research group in 1999 with the synthesis and characterization of BEDT-TTF salts containing tris(oxalato)metallate. After working for The International Union of Crystallography and The Royal Society of Chemistry, Dr. Martin returned to the Davy Faraday Research Laboratory at the Royal Institution of Great Britain as a postdoctoral research assistant in Peter Day's group. Dr. Martin took up his first lecturer role in 2007 and has worked at the University of Hertfordshire and Nottingham Trent University. Dr. Martin currently works on chiral molecular materials.

Scott Turner

Dr. Scott Turner (Senior Lecturer in Solid State Chemistry) was awarded a PhD from the University of Manchester, UK, working on the EPR of vanadyl oligomers. He then completed a NATO postdoc with the late Prof. Olivier Kahn at the Université de Paris-Sud and the Institut de Chimie Matière Condensée de Bordeaux (ICMCB). Another post-doc from 1996 at the Royal Institution in London was when he first worked with Prof. Peter Day. Dr. Turner enjoyed working with Prof. Peter so much that he stayed for 8 years, prior to an academic position at the University of Exeter, UK. With Prof. Peter, he worked with BEDT-TTF-based (super)conductors and TTF-containing magnets. Academic posts at the UK Universities of Warwick and Nottingham followed, before he moved to the University of Surrey (2009). Dr. Turner currently works with molecular materials, with an emphasis on spin-transition chemistry. He is continually inspired by the work of Prof. Peter Day.

John Wallis

Prof. John Wallis has been Emeritus Professor of Organic Chemistry at Nottingham Trent University, UK, since 2019, following 20 years as Professor and, before that, 11 years at the University of Kent. His main interest is the synthesis of new organosulfur donors, especially those related to BEDT-TTF, including chiral ones, and the crystalline structures of radical cation salts. He also uses synthesis and X-ray crystallography to study interactions between functional groups as models for chemical reactions.

Hiroki Akutsu

Dr. Hiroki Akutsu (Associate Professor) completed his doctoral studies at the Tokyo University of Science, Japan, working on X-ray analyses of organic charge-transfer salts with Prof. Tokiko Uchida. He then completed a post-doctoral fellowship with Prof. Hayao Kobayashi at the Institute for Molecular Science (IMS), Okazaki, Japan. A second postdoc was at the Microcalorimetry Research Center, Faculty of Science, Osaka University, with Profs. Kazuya Saito and Michio Sorai. He took up his first research associate position in 1999 and worked at the University of Hyogo with Prof. Jun-ichi Yamada and Prof. Shin'ichi Nakatsuji, where he was awarded a grant from the JSPS Researcher Exchange program, enabling him to work at the Royal Institution of Great Britain with Prof. Peter Day from April 2001 to March 2002 with his wife, Dr. Akane Akutsu-Sato, a JSPS Research fellow at the Tokyo Institute of Technology with Prof. Takehiko Mori. The collaboration continued, during which time Peter visited our laboratory in Japan four times, presenting a seminar on each visit. Since 2014, Dr. Akutsu has been an associate professor at Osaka University with Prof. Yasuhiro Nakazawa. He currently works on novel organic conductors with polar organic counterions.

Carlos J. Gómez García

Prof. Carlos J. Gómez García (Professor in Inorganic Chemistry) completed his Ph.D. in 1991 with Prof. E. Coronado at the University of Valencia. In 1993, he performed a first post-doctoral stay at the University of Rennes (France) with Prof. Lahcène Ouahab, and in 1994–1995, he did a second post-doctoral stay with Dr. Pierre Delhaès at the CRPP in Bordeaux (France) with a Human Capital and Mobility grant from the European Union. In 1998, he became an Assistant Professor, and, after a national habilitation in 2006, he became a Full Professor in 2007 at the University of Valencia. Dr. Gómez-García met Prof. Peter Day for the first time at a Molecular Materials School in Spain in 1992. Since then, he has met Prof. Peter Day at countless congresses and conferences around the globe and during the long sabbatical stays that Prof. Peter performed at the University of Valencia. Prof. Peter's work and lessons were, are, and will be a source of inspiration for him.

 *magnetochemistry*

Editorial

Perspectives on Molecular Materials—A Tribute to Professor Peter Day

Lee Martin [1,*], Scott S. Turner [2,*], John D. Wallis [1,*], Hiroki Akutsu [3,*] and Carlos J. Gómez-García [4,*]

1. Department of Chemistry and Forensics, School of Science and Technology, Nottingham Trent University, Clifton Lane, Nottingham NG11 8NS, UK
2. Chemistry Department, Faculty of Engineering and Physical Sciences, University of Surrey, Guildford GU2 7XH, UK
3. Department of Chemistry, Graduate School of Science, Osaka University, Osaka 560-0043, Japan
4. Departamento de Química Inorgánica, Universidad de Valencia, C/Dr. Moliner 50, Burjasot, 46100 Valencia, Spain
* Correspondence: lee.martin@ntu.ac.uk (L.M.); s.s.turner@surrey.ac.uk (S.S.T.); john.wallis@ntu.ac.uk (J.D.W.); akutsu@chem.sci.osaka-u.ac.jp (H.A.); carlos.gomez@uv.es (C.J.G.-G.)

Citation: Martin, L.; Turner, S.S.; Wallis, J.D.; Akutsu, H.; Gómez-García, C.J. Perspectives on Molecular Materials—A Tribute to Professor Peter Day. *Magnetochemistry* **2021**, *7*, 152. https://doi.org/10.3390/magnetochemistry7120152

Received: 11 November 2021
Accepted: 19 November 2021
Published: 23 November 2021

Publisher's Note: MDPI stays neutral with regard to jurisdictional claims in published maps and institutional affiliations.

Copyright: © 2021 by the authors. Licensee MDPI, Basel, Switzerland. This article is an open access article distributed under the terms and conditions of the Creative Commons Attribution (CC BY) license (https://creativecommons.org/licenses/by/4.0/).

Editorial

Professor Peter Day FRS was born on 20 August 1938 in Kent (UK) and attended Maidstone Grammar School.

Peter completed his undergraduate studies at Wadham College, Oxford, followed by a DPhil in chemistry in 1965 under the supervision of Bob Williams. His thesis concerned Light Induced Charge Transfer in Solids. After his DPhil, Peter became a junior research fellow at St John's College, Oxford and later became an official fellow, tutor and university lecturer.

In 1988, Peter became the Director of the Institut Laue-Langevin in Grenoble (France). Peter was Director of the Royal Institution of Great Britain from 1991 to 1998, following illustrious predecessors, such as Humphry Davy (1801), Michael Faraday (1825), John Tyndall (1867) and James Dewar (1887). Peter continued the proud traditions of the Royal Institution with an extensive outreach program popularizing science to both adults and children, including the Royal Institution Christmas Lectures, which were televised each year in the UK and Japan.

At this time, Peter was also the Director of the Davy Faraday Research Laboratory at the RI where internationally leading research in solid-state chemistry produced many high-impact publications. As Fullerian Professor of Chemistry from 1991 to 2008, and subsequently Emeritus Professor at University College London, Peter continued to perform research in materials chemistry.

Peter received many honors and awards, including fellowship of the Royal Society in 1986 for his pioneering research on mixed valence compounds. The Royal Society of Chemistry introduced the Peter Day award in Materials Chemistry, which is awarded each year.

Maybe some of the readers will still remember an anecdote that reflects his very fast mind and his sense of humor. It took place in the last days of September 1995, during the gala dinner of the NATO Advanced Workshop: *Magnetism, a Supramolecular Function*, organized by his good friend Olivier Kahn, in Carcans (France). The dinner took place at the wine cellar Chateau Maucaillou in Medoc, near Bordeaux. At the end of the dinner, after several cups of good wine, Olivier Kahn stood up and proposed a toast suggesting that researchers should only sign our articles when the results are really good; otherwise, we should use a nickname. Peter stood up and asked: "Olivier, which is your true name?"

Peter passed away aged 81 on 19 May 2020.

Peter will be greatly missed by the guest editors of this Special Issue who were privileged to work closely with him. Peter's contributions to materials chemistry are well documented, but his friends and colleagues will remember the kindness of an inspirational mentor and a dear friend.

To pay tribute to his countless key contributions in many different domains in Materials Chemistry, the journal *Magnetochemistry* has decided to publish this memorial Special Issue (SI) entitled *Perspectives on Molecular Materials—A Tribute to Professor Peter Day*, including a total of twenty-one contributions from some of his collaborators and friends.

Among these contributions, there is a personal editorial written by *S. Álvarez*, from the University of Barcelona (Spain) [1], showing that Peter Day was not only an outstanding materials chemist, but he was also passionate about time and space. Prof. Álvarez shows the interest of Peter Day in the history of chemistry and of those institutions he belonged to, as well as in the many places he visited during his more than half a century-long scientific career.

Following chronological order, the first research contribution, by *F. Setifi* et al. at the University Ferhat Abbas of Sétif (Algeria), *J. Reedijk* at Leiden University (The Netherlands), *C. I. Yang* at Tunghai University (Taiwan), *S. Bernés* at Autónoma University of Puebla (Mexico), *D. K. Geiger* at SUNY college at Geneseo (USA) and *G. Süheyla Kürkçüoğlu* at Eskişehir Osmangazi University (Turkey), ref. [2], presents the first Mn(II) chain with alternating double end-to-end and end-on azido bridges showing coexistence of spin-canted antiferromagnetic order (at T_N = 3.4 K) and metamagnetism (with a critical field of ca. 30 mT). The analogous Co(II) compound shows weak ferromagnetic order below 16.2 K.

The contribution by *N. D. Kushch, E. B. Yagubskii* et al. at the Institute of Problems of Chemical Physics (Russia), *V. N. Zverev* at the Institute of Solid-State Physics (Russia), *E. Canadel* at the Institute of Materials Science of Barcelona (Spain) and *J. Yamada* at the University of Hyogo (Japan), ref. [3], shows three new radical salts of a fused bis-TTF donor (BDH-TTP) with the octahedral paramagnetic $[ReF_6]^{2-}$ anion and the tetrahedral diamagnetic one $[ReO_4]^-$. The two salts with the $[ReF_6]^{2-}$ anion are metallic down to low temperatures, whereas the salt with the $[ReO_4]^-$ anion is a semiconductor. No slow relaxation of magnetization was observed in the $[ReF_6]^{2-}$ salts.

The contribution by *D. Dragancea* at the Institute of Chemistry in Chișinău (Moldova), *M. Andruh* et al. at the University of Bucharest (Romania) and *G. Novitchi* at the University Grenoble Alpes (France), ref. [4], reports three new cyano-bridged 3d-4f chain compounds containing alternating Gd-Fe, Dy-Fe and Dy-Co ions. The two Dy-containing compounds show slow relaxation of the magnetization, and the Dy-Co polymer is one the few known examples of chains of Single-Ion Magnets.

The contribution by *T. Sugano* at Meiji Gakuin University (Japan), *H. Mori* at the University of Tokyo (Japan) and *S. J. Blundell* and *W. Hayes* at the University of Oxford (UK), ref. [5], presents the magnetic properties of nitronyl nitroxide (NN) and iminonitroxide (IN) functionalized organic radicals. Among other nitroxyl radicals, they show that the radical 2-benzo[b]thienyl-NN presents two magnetic dimers, one exhibiting ferro- and the other antiferromagnetic intermolecular interactions, in agreement with the crystal structure. They also show that the 4-(2′-thienyl)phenyl-NN radical behaves as an alternating 1D antiferromagnet, also in agreement with its crystal structure.

The contribution by *N. Marino* and *G. De Munno* at the University of Calabria (Italy) and *F. Lloret, M. Julve* et al. at the University of Valencia (Spain), ref. [6], presents a very original novel Cu(II) oxalato-based 2D coordination polymer where the oxalato ligand shows two different coordination modes (bis-bidentate and bi/mono-dentate). This cyan 2D polymer shows an external control of the optical and magnetic properties, since it exhibits fast and selective adsorption of methylamine, which leads to a deep blue adsorbate that further transforms into a third polymer under vacuum. The magnetic properties of the three polymers show a unique switching from strong to weak antiferromagnetic interactions.

The contribution by *B. Tsukerblat* at Ben-Gurion University of the Negev (Israel) with *A. Palii* and *S. Aldoshin* at the Institute of Problems of Chemical Physics (Russia), ref. [7], shows a study of the dimeric molecular mixed-valence cell for quantum cellular automata

(QCA) using the two-mode vibronic model. The authors consider a multielectron mixed-valence dimer of the type d^2-d^1, where the double exchange and the Heisenberg-Dirac-Van Vleck exchange interactions are operative together with the inter-center vibrational modes. The quantum-mechanical calculations show the possibility of combining the function of molecular QCA with that of spin switching in an electronic device and may guide the rational design of such multifunctional molecular electronic devices.

The contribution by *S. J. Blundell* at the University of Oxford, *T. Lancaster* at Durham University (UK), *P. J. Baker* and *F. Pratt* at Rutherford Appleton Laboratory at Oxfordshire (UK) and *K. Sato et al.* at Osaka City University (Japan), ref. [8], presents the unusual crystallization of an achiral organic radical in two chiral enantiomorphs that are mirror images of each other. The study of the magnetic properties of the crystals by muon-spin rotation experiments show the presence of long-range magnetic order below 1.10 K and of two oscillatory components showing different temperature dependences.

The contribution by *S. S. Turner* and *J. Daniell* at the University of Surrey (UK), *H. Akutsu* at Osaka University (Japan), *P. N. Horton* and *S. J. Coles* at the University of Southampton (UK) and *V. Schünemann* at the Technical University of Kaiserslautern (Germany), ref. [9], presents two novel salts with the anion $[Ni(mnt)_2]^-$ (mnt = maleonitriledithiolate) and two cationic Fe(II) complexes with derivatives of 2,6-bis(pyazolyl)pyridine or pyrazine. Both salts were characterized by variable temperature single crystal X-ray diffraction and magnetic measurements. The first salt displays an incomplete and gradual spin crossover up to 300 K and a rapid increase in the high-spin fraction between 300 and 350 K. The second salt shows a gradual and more complete SCO response centered at 250 K, confirmed by variable temperature Mössbauer spectroscopy. In both cases, the anionic moieties are isolated, and no electrical conductivity was observed.

The contribution by *F. Pratt et al.* at Rutherford Appleton Laboratory at Oxfordshire (UK), *C. E. Anson* and *A. K. Powell* at Karlsruhe Institute of Technology (Germany), *J. Tang* at Changchun Institute of Applied Chemistry (China) and *S. J. Blundell* at the University of Oxford (UK), ref. [10], reports a new molecular cluster system with 19 Fe(III) ions arranged in a disc-like structure with a total spin S = 35/2, that behaves as a single molecule magnet with an anisotropy barrier of 16 K. Below 1.2 K, the cluster presents an antiferromagnetic order due to the presence of weak inter-cluster interactions. The authors use neutron diffraction to determine the nature of the magnetic ordering and easy spin axis and inelastic neutron scattering to follow the magnetic order parameters and the magnetic excitations.

The contribution by *S. Triki et al.* at the University of Brest (France) and *S. Benmanosur* and *C. J. Gómez-García* at the University of Valencia (Spain), ref. [11], describes the synthesis and complete characterization of a rare anionic Fe(II) spin crossover (SCO) complex that shows a two-step SCO at around 170 and 298 K, as confirmed by crystallographic and magnetic studies. After complete de-solvation of the complex at around 400 K, the high temperature step shifts to lower temperatures and merges in only one gradual SCO at around 216 K.

The contribution by *B. Zhang et al.* at the Chinese Academy of Sciences at Beijing (China) and *Y. Zhang et al.* at the University of Peking (China), ref. [12], presents a nice example of a crystal-to-crystal transformation of a monomeric anionic oxalato Co(II) complex into an oxalate-bridged zigzag chain via a dehydration process. The monomer shows weak antiferromagnetic interactions mediated by hydrogen bonds, whereas the chain presents an antiferromagnetic ordering at 8.2 K.

The contribution by *T. Prokhorova, E. Yagubskii et al.* at the Institute of Problems of Chemical Physics (Russia), *A. A. Bardin* at the Russian Academy of Sciences at Chernogolovka (Russia) and *V. N. Zverev* at the Institute of Solid-State Physics (Russia), ref. [13], shows the crystal structures and physical properties of new organic (super)conductors of the β"-(BEDT-TTF)$_4$(NH$_4$)[Fe^{3+}(C$_2$O$_4$)$_3$]G family, where BEDT-TTF is bis(ethylenedithio)tetrathiafulvalene and G are different halo-pyridine derivatives. In all cases, the structures show the classical paramagnetic honeycomb anionic layers alternating with conducting layers of BEDT-TTF radical cations. One of the crystals (with 2-fluoropyridine) undergoes a monoclinic (C2/c) to triclinic (P-1) phase transition at 100–150 K, while the remaining salts keep the monoclinic phase in the temperature range 300–100 K. All the salts are metallic conductors and one of them (with 2,6-difluoropyridine) shows the onset of a superconducting transition at 3.1 K. The properties of these salts are compared with those of the known monoclinic phases of the family with different mono-halo-pyridines as solvent molecules.

The contribution by *N. Avarvari et al.* at the University of Angers (France), *P. Auban-Senzier* at the University of Paris-Saclay (France), *N. Vanthuyne* at the CNRS at Marseille (France) and *M. Almeida et al.* at the University of Lisbon (Portugal), ref. [14], presents the synthesis and structural characterization of different series of radical cation salts with the organic precursors for chiral conductors: methyl-ethylenedithio-tetrathiafulvalene (**1**) and dimethyl-ethylenedithio-tetrathiafulvalene (**2**). Thus, the authors report the salts: (**1**)$_2$AsF$_6$, (**1**)I$_3$ and (**2**)I$_3$ as racemic and as (S) and (R) enantiopure forms, the enantiomeric pure forms [(S)-**1**]AsF$_6$·C$_4$H$_8$O and [(R)-**1**]AsF$_6$·C$_4$H$_8$O and also the [(meso)-**2**]PF$_6$ and [(meso)-**2**]XO$_4$ (X = Cl, Re) salts, containing the new donor (meso)-**2**. The crystallographic study shows the presence of a different packing in the latter case, compared to the chiral form, since the two methyl substituents adopt axial and equatorial conformations. The electrical properties show a quasi-metallic conductivity in (**1**)$_2$AsF$_6$ in the high temperature regime, whereas the (meso)-**2** based salts are semiconductors.

The contribution by *L. Martin et al.* at Nottingham Trent University (UK), *H. Akutsu et al.* at Osaka University (Japan) and *S. Imajo* at the University of Tokyo (Japan), ref. [15], reports the synthesis, crystal structure and conducting properties of two new members of the Peter Day series, reported in 1995. These new salts are: the first superconductor with the diamagnetic tris(oxalato)aluminate anion (with a T$_c$ of ≈ 2.5 K) and the first example of a β″ phase with the tris(oxalato)cobaltate anion.

The contribution by *H. Akutsu et al.* at Osaka University (Japan) and *S. S. Turner* at the University of Surrey (UK), ref. [16], reports three novel radical salts prepared with the organic donors bis(ethylenedithio)tetrathiafulvalene (BEDT-TTF) and bis(ethylenedithio) tetraselenafulvalene (BETS) with the organic anion 2-bromoethanesulfonate (BrC$_2$H$_4$SO$_3^-$). The salt with BEDT-TTF presents a double β″ packing, whereas the two BETS salts present a double β″ packing or a θ-type packing in the organic donors. The BEDT-TTF salt shows a metal-insulator transition at ≈ 70 K, but the BETS salts are metallic down to 4.2 K.

The contribution by *C. Mathonière, P. Guionneau et al.* at the University of Bordeaux (France), ref. [17], reports three new ionic salts containing [M(CN)$_8$]$^{4-}$ (M = Mo^{IV} and W^{IV}) and large Cu(II) and Zn(II) complex cations containing a non-conventional motif built with the ligand tris(2-aminoethyl)amine. The three novel compounds show photomagnetic effects when irradiated with blue light at low temperatures, remarkable properties as long-lived photomagnetic metastable states for the [Mo(CN)$_8$]$^{4-}$-based compounds above 200 K and rare efficient photomagnetic properties of the [W(CN)$_8$]$^{4-}$-based compound. The contribution is completed with a comparison of the photomagnetic properties of the three reported compounds with the singlet-triplet conversion recently reported for the K$_4$[Mo(CN)$_8$]·2H$_2$O compound.

The contribution by *T. Mallah, L. Catala et al.* at the University of Paris-Saclay (France), ref. [18], shows the collective magnetic behavior of photo-switchable cyanide-bridged nanoparticles, based on the Prussian blue analogue CsCoFe, when embedded in different matrices with different concentrations. The authors study the effect of the intensity of light irradiation and show that the magnetization and AC magnetic susceptibility data suggest a collective magnetic behavior due to interparticle dipolar magnetic interactions, despite the nanoparticles having a size that places them in the superparamagnetic regime.

The contribution by *J. D. Wallis et al.* at Nottingham Trent University (UK) and *M. Pilkington et al.* at Brock University (Canada), ref. [19], reports the syntheses of new BEDT-TTF derivatives with (i) one ethynyl group (HC≡C-), (ii) two (n-heptyl), (iii) four (n-butyl) alkyl side chains, (iv) two trans acetal (-CH(OMe)$_2$) groups, (v) two trans aminomethyl (-CH$_2$NH$_2$) groups, or (vi) an iminodiacetate (-CH$_2$N(CH$_2$CO$_2^-$)$_2$ side chain. The contribution also reports the synthesis and magnetic properties of three transition metal salts from the latter donor, as well as the synthesis of three tris-donor systems bearing three BEDT-TTF derivatives with ester links to a core derived from benzene-1,3,5-tricarboxylic acid. Authors discuss the stereochemistry and molecular structure of the donors and describe the X-ray crystal structures of two BEDT-TTF donors: one with two CH(OMe)$_2$ groups and one with a -CH$_2$N(CH$_2$CO$_2$Me)$_2$ side chain.

The perspective article by *M. L. Mercuri et al.* at the University of Cagliari (Italy) and *S. A. Sahadevan* at the Royal Institute of Technology at Stockholm (Sweden), ref. [20], presents a review of coordination networks incorporating redox activity to enhance the magnetic, conducting and optical properties of these lattices. The contribution reviews the compounds prepared with quinone derivatives as redox-active linkers, since they have been widely used for electrode materials, flow batteries, pseudo-capacitors and other applications, thanks to the reversible two-electron redox reaction to form hydroquinone dianions via intermediate semiquinone radicals. Furthermore, these quinone linkers can be easily functionalized with different substituents and functional groups, making them excellent building blocks to prepare multifunctional tunable metal-organic frameworks (MOFs). The authors present an overview of the recent advances on benzoquinone-based MOFs, including key examples where magnetic and/or conducting properties are tuned/switched by playing with the redox activity.

Finally, there is a comprehensive review by *S. Benmansour* and *C. J. Gómez-García* at the University of Valencia (Spain), ref. [21], with all the reported series of (super)conducting and magnetic radical salts prepared with organic donors of the tetrathiafulvalene (TTF) family and oxalato-based metal complexes. The review starts with the Peter Day series of magnetic superconductors with the monoclinic β'' packing prepared with the donor bis(ethylenedithio)tetrathiafulvalene (BEDT-TTF) and continues with the orthorhombic pseudo-κ semiconducting polymorphs, also reported with BEDT-TTF for the first time by P. Day *et al.* The exhaustive review shows other series prepared with different oxalate-based complexes, including monomers, such as $[M^{III}(C_2O_4)_3]^{3-}$, $[Ge(C_2O_4)_3]^{2-}$ or $[Cu(C_2O_4)_2]^{2-}$, dimers, such as $[Fe_2(C_2O_4)_5]^{4-}$, trimers, such as $[M^{II}(H_2O)_2[M^{III}(C_2O_4)_3]_2]^{4-}$ and homo- or heterometallic extended 2D layers, such as $[M^{II}M^{III}(C_2O_4)_3]^{-}$ and $[M^{II}_2(C_2O_4)_3]^{2-}$. In addition to the different structural types, the review describes the magnetic properties (dia-, para-, antiferro-, ferromagnetism or long-range magnetic ordering), coexisting with electrical properties (semiconductivity, metallic conductivity or superconductivity) in these salts. The review finishes including the radical salts prepared with oxalate-based complexes and lattices with other organic donors of the TTF-type donors.

As can be seen in the previous list of contributions, Peter Day not only left a long list of key contributions in many different domains in materials chemistry, but also an endless list of collaborators and friends in many different countries. We would like to thank all of them for their high-level contributions in this Special Issue devoted to his memory. We are sure that Peter would have enjoyed reading all of them as much as we will miss him.

Funding: This research received no external funding.

Conflicts of Interest: The authors declare no conflict of interest.

References

1. Alvarez, S. Peter Day's Exploration of Time and Space. *Magnetochemistry* **2021**, *7*, 103. [CrossRef]
2. Benamara, N.; Setifi, Z.; Yang, C.; Bernès, S.; Geiger, D.K.; Kürkçüoğlu, G.S.; Setifi, F.; Reedijk, J. Coexistence of Spin Canting and Metamagnetism in a One-Dimensional Mn(II) Compound Bridged by Alternating Double End-to-End and Double End-on Azido Ligands and the Analog Co(II) Compound. *Magnetochemistry* **2021**, *7*, 50. [CrossRef]
3. Kushch, N.D.; Shilov, G.V.; Buravov, L.I.; Yagubskii, E.B.; Zverev, V.N.; Canadell, E.; Yamada, J. New Radical Cation Salts Based on BDH-TTP Donor: Two Stable Molecular Metals with a Magnetic [ReF$_6$]$^{2-}$ Anion and a Semiconductor with a [ReO$_4$]$^{-}$ Anion. *Magnetochemistry* **2021**, *7*, 54. [CrossRef]
4. Dragancea, D.; Novitchi, G.; Mădălan, A.M.; Andruh, M. New Cyanido-Bridged Heterometallic 3d-4f 1D Coordination Polymers: Synthesis, Crystal Structures and Magnetic Properties. *Magnetochemistry* **2021**, *7*, 57. [CrossRef]
5. Sugano, T.; Blundell, S.J.; Hayes, W.; Mori, H. Magnetic and Structural Properties of Organic Radicals Based on Thienyl- and Furyl-Substituted Nitronyl Nitroxide. *Magnetochemistry* **2021**, *7*, 62. [CrossRef]

6. Marino, N.; Calatayud, M.L.; Orts-Arroyo, M.; Pascual-Álvarez, A.; Moliner, N.; Julve, M.; Lloret, F.; De Munno, G.; Ruiz-García, R.; Castro, I. Magnetic Switching in Vapochromic Oxalato-Bridged 2D Copper(II)-Pyrazole Compounds for Biogenic Amine Sensing. *Magnetochemistry* **2021**, *7*, 65. [CrossRef]
7. Tsukerblat, B.; Palii, A.; Aldoshin, S. In Quest of Molecular Materials for Quantum Cellular Automata: Exploration of the Double Exchange in the Two-Mode Vibronic Model of a Dimeric Mixed Valence Cell. *Magnetochemistry* **2021**, *7*, 66. [CrossRef]
8. Blundell, S.J.; Lancaster, T.; Baker, P.J.; Pratt, F.L.; Shiomi, D.; Sato, K.; Takui, T. The Internal Field in a Ferromagnetic Crystal with Chiral Molecular Packing of Achiral Organic Radicals. *Magnetochemistry* **2021**, *7*, 71. [CrossRef]
9. Turner, S.S.; Daniell, J.; Akutsu, H.; Horton, P.N.; Coles, S.J.; Schünemann, V. New Spin-Crossover Compounds Containing the [Ni(mnt)] Anion (mnt = Maleonitriledithiolate). *Magnetochemistry* **2021**, *7*, 72. [CrossRef]
10. Pratt, F.L.; Guidi, T.; Manuel, P.; Anson, C.E.; Tang, J.; Blundell, S.J.; Powell, A.K. Neutron Studies of a High Spin Fe_{19} Molecular Nanodisc. *Magnetochemistry* **2021**, *7*, 74. [CrossRef]
11. Cuza, E.; Benmansour, S.; Cosquer, N.; Conan, F.; Gómez-García, C.J.; Triki, S. Solvent-Induced Hysteresis Loop in Anionic Spin Crossover (SCO) Isomorph Complexes. *Magnetochemistry* **2021**, *7*, 75. [CrossRef]
12. Zhang, B.; Zhang, Y.; Chang, G.; Wang, Z.; Zhu, D. Crystal-to-Crystal Transformation from $K_2[Co(C_2O_4)_2(H_2O)_2]\cdot 4H_2O$ to $K_2[Co(\mu\text{-}C_2O_4)(C_2O_4)]$. *Magnetochemistry* **2021**, *7*, 77. [CrossRef]
13. Prokhorova, T.G.; Yagubskii, E.B.; Bardin, A.A.; Zverev, V.N.; Shilov, G.V.; Buravov, L.I. Influence of the Size and Shape of Halopyridines Guest Molecules G on the Crystal Structure and Conducting Properties of Molecular (Super)Conductors of $(BEDT\text{-}TTF)_4A^+[M^{3+}(C_2O_4)_3]\cdot G$ Family. *Magnetochemistry* **2021**, *7*, 83. [CrossRef]
14. Mroweh, N.; Bogdan, A.; Pop, F.; Auban-Senzier, P.; Vanthuyne, N.; Lopes, E.B.; Almeida, M.; Avarvari, N. Chiral Radical Cation Salts of Me-EDT-TTF and DM-EDT-TTF with Octahedral, Linear and Tetrahedral Monoanions. *Magnetochemistry* **2021**, *7*, 87. [CrossRef]
15. Blundell, T.J.; Brannan, M.; Mburu-Newman, J.; Akutsu, H.; Nakazawa, Y.; Imajo, S.; Martin, L. First Molecular Superconductor with the Tris(Oxalato)Aluminate Anion, $\beta''\text{-}(BEDT\text{-}TTF)_4(H_3O)Al(C_2O_4)_3\cdot C_6H_5Br$, and Isostructural Tris(Oxalato)Cobaltate and Tris(Oxalato)Ruthenate Radical Cation Salts. *Magnetochemistry* **2021**, *7*, 90. [CrossRef]
16. Akutsu, H.; Koyama, Y.; Turner, S.S.; Nakazawa, Y. Structures and Properties of New Organic Molecule-Based Metals, $(D)_2BrC_2H_4SO_3$ [D = BEDT-TTF and BETS]. *Magnetochemistry* **2021**, *7*, 91. [CrossRef]
17. Qi, X.; Guionneau, P.; Lafon, E.; Perot, S.; Kauffmann, B.; Mathonière, C. New Photomagnetic Ionic Salts Based on $[Mo^{IV}(CN)_8]^{4-}$ and $[W^{IV}(CN)_8]^{4-}$ Anions. *Magnetochemistry* **2021**, *7*, 97. [CrossRef]
18. Trinh, L.; Rivière, E.; Mazerat, S.; Catala, L.; Mallah, T. Collective Magnetic Behavior of 11 nm Photo-Switchable CsCoFe Prussian Blue Analogue Nanocrystals: Effect of Dilution and Light Intensity. *Magnetochemistry* **2021**, *7*, 99. [CrossRef]
19. Yang, S.; Zecchini, M.; Brooks, A.C.; Krivickas, S.J.; Dalligos, D.M.; Matuszek, A.M.; Stares, E.L.; Pilkington, M.; Wallis, J.D. Synthesis of New Derivatives of BEDT-TTF: Installation of Alkyl, Ethynyl, and Metal-Binding Side Chains and Formation of Tris(BEDT-TTF) Systems. *Magnetochemistry* **2021**, *7*, 110. [CrossRef]
20. Monni, N.; Oggianu, M.; Ashoka Sahadevan, S.; Mercuri, M.L. Redox Activity as a Powerful Strategy to Tune Magnetic and/or Conducting Properties in Benzoquinone-Based Metal-Organic Frameworks. *Magnetochemistry* **2021**, *7*, 109. [CrossRef]
21. Benmansour, S.; Gómez-García, C.J. The Peter Day Series of Magnetic (Super)Conductors. *Magnetochemistry* **2021**, *7*, 93. [CrossRef]

Article

Coexistence of Spin Canting and Metamagnetism in a One-Dimensional Mn(II) Compound Bridged by Alternating Double End-to-End and Double End-On Azido Ligands and the Analog Co(II) Compound †

Nesrine Benamara [1], Zouaoui Setifi [1,2], Chen-I Yang [3,*], Sylvain Bernès [4], David K. Geiger [5], Güneş Süheyla Kürkçüoğlu [6], Fatima Setifi [1,*] and Jan Reedijk [7,*]

1. Laboratoire de Chimie, Ingénierie Moléculaire et Nanostructures (LCIMN), Université Ferhat Abbas Sétif 1, Sétif 19000, Algeria; nes_benamara@yahoo.com (N.B.); setifi_zouaoui@yahoo.fr (Z.S.)
2. Département de Technologie, Faculté de Technologie, Université 20 Août 1955-Skikda, Skikda 21000, Algeria
3. Department of Chemistry, Tunghai University, Taichung 407, Taiwan
4. Instituto de Física, Benemérita Universidad Autónoma de Puebla, 72570 Puebla, Mexico; sylvain_bernes@hotmail.com
5. Department of Chemistry, SUNY-College at Geneseo, Geneseo, NY 14387, USA; geiger@geneseo.edu
6. Department of Physics, Eskişehir Osmangazi University, 26040 Eskişehir, Turkey; gkurkcuo@ogu.edu.tr
7. Leiden Institute of Chemistry, Leiden University, P.O. Box 9502, 2300 RA Leiden, The Netherlands
* Correspondence: ciyang@thu.edu.tw (C.-I.Y.); fat_setifi@yahoo.fr (F.S.); reedijk@chem.leidenuniv.nl (J.R.)
† Dedicated to the memory of Prof. Peter Day, who passed away in May 2020.

Citation: Benamara, N.; Setifi, Z.; Yang, C.-I; Bernès, S.; Geiger, D.K.; Kürkçüoğlu, G.S.; Setifi, F.; Reedijk, J. Coexistence of Spin Canting and Metamagnetism in a One-Dimensional Mn(II) Compound Bridged by Alternating Double End-to-End and Double End-On Azido Ligands and the Analog Co(II) Compound . *Magnetochemistry* 2021, 7, 50. https://doi.org/10.3390/magnetochemistry7040050

Academic Editor: Lee Martin

Received: 26 February 2021
Accepted: 30 March 2021
Published: 6 April 2021

Publisher's Note: MDPI stays neutral with regard to jurisdictional claims in published maps and institutional affiliations.

Copyright: © 2021 by the authors. Licensee MDPI, Basel, Switzerland. This article is an open access article distributed under the terms and conditions of the Creative Commons Attribution (CC BY) license (https://creativecommons.org/licenses/by/4.0/).

Abstract: Two new compounds of general formula [M(N$_3$)$_2$(dmbpy)] in which dmbpy = 5,5'-dimethyl-2,2'-bipyridine, and M = Mn(II) or Co(II), have been solvothermally synthesized and characterized structurally and magnetically. The structures consist of zig-zag polymeric chains with alternating bis-μ(azide-N1)$_2$M and bis-μ(azide-N1,N3)$_2$M units in which the cis-octahedrally based coordination geometry is completed by the N,N'-chelating ligand dmbpy. The molecular structures are basically the same for each metal. The Mn(II) compound has a slightly different packing mode compared to the Co(II) compound, resulting from their different space groups. Interestingly, relatively weak interchain interactions are present in both compounds and this originates from π–π stacking between the dmbpy rings. The magnetic properties of both compounds have been investigated down to 2 K. The measurements indicate that the manganese compound shows spin-canted antiferromagnetic ordering with a Néel temperature of T_N = 3.4 K and further, a field-induced magnetic transition of metamagnetism at temperatures below the T_N. This finding affords the first example of an 1D Mn(II) compound with alternating double end-on (EO) and double end-to-end (EE) azido-bridged ligands, showing the coexistence of spin canting and metamagnetism. The cobalt compound shows a weak ferromagnetism resulting from a spin-canted antiferromagnetism and long-range magnetic ordering with a critical temperature, T_C = 16.2 K.

Keywords: azide; chain compounds; ferromagnetism; antiferromagnetism; metamagnetism; spin canting

1. Introduction

Coordination compounds with azide ligand have been studied for decades, not only for their potential use as detonation agents or explosives [1–5], but also because of the intrinsic properties of N$_3^-$ as a "pseudo halogen" [6,7] and as a subject of magnetism [8]. Many studies of stable azide coordination compounds have been reported, and the Cambridge Structural Database (2020 release) contains over 5000 items having at least one coordinated azide ligand [9].

Azide ligands can bind monodentately to metal ions in an end-on mode (N1), or they can bridge between two or more metal ions. Bridging azide between two metals can take place using one terminal nitrogen only (i.e., N1,N1, known as *end-on* mode, abbreviated in this manuscript as EO), or two terminal nitrogens (i.e., N1,N3 or *end-to-end* mode, abbreviated as EE). In the bridging mode, dinuclear compounds can be formed, like in M(azide)$_2$M species [10], as depicted in Figure 1, but also polynuclear compounds, including polymeric linear or zig-zag species of formula $\cdots$(azide)$_2$M(azide)$_2$M(azide)$_2$ $\cdots$. Mixed species with both terminal (non-bridging) and bridging azide are known in the literature [8,11–14], but are not so common.

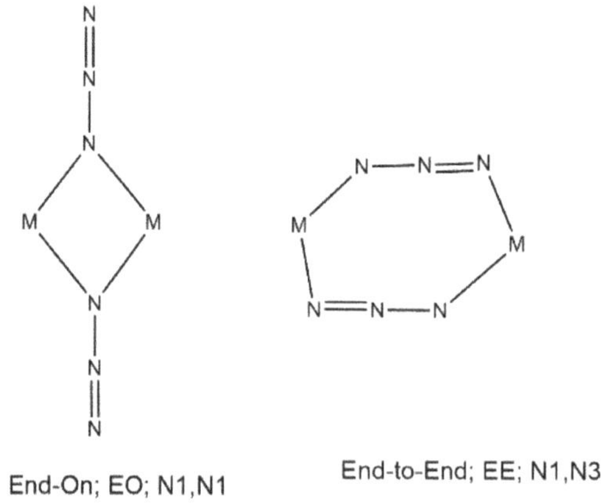

Figure 1. Schematic representation of the binding modes for bridging azido ligands between two metal ions M.

In earlier reports from our laboratories, we have given attention to metal–azide compounds as bridging ligands for dinuclear compounds, or higher aggregates up to 2D networks [15–18] and we have found both ferromagnetic and antiferromagnetic cases. The ferromagnetic cases are only known for the EO azido-bridged compounds. It appears that the choice of the non-bridging co-ligands plays a major role on the formation of specific compounds and structures, and also whether EO, EE or a combination is found. As the azido ligands are rather small or narrow, to generate relative stable coordination spheres around the metal, the co-ligands should be rather bulky, as shown from the literature examples above.

In the present paper, we report on two new compounds that have alternating modes (EE and EO) of bridging azide, forming zig-zag chains in the solid state. The compounds have interesting magnetic properties and these are studied in detail using magnetic susceptibility and magnetization studies at low-temperature (down to 2 K). We feel that the study is relevant in the search for potentially cheap, stable and useful new magnets.

2. Results and Discussion

2.1. Synthetic Efforts

Both title compounds were easily synthesized by the hydrothermal method and the synthesis was found to be reproducible. Despite several attempts, no pure crystalline materials could be isolated for the similar Fe(II) compound. In most attempts, two crystalline forms could be obtained, one isomorphous with the Mn(II) compound and the other isomorphous with the Co(II) compound. Therefore, this compound was not considered suitable for the study of the magnetic properties. In contrast, pure bulk samples for the Mn(II) and Co(II) compounds were available, as evidenced by powder diffraction data, which are consistent with patterns simulated from single-crystal data (Figure S1).

2.2. Characterization of the Compounds by IR and Elemental Analysis

Both compounds display very similar infrared spectra (Figure S2), where bands resembling the free dmbpy ligand are easily recognized. Most characteristic is the broad strong band (doublet) near 2087 cm^{-1}, typical for the coordinated bridged azide ligand [19–21]. The elemental analyses of both compounds are in full agreement with the values calculated, and with the 3D structures presented below, indicating that no secondary phases were obtained with the used synthetic methodology.

2.3. Structure Description of the Compounds

Single crystal structures were determined for both synthesized compounds (**1** and **2**) [$M(N_3)_2$(dmbpy)], with M = Mn(II) and Co(II) and dmbpy = 5,5'-dimethyl-2,2'-bipyridine ($C_{12}H_{12}N_2$). The Mn(II) compound (**1**) crystallizes in space group P-1, while the Co(II) compound (**2**) crystallizes in space group $P2_1/c$. Even though the space groups are different, both compounds share the same molecular structure (see Figure 2 for the Mn(II) compound). The metal center is coordinated by the bidentate dmbpy ligand and two azido anions. Since both pseudohalides, azido N_3^- ligands, bridge between symmetry-related metal centers in the crystal, a one-dimensional polymeric structure is formed, in which each azido anion has a different function: two anions N3/N4/N5 form a centrosymmetric double-bridge, with the µ-1,1 mode of coordination (EO). Two other anions, N6/N7/N8, also form a centrosymmetric double-bridge, but this time with the µ-1,3 coordination mode (EE).

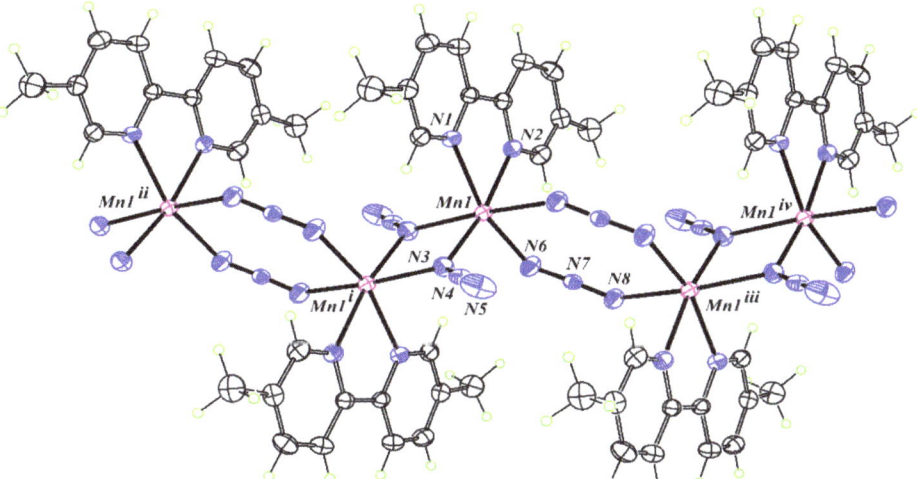

Figure 2. Part of the polymeric structure of *catena*-poly-[Mn(N$_3$)$_2$(dmbpy)] with displacement ellipsoids for non-H atoms at the 30% probability level. Mn and N atoms belonging to the asymmetric unit are labelled, as well as symmetry-related metallic centers along the chain. Symmetry codes: (**i**) 1 − x, 1 − y, 1 − z; (**ii**) −1 + x, y, z; (**iii**) 2 − x, 1 − y, 1 − z; (**iv**) 1 + x, y, z.

Both coordination modes alternate along the zig-zag polymeric chains, which run parallel to the a-axis regardless of the crystal system, triclinic (M = Mn) or monoclinic (M = Co). Relevant coordination bond lengths and angles are given in Table 1. The trend of the well-known smaller ionic radii going from Mn(II) to Co(II) is clearly visible.

Table 1. Selected bond lengths (in Å) and angles (deg.) for both compounds.

Compounds	[Mn(N$_3$)$_2$(dmbpy)]	[Co(N$_3$)$_2$(dmbpy)]
M–N1	2.272(3)	2.115(5)
M–N2	2.256(4)	2.133(5)
M–N3	2.215(4)	2.144(5)
M–N3	2.222(3) [a]	2.198(6) [c]
M–N6	2.208(3)	2.140(6)
M–N8	2.263(3) [b]	2.204(6) [d]
N1–M–N2	72.26(11)	76.7(2)
N3–M–N6	99.50(15)	95.2(2)
N1–M–N3	95.59(12) [a]	94.1(2)
N1–M–N6	162.97(15)	170.6(2)
N2–M–N6	91.83(14)	94.3(2)
N3–M–N3	79.08(12) [a]	78.6(2) [c]
N6–M–N8	90.56(12) [b]	88.8(2) [d]
N2–M–N3	168.67(12)	167.5(2)

Symmetry for the last N atom: [a] $1-x, 1-y, 1-z$; [b] $2-x, 1-y, 1-z$; [c] $-x, -y, 1-z$; [d] $1-x, -y, 1-z$.

Along the chain, long and short $M\cdots M$ separations alternate, 5.430(1) and 3.422(1) Å for M = Mn and 3.361(2) and 5.317(2) Å for M = Co. These distances are slightly shorter for the Co(II) compound as a result of the smaller ionic radii of the cations. The polynuclear zig-zag chains are packed efficiently in the crystal. With M = Mn, chains are parallel, and the dmbpy ligands of two neighboring chains are stacked in such a way that they interact with a rather short distance, 3.440 Å. For M = Co, the arrangement of chains in the crystal is slightly modified, because of the monoclinic cell symmetry. However, the relative position of two neighboring chains is essentially preserved, and the π–π interactions between the dmbpy ligands are even strengthened, with separations between ligand mean planes of 3.283 Å (Co). In Figure 3, both different packings are depicted.

In [Mn(N$_3$)$_2$(dmbpy)], the N–N bond lengths are symmetrical in the EE azido ligands but are asymmetric in the EO ligands. The bite angle exhibited by the dmbpy ligand (72.26(11)°) is the largest distortion in the geometry of the *cis*-octahedral coordination sphere. The four-membered Mn$_2$(EO-N$_3$)$_2$ ring is planar as a result of the inversion center. The eight-membered Mn$_2$(EE-N$_3$)$_2$ ring adopts a chair configuration. The dihedral angle δ, defined by the N6/Mn1/N8 plane and the (EE-N$_3$)$_2$ plane, is 9.69(3)° and Mn1 sits 0.266(7) Å out of the (EE-N$_3$)$_2$ plane. The Mn–azido–Mn torsion angle τ, defined by the dihedral angle between the mean planes of Mn1-N6-N7-N8 and Mn1-N8-N7-N6, is 20.3(6)°. Similarly, in [Co(N$_3$)$_2$(dmbpy)], the N–N bond lengths are approximately symmetrical in the EE azido ligands but are asymmetric in the EO ligands. The bite angle exhibited by the dmbpy ligand (76.7(2)°) is the largest distortion in the geometry of the *cis*-octahedral coordination sphere. The four-membered Co$_2$(EO-N$_3$)$_2$ ring is planar as a result of the inversion center. The eight-membered Co$_2$(EE-N$_3$)$_2$ ring adopts a chair configuration. The dihedral angle δ, defined by the N6/Co1/N8 plane and the (EE-N$_3$)$_2$ plane, is 25.2(4)°. Co1 sits 0.66(1) Å out of the (EE-N$_3$)$_2$ plane. The Co–azido–Co torsion angle τ, defined by the dihedral angle between the mean planes of Co1-N6-N7-N8 and Co1-N8-N7-N6, is 47.7(6)°.

The azide ligand is well known for its versatility in coordination behavior, as explained in the introduction. When involved in metal-to-metal bridges, coordination modes EO and EE are frequent; however, the EO mode is roughly ten times more common than the EE mode, based on a survey of the Cambridge Structural Database [9]. Toggling EO and EE modes along a 1D polymeric structure is rare, but not unprecedented (see Table S1). Indeed, quite similar azido-bridged structures have been described using other ancillary ligands and a variety of transition metals: Schiff bases and Mn(II) [22,23] pyridine derivatives and Mn(II) [24,25], Co(II), Ni(II) [14] or Zn(II) [21] amine/pyridine derivatives and Ni(II) [11,26,27], among others. The nearest structurally related compound is certainly [Mn(N$_3$)$_2$(2,2′-bipyridine)], which crystallizes in space group *P*-1, with unit–cell parameters quite close to those of [Mn(N$_3$)$_2$(dmbpy)] [13,28] and in fact is even isostructural with the Fe(II) and Co(II) analogues [29].

Compounds based on azido-EE/EO double bridges are of interest in the field of magnetochemistry, because the type of interaction between magnetic centers is not unexpected. Therefore, it was decided to perform a detailed magnetic analysis down to very low temperatures. The results are described in the next section.

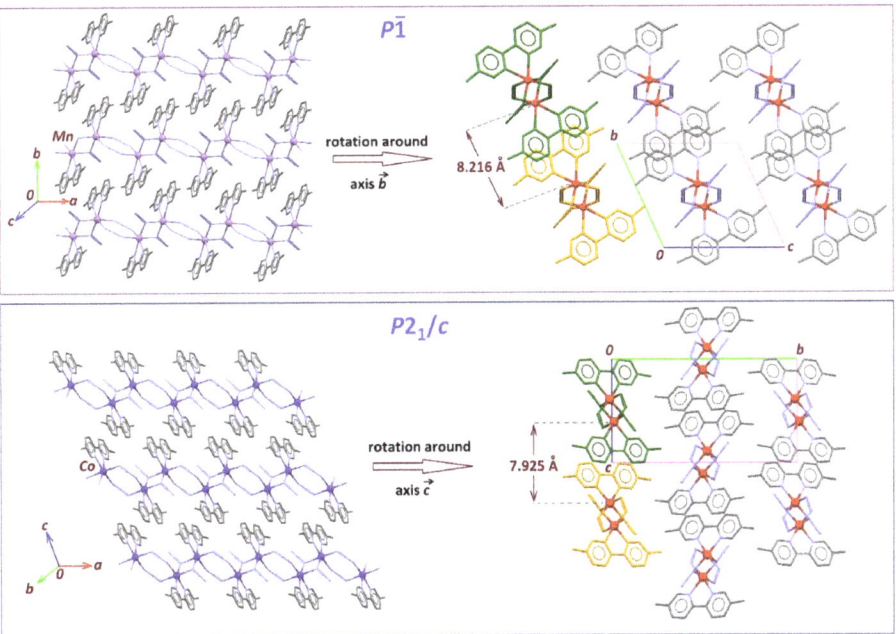

Figure 3. Projection of the packing structures for compounds **1** (top) and **2** (bottom). On the figures on the left side, crystal structures are viewed in a plane where 1D chains are parallel, emphasizing π–π interactions between aromatic rings. By rotation around a crystallographic axis, crystal structures are viewed down the polymerization direction (right side), showing the arrangement of the chains in the crystal. Green and gold parts are different parallel chains, and shortest interchain metal–metal (M–M) distances are quoted in both crystals. Closest M–M contacts are greater than 8.2 Å for **1** and greater than 7.9 Å for **2**.

2.4. Magnetic Properties

For compounds with two different azide binding modes, one can expect interesting magnetic properties, as metal–metal exchange can occur via two different pathways. It is clear that the EE bridges, with their large metal···metal separations, will always be involved in antiferromagnetic coupling (AF) [8]. The AF interaction is increased if the eight-membered metallacycle $M(N_3)_2M'$ is essentially planar. In contrast, EO bridges may promote ferromagnetism (F), provided the M–N–M angle is less than 108° [8]. Regarding the AF component, the triclinic compound (M = Mn) is expected to display a similar or even greater AF interaction than the monoclinic compounds (M = Co), since the former has an EE bridge almost flat, while the bridges in the latter have a butterfly conformation: the dihedral angle δ between the (N$_3$)$_2$ mean plane and the plane formed by M and the bonded N$_{azido}$ atoms is δ = 9.69(3)° for M = Mn, and δ = 25.2(4)° for M = Co. Regarding the F component, both compounds fulfil the requirement for potentially having $J > 0$. The observed angles at EO bridges are 100.92(12) and 101.4(2)° for M = Mn, Co, respectively.

On the other hand, the six-coordinated metal ions used in this work are located in a slightly distorted octahedral ligand field, and their electronic configurations allow for either a low, or a high-spin state, depending on the crystal field splitting Δ_o for $3d$ orbitals. In the present case, the spectrochemical series indicates that N_3^- is a relatively weak ligand, while dmbpy is a relatively strong ligand. There is thus a competition between π-donation and π-backdonation in the coordination sphere, which makes the ground spin state not straightforward to anticipate. However, crystal structures are helpful in this regard, because Δ_o is also related to the strength of any Jahn–Teller (JT) effect in a crystal field with octahedral symmetry. Assuming a small tetragonal distortion, a symmetry measure accounting for the octahedral character of the field can be computed as $S(O_h) = 5.39\Delta^2 - 0.33|\Delta|$, where Δ is the difference between the largest and the shortest coordination bond lengths. Structures with $S(O_h) < 4.42$ are closer to the octahedron than to the trigonal prism geometry [30].

For the Mn(II) compound, $S_{Mn}(O_h) \sim 10^{-3}$, reflecting a very small departure from the ideal O_h symmetry, which, in turn, confirms that the JT distortion indeed is not present, consistent with a high-spin configuration. For the Co(II) compound, the JT distortion is much more noticeable, with $S_{Co}(O_h) \sim 13 \times 10^{-3}$, in line with weak JT distortions, as expected for high-spin d^7 ions; (in particular, a low-spin configuration for Co^{2+}, with an odd number of electrons in the e_g orbitals would give rise to a strong JT effect, associated to a symmetry measure of the field $S(O_h) \gg 10^{-2}$).

It is evident that the crystal structures for the newly synthesized compounds allow one to predict the ground spin state for each one, $S = 5/2$ and $S = 3/2$ for M = Mn(II) and Co(II), respectively. On the other hand, both antiferromagnetic and ferromagnetic interactions should alternate along the 1D chains. Structural features obtained from X-ray structures are, however, not enough to confidently assess the balance between F and AF interactions in these materials, and a comprehensive experimental study of magnetic susceptibility was thus warranted.

Mn Compound (1). The temperature dependences of χ_M and $\chi_M T$ are depicted in Figure 4. At 300 K, the $\chi_M T$ value per Mn(II) ion of compound **1** is 3.80 cm^3 mol^{-1} K, which is lower than the spin-only value of 4.38 cm^3 mol^{-1} K expected for a magnetically isolated octahedral high-spin Mn(II) ion with g = 2.00. Upon cooling, the $\chi_M T$ values decrease gradually and show the cusp around 3.5 K, with a $\chi_M T$ value of 0.317 cm^3 mol^{-1} K at 3.5 K, decreasing to a value of 0.194 cm^3 mol^{-1} K at 2.0 K. The monotonic $\chi_M T$ decrease at a high temperature is indicative of the existence of antiferromagnetic coupling. Upon cooling, the χ_M increases smoothly from 0.013 cm^3 mol^{-1} at 300 K to reach a plateau of 0.035 cm^3 mol^{-1} at about 14.6 K, and then increases rapidly, reaching a maximum of 0.098 cm^3 mol^{-1} at 2.5 K, before a slight decrease to a value of 0.096 cm^3 mol^{-1} at 2.0 K. The sharp increase in χ_M at a low temperature is likely due to a small amount of paramagnetic impurities, e.g., at crystal edges and vacancies along the plane, as is known for related cases [31,32], or perhaps to some features of uncompensated spin. The temperature dependence of $1/\chi_M$ at temperatures above 110 K can be fitted by the Curie–Weiss law with C = 4.64(4) cm^3 mol^{-1} K and θ = −64.7(6) K (Figure S3). The negative Weiss constant suggests the presence of overall antiferromagnetic coupling between the adjacent Mn(II) ions.

To try to obtain intrachain magnetic couplings between Mn(II) ions through the double EO–N_3 and double EE–N_3 bridges, the magnetic susceptibility of compound **1** was fitted using the expression proposed by Cortés [13,28], for alternating chains of classical spins on the Hamiltonian $H = -J_1\Sigma S_{2i}S_{2i+1} - J_2\Sigma S_{2i+1}S_{2i+2}$.

$$\chi_M = [Ng^2\beta^2 S(S+1)/3kT][(1 + u_1 + u_2 + u_1u_2)/(1 - u_1u_2)] \quad (1)$$

where $u_i = \coth[J_i S(S+1)/kT] - kT/[J_i S(S+1)]$ (i = 1 and 2) with $S = 5/2$, J_1 and J_2 are the F and AF exchange constants through double EO–N_3 and double EE–N_3 superexchange pathways, respectively.

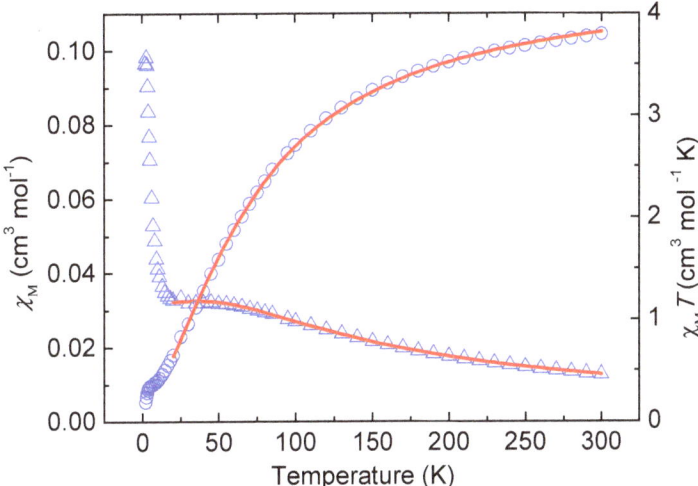

Figure 4. The temperature dependences of χ_M (triangles) and $\chi_M T$ (circles) of compound **1**. The solid line represents the best fit given in the text.

The data above 20 K were fitted and the best fit to experimental data led to J_1 = 2.87(7) cm^{-1}, and J_2 = −11.3(9) cm^{-1} with a fixed g = 2.0. The J_1 and J_2 parameters are consistent with the reported results for related compounds. As in the earlier theoretical and experimental studies, the magnetic couplings between Mn(II) ions via the double EO–N$_3$ bridge were usually ferromagnetic due to small Mn–N–Mn angles which would lead the orthogonality of the magnetic orbitals of the adjacent Mn(II) centers. In contrast, the antiferromagnetic couplings were usually dominated via the EE–N$_3$ bridge due to the well overlap of magnetic orbitals of the adjacent Mn(II) centers. The ferromagnetic interaction decreases with the increasing Mn–N$_{azide}$–Mn angle, while the antferromagnetic couplings via double EE–N$_3$ bridges were dependent on the δ angle, the dihedral angle between the N$_{azide}$–Mn–N$_{azide}$ plane and the plane defined by the two azido bridges, the magnetic coupling decreases with the increasing δ angles [13]. The obtained magnetic coupling of two Mn(II) ions is ferromagnetic for a double EO–N$_3$ bridging with a small Mn–N$_{azide}$–Mn angle of 100.92(12)°, and is antiferromagnetic for a double EE–N$_3$ bridging with a small δ angle, 9.69(3)°, which are consistent with the reported results for related compounds [13,22,23,33–38]. In Table S1, a detailed overview of the literature is given with structural details and J values for a variety of Mn compounds, including the present two new compounds.

Furthermore, the temperature dependences of $\chi_M T$ for compound **1** under 100 Oe were also collected (see Figure S4), showing similar behavior to that under 1000 Oe except for that in the low-temperature range, in which the $\chi_M T$ value slightly increases with decreasing temperature below 6.0 K to a maximum of 0.381 cm^3 mol^{-1} K at 4.5 K and then sharply decreases to 0.076 cm^3 mol^{-1} K at 2.0 K. This suggests that a possible mechanism involving weak ferromagnetic correlations is operative within compound **1** below 6.0 K and the final decrease may be due to antiferromagnetic interactions between the chains and/or saturation effects. These weak ferromagnetic correlations can be attributed to spin canting, i.e., the antiferromagnetically coupled local spins within the –Mn–(EE-N$_3$)$_2$–Mn–EO-N$_3$)$_2$–chains are not perfectly antiparallel, but are canted with respect to each other, resulting in uncompensated residual spins [13,22,23,33–37].

To further characterize the low-temperature magnetic behavior of compound **1**, ZFC/FC magnetization measurements under a field of 10 Oe were carried out. As shown in Figure S5, the ZFC/FC magnetizations were found to be non-bifurcated and show a sharp maximum at 3.4 K, suggesting the occurrence of antiferromagnetic ordering. The temperature dependence of the AC susceptibility of compound **1** was also measured at $H_{dc} = 0$ Oe and $H_{ac} = 3.5$ Oe at different frequencies (Figure S6), which shows the sharp frequency-independent value of the $\chi_M{'}$ signals with the peak maximum at 3.4 K. The absence of $\chi_M{''}$ signals confirms the onset of antiferromagnetic ordering with a conventional $T_N = 3.4$ K and implies the existence of a magnetic phase transition.

The isothermal field dependence data of the magnetization of compound **1** was collected at 1.8 K (Figure 5), in which the magnetization shows a sigmoidal shape with an abrupt increase at a field above ~0.3 kOe to reach a value of 0.36 $N\beta$ at 70 kOe. This sigmoidal magnetization clearly indicates a field-induced magnetic transition of metamagnetic nature [39–41]. In this metamagnetic transition, the net moments of spin-canting Mn–N$_3$ chains aligned antiparallel under a weak applied field by weak interchain antiferromagnetic interactions are overcome by a stronger external field and result in the state transition from antiferromagnetic (AF) to paramagnetic (P). The critical field of magnetic transition, H_C, at 1.8 K, was estimated to be about 0.48 kOe as determined by dM/dH (Figure 5, inset). The M value of 0.36 $N\beta$ at 70 kOe is far below the expected saturation value of 5.0 $N\beta$ for an isotropic high-spin Mn(II) system, confirming the antiferromagnetic nature of **1**. Moreover, at 1.8 K, a small butterfly-shaped magnetic hysteresis loop was obtained, indicating a soft magnetic behavior (see Figure S7). The spin canting angle was estimated to be about $\alpha = 0.30°$, based on the equation $\sin(\alpha) = M_R/M_S$ ($M_R = 0.026$ $N\beta$; obtained by extrapolating the high-field linear part of the magnetization curve at 1.8 K to zero field, and $M_S = 5.0$ $N\beta$) [42–45].

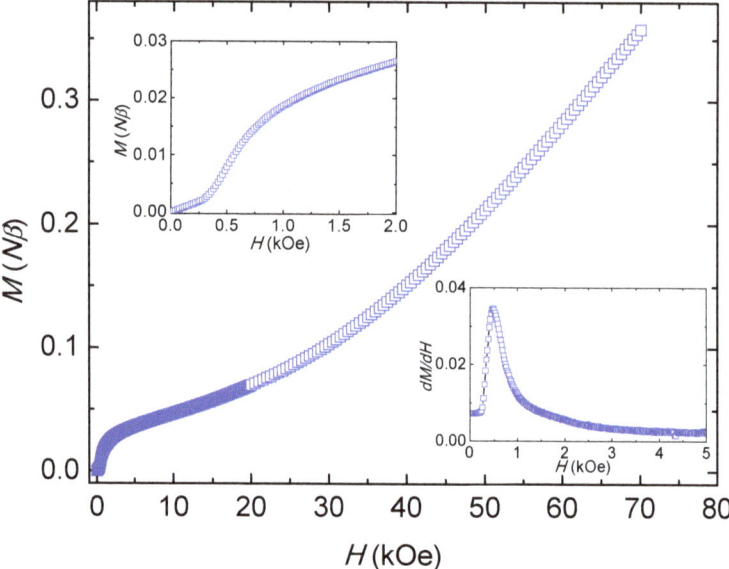

Figure 5. Isothermal magnetization of compound **1** at 1.8 K. The insets give the low field range and the derivative of M vs. H.

Generally speaking, spin canting can arise from two contributions: (i) the presence of an antisymmetric exchange Dzyaloshinsky–Moriya interaction [46] and (ii) the existence of single-ion magnetic anisotropy [47–52]. The presence of an inversion center between adjacent spin centers can result in the disappearance of the antisymmetric exchange. Hence, the lack of antisymmetric exchange in compound **1** would be expected, due to the existence of a crystallographic inversion center between the Mn(II) ions (see symmetry codes in caption of Figure 2). However, the Mn1 center in **1** displays a distorted octahedral geometry due to the small bite angle of 72.26(11)° of the dmbpy chelating ligand. Such distorted octahedral geometries were reported as originating from the weak single-ion anisotropy of the high-spin Mn(II) site [53–57]. Therefore, it is assumed that the spin-canted antiferromagnetism in **1** can be attributed to weak single-ion magnetic anisotropy by a distorted metal coordination environment at low temperature. Similar spin canting behavior has been observed in a few other Mn(II) compounds containing chains of alternating double EO and double EE bridging modes of azides [34,58].

The field-induced magnetic phase transition for compound **1** was further investigated by the measurements of various fields of the FC magnetic susceptibilities, $\chi_M(T)$, and the field dependence of the magnetizations, $M(H)$, at different temperatures. As shown in Figure S8, the maximum of $\chi_M(T)$ shifts to lower temperatures with increasing applied field, until the $\chi_M(T)$ reaches a plateau at a field larger than 600 Oe, confirming that the weak interchain antiferromagnetic interaction is overcome by a stronger external field. As shown in Figure S9, at 2.0 K, the stepwise $M(H)$ curve clearly indicates a field-induced magnetic transition from AF to P. This stepwise magnetization becomes less pronounced with increasing temperature, and the differentials of these curves show peaks that shift to lower fields with increasing temperature (see Figure S10), indicating the phase transitions of metamagnetism.

Combining $M(H)$, FCM and the frequency-independent χ_M' data, the magnetic phase (T, H) diagram has been plotted in Figure 6. The value of H_C decreases with increasing temperature and finally disappears at about 3.4 K. The solid line of $H_C(T)$ in Figure 6, on an analysis of the M–H curves, signifies a typical magnetic transition from AF to P corresponding to metamagnetic materials.

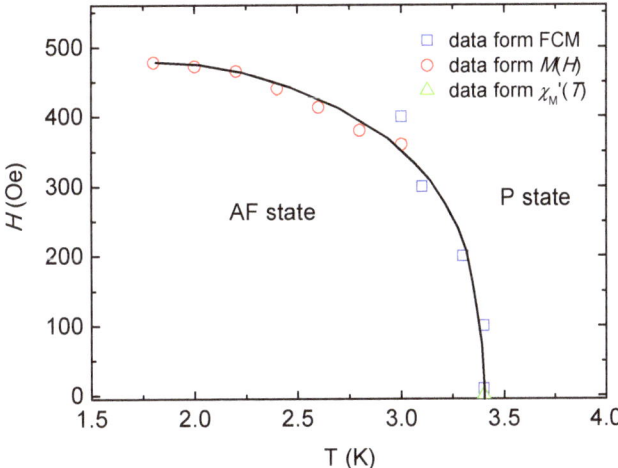

Figure 6. Magnetic phase (T, H) diagram for **1**, build up using the location of the maximum of χ_M vs. T data (open blue square), location of the maximum from dM/dH vs. H data (open red circle) and AC data (open green triangle); the solid line is a guide.

Co compound (2). The temperature dependences of χ_M and $\chi_M T$ of compound **2** are shown in Figure S11 and Figure 7, respectively. As the temperature decreases from 300 K, the χ_M value increases smoothly, reaching a rounded maximum of 0.021 cm^3 mol^{-1} at about 70 K, and then decreases slightly reaching a value of 0.015 cm^3 mol^{-1} at 20 K. Upon further cooling, the χ_M value increases rapidly to a sharp maximum of 0.017 cm^3 mol^{-1} at 12.8 K, after slightly decreasing, χ_M value increases again to 0.017 cm^3 mol^{-1} at 2.0 K. The temperature dependence of $1/\chi_M$ at temperatures above 100 K has been fitted by the Curie–Weiss law with a Curie constant C = 4.85 cm^3 mol^{-1} K and a Weiss constant $\theta = -130.5$ K (see Figure S12).

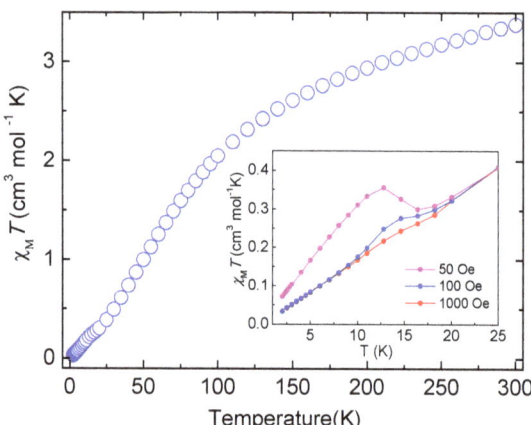

Figure 7. Plot of $\chi_M T$ vs. T of compound **2** in an applied field of 1 kOe from 2 to 300 K. The insets give the $\chi_M T$ vs. T at low temperature measured under the indicated external fields.

The large negative Weiss constant suggests the presence of strong spin-orbital coupling and/or overall antiferromagnetic interactions between the adjacent Co(II) ions. As shown in Figure 7, at 300 K, the $\chi_M T$ value per Co(II) of compound **2** is 3.38 cm^3 mol^{-1} K, which is larger than the spin-only value of 1.87 cm^3 mol^{-1} K for a magnetically isolated octahedral Co(II) ion ($S = 3/2$), with g = 2.00. Upon cooling, the value of $\chi_M T$ decreases monotonically to attain a local minimum value of 0.299 cm^3 mol^{-1} K at 18.2 K, which is indicative of the existence of antiferromagnetic coupling. After a very small increase to a maximum value of 0.277 cm^3 mol^{-1} K at 14.6 K, the $\chi_M T$ value decreases again with further cooling to 2.0 K. The increase in $\chi_M T$ below 18.2 K is field-dependent, as shown in the inset of Figure 7; this suggests that a mechanism of weak ferromagnetic correlations due to spin canting antiferromagnetism is operative within compound **2**. The final decrease in $\chi_M T$ value may be attributed to antiferromagnetic interactions between the chain and/or saturation effects. Similar to the observations in compound **1**, the lack of antisymmetric magnetic interactions in compound **2** would be expected because of the presence of inversion centers in the $P2_1/c$ crystal structure. Thus, the spin canting of compound **2** originates from the single-ion anisotropy of the Co(II) ion, which is in agreement with the reported Co(II) spin canting compounds containing the same bridging mode of azide [59–61].

In order to substantiate the low-temperature magnetic properties of compound **2**, ZFC/FC magnetization studies were carried out at 50 Oe. As shown in Figure 8, upon cooling, both ZFC and FC magnetizations increase abruptly at temperatures below 18 K and a divergence between ZFC/FC below 16.2 K is observed, suggesting the occurrence of magnetic ordering for the formation of an ordered state and the existence of an uncompensated moment below the critical temperature of T_c = 16.2 K. Upon cooling, both ZFC and FC magnetizations increase again below 5.0 K, which may be due to the spin-reorientations of the domain wall. The existence of magnetic ordering was also confirmed by the AC magnetic susceptibility measurements of compound **2** performed at H_{dc} = 0 Oe and H_{ac} = 3.5 Oe at different frequencies (Figure S13). As can be seen from Figure S13, both χ_M' and χ_M'' signals are frequency-independent, where the χ_M' signals show two peaks at ca. 14.8 K and 5.2 K with two corresponding non-zero χ_M'' signals formed at temperatures below 16.2 and 6.2 K. The presence of χ_M' and χ_M'' peaks at about 15 K are the result of the formation of an ordered state with an uncompensated moment and the peaks of χ_M' and χ_M'' at about 5 K may be caused by spin-reforestation [62,63]. These data confirm the occurrence of magnetic ordering by weak ferromagnetism due to spin canting, which is consistent with the obtained results from ZFC/FC magnetizations data. Due to the presence of weak non-zero χ_M'' signals, a coercive magnetic behavior would be expected below T_c.

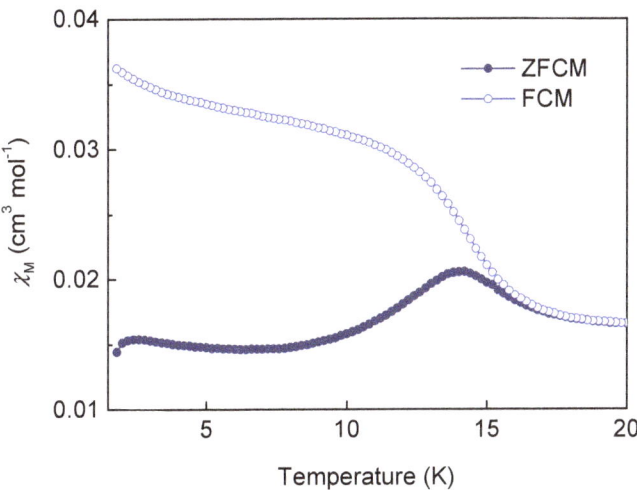

Figure 8. FC and ZFC magnetization plots of compound **2** at the field of 50 Oe.

To further study the magnetic ordering of compound **2**, the isothermal field-dependent magnetization was collected at 2.0 K. As depicted in Figure S14, the initial increase in magnetization shows a positive curvature to 0.175 $N\beta$ at 70 kOe, which is far below the theoretical value of saturation for an isotropic high-spin Co(II) system, and the absence of saturation of magnetization; this is an indication of an overall antiferromagnetic interaction between the Co(II) ions in compound **2**. In addition, when the field is less than 10 kOe, a hysteresis loop is clearly observed at 2.0 K, suggesting the soft magnet property of compound **2** (Figure S14, inset). The hysteresis loop shows a remanent magnetization (M_r) of ≈0.007 $N\beta$ and a coercive field of ≈350 Oe. Based on the value of M_r at 2.0 K, the canting angle of compound **2** is estimated to be approximately 0.20°, where M_S is 2.15 $N\beta$ for an octahedral Co(II) at 2 K with the effective spin of $S' = 1/2$ and a common value of $g' = 4.3$ [60].

Finally, to rule out any contributions of Mn(II)/Mn(III) or Co(II) oxides in the low-temperature magnetic behaviors in compounds **1** and **2**, the AC magnetic data and 2.0 K field-dependent magnetization were collected using the thermal decomposed samples of **1** and **2** that had been heated at 350 °C for two hours (Figures S15–S18), in which the disappearance of peaks in χ_M' and/or χ_M'' and the absence of magnetic hysteresis loops in field-dependent magnetization are obtained, excluding the contribution of the behavior of Mn(II)/Mn(III) or Co(II) oxides.

3. Concluding Remarks

The results presented and discussed above have shown that the new compounds of general formula [M(N₃)₂(dmbpy)] for M = Mn (**1**) and Co (**2**) show very similar and almost identical linear chain structures, with alternating azide double bridging anions (EE and EO). Using magnetic studies down to a very low T (2K), the existence of intrachain ferro- and antiferromagnetic interactions was established, and these interactions are dominated by the double EE and double EO azido ligand bridges. Overall, the compounds are found to behave as antiferromagnets, and the study is relevant in the search for potentially cheap, stable and useful new magnets.

Compounds **1**·(Mn) and **2**·(Co) exhibit spin-canted antiferromagnetism at very low-temperatures, which is ascribed to the presence of single-ion anisotropy. Furthermore, below the Néel temperature, T_N, field-induced magnetic transitions have also been observed, and these are indicative for metamagnetism in the case of **1**·(Mn). Such coexistence of spin-canted antiferromagnetism and metamagnetism in 1D Mn(II) compounds with alternating double EE and double EO azido ligands is unprecedented. Although the spin-canted antiferromagnetism might seem incompatible with the crystal structure of **2**, because of the presence of an inversion center between the bridged Mn(II) centers, the observation of the spin canting in **2** should be attributed to a structural phase transition, or a distortion in the crystal at low temperature, thereby removing the inversion center. Such types of distortions have been reported before [34,58]. The weak interchain interactions present in both compounds are ascribed to the π–π stacking interactions between the dmbpy rings.

4. Material and Methods

4.1. General Remarks

The starting materials (metal salts, sodium azide and the ligand 5,5'-dimethyl-2,2'-bipyridine, $C_{12}H_{12}N_2$), and used solvents were purchased from commercial sources (analytical reagent grade) and used without further purification. All the compounds were synthesized solvothermally under autogenous pressure. Azido derivatives are potentially explosive and should be handled with great care and prepared only in small quantities by trained persons.

4.2. Synthesis

Synthesis of catena-poly-[Mn(N$_3$)$_2$(dmbpy)] (**1**) A mixture of Mn(NO$_3$)$_2 \cdot$ 4H$_2$O (50 mg, 0.2 mmol), dmbpy (37 mg, 0.2 mmol) and NaN$_3$ (26 mg, 0.4 mmol) in H$_2$O/EtOH (3:1 v/v, 20 mL) was sealed in a Teflon-lined autoclave and heated at 130 °C for 2 days. After cooling to room temperature at a rate of 10 °C h^{-1}, yellow-colored crystals of **1** were obtained (yield 32%). Anal. Calcd. (%) for $C_{12}H_{12}MnN_8$: C, 44.59; H, 3.74; N, 34.67%. Found: C, 44.45; H, 3.92; N, 34.42%. Main IR band (KBr pellet, cm^{-1}): doublet at 2089s [ν(N$_3{}^-$)].

Synthesis of catena-poly-[Co(N$_3$)$_2$(dmbpy)] (**2**). This compound was prepared following a procedure similar to that of compound **1**, except that Co(NO$_3$)$_2 \cdot$6H$_2$O (58 mg, 0.2 mmol), was used instead of Mn(NO$_3$)$_2 \cdot$4H$_2$O. Brown-colored crystals were obtained, with a yield of 40%, containing **2**. Anal. Calcd. (%) for $C_{12}H_{12}CoN_8$: C, 44.05; H, 3.70; N, 34.24%. Found: C, 43.85; H, 3.82; N, 34.63%. Main IR band (KBr pellet, cm^{-1}): doublet at 2085s [ν(N$_3{}^-$)].

Physical Measurements. Elemental analyses of the obtained compounds (C, H and N) were performed using a Perkin–Elmer 2400 series II CHN analyzer. Infrared spectra were recorded as KBr pellets in the range 4000–400 cm^{-1} (4 cm^{-1} resolution) on a Perkin–Elmer 100 FT-IR spectrometer, which was calibrated using polystyrene and CO$_2$ bands. The temperature dependence DC and AC magnetic susceptibility measurements were performed on powdered samples, restrained in eicosane to prevent torquing, on a Quantum Design MPMS-7 SQUID (Superconducting Quantum Interference Device) and a PPMS (Physical Property Measurement System) magnetometer, equipped with 7.0 T and 9.0 T magnets (Quantum Design, San Diego, CA, USA), respectively, operated in the range of 2.0–300 K. Diamagnetic corrections were estimated from Pascal's constants [64] and subtracted from the experimental susceptibility data to obtain the molar paramagnetic susceptibility of the compounds. Powder X-ray diffraction (PXRD) measurements of **1** and **2** were carried out on a Siemens D-5000 diffractometer (Siemens, Karlsruhe, Germany) running in a step mode with a step size of 0.02° in θ and a fixed time of 10 s at 40 kV, 30 mA for Cu-Kα (λ = 1.5406 Å).

4.3. X-ray Crystallography

Diffraction data (Table 2) were collected at 200 K on a Bruker SMART X2S (Bruker AXS Inc., Madison, WI, USA) benchtop diffractometer [65], using the Mo $K\alpha$ radiation (λ = 0.71073 Å) and the structures were refined with SHELXL [66,67]. Crystals for M = Co(II) were found to be twinned by a twofold rotation about the c^* reciprocal axis. One batch scale factor was refined, which converged to 0.24. All H atoms were placed in calculated positions and refined as riding to their carrier atoms.

Table 2. Crystal data and refinement parameters.

Compound/Deposition in CCDC	1/1954571	2/1954572
Formula	$C_{12}H_{12}MnN_8$	$C_{12}H_{12}CoN_8$
Fw	323.24	327.23
Crystal size (mm^3)	$0.5 \times 0.3 \times 0.2$	$0.30 \times 0.30 \times 0.18$
Space group	$P\text{-}1$	$P2_1/c$
a (Å)	7.8058(18)	7.397(2)
b (Å)	9.538(3)	18.373(4)
c (Å)	10.601(3)	10.661(3)
α (°)	113.630(8)	-
β (°)	102.610(8)	110.571(9)
γ (°)	93.223(8)	-
V (Å^3)	696.4(3)	1356.5(6)
Z, Z'	2, 1	4, 1
Diffractometer	Bruker X2S	Bruker X2S
Radiation	Mo-$K\alpha$	Mo-$K\alpha$
T (K)	200	200
Abs. coef. (mm^{-1})	0.954	1.272
Transmission fact.	0.56–0.83	0.51–0.80
Refl. collected	6450	8922
Sinθ/λ (Å^{-1})	0.62	0.61
R_{int} (%)	4.26	8.68
Completeness (%)	97.4	99.2
Data/parameters	2658/192	2575/194
Restraints	0	0
R_1, wR_2 [$I > 2\sigma(I)$]	5.87, 16.64	6.53, 16.48
R_1, wR_2 [all data]	7.07, 17.60	8.26, 17.88
GOF on F^2	1.035	1.003

Supplementary Materials: Supplementary data (all 18 Supplementary Figures and one Table) associated with this paper can be found at https://www.mdpi.com/article/10.3390/magnetochemistry7040050/s1. XRD powder patterns and infrared spectra of the compounds **1** and **2**, in Figures S1 and S2, as well as 16 other Figures (S3–S18) with details of magnetic studies and a supplementary Table (S1), as indicated in the text. CCDC 1954571-1954572 contain the supplementary crystallographic data for compounds **1** and **2**, respectively. These data can be obtained free of charge via http://www.ccdc.cam.ac.uk/conts/retrieving.html, or from the Cambridge Crystallographic DataCentre, 12 Union Road, Cambridge CB2 1EZ, UK; e-mail: deposit@ccdc.cam.ac.uk.

Author Contributions: N.B. and Z.S. performed the coordination chemistry and crystallizations, G.S.K. performed and analyzed the spectroscopic measurements, D.K.G., F.S. and S.B. realized the single crystal X-ray diffraction experiments and refined the X-ray structures, C.-I.Y. performed and analyzed the magnetic measurements. F.S., S.B., C.-I.Y. and J.R. assisted with the interpretation and contributed to the writing of the article. All authors have read and agreed to the published version of the manuscript.

Funding: The authors acknowledge the Algerian DG-RSTD (Direction Générale de la Recherche Scientifique et du Développement Technologique) and Université Ferhat Abbas Sétif 1 for financial support. The APC was funded by the journal.

Data Availability Statement: The data presented in this study are available in Supplementary Material.

Acknowledgments: S.B. thanks MC Cándida Pastor Ramírez (BUAP, Mexico), for bringing Reference [30] to our attentions.

Conflicts of Interest: The authors declare no competing financial interest.

References

1. Khasainov, B.; Comet, M.; Veyssiere, B.; Spitzer, D. On the Mechanism of Efficiency of Lead Azide. *Propellants Explos. Pyrotech.* **2016**, *42*, 547–557. [CrossRef]
2. Xu, J.-G.; Sun, C.; Zhang, M.-J.; Liu, B.-W.; Li, X.-Z.; Lu, J.; Wang, S.-H.; Zheng, F.-K.; Guo, G.-C. Coordination Polymerization of Metal Azides and Powerful Nitrogen-Rich Ligand toward Primary Explosives with Excellent Energetic Performances. *Chem. Mater.* **2017**, *29*, 9725–9733. [CrossRef]
3. Ma, X.; Liu, Y.; Song, W.; Wang, Z.; Liu, X.; Xie, G.; Chen, S.; Gao, S. A difunctional azido-cobalt(ii) coordination polymer exhibiting slow magnetic relaxation behaviour and high-energy characteristics with good thermostability and insensitivity. *Dalton Trans.* **2018**, *47*, 12092–12104. [CrossRef] [PubMed]
4. Xu, J.-G.; Li, X.-Z.; Wu, H.-F.; Zheng, F.-K.; Chen, J.; Guo, G.-C. Substitution of Nitrogen-Rich Linkers with Insensitive Linkers in Azide-Based Energetic Coordination Polymers toward Safe Energetic Materials. *Cryst. Growth Des.* **2019**, *19*, 3934–3944. [CrossRef]
5. Zhang, W.; Li, T.; Zhang, B.; Wang, L.; Zhang, T.; Zhang, J. Planar, Energetic, π–π-Stacked Compound with Weak Interactions Resulting in a High-Impact- and Low-Friction-Sensitive, Safer, Primary Explosive. *Inorg. Chem.* **2019**, *58*, 7653–7656. [CrossRef]
6. Beck, W.; Werner, K.V. Reaktionen der Azido- und Isocyanatorhodium(I)-Komplexetrans-(Ph3P)2Rh(CO)X mit dem Nitrosyl-Kation in Gegenwart von Alkoholen. *Eur. J. Inorg. Chem.* **1973**, *106*, 868–873. [CrossRef]
7. Hu, B.-W.; Zhao, J.-P.; Yang, Q.; Liu, F.-C. Dzyaloshinski–Moriya (D–M) oriented weak ferromagnets in isomorphic coordination architectures constructed by flexible 1,2,4-triazole-1-acetate ligands with the assistance of halogen or pseudohalogen anions. *Inorg. Chem. Commun.* **2013**, *35*, 290–294. [CrossRef]
8. Ribas, J.; Escuer, A.; Monfort, M.; Vicente, R.; Cortés, R.; Lezama, L.; Rojo, T. Polynuclear NiII and MnII azido bridging complexes. Structural trends and magnetic behavior. *Coord. Chem. Rev.* **1999**, *193–195*, 1027–1068. [CrossRef]
9. Groom, C.R.; Bruno, I.J.; Lightfoot, M.P.; Ward, S.C. The Cambridge Structural Database. *Acta Crystallogr. Sect. B Struct. Sci. Cryst. Eng. Mater.* **2016**, *72*, 171–179. [CrossRef]
10. Ramírez, C.P.; Bernès, S.; Anzaldo, S.H.; Ortega, Y.R. Structure and NMR properties of the dinuclear complex di-μ-azido-κ4 N 1:N 1-bis[(azido-κN)(pyridine-2-carboxamide-κ2 N 1,O)zinc(II)]. *Acta Crystallogr. Sect. E Crystallogr. Commun.* **2021**, *77*, 111–116. [CrossRef]
11. Monfort, M.; Resino, I.; Ribas, J.; Solans, X.; Font-Bardia, M.; Rabu, P.; Drillon, M. Synthesis, structure, and magnetic properties of two new ferromagnetic/antiferromagnetic one-dimensional nickel(II) complexes. Magnetostructural correlations. *Inorg. Chem.* **2000**, *39*, 2572–2576. [CrossRef] [PubMed]
12. Monfort, M.; Resino, I.; Ribas, J.; Stoeckli-Evans, H. A Metamagnetic Two-Dimensional Molecular Material with Nickel(II) and Azide. *Angew. Chem. Int. Ed.* **2000**, *39*, 191–193. [CrossRef]
13. Cortés, R.; Drillon, M.; Solans, X.; Lezama, A.L.; Rojo, T. Alternating Ferromagnetic–Antiferromagnetic Interactions in a Manganese(II)–Azido One-Dimensional Compound: [Mn(bipy)(N3)2]. *Inorg. Chem.* **1997**, *36*, 677–683. [CrossRef]
14. Lu, Z.; Gamez, P.; Kou, H.-Z.; Fan, C.; Zhang, H.; Sun, G. Spin canting and metamagnetism in the two azido-bridged 1D complexes [Ni(3,5-dmpy)2(N3)2]n and [Co1.5(3,5-dmpy)3(N3)3]n. *CrystEngComm* **2012**, *14*, 5035–5041. [CrossRef]
15. Van Albada, G.A.; Van Der Horst, M.G.; Mutikainen, I.; Turpeinen, U.; Reedijk, J. Synthesis, Crystal Structure and Spectroscopy of catena-poly-bis(azido-N1,N1)(2-Aminopyrimidine)Copper(II). *J. Chem. Crystallogr.* **2008**, *38*, 413–417. [CrossRef]
16. Van Albada, G.A.; Van Der Horst, M.G.; Mutikainen, I.; Turpeinen, U.; Reedijk, J. Dinuclear and polynuclear Cu(II) azido-bridged compounds with 7-azaindole as a ligand. Synthesis, characterization and 3D structures. *Inorg. Chim. Acta* **2011**, *367*, 15–20. [CrossRef]
17. Van Albada, G.A.; Mutikainen, I.; Roubeau, O.; Reedijk, J. A dinuclear end-on azide-bridged copper(II) compound with weak antiferromagnetic interaction—Synthesis, characterization, magnetism and X-ray structure of bis[(μ-azido-κN1)-(azido-κN1)(1,3-bis(benzimidazol-2-yl)-2-methylpropane)copper(II)]. *J. Mol. Struct.* **2013**, *1036*, 252–256. [CrossRef]
18. Setifi, Z.; Ghazzali, M.; Glidewell, C.; Pérez, O.; Setifi, F.; Gómez-García, C.J.; Reedijk, J. Azide, water and adipate as bridging ligands for Cu(II): Synthesis, structure and magnetism of (μ4-adipato-κ-O)(μ-aqua)(μ-azido-κN1,N1)copper(II) monohydrate. *Polyhedron* **2016**, *117*, 244–248. [CrossRef]
19. Van Albada, G.A.; Mohamadou, A.; Mutikainen, I.; Turpeinen, U.; Reedijk, J. Diazidobis(2,2′-biimidazoline)nickel(II). *Acta Crystallogr. Sect. E Struct. Rep. Online* **2004**, *60*, m237–m238. [CrossRef]
20. Van Albada, G.A.; Smeets, W.J.J.; Spek, A.L.; Reedijk, J. The crystal structure and IR spectra of μ-(bipyrimidine-N1,N1′,N5,N5′)-bis[(azido-N1)(methanol)(bipyrimidine-N1,N1′)copper(II)] bis(triflate) bis(methanol). *J. Chem. Crystallogr.* **1998**, *28*, 427–432. [CrossRef]
21. Mautner, F.A.; Sudy, B.; Massoud, A.A.; Abu-Youssef, M.A.M. Synthesis and characterization of Mn(II) and Zn(II) azido complexes with halo-substituted pyridine derivative ligands. *Transit. Met. Chem.* **2013**, *38*, 319–325. [CrossRef]
22. Yue, Y.-F.; Gao, E.-Q.; Fang, C.-J.; Zheng, T.; Liang, J.; Yan, C.-H. Three Azido-Bridged Mn(II) Complexes Based on Open-Chain Diazine Schiff-base Ligands: Crystal Structures and Magnetic Properties. *Cryst. Growth Des.* **2008**, *8*, 3295–3301. [CrossRef]
23. Gao, E.-Q.; Bai, S.-Q.; Yue, Y.-F.; Wang, Z.-M.; Yan, C.-H. New One-Dimensional Azido-Bridged Manganese(II) Coordination Polymers Exhibiting Alternating Ferromagnetic–Antiferromagnetic Interactions: Structural and Magnetic Studies. *Inorg. Chem.* **2003**, *42*, 3642–3649. [CrossRef] [PubMed]

24. Abu-Youssef, M.A.M.; Drillon, M.; Escuer, A.; Goher, M.A.S.; Mautner, F.A.; Vicente, R. Topological Ferrimagnetic Behavior of Two New [Mn(L)2(N3)2]nChains with the New AF/AF/F Alternating Sequence (L = 3-Methylpyridine or 3,4-Dimethylpyridine). *Inorg. Chem.* **2000**, *39*, 5022–5027. [CrossRef] [PubMed]
25. Abu-Youssef, M.A.M.; Escuer, A.; Goher, M.A.S.; Mautner, F.A.; Reiss, G.J.; Vicente, R. Can a homometallic chain be ferromagnetic? *Angew. Chem. Int. Ed.* **2000**, *39*, 1624–1626. [CrossRef]
26. Ribas, J.; Monfort, M.; Ghosh, B.K.; Solans, X.; Font-Bardía, M. Versatility of the azido bridging ligand in the first two examples of ferro–antiferromagnetic alternating nickel(II) chains. *J. Chem. Soc. Chem. Commun.* **1995**, *23*, 2375–2376. [CrossRef]
27. Ribas, J.; Monfort, M.; Resino, I.; Solans, X.; Rabu, P.; Maingot, F.; Drillon, M. A Unique NiII Complex with Three Different Azido Bridges: Magneto-Structural Correlations in the First Triply Alternating $S = 1$ Chain. *Angew. Chem. Int. Ed.* **1996**, *35*, 2520–2522. [CrossRef]
28. Cortes, R.; Lezama, L.; Pizarro, J.L.; Arriortua, M.I.; Solans, X.; Rojo, T. Alternating Ferromagnetic and Antiferromagnetic Interactions In a MnII Chain With Alternating End-On and End-To-End Bridging Azido Ligands. *Angew. Chem. Int. Ed.* **1994**, *33*, 2488–2489. [CrossRef]
29. Viau, G.; Lombardi, M.G.; De Munno, G.; Julve, M.; Lloret, F.; Faus, J.; Caneschi, A.; Clemente-Juan, J.M. The azido ligand: A useful tool in designing chain compounds exhibiting alternating ferro- and antiferro-magnetic interactions. *Chem. Commun.* **1997**, *13*, 1195–1196. [CrossRef]
30. Alvarez, S.; Avnir, D.; Llunell, M.; Pinsky, M. Continuous symmetry maps and shape classification. The case of six-coordinated metal compounds. *New J. Chem.* **2002**, *26*, 996–1009. [CrossRef]
31. Bhowmik, P.; Biswas, S.; Chattopadhyay, S.; Diaz, C.; Gómez-García, C.J.; Ghosh, A. Synthesis, crystal structure and magnetic properties of two alternating double µ1,1 and µ1,3 azido bridged Cu(ii) and Ni(ii) chains. *Dalton Trans.* **2014**, *43*, 12414–12421. [CrossRef] [PubMed]
32. Wang, Y.-Q.; Tan, Q.-H.; Guo, X.-Y.; Liu, H.-T.; Liu, Z.-L.; Gao, E.-Q. Novel manganese(ii) and cobalt(ii) 2D polymers containing alternating chains with mixed azide and carboxylate bridges: Crystal structure and magnetic properties. *RSC Adv.* **2016**, *6*, 72326–72332. [CrossRef]
33. Fu, A.; Huang, X.; Li, J.; Yuen, T.; Lin, C.L. Controlled synthesis and magnetic properties of 2D and 3D iron azide networks infinity 2[Fe(N3)2(4,4'-bpy)] and infinity 3[Fe(N3)2(4,4'-bpy)]. *Chemistry* **2002**, *8*, 2239–2247. [CrossRef]
34. Zhang, J.-Y.; Liu, C.-M.; Zhang, D.-Q.; Gao, S.; Zhu, D.-B. Spin-canting in a 1D chain Mn(II) complex with alternating double end-on and double end-to-end azido bridging ligands. *Inorg. Chem. Commun.* **2007**, *10*, 897–901. [CrossRef]
35. Kar, P.; Drew, M.G.B.; Gómez-García, C.J.; Ghosh, A. Coordination Polymers Containing Manganese(II)-Azido Layers Connected by Dipyridyl-tetrazine and 4,4'-Azobis(pyridine) Linkers. *Inorg. Chem.* **2013**, *52*, 1640–1649. [CrossRef] [PubMed]
36. Mautner, F.A.; Berger, C.; Scherzer, M.; Fischer, R.C.; Maxwell, L.; Ruiz, E.; Vicente, R. Different topologies in three manganese-µ-azido 1D compounds: Magnetic behavior and DFT-quantum Monte Carlo calculations. *Dalton Trans.* **2015**, *44*, 18632–18642. [CrossRef] [PubMed]
37. Abu-Youssef, M.A.M.; Escuer, A.; Gatteschi, D.; Goher, M.A.S.; Mautner, F.A.; Vicente, R. Synthesis, Structural Characterization, Magnetic Behavior, and Single Crystal EPR Spectra of Three New One-Dimensional Manganese Azido Systems with FM, Alternating FM-AF, and AF Coupling. *Inorg. Chem.* **1999**, *38*, 5716–5723. [CrossRef]
38. Gao, E.-Q.; Cheng, A.-L.; Xu, Y.-X.; He, M.-Y.; Yan, C.-H. From Low-Dimensional Manganese(II) Azido Motifs to Higher-Dimensional Materials: Structure and Magnetic Properties. *Inorg. Chem.* **2005**, *44*, 8822–8835. [CrossRef]
39. Wang, X.-Y.; Wang, L.; Wang, Z.-M.; Su, G.; Gao, S. Coexistence of Spin-Canting, Metamagnetism, and Spin-Flop in a (4,4) Layered Manganese Azide Polymer. *Chem. Mater.* **2005**, *17*, 6369–6380. [CrossRef]
40. Zhang, X.-M.; Hao, Z.-M.; Zhang, W.-X.; Chen, X.-M. Dehydration-Induced Conversion from a Single-Chain Magnet into a Metamagnet in a Homometallic Nanoporous Metal–Organic Framework. *Angew. Chem. Int. Ed.* **2007**, *46*, 3456–3459. [CrossRef]
41. Cheng, X.-N.; Xue, W.; Huang, J.-H.; Chen, X.-M. Spin canting and/or metamagnetic behaviours of four isostructural grid-type coordination networks. *Dalton Trans.* **2009**, *29*, 5701–5707. [CrossRef] [PubMed]
42. Bellitto, C.; Federici, F.; Colapietro, M.; Portalone, G.; Caschera, D. X-ray Single-Crystal Structure and Magnetic Properties of Fe[CH3PO3)]·H2O: A Layered Weak Ferromagnet. *Inorg. Chem.* **2002**, *41*, 709–714. [CrossRef] [PubMed]
43. Liu, B.; Shang, R.; Hu, K.-L.; Wang, Z.-M.; Gao, S. A New Series of Chiral Metal Formate Frameworks of [HONH3][MII(HCOO)3] (M = Mn, Co, Ni, Zn, and Mg): Synthesis, Structures, and Properties. *Inorg. Chem.* **2012**, *51*, 13363–13372. [CrossRef] [PubMed]
44. Cheng, X.-N.; Xue, W.; Zhang, W.-X.; Chen, X.-M. Weak Ferromagnetism and Dynamic Magnetic Behavior of Two 2D Compounds with Hydroxy/Carboxylate-Bridged Co(II) Chains. *Chem. Mater.* **2008**, *20*, 5345–5350. [CrossRef]
45. Wang, T.-T.; Ren, M.; Bao, S.-S.; Liu, B.; Pi, L.; Cai, Z.-S.; Zheng, Z.-H.; Xu, Z.-L.; Zheng, L.-M. Effect of Structural Isomerism on Magnetic Dynamics: From Single-Molecule Magnet to Single-Chain Magnet. *Inorg. Chem.* **2014**, *53*, 3117–3125. [CrossRef]
46. Boca, R.; Herchel, R. Antisymmetric exchange in polynuclear metal complexes. *Coord. Chem. Rev.* **2010**, *254*, 2973–3025. [CrossRef]
47. Verdaguer, M.; Bleuzen, A.; Marvaud, V.; Vaissermann, J.; Seuleiman, M.; Desplanches, C.; Scuiller, A.; Train, C.; Garde, R.; Gelly, G.; et al. Molecules to build solids: High TC molecule-based magnets by design and recent revival of cyano complexes chemistry. *Coord. Chem. Rev.* **1999**, *190–192*, 1023–1047. [CrossRef]
48. Rodríguez, A.; Kivekäs, R.; Colacio, E. Unique self-assembled 2D metal-tetrazolate networks: Crystal structure and magnetic properties of [M(pmtz)2](M = Co(ii) and Fe(ii); Hpmtz = 5-(pyrimidyl)tetrazole). *Chem. Commun.* **2005**, *41*, 5228–5230. [CrossRef]

49. Li, J.-R.; Yu, Q.; Sañudo, E.C.; Tao, Y.; Bu, X.-H. An azido–CuII–triazolate complex with utp-type topological network, showing spin-canted antiferromagnetism. *Chem. Commun.* **2007**, *25*, 2602–2604. [CrossRef]
50. Wang, X.-Y.; Wei, H.-Y.; Wang, Z.-M.; Chen, Z.-D.; Gao, S. FormateThe Analogue of Azide: Structural and Magnetic Properties of M(HCOO)2(4,4′-bpy)·nH2O (M = Mn, Co, Ni; n = 0, 5). *Inorg. Chem.* **2005**, *44*, 572–583. [CrossRef]
51. Escuer, A.; Cano, J.; Goher, M.A.; Journaux, Y.; Lloret, F.; Mautner, F.A.; Vicente, R. Synthesis, Structural Characterization, and Monte Carlo Simulation of the Magnetic Properties of Two New Alternating MnIIAzide 2-D Honeycombs. Study of the Ferromagnetic Ordered Phase below 20 K. *Inorg. Chem.* **2000**, *39*, 4688–4695. [CrossRef] [PubMed]
52. Cheng, L.; Zhang, W.-X.; Ye, B.-H.; Lin, J.-B.; Chen, X.-M. Spin Canting and Topological Ferrimagnetism in Two Manganese(II) Coordination Polymers Generated by In Situ Solvothermal Ligand Reactions. *Eur. J. Inorg. Chem.* **2007**, *2007*, 2668–2676. [CrossRef]
53. Carlin, R.L.; Van Duyneveldt, A.J. Field-dependent magnetic phenomena. *Acc. Chem. Res.* **1980**, *13*, 231–236. [CrossRef]
54. Manson, J.L.; Kmety, C.R.; Palacio, F.; Epstein, A.J.; Miller, J.S. Low-Field Remanent Magnetization in the Weak Ferromagnet Mn[N(CN)2]2. Evidence for Spin-Flop Behavior. *Chem. Mater.* **2001**, *13*, 1068–1073. [CrossRef]
55. Pinkowicz, D.; Rams, M.; Nitek, W.; Czarnecki, B.; Sieklucka, B. Evidence for magnetic anisotropy of [NbIV(CN)8]4− in a pillared-layered Mn2Nb framework showing spin-flop transition. *Chem. Commun.* **2012**, *48*, 8323–8325. [CrossRef]
56. Lu, Y.-B.; Wang, M.-S.; Zhou, W.-W.; Xu, G.; Guo, G.-C.; Huang, J.-S. Novel 3-D PtS-like Tetrazolate-Bridged Manganese(II) Complex Exhibiting Spin-Canted Antiferromagnetism and Field-Induced Spin-Flop Transition. *Inorg. Chem.* **2008**, *47*, 8935–8942. [CrossRef] [PubMed]
57. Schlueter, J.A.; Manson, J.L.; Hyzer, K.A.; Geiser, U. Spin Canting in the 3D Anionic Dicyanamide Structure (SPh3)Mn(dca)3(Ph = Phenyl, dca = Dicyanamide). *Inorg. Chem.* **2004**, *43*, 4100–4102. [CrossRef]
58. Ray, U.; Jasimuddin, S.; Ghosh, B.K.; Monfort, M.; Ribas, J.; Mostafa, G.; Lu, T.-H.; Sinha, C. A New Alternating Ferro-Antiferromagnetic One-Dimensional Azido-Bridged (Arylazoimidazole)manganese(II), [Mn(TaiEt)(N3)2]n [TaiEt = 1-Ethyl-2-(p-tolylazo)imidazole], Exhibiting Bulk Weak Ferromagnetic Long-Range Ordering. *Eur. J. Inorg. Chem.* **2004**, *2004*, 250–259. [CrossRef]
59. Boonmak, J.; Nakano, M.; Youngme, S. Structural diversity and magnetic properties in 1D and 2D azido-bridged cobalt(ii) complexes with 1,2-bis(2-pyridyl)ethylene. *Dalton Trans.* **2011**, *40*, 1254–1260. [CrossRef]
60. Boonmak, J.; Nakano, M.; Chaichit, N.; Pakawatchai, C.; Youngme, S. Spin Canting and Metamagnetism in 2D and 3D Cobalt(II) Coordination Networks with Alternating Double End-On and Double End-to-End Azido Bridges. *Inorg. Chem.* **2011**, *50*, 7324–7333. [CrossRef]
61. Wang, X.-T.; Wang, X.-H.; Wang, Z.-M.; Gao, S. Diversity of Azido Magnetic Chains Constructed with Flexible Ligand 2,2′-Dipyridylamine. *Inorg. Chem.* **2009**, *48*, 1301–1308. [CrossRef] [PubMed]
62. Li, R.-Y.; Wang, X.-Y.; Liu, T.; Xu, H.-B.; Zhao, F.; Wang, Z.-M.; Gao, S. Synthesis, Structure, and Magnetism of Three Azido-Bridged Co2+Compounds with a Flexible Coligand 1,2-(Tetrazole-1-yl)ethane. *Inorg. Chem.* **2008**, *47*, 8134–8142. [CrossRef] [PubMed]
63. Bałanda, M. AC Susceptibility Studies of Phase Transitions and Magnetic Relaxation: Conventional, Molecular and Low-Dimensional Magnets. *Acta Phys. Pol. A* **2013**, *124*, 964–976. [CrossRef]
64. Bain, G.A.; Berry, J.F. Diamagnetic Corrections and Pascal's Constants. *J. Chem. Educ.* **2008**, *85*, 532–533. [CrossRef]
65. Eccles, K.S.; Stokes, S.P.; Daly, C.A.; Barry, N.M.; McSweeney, S.P.; O'Neill, D.J.; Kelly, D.M.; Jennings, W.B.; Dhubhghaill, O.M.N.; Moynihan, H.A.; et al. Evaluation of the Bruker SMART X2S: Crystallography for the nonspecialist? *J. Appl. Crystallogr.* **2010**, *44*, 213–215. [CrossRef]
66. Sheldrick, G.M. A short history of SHELX. *Acta Crystallogr. A* **2008**, *64*, 112–122. [CrossRef]
67. Sheldrick, G.M. Crystal structure refinement with SHELXL. *Acta Crystallogr.* **2015**, *71*, 3–8. [CrossRef]

Editorial

Peter Day's Exploration of Time and Space

Santiago Alvarez

Departament de Química Inorgànica i Orgànica-Secció de Química Inorgànica-and Institut de Química Teòrica i Computacional, Universitat de Barcelona, Martí i Franquès 1-11, 08028 Barcelona, Spain; santiago@qi.ub.es

Abstract: Peter Day was one of the most asiduous participants of the NoSIC (Not Strictly Inorganic Chemistry) meetings, where he showed his interest in, and knowledge of, historical, sociological and other non-scientific aspects of the research activities in the institutions led by him as well as in those he visited worldwide, both as a lecturer and as an active participant. This article tries to stress that side of his personality, reflected also in his three autobiographical books, and in his motto "the past is another country", a quotation from L.P. Hartley.

Keywords: scientific tourism; biography; conferences

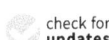

Citation: Alvarez, S. Peter Day's Exploration of Time and Space. *Magnetochemistry* **2021**, *7*, 103. https://doi.org/10.3390/magnetochemistry7070103

Received: 7 July 2021
Accepted: 8 July 2021
Published: 13 July 2021

Publisher's Note: MDPI stays neutral with regard to jurisdictional claims in published maps and institutional affiliations.

Copyright: © 2021 by the author. Licensee MDPI, Basel, Switzerland. This article is an open access article distributed under the terms and conditions of the Creative Commons Attribution (CC BY) license (https://creativecommons.org/licenses/by/4.0/).

Time and space are the two main axes of Peter Day's non-research essays and books. On the time axis, we find his interest in the history of chemistry and, in particular, the history of the institutions in which he developed his scientific activity: the Royal Institution in London, the Institute Laue–Langevin in Grenoble, and the coffee rooms. The space axis corresponds to the territories he went across during his scientific journey, the specificities of the human atmosphere in each city and town he visited, the social constructs—both political and academic—that float in their atmospheres, and the history that is required to understand how all these factors converged at the time and place of his visit. This two-fold endeavour permeates the three autobiographic books published by Peter Day [1–3], as well as the lectures and communications he gave from 2008 to 2012 in Prullans (Table 1).

Prullans is a small town of the Catalan Pyrenees overseeing the valley of La Cerdanya with the impressive walls of the Cadí mountain range in the background (Figure 1). There, between 2003 and 2018, took place a series of eight NoSIC meetings. NoSIC is a convenient acronym for an odd title, Not Strictly Inorganic Chemistry. The first call stated that the goal was "to gather in a friendly atmosphere about 50 chemists, with research activities in the area of Inorganic Chemistry, to discuss at the interface between Chemistry and other areas of Culture to enhance the permeability of the Chemistry curricula to our cultural and historical background".

Peter Day first participated in the NoSIC-2 meeting, in 2006, kindly accepting my invitation following a suggestion by Michel Verdaguer. The arrangements were facilitated because he had established a second residence in the small village of Marquixanes, in the French Roussillon, in 1997. Marquixanes is barely 85 km from Prullans, so he could easily drive through the French–Spanish border to attend the meeting. Let me note in passing that his interest for the historical and geographic context as well as for domestic details and human aspects of his environment is well reflected in the description of his life in Marquixanes that appears in a chapter of his book, "On the Cucumber Tree" [2].

He proposed as his lecture's title "The past is another country. But is it?" (L. P. Hartley). The quotation in the title comes from the novel "The Go-Between" by British writer Leslie Pole Hartley (1895–1972). If we consider the full sentence "The past is another country: they do things differently there", it becomes clear that in his mind culture and social structures differ from past to present as from one space (country) to another, and nicely summarizes the two-dimensional perspective of his scientific and human activity worldwide.

Table 1. Titles and dates of some books published, and lectures given, at the NoSIC meetings in Prullans, by Peter Day, pointing to chapters of his books related to the contents of the talks.

Year	Meeting	Book Published	Lectures and Related Book Chapters
2005		*Nature not Mocked* [1].	
2006	NoSIC-2		"The past is another country…" (L. P. Hartley) But is it? *On the Cucumber Tree*, chapter 8.
2008	NoSIC-3		Benjamin Thomson, Count Rumford–European Citizen and Cooking Expert *Nature not Mocked*, chapter 4.
2010	NoSIC-4		Davy's Batteries: the World's First Research Grant Proposal. *Nature not Mocked*, chapter 6.
2012	NoSIC-5		Conversation Rooms: Coffee and Chemistry. *Nature not Mocked*, chapter 2.
		On the Cucumber Tree [2].	
2018	NoSIC-8		Scientific Tourism: Learning About the World Through Travel. *Scientific Tourism*
2020		*Scientific Tourism* [3].	

(a)

(b)

Figure 1. (a) The Romanesque church of Prullans, 11th–12th century, (b) view of the Cadí mountain range from the NoSIC conference site.

On that occasion, he told a series of interesting stories of the Royal Institution (RI), "the most unusual scientific organisation" of which he had been the director for some years. The three most remarkable leaders of that institution went then on stage: Benjamin Thomson, Humphry Davy and Michael Faraday. He developed further their relevance and less well-known activities in subsequent NoSIC participations (Table 1). First, the physicist Sir Benjamin Thomson, made Count Rumford by the King of Bavaria, whom Day qualified as a "mover and shaker" (one who wants to organise things), and responsible for the establishment of the RI in 1799. He worked on the theory and uses of heat, "heat is nothing but motion", designed lamps and fireplaces, was interested in cooking, and authored a book titled "The Chemistry and Physics of Cooking". Remarks were made also of Davy's famous lectures on laughing gas given in 1810.

Day presented the second director of the RI, Humphry Davy, as the first researcher to write a grant proposal, asking for financial help to make a larger battery for his electrolytic experiments. With that battery, Davy was able to isolate new elements. Quoting Day, "Apart from the Lawrence Berkeley Laboratory under Glenn Seaborg, there can be no other building on the planet that has seen the isolation of so many chemical elements as 21 Albemarle Street under Davy; most of groups 1 and 2 of the periodic table and, at a further remove, chlorine and iodine were identified there".

He told us about the activity of Michael Faraday, "the young man who left his trade as a bookbinder to come to the RI as Davy's chemical assistant", who started the two famous series of popular lectures, one for adults and one for children in 1826. Less well-known activities of Faraday were also commented upon by Day, such as his contribution to the improvement of steel or the development of optical glass.

By sheer coincidence, two books have become neighbours on the shelves of my domestic library. One, written by Charles Tanford and Jacqueline Reynolds, is "The Scientific Traveler" [4]. Its younger neighbour is Peter Day's Scientific Tourism [3]. In spite of the nearly coincident titles, hardly can two books be more different in their approach to "scientific tourism". Tanford and Reynolds focus on places throughout the world where landmarks in the history of science can be visited, preceded by a general historical and geographical introduction. Peter Day's book, instead, should be considered as a memoir of a scientist on duty, who travels for scientific events and yet observes the natural, urban and human environment.

There are not many coincident geographical spots in the two books, but we can compare, for instance, the sections devoted to Prague. Tanford and Reynolds stress the figures of astronomers Tyco Brahe and Johannes Kepler, who worked under the patronage of King Rudolf II. Their proposed tour for Prague includes a sculpture of both scientists, remains of the palace used by them as residence and observatory, the sepulchre of Brahe in a church, and a plaque that marks the house where Kepler lived. I wish one could find some hints to the activity of famous alchemists contemporary of Brahe and Kepler, with whom they shared the patronage of the King.

In contrast, Peter Day visits Prague in 1969, exactly one year after the Prague Spring during which Czech Prime Minister Alexander Dubcek attempted to promote a "communism with a human face" supported by mass demonstrations of the population that ended with the invasion of the Soviet tanks and the abolition of the political liberalization. He explains several aspects of the conference attended by him within the political context of the moment, without forgetting to mention the theatre where Mozart's opera Don Giovanni was first performed or the simultaneity of the conference organized by the Czechoslovak Academy of Science and the meeting of the Central Committee of the Communist Party that was to decide on the fate of Dubcek. These and other stories were transmitted by Day in Prullans before they got published as a book, with his clear diction, tidy transparencies, and even with his long-fingered hands (Figure 2). To get the flavour of this lecture, the reader is urged to search for Peter Day's quotation of Chaucer's "Canterbury Tales" in his book "Scientific Tourism".

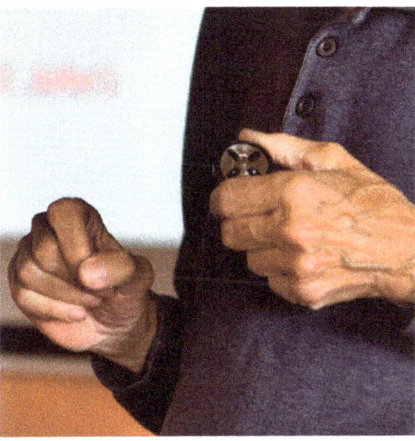

Figure 2. Peter Day's hands during a lecture on Scientific Tourism at NoSiC-8, 8 June 2018. Photo: S. Alvarez.

The travel chronicles of Peter Day have famous precedents in the history of chemistry. Ludwig Boltzmann, professor of theoretical physics at the University of Vienna, published an account of his travel to California in the summer of 1905 under the title "Reise eines deutschen Professors ins Eldorado", of which two English translations are available [5,6]. The 61-year old scientist travelled by train to Leipzig, continuing to Bremen where he took a steamship to New York. Then, after four days and nights on a railroad, with eventual visits to the smoking car, he made it to San Francisco. While the purpose of the travel was to give a summer course at Berkeley, Boltzmann seems more interested in keeping a record of the social events, meals and drinks than in the scientific aspects of his tour. Two quotations show his gastronomic interest the word "stomach" appears 11 times in the 25 pages of the article:

> «Journey to California is champagne, *Veuve Clicquot*, and oysters. Nobody with any experience in traveling will be surprised when I talk about eating and drinking. It is not only an important factor, it is the central point. It is most important during travel to keep the body healthy in face of the large variety of unaccustomed influences, most of all the stomach, and in particular the pampered stomach of a Viennese [6].»

Even when he comments on the rights of women at the university, his epicureanism emerges:

> «It goes without saying that male and female students have equal rights at universities such as these. The same is true of the faculty. I just want to present one drastic example of the far-reaching dominance achieved by the female element. One of my faculty colleagues, Miss Lilian Seraphine Hyde, a not unworthy lady whose name I committed to memory, gave a course of lectures on the preparation of salads and desserts. It was announced in the catalogue just like the course I was giving. I have kept that catalogue as evidence [6].»

Scientific tourism is of a totally different style in the pen of Humphry Davy, who published a book titled "Consolation in Travel" [7]. It describes imagined travels to a few archaeological and natural sites, and his interest is more on spiritual, historical, chemical or cultural issues, often unrelated to the touristic site. The book was written after Davy's recovery "from a long and dangerous illness", "under the same unfavourable and painful circumstances, and at a period when the constitution of the author suffered from new attacks." It was finished on 21 February 1829, but Davy passed away three months later, and the book was published posthumously in 1830.

The book presents Davy's six fictitious dialogues with two friends in Rome, Ambrosio and Onuphrio, in which they deal with history, religion and philosophy. After the first dialogue with his friends, inspired by the impressive ruins of the Roman Colosseum, Davy is immersed in deep reflections and has a vision in which a genius guides him through the evolution of the civilisation. He saw, for instance:

> « ... that in the place of the rolls of papyrus, libraries were now filled with books. "Behold", the Genius said, "the printing press; by the invention of Faust the productions of genius are, as it were, made imperishable, capable of indefinite multiplication, and rendered an unalienable heritage of the human mind.»

He saw also:

> «in the laboratories, alchemists searching for a universal medicine, an elixir of life, and for the philosopher's stone, or a method of converting all metals into gold.»

The subsequent dialogues are in a similar spirit, and they take place during visits to the Vesuvius, the temples of Paestum, the Austrian Alps (this time he travels by himself, and the conversation takes place with an unknown, a philosopher, and Eubathes, a travel companion), and Pola in the peninsula of Istria (Croatia).

There were, however, the short travels of Peter Day from Marquixanes to Prullans that have been important to me and left a deep imprint on me and other colleagues. After Day's first participation in a NoSIC, he was enthusiastic and highly encouraging about the style and contents of the meeting and participated in all subsequent editions except when health or family problems prevented him to do so. I have memories of him chatting with whoever happened to be next to him at breakfast, lunch or dinner time, and participating in every discussion after a lecture, as we see in Figure 3a, where he debates with inorganic chemist and poet Àngel Terron after his lecture "Science and Poetry: Exploring the Same Abyss". He had active participation also in the hands-on workshop "Chemical Answers to Culinary Challenges", led by Pere Castells, the chemist behind many of the innovations introduced by the world-famous cook Ferran Adrià, as seen in Figure 3b where he prepares pearls of yoghurt through the *spherification* process. He formed part of the group that visited the cathedral of La Seu d'Urgell, as recorded in a photograph (Figure 4a) in which he appears in conversation with historian of Chemistry Agustí Nieto-Galán in the cloister.

Figure 3. (**a**) Àngel Terron debates with Peter Day, with Magdolna and Istvan Hargittai in the background, at NoSIC-2, 31 May 2006; (**b**) Peter Day performs the spherification of a teaspoonful of yoghurt at NoSIC-2, surrounded by Rolf Saalfrank, Mario Ruben and Oriol Rossell. Prullans, 1 June 2006. Photos: S. Alvarez.

Figure 4. (**a**) Peter Day and Agustí Nieto-Galán during a visit to the cathedral of La Seu d'Urgell, 5 June 2008, (**b**) Day lecturing on the history of magnetism at the University of Barcelona. 8 October 2008. Photos by S. Alvarez.

Remarkably, Peter Day has left us three autobiographical books, yet they are not devoted to his many achievements in science, but rather to explain his perception of the many corners of the planet that he travelled to, and of the workings of the academic and scientific institutions that he visited or worked in. One of those books has an apparently odd title, "On the Cucumber Tree" a translation of a Hungarian expression "az uberkafán" that means "on the make" or "on the climb", taken by Day from a novel by English writer William Cooper [8], "nom de plume" of Harry S. Hoff (1910–2002). There are at least four botanical species known as "cucumber tree": *Averrhoa bilimbi*, *Dendrosicyos socotranus*, *Kigelia africana* and *Magnolia acuminata*. A photograph of the latter illustrates the cover of Day's book. None of them, however, is native to Europe, nor do they produce cucumbers, but fruits of similar shape. It is also interesting to inspect the chronology of his talks at Prullans and the publication of these autobiographical books (Table 1). In some cases, he would lecture about topics already dealt with in one of his books, while on other occasions he would give a talk, equally entertaining and well documented, that would be published later. For him, giving a lecture was something as fluent and natural as talking with a friend at teatime.

Other than the NoSIC meetings, we had a memorable chance to attend a talk delivered by him on the history of magnetism at the University of Barcelona, where he was invited to give the inaugural lecture of the exhibition "Two Millennia of Magnetism" organised by the Rare Books section of the Library of Physics and Chemistry of the University of Barcelona, on 8 October 2008. The lecture "Magnetism—A Mysterious Force of Nature and Some of its Consequences" was praised equally by scientists and non-scientists in the audience. In a picture taken during that event (Figure 4b) he appears to form part of the periodic table of the elements in the front wall of the "Aula Magna Enric Casassas", commonly known as Taula Magna ("taula" is the Catalan for "table").

Before concluding this account, let me quote Peter Day's view on the social role of research institutions and universities, taken from the epilogue of his book "Nature not Mocked":

«The unique positive feature distinguishing the university is the infinite and continuously changing cohort of young people that it brings to the research adventure. It is through fresh approaches that research flourishes most vigorously, and the new generations are its guarantors.»

At the NoSIC meetings, Thursday evening was the time for a special dinner served on long tables. The dates chosen were always at the beginning of June after we finished our teaching duties at the university and before the exams took place. Those were low-season dates for the weekdays at the Hotel Muntanya of Prullans, so we could charge a modest registration fee and be practically the only guests, which facilitated the sense of belonging to a group of friends. In the edition of 2010 during that dinner, a few days before my 60th birthday, I received as a gift a book signed by all the participants. At that point, Peter Day stood up and gave a toast to me with affection and a touch of sarcasm, "Santiago, there is life beyond 60!" he said. I have not forgotten his words, and now am sure that for him there is life beyond life.

Conflicts of Interest: The author declares no conflict of interest.

References

1. Day, P. *Nature Not Mocked: Places, People and Science*; Imperial College Press: London, UK, 2005.
2. Day, P. *On the Cucumber Tree: Scenes from the Life of an Itinerant Jobbing Scientist*, 5th ed.; The Grimsay Press: Glasgow, Scotland, 2012.
3. Day, P. *Scientific Tourism: Some Places in the Way*; P. Day: London, UK, 2020.
4. Tanford, C.; Reynolds, J. *The Scientific Traveler: A Guide to the People, Places & Institutions of Europe*; John Wiley: New York, NY, USA, 1992.
5. Boltzmann, L. Journey of a German Professor to El Dorado. *Transp. Theor. Stat. Phys.* **1991**, *20*, 499–523. [CrossRef]
6. Boltzmann, L. A German Professor's Trip to El Dorado. *Phys. Today* **1992**, *45*, 44–51. [CrossRef]
7. Davy, H. *Consolations in Travel, or the Last Days of a Philosopher*; John Murray: London, UK, 1851.
8. Cooper, W. *Memoirs of a New Man*; Macmillan: London, UK, 1966.

Article

Synthesis of New Derivatives of BEDT-TTF: Installation of Alkyl, Ethynyl, and Metal-Binding Side Chains and Formation of *Tris*(BEDT-TTF) Systems [†]

Songjie Yang [1], Matteo Zecchini [1], Andrew C. Brooks [1], Sara J. Krivickas [1], Desiree M. Dalligos [1], Anna M. Matuszek [1], Emma L. Stares [2], Melanie Pilkington [2] and John D. Wallis [1,*]

[1] School of Science and Technology, Nottingham Trent University, Clifton Lane, Nottingham NG11 8NS, UK; songjie.yang@ntu.ac.uk (S.Y.); info@umbriachem.com (M.Z.); sara.krivickas@adelaide.edu.au (S.J.K.)
[2] Department of Chemistry, Brock University, 1812 Sir Isaac Brock Way, St. Catharines, ON L2S 3A1, Canada; mpilkington@brocku.ca (M.P.)
* Correspondence: john.wallis@ntu.ac.uk
† This paper is dedicated to the memory of Professor Peter Day FRS, who greatly encouraged and inspired our work in this area.

Citation: Yang, S.; Zecchini, M.; Brooks, A.C.; Krivickas, S.J.; Dalligos, D.M.; Matuszek, A.M.; Stares, E.L.; Pilkington, M.; Wallis, J.D. Synthesis of New Derivatives of BEDT-TTF: Installation of Alkyl, Ethynyl, and Metal-Binding Side Chains and Formation of *Tris*(BEDT-TTF) Systems. *Magnetochemistry* **2021**, *7*, 110. https://doi.org/10.3390/magnetochemistry7080110

Academic Editors: Marius Andruh and Zheng Gai

Received: 1 June 2021
Accepted: 22 July 2021
Published: 3 August 2021

Publisher's Note: MDPI stays neutral with regard to jurisdictional claims in published maps and institutional affiliations.

Copyright: © 2021 by the authors. Licensee MDPI, Basel, Switzerland. This article is an open access article distributed under the terms and conditions of the Creative Commons Attribution (CC BY) license (https://creativecommons.org/licenses/by/4.0/).

Abstract: The syntheses of new BEDT-TTF derivatives are described. These comprise BEDT-TTF with one ethynyl group (HC≡C-), with two (*n*-heptyl) or four (*n*-butyl) alkyl side chains, with two *trans* acetal (-CH(OMe)$_2$) groups, with two *trans* aminomethyl (-CH$_2$NH$_2$) groups, and with an iminodiacetate (-CH$_2$N(CH$_2$CO$_2^-$)$_2$) side chain. Three transition metal salts have been prepared from the latter donor, and their magnetic properties are reported. Three *tris*-donor systems are reported bearing three BEDT-TTF derivatives with ester links to a core derived from benzene-1,3,5-tricarboxylic acid. The stereochemistry and molecular structure of the donors are discussed. X-ray crystal structures of two BEDT-TTF donors are reported: one with two CH(OMe)$_2$ groups and with one a -CH$_2$N(CH$_2$CO$_2$Me)$_2$ side chain.

Keywords: organic conductors; BEDT-TTF; synthesis; magnetism; crystal structures

1. Introduction

BEDT-TTF **1** has played a significant role in the development of electroactive organic materials. A very wide range of crystalline radical cation salts, with different stoichiometries, have been prepared, some of which are semi-conductors, conductors, or low temperature superconductors [1–4]. A range of different packing modes for the donors in these salts have been identified [5–7], which includes a κ-phase in the superconducting salts, such as (BEDT-TTF)$_2$Cu(NCS)$_2$, where the donors pack in face-to-face pairs but lie roughly perpendicular to their neighbouring pairs [8–10]. BEDT-TTF has been used to prepare hybrid materials with conducting and magnetic properties [11–14], as pioneered initially by Day et al. who prepared salts with iron *tris*(oxalate) salts which also showed low temperature superconductivity [11,12].

Following these studies on BEDT-TTF, a range of substituted BEDT-TTF derivatives that include compounds **2–5** have been reported (Scheme 1) [15]. Particularly notable examples are the enantiopure tetramethyl derivative **2** which forms a range of radical cation salts [16–19], the enantiopure amide **3** which forms a 4:1 TCNQ complex which changes from an insulator to an organic metal at 283 K and stays metallic down to ca. 4 K [20], the racemic bipyridylthiomethyl derivative **4** which forms capsular structures with metal ions [21], and the *spiro* chiral derivative **5** [22]. BEDT-TTF donors with one amino-methyl or -ethyl side chain [23], one hydroxy-methyl or -ethyl side chain [24,25], or with multiple hydroxymethyl or 1,2-dihydroxyethyl chains such as **6** and **7** have also been reported [26–28], as well as systems where the BEDT-TTF unit is fused to a thiophene or furan ring [29].

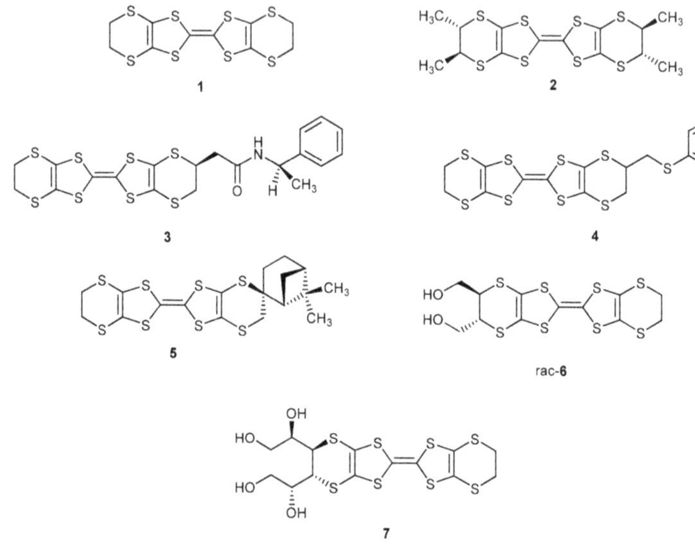

Scheme 1. Structures of donors 1–7.

Two general synthetic routes have been used to prepare the aforementioned BEDT-TTF derivatives (Scheme 2) [15]. The first is via the dithiolate 8, available from carbon disulphide and sodium, by double substitutions with dihalides or cyclic sulphate esters to give the bicyclic thione 9 [16,24]. The second is via the trithione 10, which reacts with alkenes in a 4 + 2 electrocyclic reaction to also give the bicyclic thione 9 [22–28,30–34]. In both cases, the synthesis is completed by conversion of the thione to the oxo compound 11 by treatment with mercuric acetate, and then reaction with triethyl or trimethyl phosphite to form the homo-coupled BEDT-TTF system 13. Cross-coupling of an oxo compound with a different thione is used to prepare an unsymmetrical derivative BEDT-TTF 12.

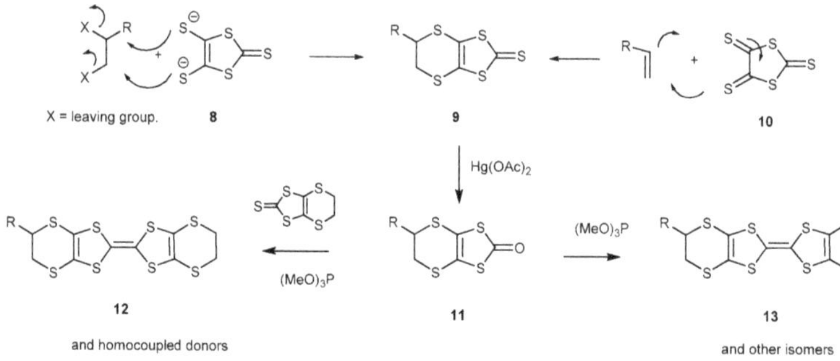

Scheme 2. Two general synthetic routes to BEDT-TTF derivatives.

In recent work, we have expanded our library of BEDT-TTF donors and now report the syntheses of a range of new racemic BEDT-TTF derivatives, prepared by the second synthetic strategy using the trithione intermediate 10. This includes BEDT-TTFs substituted with an ethynyl side chain, 38, or with metal binding groups, 48 and 53, in addition to molecules bearing three BEDT-TTF units around a benzene ring core, 26–28. Preliminary magnetic results for complexes of donor 53 with 3D transition metals are presented. Stereochemical aspects of substituted donors are also discussed. Full synthetic details are provided in Section 4 and in the Supplementary Information.

2. Results and Discussion

2.1. BEDT-TTF with Alkyl Chains

The installation of alkyl chains on a BEDT-TTF molecule should increase the solubility, which is very low for BEDT-TTF itself. Thus, the reaction of trithione **10** with *trans*-dec-5-ene gave the racemic thione **14** in 95% yield (Scheme 3). The treatment of this thione with mercuric acetate afforded the oxo compound **15** in an almost quantitative yield, which was homo-coupled using trimethyl phosphite to give the substituted BEDT-TTF in 57% yield as a mixture of stereoisomers. The configurations of the two stereogenic centres at one end of the molecule are the same but can be opposite to, or the same as, those at the other end. Thus, there are three stereoisomers: racemic (*R,R,R,R*)- and (*S,S,S,S*)-**16** and the *meso* compound (*R,R,S,S*)-**17**. Unfortunately, these are not separable by chromatography, but the structures are expected to have very similar shapes. In this respect, the conformation of the dithiin ring is typically an envelope or a half chair. Thus, the *R,R* and *S,S* configurations can position their side chains in similar pseudo-equatorial positions by adopting opposite conformations of their dithiin rings.

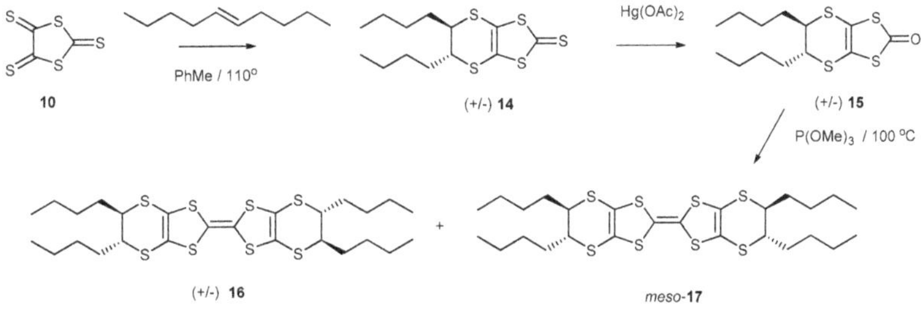

Scheme 3. Synthetic route to tetra-*n*-butyl-BEDT-TTF derivatives **16** and **17**.

Following the same synthetic strategy, the trithione **10** was reacted with non-1-ene to provide the bicyclic thione **18** with one heptyl side chain in 57% yield (Scheme 4). The thione was directly homo-coupled using trimethyl phosphite to give a mixture of two disubstituted BEDT-TTF donors (51%), each one having two diastereomers. The locations of the side chains could be at the top edge of the molecule, or one can be at the bottom edge, leading to the two differently substituted donors, each one of which has a racemic pair and a meso stereoisomer, **19–22**. Again, these were found to be inseparable. Both the **16–17** and **19–22** mixtures showed the expected two reversible oxidation peaks in their cyclic voltammograms (Table 1) at 0.47 and 0.89 V for **16–17** and 0.49 and 0.89 V for **19–22** (relative to Ag/AgCl).

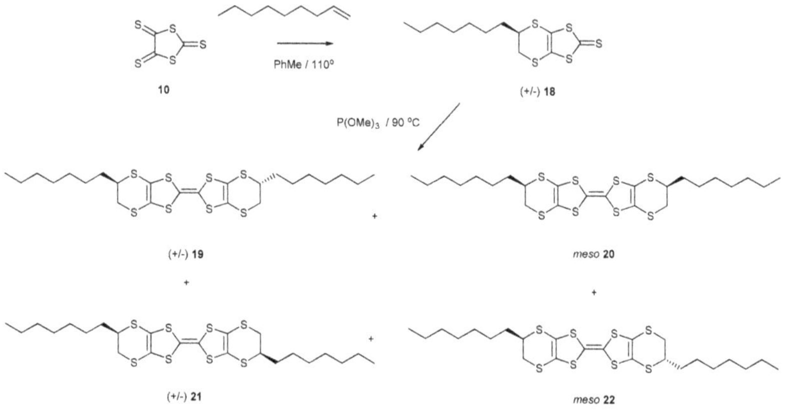

Scheme 4. Synthetic routes to di-n-heptyl-BEDT-TTF derivatives **19–22**.

Table 1. Cyclic voltammetry data for selected BEDT-TTF derivatives [a].

Donor	E_1 (V)	E_2 (V)
16–17	0.47	0.89
19–22	0.49	0.89
26	0.48, 0.55 [b]	0.88 [b]
27	0.44, 0.56 [b]	0.86 [b]
28	0.50	0.88
38	0.52	0.92
42	0.50	0.93
48	0.72 [b,c]	0.89 [b,c]
52	0.53	0.94

[a] Measured in 0.1 M tetrabutylammonium hexafluorophosphate in DCM at 20 °C, substrate concentrations 0.01 mM, and scan rate of 0.1 V s^{-1}, unless otherwise stated; [b] irreversible; [c] in THF.

2.2. Tris-(BEDT-TTF) Donors

To provide systems which will show different packing arrangements versus the classical BEDT-TTF systems and which could also act as supramolecular synthons, e.g., for forming charge transfer complexes with fullerenes, three *tris*-donor systems **26–28** were prepared containing three donors appended to a benzene core. The synthetic route to these compounds involved the coupling of benzene-1,3,5-tricarboxylic acid with three equivalents of a BEDT-TTF donor bearing a hydroxyalkyl side chain using *n*-propyl phosphonic anhydride ("T3P") or DCC (Scheme 5). Thus, donors **23** and **24** with a hydroxy-methyl or hydroxyethyl side chain afforded the *tris*-donors **26** and **27** in 22 and 33% yields, respectively. Interestingly, there are two racemic disastereomers for these two donors, one with the same configuration at all three stereogenic centres (*R,R,R*) and one with the same configuration at just two centres (*R,R,S*). However, the products proved to have low solubilities in common organic solvents, so the hydroxyethyl-BEDT-TTF donor **25**, functionalised with two *trans* oriented *n*-butyl side chains on the other ethylene bridge, was prepared and was coupled with the triacid to give the more soluble *tris*-donor **28** in 45% yield. The number of stereoisomers is increased over **26–27** because the three sets of butyl groups can have all six or just four configurations be the same. However, these stereochemical issues are unlikely to have a profound influence on the behaviour of the donor, due to the flexibilities of the dithiin rings, as discussed earlier. The structures were supported by their chemical analyses and spectral data, e.g., **26** showed a molecular ion in the mass spectrum, and **27–28** showed the three aromatic H atoms at ca. 8.8 ppm. The more soluble *tris* derivative **28** showed a reversible cyclic voltammogram with peaks at 0.50 and 0.88 V as expected of a BEDT-TTF derivative, but *tris* donors **26** and **27** showed irreversible volt-ammograms probably due to the binding of the donor to the electrode, with two oxidation peaks between 0.44 and 0.56 V and another in the range 0.86–0.88 V on the outward scan.

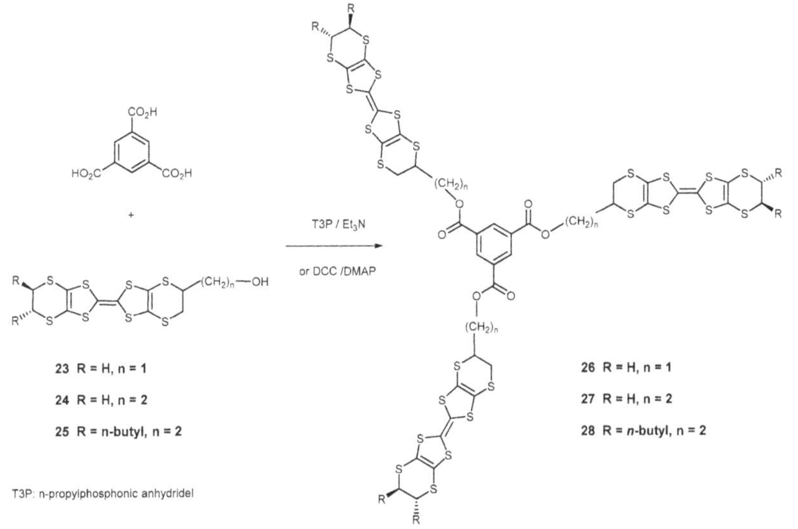

Scheme 5. Synthetic route to *tris*-BEDT-TTF derivatives **26–28**.

2.3. New Functionalised BEDT-TTFs: Ethynyl and Acetal Substituted Derivatives

There are several examples of the cyclisation of the trithione **10** with alkynes (Scheme 6), in particular with those where the triple bond bears two carbonyl or acetal functionalities, affording thiones such as **29** and **30** [35–37]. The electron deficient diarylethynes also react with the trithione **10**, giving access to thiones such as **31** and **32** [38,39]. An interesting case, therefore, would be the reaction with 4-trimethylsilyl-1-buten-3-yne **33**, which has both an alkene and an alkyne group, to determine which group preferentially reacts with the trithione **10** (Scheme 7). Refluxing the two materials together gave a 55% yield of the bicyclic trithione **34** in which the alkene, and not the alkyne, had reacted. The reaction of the trimethylsilylethynyl thione with mercuric acetate gave the oxo compound **35** in an almost quantitative yield, without affecting the triple bond, and this material was cross-coupled with the unsubstituted thione **36** to give the trimethylsilylethynyl substituted BEDT-TTF donor **37** in 45% yield. Treatment with potassium fluoride in THF/methanol gave the deprotected ethynyl-BEDT-TTF donor **38** in 88% yield. This shows the greater reactivity of the alkene over the alkyne with the trithione. This has provided an interesting donor which has the potential for connecting to larger molecular systems, either preserving the triple bond by substitution of the ethynyl H atom, e.g., by a Sonogashira reaction, or using the triple bond as a substrate for click chemistry with azide-functionalised materials.

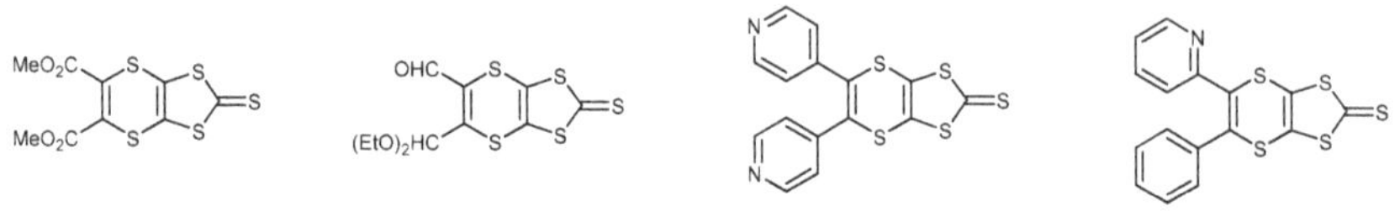

Scheme 6. Structures of thiones prepared from trithione **10** and various disubstituted alkynes.

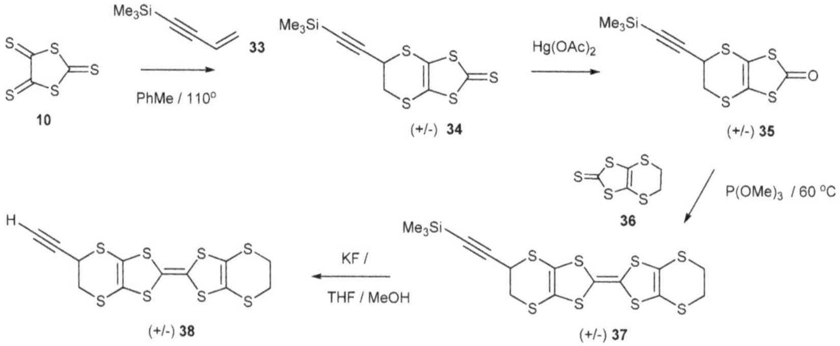

Scheme 7. Synthetic route to ethynyl-BEDT-TTF **38**.

We were interested in making BEDT-TTF with two aldehyde groups attached, which could facilitate further ring constructions or the attachment of side chains. Thus, we targeted the corresponding *bis*(dimethylacetal) containing donor **42**. The reaction of the corresponding alkene, the *bis*-dimethylacetal of fumaraldehyde **39**, with trithione **10** gave the disubstituted thione **40** in a low yield of 17%, which was converted to the oxo compound **41** in almost quantitative yield (Scheme 8). The donor **42** was obtained by the cross-coupling of oxo compound **41** with the unsubstituted thione **36** in 49% yield. However, all attempts to hydrolyse the acetal groups with HCl or tosic acid in a range of concentrations failed to give the corresponding dialdehyde. Similar experiences have been reported on organo-sulphur donors containing acetal or ketal functionality [24,40]. It may be that these donor molecules stack together in clumps and isolate themselves from the acidic environment. Nevertheless, it is quite remarkable that these groups are resistant to hydrolysis, when dilute acid is usually quite sufficient. The cyclic voltammograms of BEDT-TTF derivatives **38** and **42** showed the expected two reversible oxidation peaks (Table 1).

Single crystals suitable for X-ray diffraction of **42** were grown from CH_3CN, and the molecular structure of the donor was determined (Figure 1). The substituted dithiin ring adopts a half-chair conformation and directs the acetal groups into pseudo-axial positions. The organosulphur moiety adopts a bowed shape with inflexions about the S···S vectors across the two dithiole rings of 19.6 and 17.2°. In the crystal structure, the donors pack in centrosymmetric slipped pairs to accommodate the side chains, with the shortest C···C contact of 3.546 Å between the TTF cores.

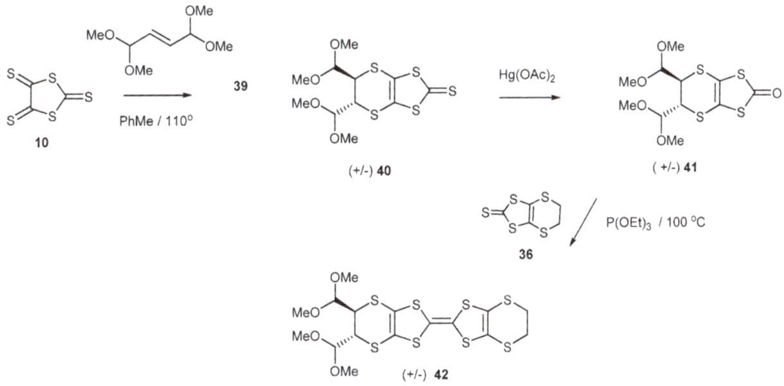

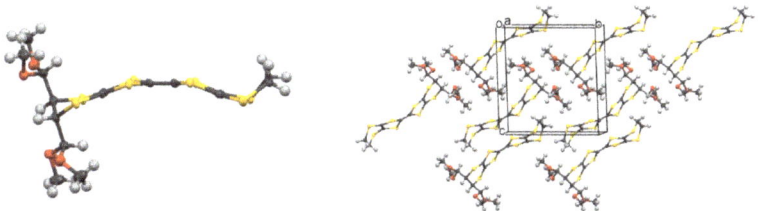

Scheme 8. Synthetic route to the BEDT-TTF derivative functionalised with two *trans*-oriented, dimethylacetal groups **42**.

Figure 1. Molecular structure of donor **42** (**left**) and its crystal packing arrangement (**right**).

2.4. Donors with Metal Ion Binding Potential and the First Salts and Their Magnetic Properties

The BEDT-TTF donor **48** with *trans* aminomethyl groups was an interesting target, both for further synthetic elaborations and for the binding of metal ions. In a first approach, which we hoped would have a wide scope, thione **10** was reacted successfully with *trans*-1,4-dibromobut-2-ene to give the *trans bis*(bromomethyl)thione **43**. However, attempts to substitute the bromide groups of this material with amines were unsuccessful, with the amine probably causing the elimination of HBr. Therefore, to synthesize the donor with two aminomethyl groups **48**, the trithione **10** was reacted with the *bis*-N-boc derivative of *trans*-but-2-en-1,4-diamine **44** to give the thione **45** in 53% yield (Scheme 9). The corresponding reaction of the trithione with the readily obtained *bis*-phthalimido derivative of *trans*-but-2-en-1,4-diamine was unsuccessful. Following the established procedure, the thione **45** was converted to the oxo compound **46** which was cross-coupled with the unsubstituted thione **36** to give the *bis*-N-Boc donor **47** in 63% yield. Deprotection of this donor with HCl in dioxane followed by basification gave the *bis*(aminomethyl)BEDT-TTF **48** in 78% yield. The cyclic voltammogram of **48** in THF showed broad irreversible peaks, probably due to the donor binding to the electrode.

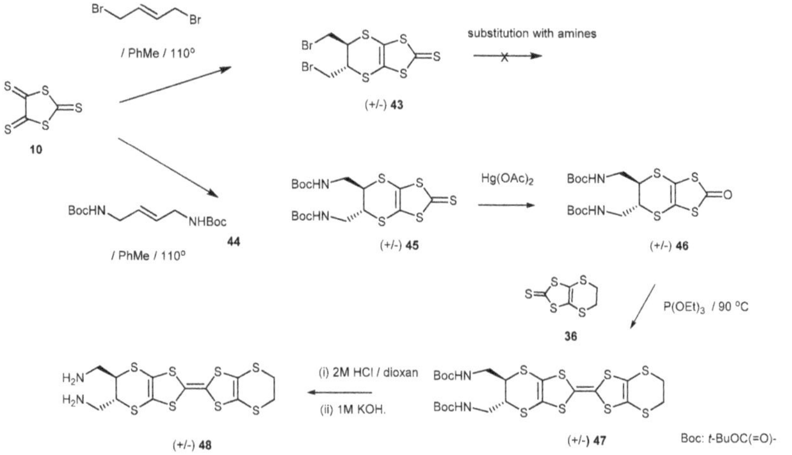

Scheme 9. Synthetic route to *trans-bis*-(aminomethyl)-BEDT-TTF **48**.

Installing an iminodicarboxylate dianion group on the side chain of BEDT-TTF could provide another BEDT-TTF donor with potential metal binding properties, so donor **53** with just a methylene between the donor and the metal binding group was prepared. This was synthesized following the synthetic strategy outlined in Scheme 10. The attachment of an allyl group to diethyl iminodiacetate gave alkene **49** which reacted with trithione **10** to give the substituted thione **50** in high yield. Conversion to the corresponding oxo compound **51** and subsequent cross-coupling with the unsubstituted thione **36** afforded the BEDT-TTF donor **52** with an imino diethyl ester group in the side chain in 34% yield.

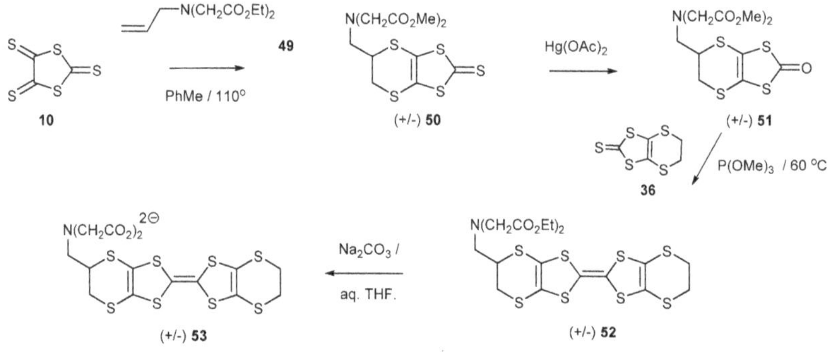

Scheme 10. Synthetic route to the disodium salt of BEDT-TTF-methylamino-N,N-dicarboxylate **53**.

Single crystals of **52** were grown from DCM, and the crystal structure was determined by single crystal X-ray diffraction (Figure 2). The BEDT-TTF moiety adopts a bowed structure, with inflexions about the two S···S vectors in the dithiole rings of 14.6 and 23.3°. The unsubstituted ethylene bridge is disordered between two half-chair conformations (62:38). On the other bridge the carbon bearing the side chain and its attached hydrogen atom are disordered between two positions (82:18) which correspond to the structures of opposite enantiomers of the donor. This illustrates how the bridge can flex to accommodate both enantiomers without affecting the positions of either the main parts of the organosulphur system or the side chain. The donors are packed in centrosymmetric pairs with the shortest C···C distance between the TTF cores of 3.42 Å.

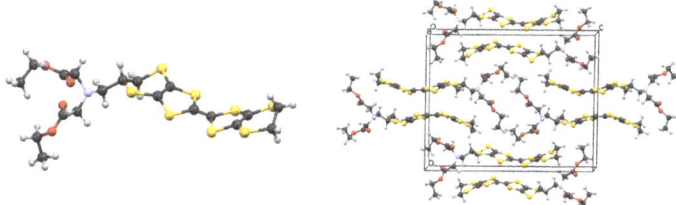

Figure 2. Crystal structure of donor **52**, showing the main molecular conformation (**left**) and the crystal packing arrangement (**right**).

Hydrolysis of the imino diethyl ester substituted donor **52** with sodium hydroxide gave the dianion **53** as a disodium salt in high yield. The reaction of this donor with a range of transition metal salts afforded precipitates which were all insoluble in a wide range of solvents, and from which single crystals could not be obtained. Thus, assignments of compositions are tentative, based only on chemical analysis, magnetic measurements, and infrared spectra, and we acknowledge that the structural topologies may be more complex. Ligand **53** can show a number of coordination modes (Figure 3). It can act as tridentate dianion ($\mathbf{L^{2-}}$) binding by two oxygens and nitrogen; be protonated once ($\mathbf{HL^-}$), on nitrogen or oxygen, and act as a bidentate or tridentate monoanion, respectively; or be protonated twice and coordinate as a neutral ligand ($\mathbf{H_2L}$). Such behaviours are seen with simpler N-alkylated iminodiacetate ligands [41–43]. It can also act as just a bridging ligand [44]. Furthermore, binding to the outer set of the BEDT-TTF unit's sulphur atoms cannot be excluded, as has been observed with copper(I) and silver(I) [45–47]. Details of three of the coordination complexes obtained are provided in Table 2. With zinc triflate, the CHN data are consistent with the product **54** containing two equivalents of the $\mathbf{HL^-}$ monoanion coordinated to Zn(II). With MnCl$_2$, the product **55** contains two neutral $\mathbf{H_2L}$ ligands along with MnCl$_2$ and two waters. For the MnCl$_2$ complex, at room temperature, the value of the χT product (4.4 cm^3 K mol^{-1}) is in excellent agreement with the theoretical expected value for an isolated HS octahedral Mn(II) ion (4.375 cm^3 K mol^{-1}) with a g value of 2.00. Similarly, for the second Mn(II) complex **56** prepared from Mn(hfac)$_2$, the room temperature χT product of 4.5 cm^3 K mol^{-1} is again very reasonable for an isolated HS Mn(II) ion in octahedral geometry, where compared to complex **55**, the hfac ions replace the chlorides, though with one less $\mathbf{H_2L}$ ligand and one less water molecule. For both Mn(II) complexes, dc magnetic studies reveal that the χT product is temperature independent down to ca. 50 K, after which time it rapidly decreases, which is consistent with the presence of antiferromagnetic interactions.

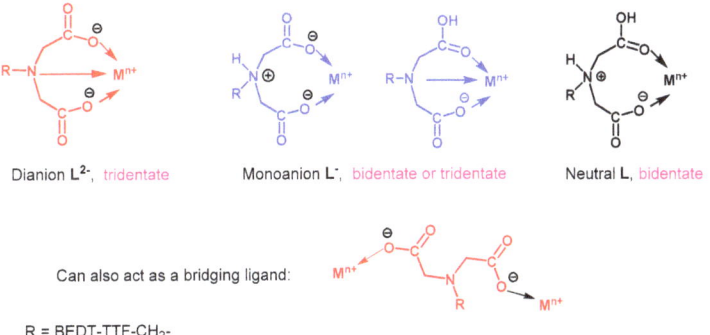

Figure 3. Possible binding modes of donor **53** to a metal ion, depending on its degree of protonation.

Table 2. Proposed compositions, magnetic data, and chemical analysis for the reaction of ligand **53** with Zn(II) and Mn(II) salts.

Metal salt, Colour.	Proposed Composition, Compound number, RMM.	Magnetic properties: [a] χT (cm^3 K mol^{-1}); C (cm^3 K mol^{-1}); θ (K).	Found CHN (%)	Calculated CHN (%)
Zn(II)triflate Red/pink.	Zn(HL)$_2$ **54** RMM 1121.4		C: 31.95 H: 2.41 N: 2.64	C: 32.09 H: 2.51 N: 2.49
MnCl$_2$ Red.	Mn(H$_2$L)$_2$Cl$_2$.2H$_2$O **55** RMM 1220	χT: 4.4 C: 4.47 θ: −6.07	C: 29.28 H: 2.57 N: 2.58	C: 29.50 H: 2.81 N: 2.29
Mn(hfac)$_2$ Orange/brown.	Mn(H$_2$L)(hfac)$_2$.H$_2$O **56** RMM 1016.8	χT: 4.5 C: 4.66 θ: −7.72	C: 29.58 H: 1.88 N: 1.48	C: 29.53 H: 1.88 N: 1.38

[a] Plots of $1/\chi$ vs. T, and χT vs. T are provided in the ESI.

3. Conclusions

The synthesis of a range of new BEDT-TTF derivatives, which are substrates for the formation of charge transfer salts as well as for incorporation in more complex systems, has been described. The development of conducting and hybrid materials is highly dependent on the availability of new donor systems, and such syntheses are often not straightforward. For example, the trithione **10** does not react with all alkenes, and the presence of the BEDT-TTF unit can disrupt apparently standard synthetic manipulations to side chain functionalities. The possibility of forming hybrid materials between the donor bearing an iminodiacetate function **53** and metal ions has been demonstrated and will be extended in the future to the *bis*(diaminomethyl) donor **48**, though methods for forming crystalline products, e.g., by hydrothermal synthesis, need to be developed. Furthermore, to produce conducting materials, an oxidation step needs to be included. Of particular note is that the flexibility of the ethylene bridges of BEDT-TTF donors results in both enantiomers of the substituted BEDT-TTF being close to superimposable, as shown in the crystal structure of **52**. This indicates that to prepare systems with chiral packing arrangements, the stereogenic feature might be more effective in the side chain. However, we note that the first observation of magnetochiral anisotropy in organic conductors was in the perchlorate salts of enantiomers of dimethyl EDT-TTF, where the chirality was on the donor [48].

4. Materials and Methods

General. Solution NMR spectra were measured on either a Jeol ECLIPSE ECX or ECZ spectrometer operating at 400 MHz for ^{1}H and at 100.6 MHz for ^{13}C, using CDCl$_3$ as solvent and tetramethylsilane (TMS) as standard unless otherwise stated, and measured in p.p.m. downfield from TMS with coupling constants reported in hertz. IR spectra were recorded on a Perkin Elmer Spectrum 100 FT-IR Spectrometer using attenuated total reflection sampling on solids or oils and are reported in cm^{-1}. Mass spectra were recorded at the EPSRC Mass Spectrometry Centre at the University of Swansea or on a Waters Xevo QTOF G2 XS using an ESI source with a Waters Acquity UPLC system at NTU. Chemical analysis data were obtained from London Metropolitan University and Nottingham University, UK. Flash chromatography was performed on 40–63 µm silica gel obtained from Fluorochem Ltd. (Hadfield, Glossop, Derbyshire, UK).

Magnetic Measurements: Variable-temperature dc magnetic measurements were performed on a Quantum Design SQUID MPMS magnetometer in an applied field of 0.1 T, from 2 to 300 K. The experimental data were corrected for the diamagnetism and signal of the sample holder.

Full details of the synthetic procedures and further details of the magnetic data are provided in the ESI. The syntheses of **49–56** are given below to illustrate the general methods.

Diethyl allylamino-N,N-diacetate, **49**: To a stirred solution of diethyl amino-N,N-diacetate (3.04 g, 16.1 mmol) in THF (50 mL), allyl bromide (2.20 mL, 25.3 mmol) was slowly added followed by addition of K_2CO_3 (3.42 g, 24.7 mmol), and the resulting suspension was warmed up to reflux and left to stir for 20 h. The reaction mixture was cooled to room temperature, and THF was evaporated under reduced pressure. DCM was added to the residue (100 mL) and the organic layer was washed with water (100 mL) and brine (100 mL) and dried over Na_2SO_4. Evaporation of DCM afforded diethyl allyl-amino-N,N-diacetate **49** (3.16 g, 86%) as a pale yellow oil; δ_H (400 MHz, $CDCl_3$): 5.88 (1H, m, -CH=CH_2), 5.22 (1H, dd, J = 18.8, 1.7 Hz, -CH=CH_{trans}H), 5.17 (1H, dd, J = 10.1, 1.8 Hz, -CH=CHH_{cis}), 4.17 (4H, q, J = 7.2 Hz, -O-CH_2-CH_3), 3.57 (4H, s, 2 × –N(CH_2COOEt)$_2$), 3.38 (2H, d, J = 6.7 Hz, -CH_2-CH=), 1.27 (6H, t, J = 7.0 Hz, 2 × -CH_3); δ_C: (100 MHz, $CDCl_3$): 170.6 (2 × -C=O), 134.9 (=CH), 118.0 (=CH_2), 60.0 (2 × -O-CH_2-CH_3), 57.0 (=CH-CH_2-N-), 53.8 (2 × -N-CH_2CO), 13.8 (2 x-CH_3); ν_{max}: 2981, 1737, 1674, 1399, 1333, 1190, 714; HRMS: (ASAP) found: 230.1383 (100%), $C_{11}H_{20}NO_4$ +H: requires: 230.1387; found C, 57.69; H, 8.36; N, 6.19%; $C_{11}H_{20}NO_4$ requires C, 57.64; H, 8.29; N, 6.11%.

Diethyl (+/−)-5,6-dihydro-2-thioxo-1,3-dithiolo[4,5-b]1,4-dithiin-5-methylamine-N,N-diacetate, **50**: Diethyl allyl-amino-N,N-diacetate **49** (1.17 g, 5.0 mmol) was added to a suspension of trithione **10** (3.25 g, 17.0 mmol) in toluene (100 mL), and the resulted suspension was warmed up to reflux and left to stir overnight under a nitrogen atmosphere. The solid formed during the reaction was filtered off and washed with $CHCl_3$ until washes ran clear. The combined filtrates were evaporated under reduced pressure to give the desired thione **50** as a dark red oil (2.09 g, 96%). δ_H (400 MHz, $CDCl_3$): 4.10 (4H, q, J = 7.0 Hz, -O-CH_2-CH_3), 3.72 (1H, m, 5-*H*), 3.51 (4H, s, 2 × (-N-CH_2-C=O), 3.43 (1H, dd, J = 13.4, 6.1 Hz, 5-(CH_αH)-N), 3.29 (1H, dd, J = 13.3, 2.8 Hz, 5-(CHH_β)-N)), 3.13 (1H, dd, J = 13.8, 8.7 Hz, 6-H_α), 3.05 (1H, dd, J = 13.8, 6.5 Hz, 6-H_β), 1.20 (6H, t, J = 7.2 Hz, 2 × -CH_3); δ_C: (100 MHz, $CDCl_3$): 208.0 (C=S), 170.9 (2 × C=O), 123.6, 122.2 (3a-, 7a-C), 60.8 (2 x-O-CH_2), 58.7 (5-CH_2-N), 56.2 (2 x-N-CH_2-C=O), 42.6 (5-C), 31.9 (6-C), 14.2 (2 x-CH_3); ν_{max}: 2975, 1729, 1483, 1368, 1259, 1180, 1140, 1054, 883, 799, 512; HRMS: (ASAP) found: 425.9995 (100%), $C_{14}H_{19}NO_4S_5$ + H: requires: 425.9990; found C, 39.55; H, 4.51; N, 3.35%, $C_{14}H_{19}NO_4S_5$ requires C, 39.53; H, 4.47; N, 3.29%.

Diethyl (+/−)-5,6-dihydro-2-oxo-1,3-dithiolo[4,5-b]1,4-dithiin-5-methylamine-N,N-diacetate, **51**: To a solution of thione **50** (1.88 g, 4.40 mmol) in $CHCl_3$ (100 mL), mercury (II) acetate (3.54 g, 11.0 mmol) was added, and the suspension was left stirring for 3 h at room temperature under a nitrogen atmosphere. The solid formed during the reaction was filtered off and washed with $CHCl_3$ until washes ran clear. The combined filtrates were washed with saturated sodium hydrogen carbonate solution (5 × 50 mL).The organic layer was washed with water (50 mL) and brine (50 mL) and dried over $MgSO_4$. Evaporation of the chloroform yielded the desired oxo-compound **51** as a brown oil (1.71 g, 91%). δ_H (400 MHz, $CDCl_3$): 4.17 (4H, q, J = 7.1 Hz, -O-CH_2CH_3), 3.82 (1H, m, 5-*H*), 3.59 (4H, s, 2 × -N-CH_2-C=O), 3.49 (1H, dd, J = 13.3, 6.2 Hz, 5-(CH_α)N-), 3.39 (1H, dd, J = 13.3, 3.1 Hz, 5-(CH_βH)N-), 3.23 (1H, dd, J = 13.8, 8.7 Hz, 6-H_α), 3.13 (1H, dd, J = 13.8, 6.4 Hz, 6-H_β), 1.28 (6H, t, 2 × -CH_3); δ_C (100 MHz, $CDCl_3$): 188.8 (C=O), 171.0 (2 × -C=O(OEt)), 113.7, 111.8 (3a-, 7a-C), 60.8 (2 × -O-CH_2-CH_3), 58.8 (-CH_2-N-), 56.2 (2 × -N-CH_2C=O), 44.1 (5-C), 32.9 (6-C), 14.2 (2 × -CH_3); ν_{max}: 2978, 1731, 1670, 1443, 1411, 1368, 1186, 1139, 1024, 762, 463; HRMS: (ASAP) found: 410.0208, $C_{14}H_{19}NO_5S_4$+H: requires: 410.0219; found C, 40.94; H, 4.72; N, 3.50%, $C_{14}H_{19}NO_5S_4$ requires C, 41.07; H, 4.64; N, 3.42%.

Diethyl (+/−)-BEDT-TTF-methylamino-N,N-diacetate, **52**: Oxo compound **51** (2.01 g, 5.0 mmol) and 5,6-dihydro-dithiolo[4,5-b]dithiin-2-thione **36** (1.79 g, 8.0 mmol) were heated together in freshly distilled triethyl phosphite (50 mL) at 90 °C for 6.5 h under a nitrogen atmosphere. The solid formed was filtered and washed with $CHCl_3$, and the combined filtrates were concentrated under reduced pressure. The triethyl phosphite was distilled off using a Kugelrohr apparatus. The residue was purified by flash chromatography (6:1 = cyclohexane: ethyl acetate) to give the substituted BEDT-TTF **52** as an orange solid (0.99 g, 34%), m.p. 107–108 °C; δ_H (400 MHz, $CDCl_3$): 4.09 (4H, q, J = 7.1 Hz, 2 × -O-

CH_2-CH_3), 3.64 (1H, m, 5-*H*), 3.50 & 3.51 (2 × 2H, 2 × s, 2 × (-N-CH_2-C(O)), 3.33 (1H, dd, J = 13.0, 6.0 Hz, 5-(CH_αH)-N, 3.21 (4H, s, (5'-, 6'-CH_2)), 3.20 (1H, dd, J = 13.0, 3.1 Hz, 5-(CH_βH)-N), 3.03 (2H, m, 6-H_2), 1.20 (6H, t, J = 7.2 Hz, 2 × -CH_3); δ_C (100 MHz, CDCl$_3$): 171.0 (2 × C=O), 113.8, 113.7, 112.5, 111.5, 109.8 (2-, 2'-, 3a-, 7a-, 3'a-, 7'a-C), 60.6 (2 x-O-CH_2-CH_3), 58.7 (5-CH_2-N-), 56.1 (2 x-N-CH_2C=O), 42.6 (5-C), 32.5 (6-C), 30.1 (5'-,6'-C), 14.1 (2 x-CH_3); ν_{max}: 2921, 1733, 1449, 1412, 1372, 1190, 1022, 770; found C, 39.16; H, 4.00; N, 2.29%; C$_{19}$H$_{23}$NO$_4$S$_8$ requires C, 38.97; H, 3.93; N, 2.39%.

Disodium (+/−)-BEDT-TTF-methylamino-N,N-dicarboxylate **53** *as a tetrahydrate, Na$_2$L.4H$_2$O*: The di-ester donor **52** (0.37 g, 6.40 mmol) was dissolved in THF (10 mL), and an aqueous solution of NaOH (5 mL, 0.256 M, 12.8 mmol) was added. The suspension was warmed to 50 °C and left stirring overnight. The THF was evaporated, and the solid filtered and washed successively with DCM (3 × 5 mL), water (3 × 5 mL), and ether (3 × 10 mL). Filtration gave **53** (0.41 g, quantitative) as a highly insoluble red solid, m.p. 240 °C (dec). ν_{max}: 2917, 1582, 1400, 1326, 1122, 995, 904, 771, 669; found C, 28.16; H, 3.03; N, 2.38%, C$_{15}$H$_{13}$NO$_4$S$_8$Na$_2$.4H$_2$O required C, 27.90; H, 3.28; N, 2.17%.

Preparation of metal salts of **53**: (a) *with zinc(II) triflate to give* **54**: To a suspension of sodium salt **53** (77 mg, 0.12 mmol) in dry MeOH (8 mL) at room temperature under a nitrogen atmosphere, zinc triflate (25 mg, 0.069 mmol) was added, and immediately after the addition, a brown suspension was formed. The reaction mixture was left to stir for 1 h, and then the solid was filtered off, washed with diethyl ether, and left to dry in air to give Zn(**HL**)$_2$, **54**, (45 mg, 58%); m.p. 243–244 °C (dec.), ν_{max}: 2950, 2854, 1596 (C=O), 1422, 1398, 1342, 1143, 996, 893, 766, 748; found C, 31.94; H, 2.41; N, 2.64%; C$_{30}$H$_{28}$N$_2$O$_8$S$_{16}$Zn requires C, 32.09; H, 2.51; N, 2.49%.

(b) *with manganese(II) chloride to give* **55**: To a suspension of sodium salt **53** (0.212 g, 0.33 mmol) in distilled water (5 mL) at room temperature, MnCl$_2$.4H$_2$O (0.035 g, 0.18 mmol) was added, and immediately after the addition, a red precipitate was formed. The obtained solid was filtered and washed with diethyl ether and left to dry in air to give Mn(H$_2$L)$_2$Cl$_2$.2H$_2$O, **55**, (147 mg, 68%), m.p. 206–207 °C (dec). ν_{max}: 3321, 2914, 1595 (C=O), 1399, 1364, 1194, 1143, 996, 892; found C, 29.28; H, 2.57; N, 2.58%; C$_{30}$H$_{30}$O$_8$S$_{16}$N$_2$Cl$_2$Mn.2H$_2$O required C, 29.50; H, 2.81; H, 2.29%.

(c) *with manganese(II) (hfac)$_2$ to give* **56**: To a suspension of sodium salt **53** (26.0 mg, 0.052 mmol) in dry MeOH (6 mL) at room temperature under a nitrogen atmosphere, Mn(hfac)$_2$.3H$_2$O (24 mg, 0.046 mmol) was added. Immediately after the addition, a brown solid precipitated out from the mixture. The reaction was left to stir for 1 h, and then the solid was filtered. The brown precipitate was collected, washed with diethyl ether, and left to dry in air to give Mn(H$_2$**L**)(hfac)$_2$.H$_2$O, **56**, (22 mg, 48%), m.p. 137–140 °C; ν_{max}: 3296, 1601, 1401, 1319, 1054, 994, 902, 771; found C, 29.61; H, 1.76; N, 1.81%; C$_{25}$H$_{17}$NO$_8$F$_{12}$S$_8$Mn.H$_2$O requires C, 29.53; H, 1.88; N, 1.38%.

X-ray Crystallography. X-ray diffraction data (MoKα) were measured at low temperature for **42** (120 K) and **52** (150 K). Structures were solved and refined using the SHELXS and SHELXL suite of programs [49,50] using the XSEED interface [51]. Molecular illustrations and geometric analysis were made with Mercury [52]. Data are deposited at the Cambridge Crystallographic Data Centre with code numbers CCDC 2083578–2083579.

Crystal data for **42**: C$_{16}$H$_{20}$O$_4$S$_8$, M_r = 532.80, triclinic, a = 6.3985(4), b = 12.6825(11), c = 13.4735(10) Å, α = 89.198(4), β = 81.424(5), γ = 87.025(5)°, V = 1079.66(14) Å^3, Z = 2, P-1, D_c = 1.64 g cm^{-3}, μ = 0.849 mm^{-1}, T = 120(2) K, 4995 unique reflections (R_{int} = 0.078), 3422 with $F^2 > 2\sigma$, $R(F, F^2 > 2\sigma)$ = 0.059, R_w (F^2, all data) = 0.13. Crystal from acetonitrile.

Crystal data for **52**: C$_{19}$H$_{23}$NO$_4$S$_8$, M_r = 585.8, monoclinic, a = 6.5481(4), b = 17.6090(16), c = 22.2441(14) Å, β = 97.298(5)°, V = 2544.1(3) Å^3, Z = 4, $P2_1/c$, D_c = 1.53 g cm^{-3}, μ = 0.729 mm^{-1}, T = 150(2) K, 5831 unique reflections (R_{int} = 0.090), 4196 with $F^2 > 2\sigma$, $R(F, F^2 > 2\sigma)$ = 0.085, R_w (F^2, all data) = 0.23. Crystal from dichloromethane.

Supplementary Materials: The following are available online at https://www.mdpi.com/article/10.3390/magnetochemistry7080110/s1, S1: full experimental details for the syntheses not described in Section 4; S2: magnetic data for **55** and **56**; S3: references for S1–S2.

Author Contributions: Conceptualization, J.D.W.; synthetic methodology, S.Y., M.Z., A.C.B., S.J.K., D.M.D., and A.M.M.; magnetic measurements, E.L.S. and M.P.; crystallography J.D.W. and A.C.B.; manuscript preparation, J.D.W. and M.P. All authors have read and agreed to the published version of the manuscript.

Funding: NSERC (DG 2018-04255) (MP).

Institutional Review Board Statement: Not relevant.

Informed Consent Statement: Not applicable.

Data Availability Statement: Crystallographic information files for **42** and **52** are available at The Cambridge Crystallographic Data Centre, 12 Union Road, Cambridge, CB2 1EZ, UK, with numbers: CCDC 2083578-2083579.

Acknowledgments: J.D.W. thanks Nottingham Trent University for PhD scholarships (M.Z., A.C.B.) and for financial support. J.D.W. also thanks the EPSRC Mass Spectrometry Service, Cardiff University, Wales, UK, for data. M.P. acknowledges financial support from NSERC (DG-2018-04255).

Conflicts of Interest: The authors declare no conflict of interest.

Dedication: This work was stimulated by our interaction with Peter Day FRS whom we first met when he was the Director of the Royal Institution in London. Most of the authors have benefited from contact with Professor Day, and some have worked closely with him. We very much appreciated his great interest and strong encouragement to synthesize new BEDT-TTF derivatives for study and his continuing interest over the years after his "retirement".

References

1. Singleton, J.; Mielke, C. Quasi-Two-Dimensional Organic Superconductors. *Contemp. Phys.* **2002**, *43*, 63–96. [CrossRef]
2. Singleton, J.; Mielke, C. Superconductors Go Organic. *Phys. World* **2002**, *15*, 35–39. [CrossRef]
3. Day, P. BEDT-TTF Charge Transfer Salts: New Structures, New Functionalities. *Compt. Rend. Chim.* **2003**, *6*, 301–308. [CrossRef]
4. Mori, H. Introduction to Organic Superconducting Materials. *Opt. Sci. Eng.* **2008**, *133*, 263–285.
5. Mori, T. Structural Genealogy of BEDT-TTF-Based Organic Conductors. I. Parallel Molecules: β and β″ Phases. *Bull. Chem. Soc. Jpn.* **1998**, *71*, 2509–2526. [CrossRef]
6. Mori, T.; Mori, H.; Tanaka, S. Structural Genealogy of BEDT-TTF-Based Organic Conductors. II. Inclined Molecules: θ, α and κ Phases. *Bull. Chem. Soc. Jpn.* **1999**, *72*, 179–197. [CrossRef]
7. Mori, T. Structural Genealogy of BEDT-TTF-Based Organic Conductors. III. Twisted Molecules: δ and α′ Phases. *Bull. Chem. Soc. Jpn.* **1999**, *72*, 2011–2027. [CrossRef]
8. Williams, J.M.; Wang, H.H.; Kini, A.M.; Carlson, K.D.; Beno, M.A.; Geiser, U.; Whangbo, M.-H.; Jung., D.; Evain, M.; Novoa, J.J. Recent Progress in the Development of Structure-Property Correlations for κ-Phase Organic Superconductors. *Mol. Crys. Liq. Crys.* **1999**, *181*, 59–64. [CrossRef]
9. Ishiguro, T.; Yamaji, K.; Saito, G. *Organic Superconductors*; Springer: Berlin, Germany, 1998.
10. Mori, H. Materials Viewpoint of Organic Superconductors. *J. Phys. Soc. Jpn.* **2006**, *75*, 051003-15. [CrossRef]
11. Kurmoo, M.; Graham, A.W.; Day, P.; Coles, S.J.; Hursthouse, M.B.; Caulfield, J.L.; Singleton, J.; Pratt, F.L.; Hayes, W.; Ducasse, L.; et al. Superconducting and Semiconducting Magnetic Charge Transfer Salts: (BEDT-TTF)$_4$AFe(C$_2$O$_4$)$_3$·C$_6$H$_5$CN (A = H$_2$O, K, NH$_4$). *J. Am. Chem. Soc.* **1995**, *117*, 12209–12217. [CrossRef]
12. Coronado, E.; Day, P. Magnetic molecular conductors. *Chem. Rev.* **2004**, *104*, 5419–5448. [CrossRef]
13. Martin, L. Molecular Conductors of BEDT-TTF with Tris(oxalato)metallate Anions. *Coord. Chem. Rev.* **2018**, *376*, 277–291. [CrossRef]
14. Sahadevan, S.A.; Abherve, A.; Monni, N.; Auban-Senzier, P.; Cano, J.; Lloret, F.; Julve, M.; Cui, H.; Kato, R.; Canadell, E.; et al. Magnetic Molecular Conductors Based on Bis(ethylenedithio)tetrathiafulvalene (BEDT-TTF) with the Tris(chlorocyananilato)ferrate(III) Complex. *Inorg. Chem.* **2019**, *58*, 15359–15370. [CrossRef] [PubMed]
15. Griffiths, J.-P.; Wallis, J.D. Substituted BEDT-TTF Derivatives: Synthesis, Chirality, Properties and Potential Applications. *J. Mater. Chem.* **2005**, *15*, 347–365.
16. Wallis, J.D.; Karrer, A.; Dunitz, J.D. Chiral Metals? A Chiral Substrate for Organic Conductors and Superconductors. *Helv. Chim. Acta* **1986**, *69*, 69–70. [CrossRef]
17. Coronado, E.; Galán-Mascarós, J.R.; Coldea, A.I.; Goddard, P.; Singleton, J.; Wallis, J.D.; Coles, S.J.; Alberola, A. A Chiral Ferromagnetic Molecular Metal. *J. Am. Chem. Soc.* **2010**, *132*, 9271–9273.
18. Pop, F.; Laroussi, S.; Cauchy, T.; Gomez-Garcia, C.J.; Wallis, J.D.; Avarvari, N. Tetramethyl-Bis(ethylenedithio)-Tetrathiafulvalene (TM-BEDT-TTF) Revisited: Crystal Structures, Chiroptical Properties, Theoretical Calculations and a Complete Series of Conducting Radical Cation Salts. *Chirality* **2013**, *25*, 466–474. [CrossRef]

19. Pop, F.; Meziere, C.; Allain, M.; Auban-Senzier, P.; Tajima, N.; Hirobe, D.; Yamamoto, H.; Canadell, E.; Avarvari, N. Unusual Stoichiometry, Band Structure and Band Filling in Conducting Enantiopure Radical Cation Salts of TM-BEDT-TTF Showing Helical Packing of the Donors. *J. Mater. Chem. C.* **2021**. online. [CrossRef]
20. Short, J.; Blundell, T.J.; Krivickas, S.J.; Yang, S.; Wallis, J.D.; Akutsu, H.; Nakazawa, Y.; Martin, L. Chiral Molecular Conductor with an Insulator–Metal Transition Close to Room Temperature. *Chem. Commun.* **2020**, *56*, 9497–9500. [CrossRef]
21. Wang, Q.; Martin, L.; Blake, A.J.; Day, P.; Akutsu, H.; Wallis, J.D. Coordination Chemistry of 2,2′-Bipyridyl- and 2,2′: 6′,2″-Terpyridyl-substituted BEDT-TTFs: Formation of a Supramolecular Capsule Motif by the Iron (II) Tris Complex of 2,2′-Bipyridine-4-thiomethyl-BEDT-TTF. *Inorg. Chem.* **2016**, *55*, 8543–8551. [CrossRef] [PubMed]
22. Griffiths, J.-P.; Nie, H.; Brown, R.J.; Day, P.; Wallis, J.D. Synthetic Strategies to Chiral Organosulfur Donors related to Bis(ethylenedithio)tetrathiafulvalene. *Org. Biomolec. Chem.* **2005**, *3*, 2155–2166. [CrossRef]
23. Griffiths, J.; Arnal, A.A.; Appleby, G.; Wallis, J.D. Synthesis and Reactivity of Amino-substituted BEDT-TTF Donors as Building Blocks for Bifunctional Materials. *Tetrahedron Lett.* **2004**, *45*, 2813–2816. [CrossRef]
24. Saygili, N.; Brown, R.J.; Day, P.; Hoelzl, R.; Kathirgamanathan, P.; Mageean, E.R.; Ozturk, T.; Pilkington, M.; Qayyum, M.M.B.; Turner, S.S.; et al. Functionalised Organosulfur Donor Molecules: Synthesis of Racemic Hydroxymethyl-, Alkoxymethyl- and Dialkoxymethyl-bis(ethylenedithio)tetrathiafulvalenes. *Tetrahedron* **2001**, *57*, 5015–5026. [CrossRef]
25. Wang, Q.; Zecchini, M.; Wallis, J.D.; Wu, Y.; Rawson, J.M.; Pilkington, M. A Family of Unsymmetrical Hydroxyl-substituted BEDT-TTF Donors: Syntheses, Structures and Preliminary Thin Film Studies. *RSC Adv.* **2015**, *5*, 40205–40218. [CrossRef]
26. Brown, R.J.; Brooks, A.C.; Griffiths, J.-P.; Vital, B.; Day, P.; Wallis, J.D. Synthesis of Bis(ethylenedithio)tetrathiafulvalene (BEDT-TTF) Derivatives with Two, Four or Eight Hydroxyl Groups. *Org. Biomolec. Chem.* **2007**, *5*, 3172–3182. [CrossRef] [PubMed]
27. Li, H.; Zhang, D.; Zhang, B.; Yao, Y.; Xu, W.; Zhu, D.; Wang, Z. Synthesis of New Electron Donors with Hydroxymethyl Groups and Studies on Their Cation-Radical Salts. *J. Mater. Chem.* **2000**, *10*, 2063–2067. [CrossRef]
28. Li, H.; Zhang, D.; Xu, W.; Fan, L.; Zhu, D. New Electron Donor: Bis(ethylenedithio)tetrathiafulvalene Derivative with Four Hydroxyl Groups. *Syn. Met.* **1999**, *106*, 111–114. [CrossRef]
29. Berridge, R.; Serebryakov, I.M.; Skabara, P.J.; Orti, E.; Viruela, R.; Pou-Amerigo, R.; Coles, S.J.; Hursthouse, M.B. A New Series of π-Extended Tetrathiafulvalene Derivatives Incorporating Fused Furanodithiino and Thienodithiino Units: A Joint Experimental and Theoretical Study. *J. Mater. Chem.* **2004**, *14*, 2822–2830. [CrossRef]
30. Aoyagi, I.; Katsuhara, M.; Mori, T. Synthesis and Structure of Highly Soluble Bis(ethylenedithio)tetrathiafulvalene Molecules with Alkyl Chains. *Sci. Tech. Adv. Mater.* **2004**, *5*, 443–447. [CrossRef]
31. Kini, A.M.; Parakka, J.P.; Geiser, U.; Wang, H.-H.; Rivas, F.; DiNiro, E.; Thomas, S.; Dudek, J.D.; Williams, J.M. Tetra-alkyl and Di-alkyl Substituted BEDT-TTF Derivatives and Their Cation-Radical Salts: Synthesis, Structure and Properties. *J. Mater. Chem.* **1999**, *9*, 883–892. [CrossRef]
32. Troitsky, V.I.; Berzina, T.S.; Katsen, Y.Y.; Neilands, O.Y.; Nicolini, C. Conducting Langmuir-Blodgett Films of Heptadecylcarboxymethyl-BEDT-TTF. *Syn. Metals* **1995**, *74*, 1–6. [CrossRef]
33. Goldenberg, L.M.; Khodorkovsky, V.Y.; Vladimir, Y.; Becker, J.Y.; Lukes, P.J.; Bryce, M.R.; Petty, M.C.; Yarwood, J. Highly Conducting Langmuir-Blodgett Films of an Amphiphilic Bis(ethylenedithio)tetrathiafulvalene (BEDT-TTF) Derivative: BEDT-TTF-$C_{18}H_{37}$. *Chem. Mater.* **1994**, *6*, 1426–1431. [CrossRef]
34. Khodorkovskii, B.Y.; Pukitis, G.; Puplovskii, A.Y.; Edzina, A.; Neilands, O.Y. Synthesis and Properties of the Hexadecyl Derivative of Bis(ethylenedithio)tetrathiafulvalene. *Khim. Geterotsikl. Soedin.* **1990**, 131–132.
35. Yang, X.; Rauchfuss, T.B.; Wilson, S. The Chemistry of C_6S_{10}: A Channel Structure for $C_6S_{10}.(CS_2)_{0.5}$ and Access to the Versatile DMAD-C_3S_4O. *Chem. Commun.* **1990**, 34–36. [CrossRef]
36. Leriche, P.; Gorgues, A.; Jubault, M.; Becher, J.; Orduna, J.; Garin, J. Cycloaddition of Acetylenedicarbaldehyde Monoacetal and 2,4,5-Trithioxo-1,3-dithiole: Ready Access to Novel Highly Extended and Sulfur-rich Analogues of Tetrathiafulvalene (TTF). *Tetrahedron Lett.* **1995**, *36*, 1275–1278. [CrossRef]
37. Ishikawa, Y.; Miyamoto, T.; Yoshida, A.; Kawada, Y.; Nakazaki, J.; Izuoka, A.; Sugawara, T. New Synthesis of 2-(1,3-Dithiol-2-ylidene)-5,6-dihydro-1,3-dithiolo[4,5-b][1,4]dithiiins with Formyl Group on Fused Benzene, [1,4]dithiin or Thiophene Ring. *Tetrahedron Lett.* **1999**, *40*, 8819–8822. [CrossRef]
38. Brooks, A.C.; Day, P.; Dias, S.I.G.; Rabaca, S.; Santos, I.C.; Henriques, R.T.; Wallis, J.D.; Almeida, M. Pyridine-functionalised (Vinylenedithio)tetrathiafulvalene (VDT-TTF) Derivatives and Their Dithiolene Analogues. *Eur. J. Inorg. Chem.* **2009**, 3084–3093. [CrossRef]
39. Niu, Z.-G.; He, L.-R.; Li, L.; Cheng, W.-F.; Li, X.-Y.; Chen, H.-H.; Li, G.-N. Synthesis, Characterisation and DFT Studies of Two New π-Conjugated Pyridine-based Tetrathiafulvalene Derivatives. *Acta Chim. Sloven.* **2014**, *61*, 786–791.
40. Marshallsay, G.J.; Bryce, M.R.; Cooke, G.; Joergensen, T.; Becher, J.; Reynolds, C.D.; Wood, S. Functionalised Tetrathiafulvalene (TTF) Systems Derived from 4, 5-(Propylenedithio)-1, 3-dithiole Units. *Tetrahedron* **1993**, *49*, 6849–6862. [CrossRef]
41. Arenzano, J.A.; Virues, J.O.; Colorado-Peralta, E.; Ramirez-Montes, P.I.; Santillan, R.; Sanchez, M.; Rivera, J.M. Heterometallic Coordination Framework by Sodium Carboxylate Subunits and Cobalt(III) Centres Obtained from a Highly Hydrogen Bonding Stabilized Cobalt(II) Monomeric Complex. *Inorg. Chem. Commun.* **2015**, *51*, 55–60. [CrossRef]
42. Puentes, R.; Torres, J.; Kremer, C.; Cano, J.; Lloret, F.; Capucci, D.; Bacchi, A. Mononuclear and Polynuclear Complexes Ligated by an Iminodiacetic Acid Derivative: Synthesis, Structure, Solution Studies and Magnetic Properties. *Dalton Trans.* **2016**, *45*, 5356–5373. [CrossRef]

43. Yousuf, I.; Zeeshan, M.; Arjmand, F.; Rizvi, M.A.; Tabassum, S. Synthesis, Structural Investigations and DNA Cleavage Properties of a New Water Soluble Cu(II)-iminodiacetate Complex. *Inorg. Chem. Commun.* **2019**, *106*, 48–53. [CrossRef]
44. Puentes, R.; Torres, J.; Faccio, R.; Bacchi, A.; Kremer, C. Lanthanide Coordination Polymers Based on Flexible Ligands Derived from Iminodiacetic acid. *Polyhedron* **2019**, *170*, 683–689. [CrossRef]
45. Kanehama, R.; Umemiya, M.; Iwahori, F.; Miyasaka, H.; Sugiura, K.-I.; Yamashita, M.; Yokochi, Y.; Ito, H.; Kuroda, S.-I.; Kishida, H.; et al. New ET-Coordinated Copper(I) Complexes: Synthesis, Structures and Physical Properties. *Inorg. Chem.* **2003**, *42*, 7173–7181. [CrossRef]
46. Jia, C.; Zhang, D.; Liu, C.-M.; Xu, W.; Hu, H.; Zhu, D. Novel Silver(I) Complexes Derived From Tetrakis(methylthio)tetrathiafulvalene and Bis(ethylenedithio)tetrathiafulvalene with 3D and 1D Structures. *New J. Chem.* **2002**, *26*, 490–494. [CrossRef]
47. Inoue, M.B.; Inoue, M.; Bruck, M.A.; Fernando, Q. Structure of Bis(ethylenedithio)tetrafulvalenium Tribromodicuprate(I), (BEDT-TTF$^+$)Cu(I)$_2$Br$_3$: Coordination of the Organic Radical Cation to the Metal Ions. *J. Chem. Soc. Chem. Commun.* **1992**, 515–516. [CrossRef]
48. Pop, F.; Auban-Senzier, P.; Canadell, E.; Rikken, G.L.R.A.; Avarvari, N. Electrical Magnetochiral Anisotropy in a Bulk Chiral Molecular Conductor. *Nat. Commun.* **2014**, *5*, 3757. [CrossRef]
49. Sheldrick, G.M. A Short History of SHELX. *Acta Crystallogr. Sect. A* **2008**, *64*, 112–122. [CrossRef] [PubMed]
50. Sheldrick, G.M. Crystal Structure Refinement with SHELXL. *Acta Crystallogr. Sect. C* **2015**, *71*, 3–8. [CrossRef]
51. Barbour, L.J. X-Seed—A Software Tool for Supramolecular Crystallography. *J. Supramol. Chem.* **2001**, *1*, 189–191. [CrossRef]
52. Macrae, C.F.; Bruno, I.J.; Chisholm, J.A.; Edgington, P.R.; McCabe, P.; Pidcock, E.; Wood, P.A. Mercury: Visualization and Analysis of Crystal Structures. *J. Appl. Crystallogr.* **2006**, *39*, 453–457. [CrossRef]

Perspective

Redox Activity as a Powerful Strategy to Tune Magnetic and/or Conducting Properties in Benzoquinone-Based Metal-Organic Frameworks

Noemi Monni [1,2,*], Mariangela Oggianu [1,2], Suchithra Ashoka Sahadevan [3] and Maria Laura Mercuri [1,2,*]

[1] Department of Chemical and Geological Sciences, University of Cagliari, Highway 554, Crossroads for Sestu, I09042 Monserrato (CA), Italy; mari.oggianu@gmail.com
[2] National Interuniversity Consortium of Materials Science and Technology, INSTM, Street Giuseppe Giusti, 9, I50121 Florence, Italy
[3] Applied Physical Chemistry, Centre of Molecular Devices, Department of Chemistry, KTH Royal Institute of Technology, SE-10044 Stockholm, Sweden; suchithraiiserk@gmail.com
* Correspondence: noemi.monni@unica.it (N.M.); mercuri@unica.it (M.L.M.)

Abstract: Multifunctional molecular materials have attracted material scientists for several years as they are promising materials for the future generation of electronic devices. Careful selection of their molecular building blocks allows for the combination and/or even interplay of different physical properties in the same crystal lattice. Incorporation of redox activity in these networks is one of the most appealing and recent synthetic strategies used to enhance magnetic and/or conducting and/or optical properties. Quinone derivatives are excellent redox-active linkers, widely used for various applications such as electrode materials, flow batteries, pseudo-capacitors, etc. Quinones undergo a reversible two-electron redox reaction to form hydroquinone dianions via intermediate semiquinone radical formation. Moreover, the possibility to functionalize the six-membered ring of the quinone by various substituents/functional groups make them excellent molecular building blocks for the construction of multifunctional tunable metal-organic frameworks (MOFs). An overview of the recent advances on benzoquinone-based MOFs, with a particular focus on key examples where magnetic and/or conducting properties are tuned/switched, even simultaneously, by playing with redox activity, is herein envisioned.

Keywords: metal-organic frameworks; redox; magnetism; conductivity; benzoquinones; semiquinones

1. Introduction

Over several decades, metal-organic frameworks (MOFs) have been extensively studied [1] due to their unique supramolecular architectures, which lead to high porosity and interesting properties in magnetism [2], conductivity [3], photochromism [4,5], luminescence [6–9], etc. MOFs are coordination compounds formed by metal ions linked to organic ligands, forming an infinite array in one, two, or three dimensions (1D, 2D, and 3D) and offering a plethora of applications in different fields including gas storage, separation, catalysis, energy storage, sensing, biomedical applications, etc. [10–14]. Depending on a careful choice of metal ions/linkers [15], MOFs can also show a combination and/or even interplay of physical properties. A relatively new strategy [10,16,17] to enhance their physical properties, in particular, magnetism and conductivity, is the incorporation of redox activity. Redox activity can be promoted via various methods such as (*i*) a rational design of redox-active metals centers or linkers [18], (*ii*) post-synthetic modifications of metal ions or through ligand exchange, and (*iii*) encapsulation of redox-active guest ions in the pores of MOFs [19], leading to the formation of radical species, affecting the electronic properties of the organic linkers and, therefore, the physical properties of the related networks [20,21].

Among the redox-active linkers, derivatives of pyrazine [22], dithiolenes [23], triphenylamine [24], N,N-Dipyridil Naphthalenediimide [20], hexaaminobenzene [25], por-

phyrin [26], bipyridine [27], etc., are most commonly used in MOFs, while multidentate redox- active linkers, providing different possibilities to tune redox activity, still represent a challenge up to now [20,21,28].

The benzoquinone/hydroquinone linkers represent a redox-active couple remarkably studied in technologically important materials such as electrodes [6,29], flow batteries [30], pseudo-capacitors [31], and materials used in artificial photosynthesis [32]. Benzoquinones are a class of naturally occurring organic compounds that possess two carbonyl groups C=O in the 1 and 4 position in an unsaturated six-membered ring [33]. Benzoquinones are usually electron deficient, and their benzoquinoid-like configuration undergoes a monoelectron reversible reduction to produce the para semiquinone radical species, which, in turn, could be further reduced to form the aromatic hydroquinone dianion (*vide infra*), as described in Scheme 1.

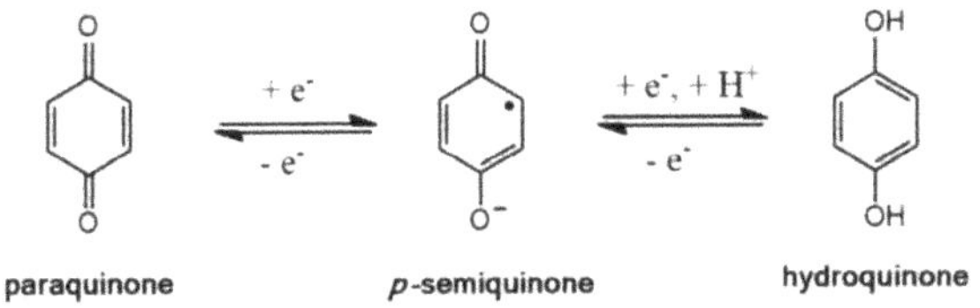

Scheme 1. Reversible redox reactions for *p*–quinone/hydroquinone couple.

The stability of the semiquinoid form could be influenced by different factors, such as the nature of the ring substituents, intra/intermolecular hydrogen bonding, solvent polarity, the presence of acidic or basic additives, and protonation [34–39]. Particularly, chemical tailoring of benzoquinone linkers, through the functionalization of the six-membered ring by various substituents (halo, nitro, amino, methoxy, etc.), can stabilize the radical anion [40], depending on the steric and electronic nature of the substituents, thus modulating the physical properties of the benzoquinone-based networks at a molecular level [41]. In fact, the presence of different substituents on the benzoquinoid ring can tune the one-electron reduction potential, as shown in Table 1. When the benzoquinones are functionalized with electron-donating substituents, i.e., methyl groups, the reduction potential is more negative depending on how many methyl groups are present, making these species more difficult to reduce. On the contrary, it is possible to observe an opposite trend when electron-withdrawing substituents, such as chlorine, are present. Therefore, benzoquinones containing chlorine substituents in the ring show more positive reduction potentials and consequently are easier to reduce [33].

Among the parabenzoquinones derivatives, 2,5-dihydroxy-1,4-benzoquinones (dhbq) have attracted ever-growing interest in material chemistry due to their versatile coordination modes (Scheme 2). Furthermore, when hydrogens at the 3 and 6 position are replaced with different substituents (halogen atoms or functional groups), they are better known in the literature as anilic acids, formulated as $H_2X_2C_6O_4$ (H_2X_2An), where X indicates the substituent and C_6O_4 indicates the anilate moiety (An) [2,42,43]. When the anilic acids are in their dianionic form, i.e., anilates, they act as valuable linkers for transition [44–51] and lanthanide metal nodes [52–58] to build materials showing a combination of fascinating physical properties and redox states (*vide infra*). All these features make them interesting molecular building blocks for constructing a large variety of novel supramolecular frameworks [49,59], as shown in Figure 1. Particularly, most of the reported structures based on parabenzoquinone derivatives are 2D, paving the way for a further exfoliation of bulk MOFs to fabricate metal-organic nanosheets (MON) that can show peculiar redox behavior [60].

Table 1. Theoretical one-electron reduction potentials of selected benzoquinone derivatives (mV vs. NHE in CH$_3$CN) [61] showing the influence of substituents.

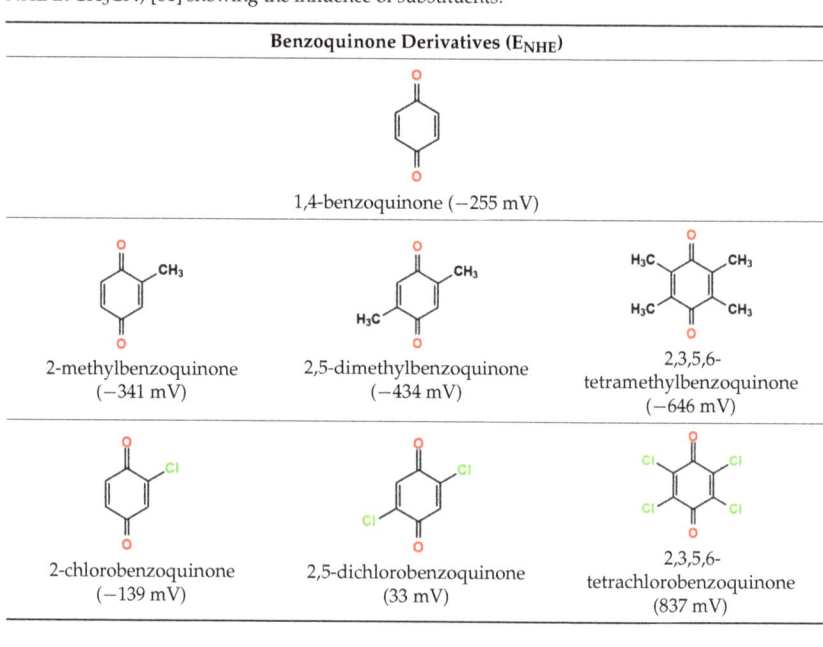

Benzoquinone Derivatives (E$_{NHE}$)		
	1,4-benzoquinone (−255 mV)	
2-methylbenzoquinone (−341 mV)	2,5-dimethylbenzoquinone (−434 mV)	2,3,5,6-tetramethylbenzoquinone (−646 mV)
2-chlorobenzoquinone (−139 mV)	2,5-dichlorobenzoquinone (33 mV)	2,3,5,6-tetrachlorobenzoquinone (837 mV)

X = H, Cl, Br, I, CN, *etc.*

3,6-disubstituted-2,5-dihydroxy-1,4-benzoquinone

1,2-bidentate   bis-1,2-bidentate   1,2-bidentate/monodentate

Scheme 2. Chemical structure of anilates and the most common coordination modes.

On this basis, MOFs formed by metal nodes and benzoquinoid-based ligands, especially anilates, feature an ideal platform for the construction of porous redox materials with switchable conducting/magnetic properties [62], due to the changeover to the semiquinoid form.

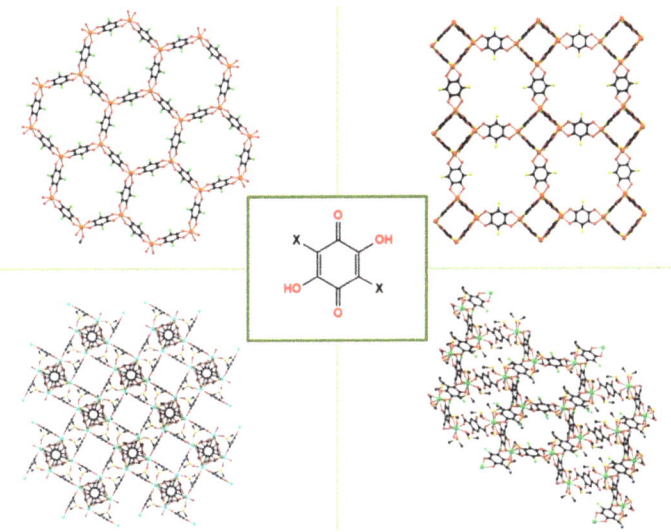

Figure 1. Extended networks based on anilates.

A pioneering study on a dhbq based-framework, reported by Abrahams et al. [63], revealed (*i*) the potential of benzoquinone to act as a suitable building block for constructing extended frameworks [64] and (*ii*) its capability to receive/lose electrons while keeping unchanged the supramolecular architecture. This study opened up unprecedented routes to tune the physical properties of extended frameworks through redox activity.

The present work focuses on key examples of the latest developments (from 2015 to date) on redox-active benzoquinone-based MOFs showing conducting and/or magnetic properties. The aim is to evidence the enhancement/switching of magnetism and/or conductivity due to a fine modulation of benzoquinone redox properties, highlighting the extreme versatility of this class of redox-active linkers in tailoring the physical properties of extended frameworks.

2. Semiquinone-Based MOFs

The incorporation of a semiquinone in a solid framework could be achieved through in situ or post-synthetic reduction of the benzoquinone derivatives. In 2016, Stock et al. [65] reported on the first example of permanently porous Al^{III}-MOFs, of the formulas $(CH_3)_2NH_2)_3$ $[Al_4(dhbq)_3(dhbq^{\bullet})_3]\cdot 3DMF$ (**1**) and $((CH_3)_2NH_2)_3[Al_4(Cl_2An)_3(Cl_2An^{\bullet})_3]\cdot 9DMF$ (**2**), containing ligands in both their dianionic form ($dhbq^{2-}$) as well as in the semiquinonic form ($dhbq^{3-}$). Interestingly, MOFs **1** and **2** were obtained by in situ reduction, using high-throughput methods, which consist of an automated solvothermal equipment with different ligand/Al stoichiometric ratios, in DMF solvent, for optimizing the synthesis conditions. MOFs **1** and **2** show specific surface areas of 1440 and 1430 m^2g^{-1}, respectively [65]. A similar MOF was also reported by Harris et al. [62,66], by combining Fe^{II} ions with H_2Cl_2An through a solvothermal reaction in DMF, leading to a novel porous semiquinoid antiferromagnet formulated as $((CH_3)_2NH_2)_2[Fe_2(Cl_2An)_3]\cdot 2H_2O\cdot 6DMF$ (**3**). The chloranilate bridging ligand is simultaneously present in both benzoquinoid and semiquinoid forms, resulting from a spontaneous electron transfer from Fe^{II} to Cl_2An^{2-}, giving rise to a mixed-valence layered MOF [62]. The ligand coordinates in its bis-bidentate mode, which generates anionic layers where six metal ions are coordinate by the ligand forming a hexagonal motif, the typical honeycomb packing. The $(CH_3)_2NH_2^+$ cations both balance the charge and orient the anionic layers to an eclipsed structure, forming 1D hexagonal channels, which show a Brunauer–Emmett–Teller (BET) surface area of

1175 m^{2}g^{-1}. MOF **3** shows antiferromagnetic interactions with a spontaneous magnetization below 80 K in its solvated form, while the magnetic ordering temperature decreases to 26 K in the desolvated form (**3_desolv**), which shows a BET surface area of 885 m^{2}g^{-1} (Figure 2) and a fully reversible structural contraction consistent with a "breathing" behavior. The high value of magnetic ordering temperature compared to other extended systems containing the same bridging ligand in its dianionic form [45] highlights the ability of semiquinone ligands to form porous magnets with enhanced magnetic coupling between metal ions (*vide supra*) [62]. Moreover, MOFs **3** and **3_desolv** show conductivity values of σ = 1.4(7) × 10^{-2} S/cm (Ea = 0.26(1) eV) and 1.0(3) × 10^{-3} S/cm (Ea = 0.19(1) eV), respectively, proving the ability of benzoquinone derivatives to construct multifunctional MOFs, in which porosity, magnetism, and conductivity coexist [62].

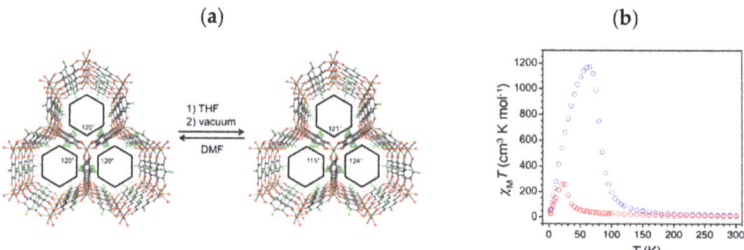

Figure 2. (**a**) Perspective view along the *c* axis of crystal structures of MOFs **3** (left) and **3_desolv** (right)—Fe, Cl, O, and C atoms are shown in orange, green, red, and grey, respectively. (**b**) Thermal variation of $\chi_M T$ for MOFs **3** (blue) and **3_desolv** (red) (applied *dc* field of 1000 Oe). Reprinted with permission from Reference [62]. Copyright © 2015, American Chemical Society.

Furthermore, H$_2$Cl$_2$An in MOF **3** could be fully reduced to its semiquinoid form via a post-synthetic approach by using cobaltocene (Cp$_2$Co), which allows a single-crystal-to-single-crystal chemical (one-electron) reduction, due to its porous crystalline structure, affording a 2D MOF, formulated as (Cp$_2$Co)$_{1.43}$((CH$_3$)$_2$NH$_2$)1.57[Fe$_2$ Cl$_2$An $_3$]·4.9DMF (**4**). Remarkably, the Tc can increase up to 105 K, a rare value for MOFs, attributable to the strong magnetic exchange interactions between metal ions mediated by the semiquinone radical form. Variable-field measurements show a magnetic hysteresis up to 100 K, which is consistent with the high Tc value (see Figure 3).

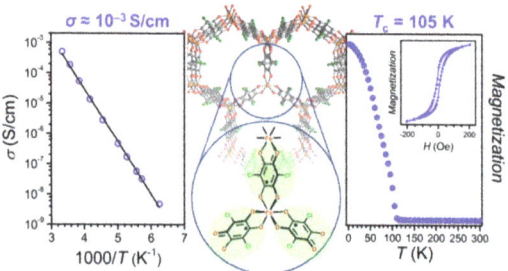

Figure 3. Magnetic and transport properties of 2D semiquinoid–based MOF **4**. Reprinted with permission from [66]. Copyright © 2017, American Chemical Society.

The r.t. conductivity of MOF **4**, on a pressed-pellet, has a value of σ = 5.1(3) × 10^{-4} Scm^{-1}, probably due to the complete ligand reduction that remove the mixed-valence character of the MOF [66]. Variable-temperature conductivity data in the 300–160 K range fit well the Arrhenius law with E$_a$ = 0.34(1) eV, in agreement with the observed conductivity value of MOF **4** and lower than MOFs **3** and **3_desolv**, supporting the complete chemical reduction.

Noteworthily, the coexistence of high magnetic ordering and electrical conductivity in the same material is rather unusual, as well as the capability of the quinoid MOF to retain its crystalline structure upon post-synthetic chemical reduction, demonstrating the potential of quinoid-based MOFs to provide a new generation of redox-active conducting magnets for future spintronics applications (*vide infra*).

In 2015, Long et al. reported the first example of a 3D dhbq-based MOF, obtained by a solvothermal reaction and formulated as $(NBu_4)_2Fe^{III}_2(dhbq)_3$ (**5**) [67]. This MOF shows a very rare topology for $dhbq^{2-}$-based coordination compounds [68–70] with two interpenetrated (10,3)-*a* networks of opposing chiralities (Figure 4), generating a topology that differs from the classic honeycomb structure frequently observed for anilates [71]. This material behaves as an Arrhenius semiconductor with a very high r.t. conductivity of 0.16(1) S/cm and an Ea of 110 meV. Mössbauer spectroscopy confirms the presence of high-spin Fe^{III} metal ions, and the high conductivity value can be ascribed to the presence of mixed-valency due to $dhbq^{3-}$ radicals, remarkably Class II/III according to Robin-Day, as evidenced by diffuse reflectance measurements in the UV–Vis–NIR range. Noteworthily, MOF **5**, where a Fe^{II}-semiquinoid transition occurred, provides a challenging scaffold for constructing tunable long-range electronic communication in MOF [72].

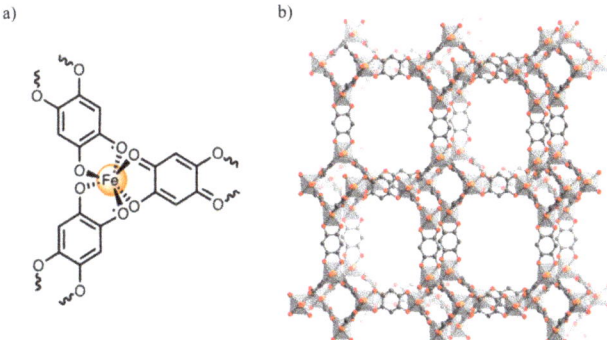

Figure 4. (a) Molecular structure of a single Fe^{III} center in MOF **5**, showing that two radical (H_2An^{3-}) bridging ligands and one diamagnetic (H_2An^{2-}) bridging ligand are coordinated to each metal site. (b) View of the porous 3D crystal structure formed by $dhbq^{n-}$-bridged Fe^{III} centers, giving the interpenetrated (10,3)-a nets. Reprinted with permission from Reference [67]. Copyright © 2015, American Chemical Society.

Interestingly, it is possible to tune the electrical conductivity by combining dhbq or H_2Cl_2An with transition metal ions having diffuse 3D orbitals. In 2018, Long et al. reported on 2D semiquinoid-bridged frameworks based on titanium, chromium, and vanadium, formulated as $((CH_3)_2NH_2)_2Ti_2(Cl_2An)_3 \cdot 4.7DMF$ (**6**), $((CH_3)_2NH_2)_{1.5}Cr_2(dhbq)_3 \cdot 4.4DMF$ (**7**), and $((CH_3)_2NH_2)_2V_2(Cl_2An)_3 \cdot 6.4DMF$ (**8**), respectively [73]. MOFs **6–8** show a honeycomb-type structure similar to MOFs **1–4**, which were studied electrochemically by using solid-state cyclic voltammetry, which pointed out only ligand-based redox processes for compounds of MOFs **6–7** and a combination of ligand- and metal-based redox activity for MOF **8**. Given their mixed-valence character, they all show electronic conductivity of values $2.7(2) \times 10^{-3}$ S/cm, $1.2(1) \times 10^{-4}$ S/cm, and remarkably 0.45(3) S/cm, respectively, for MOFs **6–8**, following the expected trend based on the correlation with the electronic structure of the frameworks and, in the case of MOF **8**, consistent with the observed metal−ligand covalency [73].

These results show that the incorporation of semiquinoid ligands in an extended scaffold is a valuable strategy for developing multifunctional MOFs with improved electrical conductivity and temperature magnetic ordering, making quinone derivatives excellent candidates for constructing next-generation data processing and storage systems.

3. Benzoquinone-Based MOFs

Very recently, Miyasaka et al. reported on post-synthetic generation of radical species via solid-state bulk electrochemistry technique, in a MOF containing diamagnetic benzoquinone derivatives, at the cathode of a lithium-ion battery system (LIB), producing a radical spin in the benzoquinone moiety and Li$^+$ insertion for preserving neutrality [74]. In this case, porosity is a fundamental requirement for host Li$^+$ ions. The precursor, formulated as $(H_3O)_2(phz)_3[Fe_2(Cl_2An)_3]$ (**9**), was obtained by the desolvatation of $(H_3O)_2(phz)_3[Fe_2(Cl_2An)_3]\cdot(CH_3COCH_3)_n\cdot(H_2O)_n$ (**9_solv**). MOF **9** shows the typical honeycomb packing shown by anilates, with alternating anionic/cationic layers, where the counter cations $[(H_3O)_2(phz)_3]^{2+}$ are placed between the layers, acting as a templating agent that leads to 1D hexagonal channels along the c axis, conferring porosity to the network. MOF **9** shows paramagnetic behavior and short-range ferromagnetic correlations among Fe^{II} ions, through a chloranilate linker, in the layered framework, as can be seen by lack of hysteresis in the magnetization vs. field (M-H) measurements, even at 5 K. With the ligand reduction due to the insertion of Li$^+$ ions, antiferromagnetic superexchange interactions between the radical anion Cl$_2$An$\bullet^{3-}$ and Fe^{II} ions took place, and the reduced MOF, formulated as $(Li)_3(H_3O)_2(phz)_3[(Fe)_2(Cl_2An)_3]$ (**9_red**), shows one of the higher T$_c$ values reported so far, T$_c$ = 128 K [74] (Figure 5). The formation of the radical ligand Cl$_2$An$\bullet^{3-}$, starting from Cl$_2$An^{2-}, leads to a long-range magnetic ordering in the MOF, making this MOF a potential cathode material of a LIB.

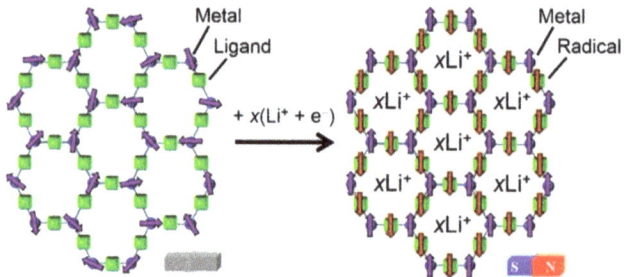

Figure 5. Schematic view of spin variation during the discharge process MOF → electron-reduced MOF, (Li$^+$)$_x$[MOF]$^{x-}$. Reprinted with permission from Reference [74].

Remarkably, Harris et al. [75] reported on the simultaneous switching of magnetic and conducting properties induced by post-synthetic chemical reduction in a MOF of formula $(Me_4N)_2[Mn_2(Cl_2An)_3]$x.DMF, containing the diamagnetic chloranilate linker (**10**). When MOF **10** is soaked in a THF equimolar reducing solution of sodium naphthalenide and 1,2-dihydroacenaphthylene for several days, a semiquinoid-based MOF, of formula Na$_3$(Me$_4$N)$_2$[Mn$_2$(Cl$_2$An)$_3$]3.9THF (**11**), is obtained. The reduction mechanism occurs via single-crystal-to-single-crystal process (*vide infra*), provoking the formation of Cl$_2$An$\bullet^{3-}$ semiquinone radical form starting from diamagnetic Cl$_2$An^{2-}, while the oxidation state of Mn^{II} remains unchanged. Upon the conversion in the radical form of Cl$_2$An^{2-}, a simultaneous change in both conductivity and magnetic properties is observed. Indeed MOF **10** shows a paramagnetic behavior above 1.8 K and a r.t. conductivity value of σ = 1.14(3) × 10^{-13} Scm^{-1} (E$_a$ = 0.74(3) eV), whereas MOF **11** shows antiferromagnetic interactions between Mn^{II} ions below 41 K, mediated by the semiquinone, and a r.t. conductivity value of σ = 2.27(1)× 10^{-8} Scm^{-1} (E$_a$ = 0.489 (8) eV), a value 200,000 times higher than the respective benzoquinoid framework. Furthermore, by soaking MOF **11** in ferrocene (Cp$_2$Fe$^+$) solution, a compound, formulated as Na((CH$_3$)$_4$N)[Mn$_2$(Cl$_2$An)$_3$] 5.5THF 0.8CH$_3$CN (**12**), is afforded, showing similar values of T$_c$ and σ as MOF **10** (oxidized compound), highlighting the reversibility of the redox process [75], as reported in Figure 6.

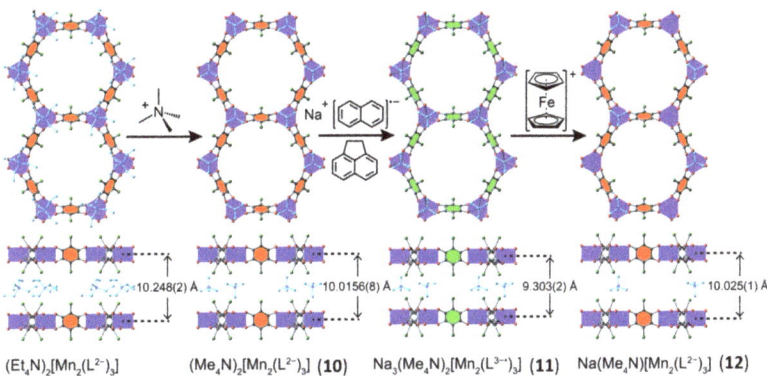

Figure 6. Schematic representation of reversible redox mechanism via single-crystal-to-single-crystal (SC-SC) process in MOFs **10–12**. Reproduced from Reference [75] with permission from the Royal Society of Chemistry.

The capability of benzoquinone-based MOFs to undergo reversible redox processes, which provoke a simultaneous switching of magnetic and transport properties, are worth being highlighted, as they may be considered suitable materials for future spintronic technologies.

4. Conclusions and Perspectives

The recent developments on redox-active benzoquinone-based MOFs, herein discussed, contribute to the ongoing research on the use of these materials for technologically relevant applications. It turns out that benzoquinoid/semiquinoid redox activity is a powerful strategy to tune their physical properties, in particular magnetism and/or conductivity. Porosity is a remarkable additional property and porous channels in benzoquinoid MOFs, allowing for the facile insertion/extraction of the electrolytes, which makes them promising materials for electrodes and rechargeable energy storage systems [76,77]. Furthermore, it has been highlighted that post-synthetic chemical redox reactions are a promising strategy to control ligand redox states in the MOF and the related changes of its conducting and magnetic properties, while the scaffold with metal–semiquinoid transitions provides tunable and delocalized electronic structures. Therefore, the whole redox control over MOFs is a very challenging task, specifically for their applications in the electronic devices realm. In more detail, redox control requires open-shell ligands and metals, frontier orbitals with similar energies, and maximal overlap to favor charge delocalization, and benzoquinones successfully match these requirements. Finally, the fabrication of semiquinoid MOFs, showing coexistence of simultaneously switchable conducting and magnetic properties, represents a forefront challenge for their potential applications in next-generation spintronic technologies, as magnetic transistors, terahertz information, and multifunctional chips, where data storage and information processing can occur at the same location.

Author Contributions: N.M., M.O. and S.A.S. contributed equally in terms of accurate literature searching and paper writing; M.L.M. supervised and wrote the paper with the help of N.M. All authors have read and agreed to the published version of the manuscript.

Funding: (a) Fondazione di Sardegna e gli Atenei Sardi, Regione Sardegna-L.R. 7/2007 annualità 2018-DGR 28/21 del 17.05.2015, project F74I19000940007; (b) CESA–RAS-Piano SULCIS (E58C16000080003) are acknowledged for the PhD grant of M.O.

Conflicts of Interest: The authors declare no conflict of interest.

References

1. Tran, M.; Kline, K.; Qin, Y.; Shen, Y.; Green, M.D.; Tongay, S. 2D coordination polymers: Design guidelines and materials perspective. *Appl. Phys. Rev.* **2019**, *6*, 041311. [CrossRef]
2. Mercuri, M.L.; Congiu, F.; Concas, G.; Ashoka Sahadevan, S. Recent Advances on Anilato-Based Molecular Materials with Magnetic and/or Conducting Properties. *Magnetochemistry* **2017**, *3*, 17. [CrossRef]
3. Coronado, E.; Galán-Mascarós, J.R.; Gómez-García, C.J.; Laukhin, V. Coexistence of ferromagnetism and metallic conductivity in a molecule-based layered compound. *Nature* **2000**, *408*, 447–449. [CrossRef]
4. Haldar, R.; Heinke, L.; Wöll, C. Advanced Photoresponsive Materials Using the Metal–Organic Framework Approach. *Adv. Mater.* **2019**, *1905227*, 1905227. [CrossRef]
5. Bénard, S.; Yu, P.; Audière, J.P.; Rivière, E.; Clément, R.; Guilhem, J.; Tchertanov, L.; Nakatani, K. Structure and NLO properties of layered bimetallic oxalato-bridged ferromagnetic networks containing stilbazolium-shaped chromophores. *J. Am. Chem. Soc.* **2000**, *122*, 9444–9454. [CrossRef]
6. Han, C.; Li, H.; Shi, R.; Zhang, T.; Tong, J.; Li, J.; Li, B. Organic quinones towards advanced electrochemical energy storage: Recent advances and challenges. *J. Mater. Chem. A* **2019**, *7*, 23378–23415. [CrossRef]
7. Pamei, M.; Puzari, A. Luminescent transition metal–organic frameworks: An emerging sensor for detecting biologically essential metal ions. *Nano-Struct. Nano-Objects* **2019**, *19*, 100364–100386. [CrossRef]
8. Ashoka Sahadevan, S.; Monni, N.; Abhervé, A.; Marongiu, D.; Sarritzu, V.; Sestu, N.; Saba, M.; Mura, A.; Bongiovanni, G.; Cannas, C.; et al. Nanosheets of Two-Dimensional Neutral Coordination Polymers Based on Near-Infrared-Emitting Lanthanides and a Chlorocyananilate Ligand. *Chem. Mater.* **2018**, *30*, 6575–6586. [CrossRef]
9. Kuznetsova, A.; Matveevskaya, V.; Pavlov, D.; Yakunenkov, A.; Potapov, A. Coordination polymers based on highly emissive ligands: Synthesis and functional properties. *Materials* **2020**, *13*, 2699. [CrossRef]
10. He, Y.; Chen, F.; Li, B.; Qian, G.; Zhou, W.; Chen, B. Porous metal–organic frameworks for fuel storage. *Coord. Chem. Rev.* **2018**, *373*, 167–198. [CrossRef]
11. Zhang, X.; Chen, A.; Zhong, M.; Zhang, Z.; Zhang, X.; Zhou, Z.; Bu, X.-H. *Metal–Organic Frameworks (MOFs) and MOF-Derived Materials for Energy Storage and Conversion*; Springer: Singapore, 2019; Volume 2, ISBN 0123456789.
12. Koo, W.-T.; Jang, J.-S.; Kim, I.-D. Metal-Organic Frameworks for Chemiresistive Sensors. *Chem* **2019**, *5*, 1938–1963. [CrossRef]
13. Baumann, A.E.; Burns, D.A.; Liu, B.; Thoi, V.S. Metal-organic framework functionalization and design strategies for advanced electrochemical energy storage devices. *Commun. Chem.* **2019**, *2*, 1–14. [CrossRef]
14. Oggianu, M.; Monni, N.; Mameli, V.; Cannas, C.; Sahadevan, S.A.; Mercuri, M.L. Designing Magnetic NanoMOFs for Biomedicine: Current Trends and Applications. *Magnetochemistry* **2020**, *6*, 39. [CrossRef]
15. Mínguez Espallargas, G.; Coronado, E. Magnetic functionalities in MOFs: From the framework to the pore. *Chem. Soc. Rev.* **2018**, *47*, 533–557. [CrossRef]
16. Wang, H.; Cui, L.; Xie, J.; Leong, C.F.; Alessandro, D.M.D.; Zuo, J. Functional coordination polymers based on redox-active tetrathiafulvalene and its derivatives. *Coord. Chem. Rev.* **2017**, *345*, 342–361. [CrossRef]
17. Degayner, J.A.; Wang, K.; Harris, T.D. A Ferric Semiquinoid Single-Chain Magnet via Thermally-Switchable Metal-Ligand Electron Transfer. *J. Am. Chem. Soc.* **2018**, *140*, 6550–6553. [CrossRef]
18. Sun, L.; Hendon, C.H.; Park, S.S.; Tulchinsky, Y.; Wan, R.; Wang, F.; Walsh, A.; Dinca, M. Is iron unique in promoting electrical conductivity in MOFs? *Chem. Sci.* **2017**, *8*, 4450–4457. [CrossRef] [PubMed]
19. Halls, J.E.; Jiang, D.; Burrows, A.D.; Kulandainathan, M.A.; Marken, F. Electrochemistry within metal-organic frameworks. *Electrochemistry* **2013**, *12*, 187–210. [CrossRef]
20. Calbo, J.; Golomb, M.J.; Walsh, A. Redox-active metal-organic frameworks for energy conversion and storage. *J. Mater. Chem. A* **2019**, *7*, 16571–16597. [CrossRef]
21. D'Alessandro, D.M. Exploiting redox activity in metal-organic frameworks: Concepts, trends and perspectives. *Chem. Commun.* **2016**, *52*, 8957–8971. [CrossRef] [PubMed]
22. Pedersen, K.S.; Perlepe, P.; Aubrey, M.L.; Woodruff, D.N.; Reyes-Lillo, S.E.; Reinholdt, A.; Voigt, L.; Li, Z.; Borup, K.; Rouzières, M.; et al. Formation of the layered conductive magnet CrCl$_2$ (pyrazine)$_2$ through redox-active coordination chemistry. *Nat. Chem.* **2018**, *10*, 1056–1061. [CrossRef] [PubMed]
23. Yan, Y.; Keating, C.; Chandrasekaran, P.; Jayarathne, U.; Mague, J.T.; Debeer, S.; Lancaster, K.M.; Sproules, S.; Rubtsov, I.V.; Donahue, J.P. Ancillary ligand effects upon dithiolene redox noninnocence in tungsten Bis(dithiolene) complexes. *Inorg. Chem.* **2013**, *52*, 6743–6751. [CrossRef]
24. Ellis, H.; Eriksson, S.K.; Feldt, S.M.; Gabrielsson, E.; Lohse, P.W.; Lindblad, R.; Sun, L.; Rensmo, H.; Boschloo, G.; Hagfeldt, A. Linker unit modification of triphenylamine-based organic dyes for efficient cobalt mediated dye-sensitized solar cells. *J. Phys. Chem. C* **2013**, *117*, 21029–21036. [CrossRef]
25. Li, C.; Shi, L.; Zhang, L.; Chen, P.; Zhu, J.; Wang, X.; Fu, Y. Ultrathin two-dimensional π–d conjugated coordination polymer Co3(hexaaminobenzene)2 nanosheets for highly efficient oxygen evolution. *J. Mater. Chem.* **2020**, *8*, 369–379. [CrossRef]
26. Campbell, W.M.; Jolley, K.W.; Wagner, P.; Wagner, K.; Walsh, P.J.; Gordon, K.C.; Schmidt-Mende, L.; Nazeeruddin, M.K.; Wang, Q.; Grätzel, M.; et al. Highly efficient porphyrin sensitizers for dye-sensitized solar cells. *J. Phys. Chem. C* **2007**, *111*, 11760–11762. [CrossRef]

27. Xiang, W.; Huang, F.; Cheng, Y.B.; Bach, U.; Spiccia, L. Aqueous dye-sensitized solar cell electrolytes based on the cobalt(ii)/(iii) tris(bipyridine) redox couple. *Energy Environ. Sci.* **2013**, *6*, 121–127. [CrossRef]
28. Su, J.; Hu, T.H.; Murase, R.; Wang, H.Y.; D'Alessandro, D.M.; Kurmoo, M.; Zuo, J.L. Redox Activities of Metal-Organic Frameworks Incorporating Rare-Earth Metal Chains and Tetrathiafulvalene Linkers. *Inorg. Chem.* **2019**, *58*, 3698–3706. [CrossRef]
29. Wu, Y.; Zeng, R.; Nan, J.; Shu, D.; Qiu, Y.; Chou, S.L. Quinone Electrode Materials for Rechargeable Lithium/Sodium Ion Batteries. *Adv. Energy Mater.* **2017**, *7*, 1700278–1700304. [CrossRef]
30. Tabor, D.P.; Gómez-Bombarelli, R.; Tong, L.; Gordon, R.G.; Aziz, M.J.; Aspuru-Guzik, A. Mapping the frontiers of quinone stability in aqueous media: Implications for organic aqueous redox flow batteries. *J. Mater. Chem. A* **2019**, *7*, 12833–12841. [CrossRef]
31. Boota, M.; Chen, C.; Be, M.; Miao, L.; Gogotsi, Y. Pseudocapacitance and excellent cyclability of 2,5-dimethoxy-1,4-benzoquinone on graphene. *Energy Environ. Sci.* **2016**, *9*, 2586–2594. [CrossRef]
32. Schon, T.B.; Mcallister, B.T.; Li, P.; Seferos, D.S. The rise of organic electrode materials for energy storage. *Chem. Soc. Rev.* **2016**, *45*, 6345–6404. [CrossRef]
33. Patai, S.; Rappoport, Z. *The Chemistry of the Quinonoid Compounds*; Bath, John Wiles & Sons Ltd.: Hoboken, NJ, USA, 1988; Volume 2.
34. Guin, P.S.; Das, S.; Mandal, P.C. Electrochemical Reduction of Quinones in Different Media: A Review. *Int. J. Electrochem.* **2011**, *2011*, 1–22. [CrossRef]
35. Lehmann, M.W.; Evans, D.H. Mechanism of the electrochemical reduction of 3,5-di-tert-butyl-1,2-benzoquinone. Evidence for a concerted electron and proton transfer reaction involving a hydrogen-bonded complex as reactant. *J. Phys. Chem. B* **2001**, *105*, 8877–8884. [CrossRef]
36. Gupta, N.; Linschitz, H. Hydrogen-bonding and protonation effects in electrochemistry of quinones in aprotic solvents. *J. Am. Chem. Soc.* **1997**, *119*, 6384–6391. [CrossRef]
37. Given, P.H.; Peover, M.E. Polarographic Reduction of Aromatic Hydrocarbons and Carbonyl Compounds in Dimethylformamide in the Presence of Proton-donors. *J. Chem. Soc.* **1960**, 385–393. [CrossRef]
38. Wightman, R.M.; Cockrell, J.R.; Murray, R.W.; Burnett, J.N.; Jones, S.B. Protonation Kinetics and Mechanism for 1,8-Dihydroxyanthraquinone and Anthraquinone Anion Radicals in Dimethylformamide Solvent. *J. Am. Chem. Soc.* **1976**, *98*, 2562–2570. [CrossRef]
39. Garza, J.; Vargas, R.; Gómez, M.; González, I.; González, F.J. Theoretical and Electrochemical Study of the Quinone-Benzoic Acid Adduct Linked by Hydrogen Bonds. *J. Phys. Chem. A* **2003**, *107*, 11161–11168. [CrossRef]
40. Yao, M.; Senoh, H.; Yamazaki, S.I.; Siroma, Z.; Sakai, T.; Yasuda, K. High-capacity organic positive-electrode material based on a benzoquinone derivative for use in rechargeable lithium batteries. *J. Power Sources* **2010**, *195*, 8336–8340. [CrossRef]
41. Khattak, A.M.; Ghazi, Z.A.; Liang, B.; Khan, N.A.; Iqbal, A.; Li, L.; Tang, Z. A redox-active 2D covalent organic framework with pyridine moieties capable of faradaic energy storage. *J. Mater. Chem. A* **2016**, *4*, 16312–16317. [CrossRef]
42. Atzori, M.; Pop, F.; Cauchy, T.; Mercuri, M.L.; Avarvari, N. Thiophene-benzoquinones: Synthesis, crystal structures and preliminary coordination chemistry of derived anilate ligands. *Org. Biomol. Chem.* **2014**, *12*, 8752–8763. [CrossRef]
43. Atzori, M.; Artizzu, F.; Sessini, E.; Marchiò, L.; Loche, D.; Serpe, A.; Deplano, P.; Concas, G.; Pop, F.; Avarvari, N.; et al. Halogen-bonding in a new family of tris(haloanilato)metallate(III) magnetic molecular building blocks. *Dalton Trans.* **2014**, *43*, 7006–7019. [CrossRef]
44. Abhervé, A.; Mañas-Valero, S.; Clemente-León, M.; Coronado, E. Graphene related magnetic materials: Micromechanical exfoliation of 2D layered magnets based on bimetallic anilate complexes with inserted [Fe^{III} (acac 2 -trien)] + and [Fe^{III} (sal 2 -trien)] + molecules. *Chem. Sci.* **2015**, *6*, 4665–4673. [CrossRef]
45. Atzori, M.; Benmansour, S.; Mínguez Espallargas, G.; Clemente-León, M.; Abhervé, A.; Gómez-Claramunt, P.; Coronado, E.; Artizzu, F.; Sessini, E.; Deplano, P.; et al. A family of layered chiral porous magnets exhibiting tunable ordering temperatures. *Inorg. Chem.* **2013**, *52*, 10031–10040. [CrossRef]
46. Benmansour, S.; Abhervé, A.; Gómez-Claramunt, P.; Vallés-García, C.; Gómez-García, C.J. Nanosheets of Two-Dimensional Magnetic and Conducting Fe(II)/Fe(III) Mixed-Valence Metal-Organic Frameworks. *ACS Appl. Mater. Interfaces* **2017**, *9*, 26210–26218. [CrossRef]
47. Sahadevan, S.A.; Abhervé, A.; Monni, N.; Sáenz De Pipaón, C.; Galán-Mascarós, J.R.; Waerenborgh, J.C.; Vieira, B.J.C.; Auban-Senzier, P.; Pillet, S.; Bendeif, E.-E.; et al. Conducting Anilate-Based Mixed-Valence Fe(II)Fe(III) Coordination Polymer: Small-Polaron Hopping Model for Oxalate-Type Fe(II)Fe(III) 2D Networks. *J. Am. Chem. Soc.* **2018**, *140*, 12611–12621. [CrossRef] [PubMed]
48. Atzori, M.; Pop, F.; Auban-Senzier, P.; Gómez-García, C.J.; Canadell, E.; Artizzu, F.; Serpe, A.; Deplano, P.; Avarvari, N.; Mercuri, M.L. Structural diversity and physical properties of paramagnetic molecular conductors based on bis(ethylenedithio) tetrathiafulvalene (BEDT-TTF) and the tris(chloranilato)ferrate(III) complex. *Inorg. Chem.* **2014**, *53*, 7028–7039. [CrossRef]
49. Kitagawa, S.; Kawata, S. Coordination compounds of 1,4-dihydroxybenzoquinone and its homologues. Structures and properties. *Coord. Chem. Rev.* **2002**, *224*, 11–34. [CrossRef]

50. Nielson, K.V.; Zhang, L.; Zhang, Q.; Liu, T.L. A strategic high yield synthesis of 2,5-dihydroxy-1,4-benzoquinone Based MOFs. *Inorg. Chem.* **2019**, *58*, 10756–10760. [CrossRef] [PubMed]
51. Poschmann, M.P.M.; Reinsch, H.; Stock, N. [M$_2$(μ-OH)$_2$(DHBQ)$_3$] (M = Zr, Hf)—Two New Isostructural Coordination Polymers based on the Unique M$_2$O$_{14}$ Inorganic Building Unit and 2,5-Dioxido-p-benzoquinone as Linker Molecule. *Z. Anorg. Und Allg. Chem.* **2021**, *647*, 436–441. [CrossRef]
52. Benmansour, S.; Hernández-Paredes, A.; Gómez-García, C.J. Effect of the lanthanoid-size on the structure of a series of lanthanoid-anilato 2-D lattices. *J. Coord. Chem.* **2018**, *71*, 845–863. [CrossRef]
53. Gómez-Claramunt, P.; Benmansour, S.; Hernández-Paredes, A.; Cerezo-Navarrete, C.; Rodríguez-Fernández, C.; Canet-Ferrer, J.; Cantarero, A.; Gómez-García, C.J. Tuning the Structure and Properties of Lanthanoid Coordination Polymers with an Asymmetric Anilato Ligand. *Magnetochemistry* **2018**, *4*, 6. [CrossRef]
54. Benmansour, S.; Pérez-Herráez, I.; López-Martínez, G.; Gómez García, C.J. Solvent-modulated structures in anilato-based 2D coordination polymers. *Polyhedron* **2017**, *135*, 17–25. [CrossRef]
55. Kingsbury, C.J.; Abrahams, B.F.; Auckett, J.E.; Chevreau, H.; Dharma, A.D.; Duyker, S.; He, Q.; Hua, C.; Hudson, T.A.; Murray, K.S.; et al. Square Grid Metal—Chloranilate Networks as Robust Host Systems for Guest Sorption. *Chem. Eur. J.* **2019**, *25*, 5222–5234. [CrossRef]
56. Sahadevan, S.A.; Monni, N.; Abhervé, A.; Cosquer, G.; Oggianu, M.; Ennas, G.; Yamashita, M.; Avarvari, N.; Mercuri, M.L. Dysprosium Chlorocyanoanilate-Based 2D-Layered Coordination Polymers. *Inorg. Chem.* **2019**, *58*, 13988–13998. [CrossRef]
57. Benmansour, S.; Gómez-García, C.J. Lanthanoid-anilato complexes and lattices. *Magnetochemistry* **2020**, *6*, 71. [CrossRef]
58. Benmansour, S.; Hernández-Paredes, A.; Bayona-Andrés, M.; Gómez-García, C.J. Slow Relaxation of the Magnetization in Anilato-Based Dy(III) 2D Lattices. *Molecules* **2021**, *26*, 1190. [CrossRef] [PubMed]
59. Kawata, S.; Kitagawa, S.; Kumagai, H.; Kudo, C.; Kamesaki, H.; Ishiyama, T.; Suzuki, R.; Kondo, M.; Katada, M. Rational Design of a Novel Intercalation System. Layer-Gap Control of Crystalline Coordination Polymers, {[Cu(CA)(H2O)m](G)}n (m = 2,5 = 2,5-Dimethylpyrazine and Phenazine; m = 1, G = 1,2,3,4,6,7,8,9-Octahydrophenazine). *Inorg. Chem.* **1996**, *35*, 4449–4461. [CrossRef] [PubMed]
60. Maka, V.K.; Mukhopadhyay, A.; Jindal, S.; Moorthy, J.N. Redox-Reversible 2D Metal–Organic Framework Nanosheets (MONs) Based on the Hydroquinone/Quinone Couple. *Chem. A Eur. J.* **2019**, *25*, 3835–3842. [CrossRef]
61. Zhu, X.Q.; Wang, C.H. Accurate estimation of the one-electron reduction potentials of various substituted quinones in DMSO and CH3CN. *J. Org. Chem.* **2010**, *75*, 5037–5047. [CrossRef] [PubMed]
62. Jeon, I.R.; Negru, B.; Van Duyne, R.P.; Harris, T.D. A 2D Semiquinone Radical-Containing Microporous Magnet with Solvent-Induced Switching from Tc = 26 to 80 K. *J. Am. Chem. Soc.* **2015**, *137*, 15699–15702. [CrossRef]
63. Abrahams, B.F.; Bond, A.M.; Le, T.H.; McCormick, L.J.; Nafady, A.; Robson, R.; Vo, N. Voltammetric reduction and re-oxidation of solid coordination polymers of dihydroxybenzoquinone. *Chem. Commun.* **2012**, *48*, 11422–11424. [CrossRef]
64. Kharitonov, A.D.; Trofimova, O.Y.; Meshcheryakova, I.N.; Fukin, G.K.; Khrizanforov, M.N.; Budnikova, Y.H.; Bogomyakov, A.S.; Aysin, R.R.; Kovalenko, K.A.; Piskunov, A.V. 2D-metal-organic coordination polymers of lanthanides (La(III), Pr(III) and Nd(III)) with redox-active dioxolene bridging ligands. *CrystEngComm* **2020**, *22*, 4675–4679. [CrossRef]
65. Halis, S.; Inge, A.K.; Dehning, N.; Weyrich, T.; Reinsch, H.; Stock, N. Dihydroxybenzoquinone as Linker for the Synthesis of Permanently Porous Aluminum Metal-Organic Frameworks. *Inorg. Chem.* **2016**, *55*, 7425–7431. [CrossRef]
66. DeGayner, J.A.; Jeon, I.R.; Sun, L.; Dincă, M.; Harris, T.D. 2D Conductive Iron-Quinoid Magnets Ordering up to Tc = 105 K via Heterogenous Redox Chemistry. *J. Am. Chem. Soc.* **2017**, *139*, 4175–4184. [CrossRef]
67. Darago, L.E.; Aubrey, M.L.; Yu, C.J.; Gonzalez, M.I.; Long, J.R. Electronic Conductivity, Ferrimagnetic Ordering, and Reductive Insertion Mediated by Organic Mixed-Valence in a Ferric Semiquinoid Metal-Organic Framework. *J. Am. Chem. Soc.* **2015**, *137*, 15703–15711. [CrossRef]
68. Benmansour, S.; Vallés-García, C.; Gómez-Claramunt, P.; Mínguez Espallargas, G.; Gómez-García, C.J. 2D and 3D Anilato-Based Heterometallic M(I)M(III) Lattices: The Missing Link. *Inorg. Chem.* **2015**, *54*, 5410–5418. [CrossRef]
69. Barltrop, J.A.; Burstall, M.L. The synthesis of Tetracyclines. Part I. Some Model Diene Reactions. *J. Chem. Soc.* **1959**, 2183–2186. [CrossRef]
70. Abrahams, B.F.; Hudson, T.A.; McCormick, L.J.; Robson, R. Coordination polymers of 2,5-dihydroxybenzoquinone and chloranilic acid with the (10,3)- A topology. *Cryst. Growth Des.* **2011**, *11*, 2717–2720. [CrossRef]
71. Luo, T.-T.; Liu, Y.-H.; Tsai, H.-L.; Su, C.-C.; Ueng, C.-H.; Lu, K.-L. A Novel Hybrid Supramolecular Network Assembled from Perfect p-p Stacking of an Anionic Inorganic Layer and a Cationic Hydronium-Ion-Mediated Organic Layer. *Eur. J. Inorg. Chem.* **2004**, *2004*, 4253–4258. [CrossRef]
72. Miller, J.S. Magnetically ordered molecule-based materials. *Chem. Soc. Rev.* **2011**, *40*, 3266–3296. [CrossRef]
73. Ziebel, M.E.; Darago, L.E.; Long, J.R. Control of Electronic Structure and Conductivity in Two-Dimensional Metal-Semiquinoid Frameworks of Titanium, Vanadium, and Chromium. *J. Am. Chem. Soc.* **2018**, *140*, 3040–3051. [CrossRef] [PubMed]
74. Chen, J.; Taniguchi, K.; Sekine, Y.; Miyasaka, H. Electrochemical development of magnetic long-range correlations with Tc = 128 K in a tetraoxolene-bridged Fe-based framework. *J. Magn. Magn. Mater.* **2020**, *494*, 165818. [CrossRef]
75. Liu, L.; Degayner, J.A.; Sun, L.; Zee, D.Z.; Harris, T.D. Reversible redox switching of magnetic order and electrical conductivity in a 2D manganese benzoquinoid framework. *Chem. Sci.* **2019**, *10*, 4652–4661. [CrossRef] [PubMed]

76. Taniguchi, K.; Chen, J.; Sekine, Y.; Miyasaka, H. Magnetic Phase Switching in a Tetraoxolene-Bridged Honeycomb Ferrimagnet Using a Lithium Ion Battery System. *Chem. Mater.* **2017**, *29*, 10053–10059. [CrossRef]
77. Jiang, Q.; Xiong, P.; Liu, J.; Xie, Z.; Wang, Q.; Yang, X.Q.; Hu, E.; Cao, Y.; Sun, J.; Xu, Y.; et al. A Redox-Active 2D Metal–Organic Framework for Efficient Lithium Storage with Extraordinary High Capacity. *Angew. Chem. Int. Ed.* **2020**, *59*, 5273–5277. [CrossRef]

Article

Collective Magnetic Behavior of 11 nm Photo-Switchable CsCoFe Prussian Blue Analogue Nanocrystals: Effect of Dilution and Light Intensity

Linh Trinh, Eric Rivière, Sandra Mazerat, Laure Catala and Talal Mallah *

Institut de Chimie Moléculaire et des Matériaux d'Orsay, Université Paris-Saclay, CNRS 15, Rue Georges Clémenceau, CEDEX 09, 91405 Orsay, France; trinh.linh@universite-paris-saclay.fr (L.T.); eric.riviere@universite-paris-saclay.fr (E.R.); sandra.mazerat@universite-paris-saclay.fr (S.M.); laure.catala@universite-paris-saclay.fr (L.C.)
* Correspondence: talal.mallah@universite-paris-saclay.fr

Abstract: The collective magnetic behavior of photoswitchable 11 nm cyanide-bridged nanoparticles based of the Prussian blue analogue CsCoFe were investigated when embedded in two different matrices with different concentrations. The effect of the intensity of light irradiation was studied in the less concentrated sample. Magnetization studies and alternating magnetic susceptibility data are consistent with a collective magnetic behavior due to interparticle dipolar magnetic interaction for the two compounds, even though the objects have a size that place them in the superparamagnetic regime.

Keywords: Prussian blue analogue; photomagnetism; nanocrystals; photo-switchable; magnetic properties; dipolar interaction

Citation: Trinh, L.; Rivière, E.; Mazerat, S.; Catala, L.; Mallah, T. Collective Magnetic Behavior of 11 nm Photo-Switchable CsCoFe Prussian Blue Analogue Nanocrystals: Effect of Dilution and Light Intensity. *Magnetochemistry* **2021**, *7*, 99. https://doi.org/10.3390/magnetochemistry7070099

Academic Editors: Carlos J. Gómez García, John Wallis, Lee Martin, Scott Turner and Hiroki Akutsu

Received: 7 May 2021
Accepted: 6 July 2021
Published: 8 July 2021

Publisher's Note: MDPI stays neutral with regard to jurisdictional claims in published maps and institutional affiliations.

Copyright: © 2021 by the authors. Licensee MDPI, Basel, Switzerland. This article is an open access article distributed under the terms and conditions of the Creative Commons Attribution (CC BY) license (https://creativecommons.org/licenses/by/4.0/).

1. Introduction

Prussian blue analogues are cyanide-bridged coordination networks with a face centered cubic (fcc) structure and general formula $A_y\{M[M'(CN)_6]_{1-x}\square_x\}\bullet zH_2O$, where A is an alkali metal ion (Na, Rb or Cs in most cases), M and M' are transition metal ions of the first series with in most cases oxidation states II and III, respectively, and $\square$ stands for metallocyanide vacancies. The cell parameter is close to 10 Å and corresponds to the distance between two metal ions of the same nature. Water molecules occupy the tetrahedral sites of the fcc structure and can also be coordinated to M when vacancies are present. When M' = Fe and M = Co, the two states $Co^{II}Fe^{III}$ ($S_{Co^{II}} = 3/2$, $S_{Fe^{III}} = 1/2$) and $Co^{III}Fe^{II}$ ($S_{Co^{III}} = 0$, $S_{Fe^{II}} = 0$) are close in energy and it is possible to switch from one to another thermally or by light irradiation at low temperatures. The electron transfer is accompanied by a spin crossover on Co (from high spin Co^{II} ($S_{Co^{II}} = 3/2$) to low spin Co^{III} ($S_{Co^{III}} = 0$)), inducing a large change in the magnetic response as first demonstrated by Hashimoto [1]. More precisely, depending on the nature of the alkali metal ions that occupy the tetrahedral sites or its absence and on the vacancies concentration, three situations may be encountered: (i) the magnetic state $Co^{II}Fe^{III}$ is present at high temperature and no electron transfer occurs upon cooling (this is the case in the absence or for low contents of alkali ions); (ii) the diamagnetic state $Co^{III}Fe^{II}$ is present for the whole temperature range, but no light induced electron transfer occurs at low temperature (this is the case when Cs^+ occupy the tetrahedral sites. For this case, the compound contains generally some amount of the $Co^{II}Fe^{III}$ magnetic phase); and (iii) the magnetic state is present at high temperature and upon cooling down a thermally assisted electron transfer occurs leading to the diamagnetic state $Co^{III}Fe^{II}$ that may be transformed upon light irradiation to $Co^{II}Fe^{III}$ at low temperature (this occurs usually when A = Rb or when the concentration of Cs is well below 1 (see formulae above)). These different cases were thoroughly investigated and rationalized for the bulk materials [2–7], made of aggregates of nanoparticles in the hundred nanometers size range.

Recently we reported [8], the photoswitching behavior of nanocrystals in the sub-15 nm size containing a large concentration of Cs and almost no vacancies i.e., $(CTA)_{0.4}[Cs_{0.7}$-$Co\{Fe(CN)_6\}_{0.9}]\bullet H_2O$ that have a majority of the diamagnetic phase at room temperature and present a light induced switching to the magnetic state at low temperatures. The photoswitching behavior was studied when the particles were embedded in CetylTrimethylAmmonium (CTA^+) which serves as counter-cation for the nanocrystals and for the objects surrounded by the organic polymer polyvinylpyrrolidone (PVP). However, we did not report the collective magnetic behavior of the nanocrystals in the photo-induced state. Indeed, upon irradiation the diamagnetic ions become paramagnetic and, due to an antiferromagnetic exchange coupling interaction, a collective ferrimagnetic behavior is observed in the bulk with a critical temperature close to 20 K [1]. In nanoparticles, the situation may be different since the magnetic correlation length is limited by the particles' size if the interparticle magnetic coupling is absent. The nanocrystals we reported are stabilized as colloids in water in the absence of stabilizing agents so that they can be embedded in different matrices and with different concentrations [9]. The objective of this paper is to investigate the effect of dilution and intensity of light irradiation on the magnetic behavior of the nanocrystals in the photoinduced metastable state when embedded in CTA and PVP.

2. Results and Discussion

2.1. Materials and Methods

The preparation and full characterization of the nanocrystals were recently reported [8]. The nanocrystals were prepared in water as a stable colloidal solution then recovered by CTA^+ and by PVP to give two samples, namely CsCoFe_CTA and CsCoFe_PVP. The two materials were prepared as follows.

CsCoFe_CTA. 200 mL of distilled water containing 673 mg of CsCl (4 mM) and 476 mg (2 mM) of $[Co(H_2O)_6]Cl_2$ were quickly added to 200 mL of distilled water containing 658 mg (2 mM) of $K_3[Fe(CN)_6]$. The solution was vigorously mixed for 30 min. A methanolic solution (600 mL) containing 1.10 g (6 mM) of cetyltrimethylammonium bromide (CTABr) was added dropwise to 200 mL (half) of the aqueous solution containing the nanoparticles. A precipitate formed during the addition, and it was recovered by centrifugation (9000 rpm for 20 min) washed with a small amount of water and dried under vacuum overnight. Elemental analysis for $Cs_{0.7}(C_{19}H_{42}N)_{0.4}Co[Fe(CN)_6]_{0.9}(H_2O)$, exp. (calc.) C: 20.10 (20.60), H: 2.43 (3.79), N: 17.70 (17.12).

CsCoFe_PVP. The remaining 200 mL of the colloidal solution containing the nanoparticles was added dropwise for 3 h to 20 mL of an aqueous solution containing 12 g of PVP. Then, 900 mL of acetone is added. A precipitate formed, and it was recovered by centrifugation (9000 rpm for 20 min), washed with a small amount of acetone, and dried under vacuum overnight.

A Transmission Electron Microscopy (Jeol 1400, Jeol, Tokyo, Japan) image of the CTA materials showed objects with an average size close to 11 nm (Figure 1). The X-ray power diffraction (Panalytical X-Pert Pro MPD, Malvern Panalytical, Malvern, UK) diagram of the same material was consistent with face centered cubic structure as expected for Prussian blue analogues with a cell parameter $a = 10.01$ Å (Figure 1) [8]. The infrared spectrum (Perkin Elmer Spectrum 100, PerkinElmer Inc., Norwalk, Connecticut, USA) (Figure 1) in the 2300–1900 cm^{-1} regions showed the asymmetric vibration mode of the cyanide at 2115 cm^{-1} characteristics of cyanide bridge corresponding to a mixture of Fe(II)-CN-Co(III) and Fe(II)-CN-Co(III) sequences [8].

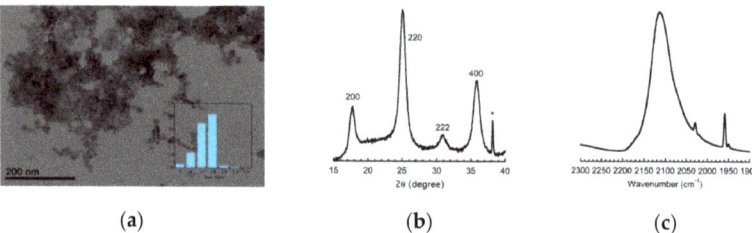

Figure 1. (a) Transmission Electron Microscopy imaging with count as a function of size in the inset; (b) Powder X-ray diffraction pattern; and (c) infra-red spectrum of the CsCoFe_CTA nanoparticles.

Combined powder X-ray diffraction, magnetic, Electron Paramagnetic Resonance, and X-ray photoelectron spectroscopy studies are consistent with the presence at room temperature of the different phases $Co^{II}Fe^{II}$, $Co^{II}Fe^{III}$, and $Co^{III}Fe^{II}$, the latter contributing to 70% of the overall concentration of each nanocrystal [8]. Because of their stability as isolated objects in water, they were used to unravel the mechanisms of charge transfer and spin crossover after light irradiation [10]. They were also deposited on graphite and their conductance was measured, showing a long range electron transport with relatively weak attenuation [11].

In order to assess the relative concentration in nanoparticles for the two materials, we compared their magnetization (Quantum Design XL7, Quantum Design, San Diego, CA, USA) values at saturation (M_{sat}) and at low temperature (T = 2 K and B = 6 T) that are proportional to the amount magnetic species within the material (Figure 2). A ratio close to 20 was found for $M_{sat(CTA)}/M_{sat(PVP)}$, showing that the concentration in nanoparticles of the PVP containing materials was 20 times lower than that of the CTA one.

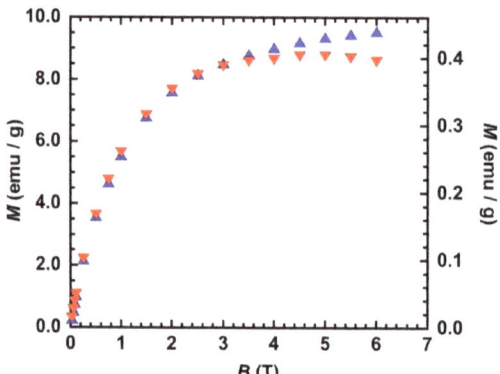

Figure 2. M = f(B) at T = 2 K before irradiation for CsCoFe_CTA and CsCoFe_PVP. ▲: left scale; ▼: right scale.

The magnetic behavior of the two samples CsCoFe_CTA and CsCoFe_PVP was studied using a SQUID magnetometer in the dc mode by measuring the magnetization in the Field Cooled (FCM) and the Zero-Field Cooled (ZFCM) modes and the magnetic hysteresis loops at T = 2 K and in the ac mode by measuring the thermal dependence of the in-phase and out-of-phase magnetic susceptibilities in zero dc applied magnetic field for different frequencies of the alternating magnetic field equal to 3 Oe. The irradiation was carried out using a laser diode connected to an optical fiber at a wavelength of 635 nm with the values of the laser power specified below.

2.2. Magnetic Behavior of CsCoFe_CTA

Before irradiation, the FCM plot has the feature of paramagnetic species with only a slight increase below T = 4 K, consistent with the presence of the paramagnetic species $Co^{II}Fe^{II}$ and $Co^{II}Fe^{III}$ (Figure 3a). It is worth noting that the absence of a large increase of the magnetization at low temperature is in line with a very short correlation length consistent with relatively isolated $Co^{II}Fe^{III}$ pairs within the nanocrystals. The sample was irradiated at T = 2 K, heated up to 25 K, and the FCM was measured. After irradiation, the shape of the FCM curve suggests a behavior due to a magnetic order or to a blocking of the magnetization. The ZFCM curve after irradiation shows a maximum at T = 11.8 K and meets the FCM curve at T = 14.5 K, which is usually associated with the blocking of particles with different sizes.

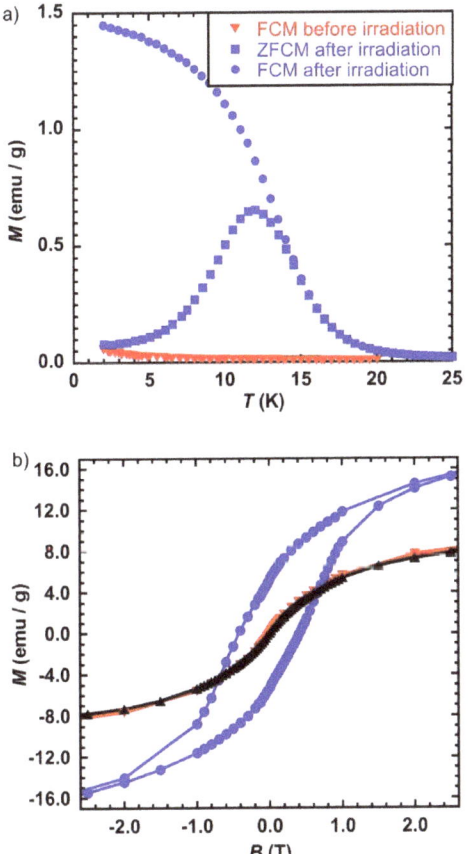

Figure 3. (a) M = f(T) in the form of Field Cooled before (▼) and after (●) irradiation and Zero-Field Cooled after irradiation (■) and (b) Magnetic hysteresis loop at T = 2 K before (▼) and after (●) irradiation and after relaxation (▲), for CsCoFe_CTA; laser power = 50 mW/cm².

The magnetic hysteresis loop measured before irradiation was characteristic of a paramagnetic behavior (Figure 3b), without an opening at zero field. After irradiation, a hysteresis loop appeared with a coercive field H_C = 0.41 T, consistent with a relatively large magnetic anisotropy for the photoinduced state as expected from the presence of Co^{II} within the nanoparticles [12]. In order to check the reversibility of the photoswitching, the sample was heated to T = 250 K (above its relaxation temperature) [8] and then cooled down to 2 K. The magnetization after relaxation was superimposable to its trace before irradiation, confirming the total recovery of the ground state.

We performed ac susceptibility studies in order to get more insight into the magnetism of the system. Before irradiation, there was no maxima in the temperature dependence of the in-phase (χ') susceptibility responses for frequencies (ν) ranging from 0.1 to 300 Hz (ac magnetic field of 3 Oe and zero dc magnetic field). After irradiation, maxima of the $\chi' = f(T)$ and the $\chi'' = f(T)$ (χ'' is the out-of-phase susceptibility) curves appeared (Figure 4), which are typical for a system with a blocking temperature that can be associated with the isolated objects or due to a spin glass like behavior. The value of the maximum of the $\chi'' = f(T)$ curve at the lowest frequency available ($\nu = 0.1$ Hz) was 11.2 K, consistent with temperature of the maximum of the ZFCM curve. The analysis of the out-of-phase data was performed by plotting the $\ln(\tau) = f(1/T)$ (with $\tau(1/2\pi\nu) = \tau_0 \exp(\Delta E/kT)$, where τ_0 is the attempt time, ΔE the anisotropy barrier, k the Boltzmann constant, and T the temperature of the maximum of the $\chi'' = f(T)$ curve at a given frequency. The linear fit of the data (not shown) led to $\tau_0 = 1.2 \times 10^{-27}$ s and $\Delta E = 690$ K. For single magnetic domain nanoparticles without (or with very weak) dipolar interactions, values for τ_0 close or larger than 10^{-12} s were expected [13], and were found for isolated CsNiCr cyanide-bridged nanoparticles [14]. The τ_0 value obtained from the fit had no physical meaning for isolated objects and it can be assumed that the observed behavior was due to magnetic interactions among the particles. The Mydosh parameter ϕ allows for discriminating among various magnetic behaviors where $\phi = (T_{max} - T_{min})/(T_{max}(\log \nu_{max} - \log \nu_{min}))$ with T_{max} and T_{min} being the temperatures of the maxima of the $\chi'' = f(T)$ curves for the two extreme applied frequencies ν_{max} and ν_{min} respectively, ϕ values close or larger than 0.12 are expected for nearly isolated magnetic nanoparticles [13,15]. For the present case, ϕ (0.03) was much smaller than 0.12, which indicates together with the very small τ_0 value the presence of magnetic interactions (dipolar) among the particles and not to a blocking of the magnetization of single domain isolated objects. It is possible to fit the relaxation time by introducing a parameter that accounts for the interaction using the modified Arrhenius law $\tau = \tau_0 \exp(\Delta E/k(T - T_0))$, where T_0 considers the interaction among the nanoparticles that leads to more physically acceptable values for τ_0 (2.2×10^{-12} s) and an average anisotropy barrier $\Delta E = 114$ K with $T_0 = 7$ K.

2.3. Magnetic Behavior of CsCoFe_PVP

Here, the nanocrystals were embedded in PVP with a concentration 20 times less. They are, therefore, spatially more separated than in CsCoFe_CTA.

The maximum of the ZFCM curve was found at $T = 4.4$ K, a value smaller than for CsCoFe_CTA (11.8 K) (Figure 5a). The FCM and ZFCM curves joined at $T = 5$ K. The difference in the temperature between the maximum of the ZFCM and the temperature where the two curves join was 0.6 K, which was lower than 2.7 K found for CsCoFe_CTA. This result means that the difference in the irreversibility temperature and the maximum of the ZFCM curve for CsCoFe_CTA was not due to the blocking of particles with different sizes, otherwise we would have had the same difference for CsCoFe_PVP, since the same objects were present in the two materials. Therefore, magnetic dipolar interactions seem to affect the irreversibility temperature of the magnetization curves in the case of the CsCoFe nanoparticles.

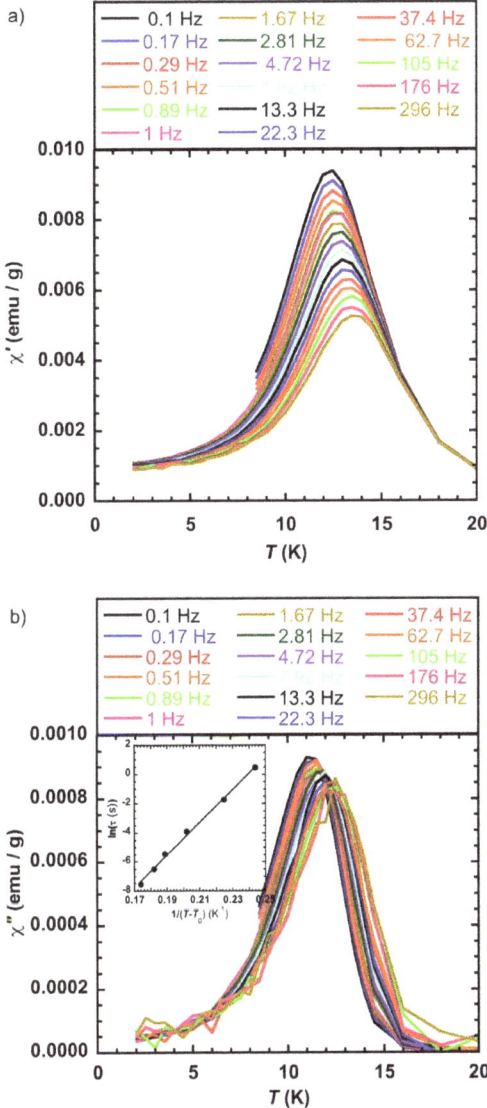

Figure 4. (a) $\chi' = f(T)$ and (b) $\chi'' = f(T)$ at different frequencies of the alternating magnetic field (3 Oe), inset $\ln(\tau) = f(1/(T-T_0))$ and linear fit with values in the text, CsCoFe_CTA; laser power = 50 mW/cm^2.

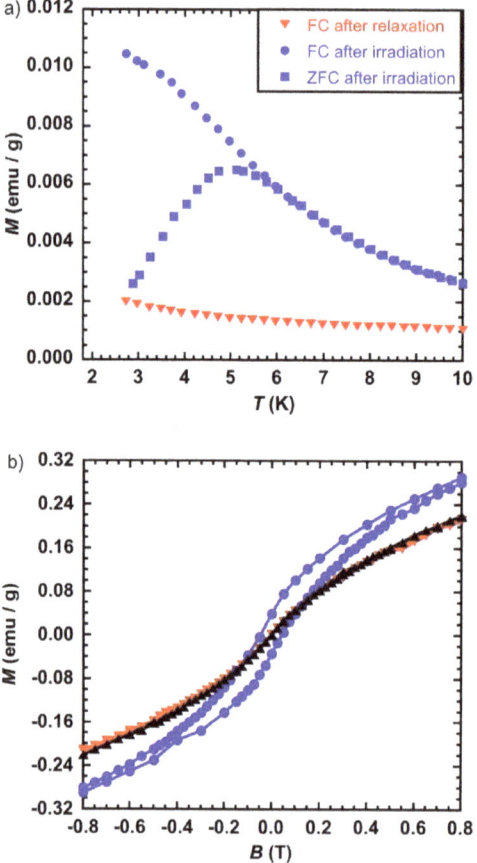

Figure 5. (a) $M = f(T)$ in the form of Field Cooled before (▼) and after (●) irradiation and Zero-Field Cooled after irradiation (■), for CsCoFe_PVP and (b) Magnetic hysteresis loop at T = 2 K before (▼) and after (●) irradiation and after relaxation (▲); laser power 65 mW/cm^2.

The magnetic hysteresis loop opens after light irradiation with a coercive field of 0.053 T (Figure 5b), one order of magnitude weaker than for the non-diluted CsCoFe_CTA compound. The magnetization curves before irradiation and after relaxation are superimposable, as expected, and are consistent with the reversibility of the process.

The susceptibility measurements with an alternating magnetic field of 3 Oe gave a very weak signal before irradiation, as expected. After irradiation, the light induced data were not of very good quality because we reduced, as much as possible, the thickness of the sample in order to have a maximal light penetration. However, we observed the maxima of the $\chi' = f(T)$ curves at different frequencies that do not coincide (Figure 6a). The temperature dependence of the out-of-phase susceptibility curves measured at 1 and 2.81 Hz shows slightly different temperature maxima (Figure 6b), even though there is some uncertainty concerning these values because of the very weak signal.

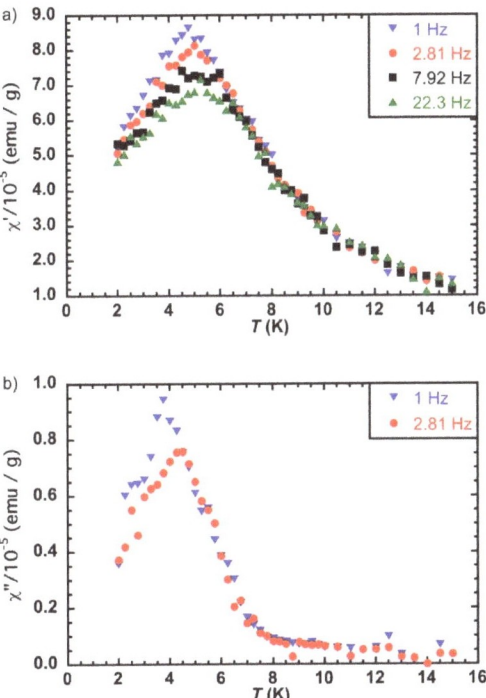

Figure 6. (a) $\chi' = f(T)$ and (b) $\chi'' = f(T)$ at different frequencies of the alternating magnetic field (3 Oe) for CsCoFe_PVP; laser power 65 mW/cm^2.

A photomagnetic effect leading to an opening of the hysteresis loop was observed as for CsCoFe_CTA, suggesting a similar behavior. However, the maximum of the ZFCM curve is at lower temperature and the out-of-phase susceptibility signals are too low to reach a conclusion of the nature of the magnetic behavior of CsCoFe_PVP. We therefore studied the effect of light intensity on the magnetic behavior of the materials.

2.4. Effect of the Power of Light Irradiation

Another way to sense the effect of magnetic dipolar interactions is to keep the same concentration of the nanoparticles and gradually increase their magnetic moments. Since the magnitude of the magnetic dipolar interactions is proportional to the value of the magnetic moments of the interacting objects, increasing the magnetic moment should lead to an upward shift of the critical temperature (everything else being equal). In order to optimize light penetration and obtain the maximum light-induced response, we used a thin film of few microns of PVP containing the nanoparticles. We first measured the temporal response to light by irradiating the sample at different laser powers at $T = 10$ K and $B = 0.5$ T. Figure 7a illustrates the variation of the magnetization (M) with time for a power of 20 mW/cm^2. It shows that 50% of the saturation is obtained within 8 min and 90% within 30 min. When irradiating at a power of 100 mW/cm^2 (not shown here), 90% of the magnetization saturation is reached within 8 min and 50% within 1.5 min, illustrating the relatively fast response of the sample to light irradiation. The jump at $t = 97$ min observed when the laser is switched off (Figure 7a) is due to the thermalization of the sample (to 10 K) that was heated up during the irradiation process. If we assume that the process takes place in the paramagnetic regime, the temperature during the irradiation process should be around 20 K. Figure 7b depicts the variation of the reduced magnetization with the laser power. It shows that at a laser power of 150 mW/cm^2 the magnetization of the sample is multiplied by around 2.5 due to an increase of the magnetic moments of the individual objects upon transformation of the diamagnetic $Co^{III}Fe^{II}$ pairs to magnetic $Co^{II}Fe^{III}$.

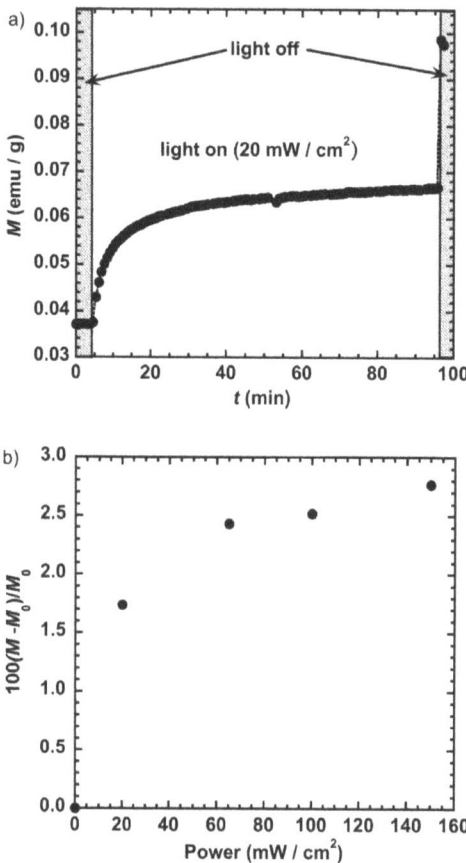

Figure 7. (**a**) Variation of the magnetization (M) with time (t) for a laser power of 20 mW/cm^2 and (**b**) reduced magnetization (($M-M_0$)/M_0)) performed at $T = 10$ K and $B = 0.5$ T showing the rate of increase of the magnetization after irradiation as a function of the laser power.

In order to assess the increase of the nanoparticles' magnetic moments with the intensity of light on the overall magnetic behavior, we measured the ZFCM curves after irradiation at $T = 2$ K and $B = 0.005$ T for laser power values of 20, 65, and 150 mW/cm^2 (Figure 8). The value of the maximum of the ZFCM curves shifts from 4.0 to 5.1 K when going from 20 to 150 mW/cm^2. This behavior shows that, everything else being equal, the increase of the critical temperature is directly related to the increase of the local magnetic moments within the sample. This behavior can be due either to interparticle magnetic dipolar interactions that increase upon an increase of the objects' magnetic moments, or to an increase of the blocking temperature of the isolated objects if they were in the superparamagnetic regime. The reversibility of the process was examined by measuring the FCM curves after relaxation (heating up to $T = 250$ K) for the different laser power values. They were all found identical to those before irradiation, showing that no damage occurs during the irradiation process even at a power of 150 mW/cm^2.

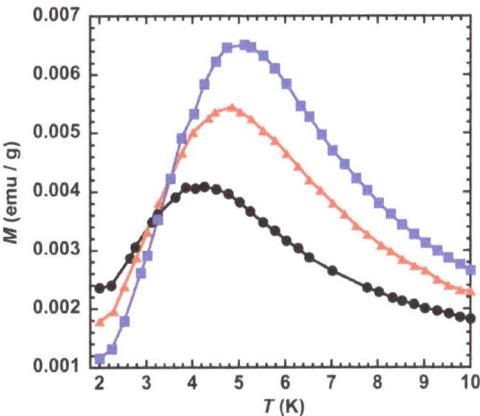

Figure 8. ZFCM after irradiation with laser power values of 20 (●), 65 (▲), and 150 (■) mW/cm^2, for CsCoFe_PVP.

A spin glass-like behavior was reported by us for the 3 nm Ni^{II}Fe^{III} cyanide-bridged nanoparticles [16,17]. It was also observed for CsCoFe nanoparticles, but was assigned to a size effect [18] and, later, CoFe particles embedded in mesoporous silica were investigated with the conclusion that interparticle dipolar interactions are present [19,20]. However, because of the nature of the materials (nanoparticles embedded in silica), it was not possible to investigate the effect of dilution in order to confirm that the magnetic behavior is indeed due to dipolar magnetic interactions and not to the intrinsic behavior of isolated particles. Indeed, if the particles have a size larger than that of the critical magnetic single domain size, they will have a behavior similar to that of superparamagnetic particles (single magnetic domain) feeling magnetic dipolar interactions. Moreover, it is usually difficult to reach a definite conclusion without highly diluting the nanoobjects.

We have already investigated the magnetic properties of highly diluted CsNiCr PBA particles where we demonstrated that successive dilution in PVP allows isolating the objects and observing the single domain regime with a Néel–Brown behavior and a critical size (D) of 22 nm [14]. It is possible to use this result to estimate the critical size for the CsCoFe network by comparing the magnitude of the exchange coupling interaction and the magnetic moments of the two networks. For cubic particles, D is given by $7.2(A/0.5\mu_0 M^2)^{1/2}$, where A is proportional to the exchange coupling between two ions of the network and M is the magnetic moment of a pair of interacting metal ions [21]. Actually, the Co^{II}-Fe^{III} exchange coupling interaction for the CsCo^{II}Fe^{III} network is about four times weaker ($|J_{CoFe}| \approx 3.7$ cm^{-1}) than that of CsNiCr ($J_{NiCr} \approx 16$ cm^{-1}) [22] since its T_C is equal to 21 K [1] while that of CsNiCr is 90 K [1,22,23]. The magnetic moment of a Co^{II}Fe^{III} pair is equal to 2 Bohr Magnetons due to the antiferromagnetic coupling between $S_{Co^{II}}$ (3/2) and $S_{Fe^{III}}$ (1/2), while this value is 5 Bohr Magnetons for the CsNiCr network due to the *ferro*magnetic exchange coupling between Ni^{II} ($S_{Ni^{II}}$ = 1 and Cr^{III} ($S_{Cr^{III}}$ = 3/2). Using these values and the expression of the critical size of a single magnetic domain leads to D = 27.5 nm for the CsCoFe network, a value much larger than the size of the nanocrystals at hand (a maximum of 11 nm if light fully converts the particles from the ground diamagnetic to the metastable magnetic state), ensuring that the investigated objects are in the superparamagnetic regime. In the superparamagnetic regime, a blocking of the magnetization of the isolated objects may be observed if the magnetic correlation length is large enough for a single object to behave as single giant spin. This cannot be the case for the present nanoparticles since the Co^{II}-Fe^{III} exchange coupling interaction value (≈ 3.7 cm^{-1}) is very close to the temperature maxima of the ZFCM curves (Figure 8). Consequently, the shift up of these temperature maxima upon increasing the laser power can only be due to interparticle dipolar magnetic interactions.

3. Concluding Remarks

The collective magnetic behavior of CsCoFe photoswitchable 11 nm cyanide-bridged nanoparticles was investigated by examining the effect of their dilution in two different matrices with a concentration ratio around 20 between the two compounds. The magnetic data performed in static magnetic field (FCM, ZFCM) and the susceptibility data in the presence of an alternating field are consistent with a spin glass-like behavior, compatible with interparticle magnetic dipolar interactions thanks to the analysis of the out-of-phase susceptibility data of the more concentrated sample (in CTA). For the less concentrated sample (in PVP), increasing the light intensity shifts up the temperature maximum of the ZFCM curves. The analysis of the critical single magnetic domain size and the magnetic correlation length within the objects is consistent with an interparticle dipolar coupling rather than an increase of the blocking temperature of single domain objects.

Author Contributions: Conceptualization, L.C. and T.M.; Preparation and Characterization, L.T., S.M. and E.R.; data curation, T.M.; writing—review and editing, L.C. and T.M. All authors have read and agreed to the published version of the manuscript.

Funding: This research was partially funded by the Vietnamese government for Scholarship to L.T., thesis code MN31.

Institutional Review Board Statement: Not applicable.

Informed Consent Statement: Not applicable.

Data Availability Statement: Data supporting reported results are available from the authors on request.

Conflicts of Interest: The authors declare no conflict of interest.

References

1. Sato, O.; Iyoda, T.; Fujishima, A.; Hashimoto, K. Photoinduced magnetization of a cobalt-iron cyanide. *Science* **1996**, *272*, 704–705. [CrossRef]
2. Sato, O.; Einaga, Y.; Iyoda, T.; Fujishima, A.; Hashimoto, K. Cation-driven electron transfer involving a spin transition at room temperature in a cobalt iron cyanide thin film. *J. Phys. Chem. B* **1997**, *101*, 3903–3905. [CrossRef]
3. Yokoyama, T.; Ohta, T.; Sato, O.; Hashimoto, K. Characterization of magnetic CoFe cyanides by x-ray-absorption fine-structure spectroscopy. *Phys. Rev. B* **1998**, *58*, 8257–8266. [CrossRef]
4. Sato, O.; Einaga, Y.; Fujishima, A.; Hashimoto, K. Photoinduced long-range magnetic ordering of a cobalt-iron cyanide. *Inorg. Chem.* **1999**, *38*, 4405–4412. [CrossRef]
5. Champion, G.; Escax, V.; Moulin, C.C.D.; Bleuzen, A.; Villain, F.O.; Baudelet, F.; Dartyge, E.; Verdaguer, N. Photoinduced ferrimagnetic systems in prussian blue analogues (CxCo4)-Co-I[Fe(CN)(6)](y) (C-I = alkali cation). 4. Characterization of the ferrimagnetism of the photoinduced metastable state in Rb1.8Co4[Fe(CN)(6)](3.3)center dot 13H(2)O by K edges X-ray magnetic circular dichroism. *J. Am. Chem. Soc.* **2001**, *123*, 12544–12546. [PubMed]
6. Escax, V.; Bleuzen, A.; Moulin, C.C.D.; Villain, F.; Goujon, A.; Varret, F.; Verdaguer, M. Photoinduced ferrimagnetic systems in prussian blue analogues (CxCo4)-Co-I[Fe(CN)(6)](y) (C-I = alkali cation). 3. Control of the photo- and thermally induced electron transfer by the [Fe(CN)(6)] vacancies in cesium derivatives. *J. Am. Chem. Soc.* **2001**, *123*, 12536–12543. [CrossRef] [PubMed]
7. Shimamoto, N.; Ohkoshi, S.; Sato, O.; Hashimoto, K. Control of charge-transfer-induced spin transition temperature on cobalt-iron Prussian blue analogues. *Inorg. Chem.* **2002**, *41*, 678–684. [CrossRef] [PubMed]
8. Trinh, L.; Zerdane, S.; Mazerat, S.; Dia, N.; Dragoe, D.; Herrero, C.; Riviere, E.; Catala, L.; Cammarata, M.; Collet, E.; et al. Photoswitchable 11 nm CsCoFe Prussian Blue Analogue Nanocrystals with High Relaxation Temperature. *Inorg. Chem.* **2020**, *59*, 13153–13161. [CrossRef] [PubMed]
9. Brinzei, D.; Catala, L.; Louvain, N.; Rogez, G.; Stephan, O.; Gloter, A.; Mallah, T. Spontaneous stabilization and isolation of dispersible bimetallic coordination nanoparticles of CsxNi[Cr(CN)(6)](y). *J. Mater. Chem.* **2006**, *16*, 2593–2599. [CrossRef]
10. Cammarata, M.; Zerdane, S.; Balducci, L.; Azzolina, G.; Mazerat, S.; Exertier, C.; Trabuco, M.; Levantino, M.; Alonso-Mori, R.; Glownia, J.M.; et al. Charge transfer driven by ultrafast spin transition in a CoFe Prussian blue analogue. *Nat. Chem.* **2021**, *13*, 10–14. [CrossRef]
11. Bonnet, R.; Lenfant, S.; Mazerat, S.; Mallah, T.; Vuillaume, D. Long-range electron transport in Prussian blue analog nanocrystals. *Nanoscale* **2020**, *12*, 20374–20385. [CrossRef] [PubMed]
12. Prado, Y.; Arrio, M.A.; Volatron, F.; Otero, E.; Moulin, C.C.D.; Sainctavit, P.; Catala, L.; Mallah, T. Magnetic Anisotropy of Cyanide-Bridged Core and CoreShell Coordination Nanoparticles Probed by X-ray Magnetic Circular Dichroism. *Chem. Eur. J.* **2013**, *19*, 6685–6694. [CrossRef] [PubMed]

13. Dormann, J.L.; Spinu, L.; Tronc, E.; Jolivet, J.P.; Lucari, F.; D'Orazio, F. Effect of interparticle interactions on the dynamical properties of gamma-Fe2O3 nanoparticles. *J. Magn. Magn. Mater.* **1998**, *183*, L255–L260. [CrossRef]
14. Prado, Y.; Mazerat, S.; Riviere, E.; Rogez, G.; Gloter, A.; Stephan, O.; Catala, L.; Mallah, T. Magnetization Reversal in (CsNiCrIII)-Cr-II(CN)(6) Coordination Nanoparticles: Unravelling Surface Anisotropy and Dipolar Interaction Effects. *Adv. Funct. Mater.* **2014**, *24*, 5402–5411. [CrossRef]
15. Dormann, J.L.; Cherkaoui, R.; Spinu, L.; Nogues, M.; Lucari, F.; D'Orazio, F.; Fiorani, D.; Garcia, A.; Tronc, E.; Jolivet, J.P. From pure superparamagnetic regime to glass collective state of magnetic moments in gamma-Fe_2O_3 nanoparticle assemblies. *J. Magn. Magn. Mater.* **1998**, *187*, L139–L144. [CrossRef]
16. Brinzei, D.; Catala, L.; Rogez, G.; Gloter, A.; Mallah, T. Magnetic behaviour of negatively charged nickel(II) hexacyanoferrate(III) coordination nanoparticles. *Inorg. Chim. Acta* **2008**, *361*, 3931–3936. [CrossRef]
17. Folch, B.; Guari, Y.; Larionova, J.; Luna, C.; Sangregorio, C.; Innocenti, C.; Caneschi, A.; Guerin, C. Synthesis and behaviour of size controlled cyano-bridged coordination polymer nanoparticles within hybrid mesoporous silica. *New J. Chem.* **2008**, *32*, 273–282. [CrossRef]
18. Pajerowski, D.M.; Frye, F.A.; Talham, D.R.; Meisel, M.W. Size dependence of the photoinduced magnetism and long-range ordering in Prussian blue analogue nanoparticles of rubidium cobalt hexacyanoferrate. *New J. Phys.* **2007**, *9*, 222. [CrossRef]
19. Moulin, R.; Delahaye, E.; Bordage, A.; Fonda, E.; Baltaze, J.P.; Beaunier, P.; Riviere, E.; Fornasieri, G.; Bleuzen, A. Ordered Mesoporous Silica Monoliths as a Versatile Platform for the Study of Magnetic and Photomagnetic Prussian Blue Analogue Nanoparticles. *Eur. J. Inorg. Chem.* **2017**, 1303–1313. [CrossRef]
20. Fornasieri, G.; Bordage, A.; Bleuzen, A. Magnetism and Photomagnetism of Prussian Blue Analogue Nanoparticles Embedded in Porous Metal Oxide Ordered Nanostructures. *Eur. J. Inorg. Chem.* **2018**, 259–271. [CrossRef]
21. Rave, W.; Fabian, K.; Hubert, A. Magnetic states of small cubic particles with uniaxial anisotropy. *J. Magn. Magn. Mater.* **1998**, *190*, 332–348. [CrossRef]
22. Mallah, T.; Auberger, C.; Verdaguer, M.; Veillet, P. A Heptanuclear Criiiniii(6) Complex with a Low-Lying S = 15/2 Ground-State. *J. Chem. Soc. Chem. Commun.* **1995**, 61–62. [CrossRef]
23. Gadet, V.; Mallah, T.; Castro, I.; Verdaguer, M.; Veillet, P. High-Tc Molecular-Based Magnets–A Ferromagnetic Bimetallic Chromium(Iii) Nickel(Ii) Cyanide with Tc = 90-K. *J. Am. Chem. Soc.* **1992**, *114*, 9213–9214. [CrossRef]

Article

New Photomagnetic Ionic Salts Based on [Mo^{IV}(CN)$_8$]$^{4-}$ and [W^{IV}(CN)$_8$]$^{4-}$ Anions [†]

Xinghui Qi [1], Philippe Guionneau [1,*], Enzo Lafon [1], Solène Perot [1], Brice Kauffmann [2] and Corine Mathonière [1,3,*]

1. Université de Bordeaux, CNRS, Bordeaux INP, ICMCB, UMR 5026, F-33600 Pessac, France; xinghui.qi@icmcb.cnrs.fr (X.Q.); enzo.lafon@etu.u-bordeaux.fr (E.L.); solene.perot@etu.u-bordeaux.fr (S.P.)
2. Université de Bordeaux, IECB, UMS 3033, Institut Européen de Chimie et Biologie, 2 rue Escarpit, F-33600 Pessac, France; b.kauffmann@iecb.u-bordeaux.fr
3. Université de Bordeaux, CNRS, Centre de Recherche Paul Pascal, UMR 5031, F-33600 Pessac, France
* Correspondence: philippe.guionneau@icmcb.cnrs.fr (P.G.); corine.mathoniere@u-bordeaux.fr (C.M.)
† Dedicated to Professor Peter Day, this article belongs to the Special Issue—Perspectives on Molecular Materials—A Tribute to Professor Peter Day.

Citation: Qi, X.; Guionneau, P.; Lafon, E.; Perot, S.; Kauffmann, B.; Mathonière, C. New Photomagnetic Ionic Salts Based on [Mo^{IV}(CN)$_8$]$^{4-}$ and [W^{IV}(CN)$_8$]$^{4-}$ Anions. *Magnetochemistry* 2021, 7, 97. https://doi.org/10.3390/magnetochemistry 7070097

Academic Editor: Fabrice Pointillart

Received: 28 May 2021
Accepted: 29 June 2021
Published: 6 July 2021

Publisher's Note: MDPI stays neutral with regard to jurisdictional claims in published maps and institutional affiliations.

Copyright: © 2021 by the authors. Licensee MDPI, Basel, Switzerland. This article is an open access article distributed under the terms and conditions of the Creative Commons Attribution (CC BY) license (https://creativecommons.org/licenses/by/4.0/).

Abstract: Three new ionic salts containing [M(CN)$_8$]$^{4-}$ (M = Mo^{IV} and W^{IV}) were prepared using large complex cations based on a non-conventional motif built with the tris(2-aminoethyl)amine (noted hereafter tren) ligand, [{M′(tren)}$_3$(μ-tren)]$^{6+}$ (M′ = Cu^{II} and Zn^{II}). The crystal structures of the three compounds show that the atomic arrangement is formed by relatively isolated anionic and cationic entities. The three compounds were irradiated with a blue light at low temperature, and show a significant photomagnetic effect. The remarkable properties of these compounds are (i) the long-lived photomagnetic metastable states for the [Mo(CN)$_8$]$^{4-}$-based compounds well above 200 K and (ii) the rare efficient photomagnetic properties of the [W(CN)$_8$]$^{4-}$-based compound. These photomagnetic properties are compared with the singlet-triplet conversion recently reported for the K$_4$[Mo(CN)$_8$]·2H$_2$O compound.

Keywords: coordination compounds; octacyanometalates; photomagnetism

1. Introduction

The richness of the photochemistry of K$_4$[MoCN$_8$]·2H$_2$O in solution has been known for a few decades [1,2]. Irradiation of aqueous solutions of [MoCN$_8$]$^{4-}$ in its ligand field bands (350–400 nm) allows photosubstitution reactions, with the isolation of different species, [MoCN$_7$(H$_2$O)]$^{3-}$ or [MoCN$_7$(OH)]$^{4-}$ [3], that depend on the pH of the solutions. The [Mo(CN)$_8$]$^{4-}$ complex can be used as a building block to form a prolific series of both polynuclear compounds and coordination polymers [4,5]. Several of these compounds show interesting photomagnetic properties based on either photo-induced electron transfer for systems exhibiting Metal-to-Metal Charge Transfer (MMCT) or Singlet-Triplet formation on Mo^{IV} [5].

More recently, an intriguing breakage of Mo–CN bond has been discovered in the crystalline solid state after blue light irradiation at 10 K for the K$_4$[MoCN$_8$]·2H$_2$O complex [6]. The removal of one CN ligand from the Mo coordination sphere is accompanied by the capture of the free CN group in the crystal lattice by water molecules. Interestingly, first, the photo-induced effect is accompanied by a spin change, and second, it is fully reversible through a thermal heating. This phenomenon of decoordination/coordination is quite common in solution, in particular for Ru [7,8], Fe [9] and Ni [10] complexes, but it remains rare in the solid state and even rarer in the crystalline solid state without loss of crystallinity. So far, only very few compounds show a reversible bond breaking. One example is based on a Co complex [11] that exhibits a dynamic bond switching with a modulation of the ligand field and the orbital momentum of the metal ion. The second example is shown in a family of spin crossover Fe^{II} [12,13] complexes where the decoordination/coordination process can be thermally and photo-induced, as it is the case in the K$_4$[Mo(CN)$_8$]·2H$_2$O complex [6]. To explore if this phenomenon can be extended to other Mo/W-based systems, we have started a systematic study of compounds having non-bridged [M(CN)$_8$]$^{4-}$ complexes. In this work, we will present the synthesis, the structural and magnetic characterizations of two new anionic [Mo(CN)$_8$]$^{4-}$ complexes crystallized with large coordination cations containing Cu^{2+} and Zn^{2+} ions. Additionally, we extend this study to the analogous [W(CN)$_8$]$^{4-}$ complex. For these three new systems, significant photomagnetic responses have been obtained. These properties have been analyzed with a model based on the recent report of the photo-induced single-triplet crossover [6].

2. Results and Discussion
2.1. Synthesis and Characterization

The compounds **1** (for [{Cu(tren)}$_3$(μ-tren)]$_4$[Mo(CN)$_8$]$_6$·45H$_2$O·2CH$_3$OH), **2** (for [{Zn(tren)}$_3$(μ-tren)]$_2$[Mo(CN)$_8$]$_3$·18H$_2$O) and **3** (for [{Zn(tren)}$_3$(μ-tren)]$_2$[W(CN)$_8$]$_3$·17H$_2$O) were obtained by mixing solutions containing 3d divalent metal M^{2+} cations, tren ligand and [M(CN)$_8$]$^{4-}$ anions following two different methods. Green crystals of compound **1** were prepared by a layering method by the diffusion of an aqueous solution of CuCl$_2$·2H$_2$O and tren ligand into the solution of K$_4$[Mo^{IV}(CN)$_8$]·2H$_2$O, the two solutions being separated by a layer of methanol and water. Compounds **2** and **3** were prepared by one-pot method. By mixing solutions of ZnCl$_2$, tren ligand and K$_4$[M^{IV}(CN)$_8$]·2H$_2$O (M = Mo (**2**) and W (**3**)), a clear yellow solution was obtained. Then the slow addition of about 1.5 mL of methanol led to the formation of target yellow crystals of **2** and **3** after one night.

Infrared spectra (IR) for compounds **1**, **2** and **3** are very similar (Figure S1 and Tables S1 and S2). The three compounds show the characteristics bands of the tren ligand at 1600, 1450, 1300, and 995 cm^{-1}. **1** and **2** show a broad band centered at 2098 cm^{-1}, the signature of terminal CN ligands coordinated to the Mo^{4+} by the C atoms. For compound **3**, this band appears at 2091 cm^{-1} in agreement with the coordination of the terminal CN to the W^{4+} by the C atoms. These spectra indicate that **1**, **2** and **3** are ionic salts and that the chemical environment of the [M(CN)$_8$]$^{4-}$ anions are similar in the three compounds.

UV-Vis spectra of compounds **2** and **3** are similar to that of K$_4$[Mo(CN)$_8$]·2H$_2$O and K$_4$[W(CN)$_8$]·2H$_2$O [14], while there is a broad absorption with maximum and shoulders at 877 and 657 nm for compound **1** (Figure S2 and Table S3). These bands also appear in [Cu(tren)]$^{2+}$ complexes in square pyramidal geometry [15], and suggest that the tren ligand acts in **1** as a tetradentate ligand for the Cu^{2+} ion. This means that a fifth ligand is necessary to assure for the Cu^{2+} ion a bipyramid geometry. It is worth noting that for **1** no additional transition except the transitions observed in its precursors is observed, suggesting the absence of an outer-sphere charge transfer in Cu^{2+}/Mo^{4+} pairs in **1**.

2.2. Crystal Structure Description of the Compounds
2.2.1. [{Cu(tren)}$_3$(μ-tren)]$_4$[Mo(CN)$_8$]$_6$·45H$_2$O·2CH$_3$OH (1)

Single Crystal X-ray diffraction (SCXRD) analysis shows that **1** crystallizes in the monoclinic space group $P2_1/c$. As shown by the crystallographic data in Table A1, the unit cell of **1** is very large, leading to up to 323 atoms (without H atoms) with 3D coordinates in the asymmetric unit which certainly puts this compound in the category of the giant unit-cell ones (V > 20,000 Å^3). This makes the SCXRD crystal structure determination a challenge in itself. **1** is constructed by the assembly of [{Cu(tren)}$_3$(μ-tren)]$^{6+}$ cations and [Mo(CN)$_8$]$^{4-}$ anions with no covalent bond between them. The packing diagram of **1** is shown on Figure S4. There are six [Mo(CN)$_8$]$^{4-}$ anions and four [{Cu(tren)}$_3$(μ-tren)]$^{6+}$ cations in the asymmetric unit (Figure 1). The very large number of solvent molecules in the asymmetric unit (>40 water and/or methanol molecules) and the flexible arms for the [{Cu(tren)}$_3$(μ-tren)]$^{6+}$ cations cause structural disorder, which increases the difficulty in the crystal structure determination. Nevertheless, it is important to note that the crystal structure has been solved without ambiguity and confirmed by the diffraction data investigation on several crystals from different batches. As a result, while the solvent part cannot be discussed in detail, the structural parameters and notably the 3D atomic coordinates for the anions and cations in **1** are, on the contrary, robust and can be discussed further. The crystallographic data and selected bond lengths and angles are presented in Table A1, Tables S4 and S7.

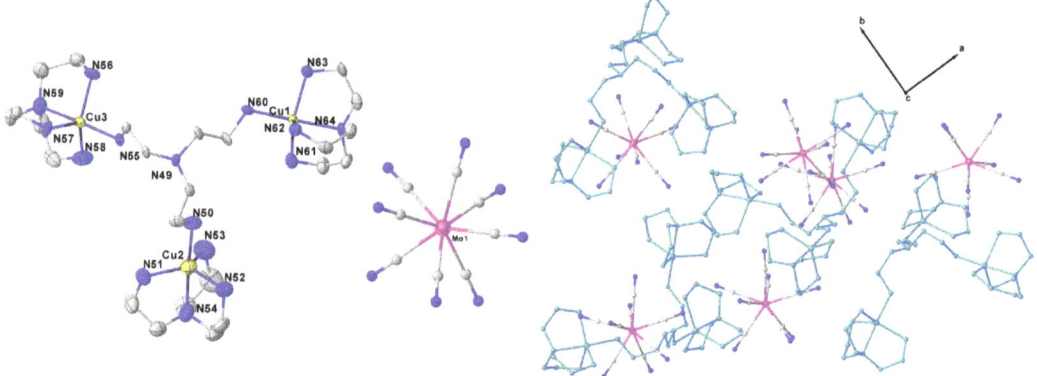

Figure 1. [{Cu(tren)}$_3$(μ-tren)]$^{6+}$ (**left**), [Mo(CN)$_8$]$^{4-}$ (**middle**) selected fragments with partial atoms labelling and asymmetric unit of **1** (without solvent entities) viewed along *c* with [{Cu(tren)}$_3$(μ-tren)]$^{6+}$ units in green (**right**). Color codes: N, blue; C, grey; Cu, yellow; Mo, pink.

The [{Cu(tren)}$_3$(μ-tren)]$^{6+}$ is constructed by one μ-tren linked to three Cu sites, where each copper site is blocked by another tren ligand. Therefore, the tren ligands serve as a tetradentate ligand for the copper sites, and as a tridentate ligand to link the three different Cu sites contained in the trimetallic cation. To the best of our knowledge, this unusual trinuclear copper complex cation has been reported only in the crystal structure of [Cu$_3$(tren)$_4$][Pt(CN)$_4$]$_3$·2H$_2$O [16]. As indicated by the continuous shape measurement (CShM) [17] values, all the copper sites adopt triangular bipyramidal geometry, with the exception of Cu9 site which corresponds to distorted square pyramidal geometry, in agreement with the UV-Visible spectra (see Table A2). For instance, the CShM value for Cu1 site is 0.420, corresponding to the triangular bipyramidal geometry and for Cu9 site is 1.491, corresponding to the square pyramidal geometry. The three copper sites in [{Cu(tren)}$_3$(μ-tren)]$^{6+}$ are arranged in the form of an irregular triangle, for example, with rather long Cu...Cu distances, such as as 7.450, 7.542, 9.156 Å for Cu1...Cu2, Cu2...Cu3 and Cu1...Cu3 distances, respectively.

The [Mo(CN)$_8$]$^{4-}$ anion is stabilized by wide-numerous N-H ... N≡C and O-H ... N≡C hydrogen bonds formed by the interaction of [{Cu(tren)$_3$(μ-tren)]$^{6+}$ or water molecules with [Mo(CN)$_8$]$^{4-}$ units, respectively. The selected bond lengths and angles for Mo sites are presented in Table S4. Average bond distances of Mo-C and C≡N are 2.179(10)/2.170(10)/2.176(14)/2.177(10)/2.153(13)/2.169(10) and 1.145(14)/1.141(14)/-1.135(6)/1.137(13)/1.157(10)/1.138(14) Å, respectively, while the average Mo-C≡N bond angles equal to 177.0(10)/176.7(10)/175.9(12)/177.5(9)/176.8(13)/177.3(10)°. All the Mo sites reveal a geometry close to the square antiprism (SAPR), as evidenced by continuous shape measurement (CShM) analysis (Table A2). The minimum Mo ... Mo distances are 9.65, 9.65, 9.77, 9.67, 9.80, 9.67 Å, which are much longer than the distance of 7.53 Å found in the reference compound K$_4$[Mo^{IV}(CN)$_8$]·2H$_2$O [6], but appear comparable to the values observed for [Ni(bipy)$_3$]$_2$[Mo(CN)$_8$]·12H$_2$O [18].

2.2.2. [{Zn(tren)}$_3$(μ-tren)]$_2$[Mo(CN)$_8$]$_3$·18H$_2$O (**2**)

Similarly to **1**, compound **2** is based on the blocks [{M'(tren)}$_3$(μ-tren)]$^{6+}$ (M' = Cu (**1**) and Zn (**2**)) and [Mo(CN)$_8$]$^{4-}$ (Figure 2), but it crystallizes in the non-centrosymmetric space group *Cc*. The unit-cell is about twice smaller than for **1**, leading to half of the asymmetric content, namely, three [Mo(CN)$_8$]$^{4-}$ anions, two [{Zn(tren)}$_3$(µ-tren)]$^{6+}$ and roughly half of the solvent (water) molecules (one formula unit). The crystal structure refinement is consequently of better quality than for **1**, reaching almost for **2** the standard criterion expected for small molecules, although here also the number of atoms in the asymmetric unit remains impressive (157 without H atoms). The crystallographic data and selected bond lengths and angles are presented in Table A1, Tables S5 and S8. In the same way as for **1**, the crystal structure of **2** contains a large number of solvent molecules that are difficult to localize, although here the determination of H atoms position could reasonably be conducted. However, it is hazardous to discuss in detail features that concern the solvent entities. The coordination metal ions are on the contrary very well defined and can be further discussed. Similar to [{Cu(tren)}$_3$(µ-tren)]$^{6+}$ found in **1**, the zinc sites adopt the triangular bipyramidal geometry, for example, with continuous shape measurement (CShM) value of 0.817 for Zn1 site (see Table A2). The coordination geometry of the Zn ions is close to the ones found in the reported trinuclear compound MoZn$_2$–tren [19]. The selected bond lengths and angles for Mo sites are presented in Table S5. Average bond distances of Mo-C and C≡N are 2.174(13)/2.166(12)/2.179(12) and 1.139(12)/1.150(12)/1.132(12) Å, respectively. The average Mo-C≡N bond angles are equal to 177.0(12)/177.7(15)/176.6(13)°. The shortest Mo . . . Mo distances are 9.48/9.46/9.46 Å, which fall in the same range as the ones found in compound **1**. The three Mo sites in **2** are also close to a square antiprism (SAPR) geometry as evidenced by the continuous shape measurement (CShM) values (Table A2).

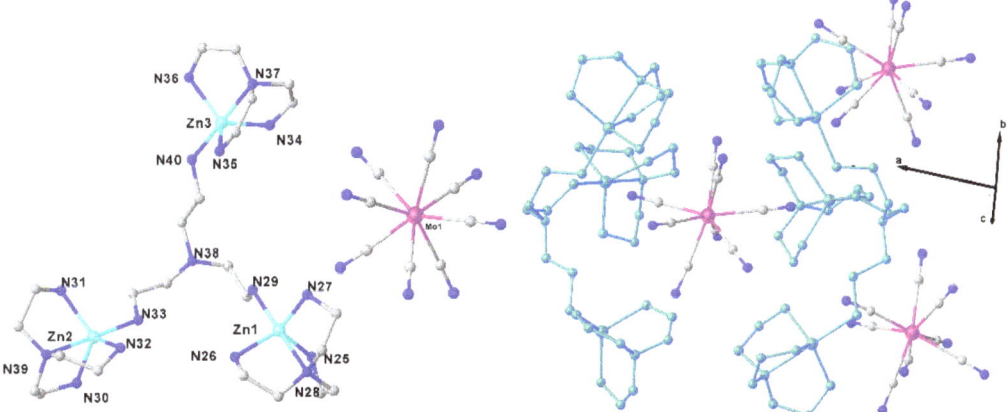

Figure 2. Selected [{Zn(tren)}$_3$(µ-tren)]$^{6+}$ (**left**), [Mo(CN)$_8$]$^{4-}$ (**middle**) fragments of **2** and asymmetric unit of **2** (without solvent entities) with [{Zn(tren)}$_3$(µ-tren)]$^{6+}$ units in green (**right**). Color codes: N, blue; C, grey; Zn, light blue; Mo, pink.

2.2.3. [{Zn(tren)}$_3$(µ-tren)]$_2$[W(CN)$_8$]$_3$·17H$_2$O (**3**)

The crystallographic data of **3** are given in Table A1. It adopts the same unit-cell and space group as **2** and atomic positions are very close in these two structures. The crystal structure of **3** is therefore similar to the one of **2** with the replacement of [Mo(CN)$_8$]$^{4-}$ anions by [W(CN)$_8$]$^{4-}$ anions. Crystal structure criteria of **3** are even slightly better than for **2**. The asymmetric unit of **3** is given in Figure S3, whereas the selected bond lengths and angles are reported in Tables S6 and S9. The [W(CN)$_8$]$^{4-}$ anions also adopt the SAPR geometry with CShM values comparable with those of the Mo sites found in **2**. The shortest W . . . W distances are 9.40/9.45/9.40 Å.

Notably, the crystal packing for **3** is quite different from that for **1** (Figure S4). Compounds **1** and **3** exhibit as two-dimensional and three-dimensional coordination polymers if considering the semi-coordination bonds, respectively.

The experimental powder X-ray diffraction (PXRD) patterns are globally consistent with the above SCXRD results but the large unit-cells combined with the low symmetry involved make the PXRD almost mute in term of reliable information (Figure S5). We can only note that the powders of **1** and **3** look more poorly crystalline than the powder of **2**.

2.3. Magnetic and Photomagnetic Properties

The magnetic and photomagnetic properties of **1**, **2** and **3** have been studied with microcrystalline powders sealed in a small PVE bag (see experimental section). Irradiations at 405 nm were selected because this wavelength fits with the energy range of one ligand field transition of the anions. For the three compounds, the magnetic properties (i.e., magnetization versus field at low temperatures and/or χT versus temperature, χ being the magnetic susceptibility and T the temperature) were first studied in the dark (curves named dark). At 10 K, the samples are irradiated and the time dependence of the magnetic properties is followed during light irradiation. After the light excitations, magnetizations versus field at low temperature are measured. Then the samples are heated again to evaluate the persistence of the photo-induced changes from 2 K to 300 K (curves named after blue irradiation). Finally, the compounds are measured again in the dark from 300 K to 10 K to check the reversibility of the photo-induced magnetic changes (curves named relaxation). All the magnetic curves shown below are normalized per Cu_2Mo for **1**, Zn_2Mo for **2** and Zn_2W for **3**. This normalization will allow an easier comparison for the discussion of the results.

2.3.1. CutrenMo Compound (**1**)

As observed from the temperature dependence of the χT and low-temperature magnetizations in the dark (Figure 3), **1** reveals a paramagnetic behavior with a χT product equal to 0.80 cm^3 mol^{-1} K in agreement with two $Cu^{2+}(3d^9)$ ions of $S = 1/2$ (per Cu_2Mo units) with a Zeeman factor of $g = 2$ (Figure S6) and one diamagnetic $Mo^{4+}(4d^2$ in square antiprism geometry) ion. The superposition of the reduced magnetizations measured at 1.8, 3 and 5 K suggests the absence of magnetic anisotropy. When **1** is irradiated with a light of 405 nm, the value of χT at 10 K increases from 0.79 to 1.4 cm^3 mol^{-1} K after 25 h of irradiation (inset, Figure 3). Then the light is switched off, and low-temperature magnetizations at 1.8, 3 and 5 K were measured. The saturation magnetization at 1.8 K is now 3.09 Nβ, significantly higher than the value of 2 Nβ found before irradiation. The non-superposed reduced magnetizations suggest anisotropy in the photo-induced state. This observation is clearly different in the ground state. The temperature dependence of the χT from 2 K to 300 K first increases to reach a plateau at 1.67 cm^3 mol^{-1} K at 30 K. When compared with the χT product before irradiation, the χT value has increased with a maximum of 0.89 cm^3 mol^{-1} K. Then, the χT product slightly decreases monotonously to reach 1.43 cm^3 mol^{-1} K at 250 K. Above this temperature, a faster decrease of the χT product is observed. At 300 K, the χT value is back to the value obtained before the light irradiation. A new plot χT vs. T (red curve on Figure 3) shows that the photo-induced process is reversible.

2.3.2. ZntrenMo Compound (**2**)

As **2** contains only diamagnetic metal ions $Zn^{2+}(3d^{10})$ and $Mo^{4+}(4d^2)$ in square antiprism geometry, the χT values measured in the dark are in agreement with the diamagnetic nature of the compound (Figure 4). Under light irradiation at 405 nm, the value of χT increases from 0 to 0.9 cm^3 mol^{-1} K after 38 h of irradiation. The reduced magnetizations measured after the light excitation are not superimposed. This is consistent with a weak magnetic anisotropy in the photoinduced state. The saturation value of magnetization at 1.8 K is about 1.15 Nβ. The temperature dependence of the χT product from 2 K to 300 K has a similar shape to the χT vs. T plot of **1**. For **2**, a maximum value of χT product of 0.89 cm^3 mol^{-1} K is reached at 15 K. Then, the χT product decreases monotonously to reach 0.7 cm^3 mol^{-1} K at 240 K. Above this temperature, a faster decrease of the χT product is observed. At 300 K, the χT value is equal at 0.3 cm^3 mol^{-1} K, well above the value obtained before the light irradiation. The lack of complete reversibility after thermal heating in **2** is quite unusual for a photomagnetic compound containing the [Mo(CN)$_8$]$^{4-}$ unit. This uncommon observation needs further structural investigations to be fully understood. Finally, a new χT vs. T plot (red curve on Figure 4) shows that the photo-induced process is not fully erased after a room-temperature treatment. A clear remaining paramagnetic signal around 0.2 cm^3 mol^{-1} K is observed.

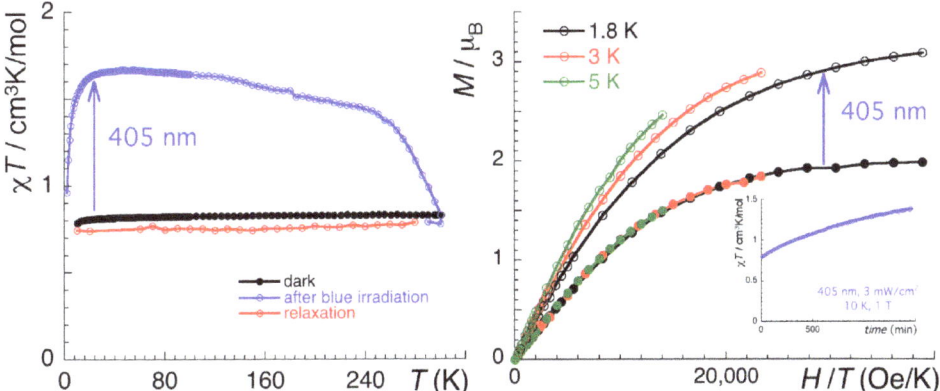

Figure 3. (**Left**) $\chi T = f(T)$ plots of **1** measured at 1 T and 0.4 K/min: in the dark before irradiation (dark points), after 405 nm irradiation (open blue points) and after a reconditioning to 300 K (red points). (**Right**) Reduced magnetizations at different temperatures (1.8 K, 3 K and 5 K) before irradiation (full points) and in the photo-excited state of **1** (open points). (Insert: Time dependence of the χT for **1** measured at 1 T and 10 K, with continuous irradiation of wavelength at 405 nm (3 mW/cm^2)).

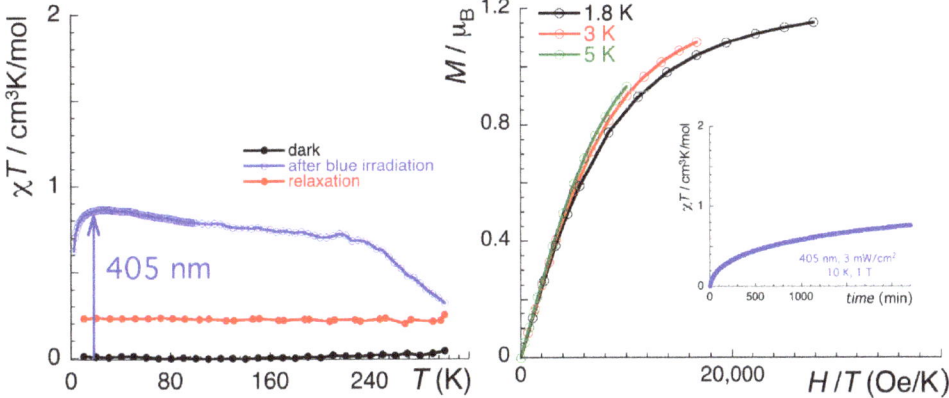

Figure 4. (**Left**) $\chi T = f(T)$ plots of **2** measured at 1 T and 0.4 K/min: in the dark before irradiation (dark points), after 405 nm irradiation (open blue points) and after reconditioning at 300 K (red points). (**Right**) Reduced magnetizations at different temperatures (1.8 K, 3 K and 5 K) in the photo-excited state of **2** (open points). (Insert: Time dependence of the χT for compound **2** measured at 1 T and 10 K, with continuous irradiation of wavelength of 405 nm (3 mW/cm^2)).

2.3.3. ZntrenW Compound (3)

As evidenced above, **2** and **3** exhibit very similar crystal structures, but they obviously contain a different octacyanometalate anion, albeit in the same geometry. As shown by the magnetic properties of **3** in Figure 5, **3** is a diamagnetic compound in agreement with the diamagnetic configuration of two Zn^{2+} ($3d^{10}$) ions and the one W^{4+} ($5d^2$ in square antiprism geometry) ion, with χT values measured in the dark close to 0. Under light irradiation at 405 nm, the value of χT increases from 0 to reach 0.49 cm^3 mol^{-1} K after 30 h of irradiation. The reduced magnetizations measured after the light excitation are almost superimposed, consistent with a very weak magnetic anisotropy in the photoinduced state. A clear saturation of magnetization at 1.8 K is observed at the value of 0.51 Nβ. The temperature dependence of the χT product from 2 K to 300 K is of similar shape to the χT vs. T plots of **1** and **2**. For **3**, a plateau is observed with a maximum value of 0.66 cm^3 mol^{-1} K around 50 K. Then, a small decrease is observed to reach 0.4 cm^3 mol^{-1} K at 200 K. Above this temperature, a faster decrease of the χT product is observed, and the χT value is back to 0 at 250 K, suggesting that the compound is back in its diamagnetic ground state. This is confirmed by a new χT vs. T plot measured after the light excitation and thermal heating of the sample.

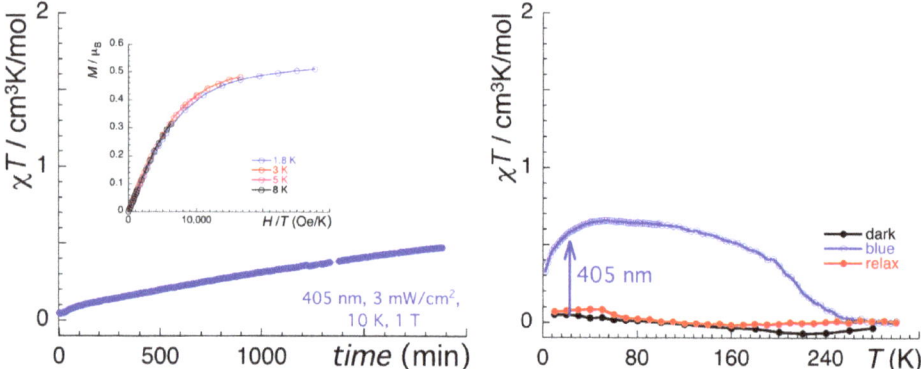

Figure 5. (**Left**) Time dependence of the χT product measured at 1 T during 405 nm light irradiation (0.3 mW/cm^2) (Inset: Reduced magnetizations at different temperatures (1.8 K, 3 K, 5 K and 8 K) in the photo-excited state of **3**). (**Right**) χT = f(T) plots at 1 T and 0.4 K/min of **3** measured in the dark before any light irradiation, after 405 nm irradiation and after reconditioning to 300 K.

3. Discussion and Conclusions

In this work, we were able to synthesize three new ionic salts containing the 4d $[Mo(CN)_8]^{4-}$ anion or its 5d analog $[W(CN)_8]^{4-}$ anion. The $[Mo(CN)_8]^{4-}$ anion is known to be involved in several polynuclear compounds exhibiting photomagnetic properties. On the other hand, only few examples of photomagnetic systems based on the $[W(CN)_8]^{4-}$ complex have been reported [20,21]. The three systems reported in this work are ionic salts and are based on large cations of formula of $[\{M'(tren)\}_3(\mu\text{-tren})]^{6+}$ containing 3d metal ions and the $[M(CN)_8]^{4-}$ anions. **1** adopts a slightly different crystal packing than the ones found in **2** and **3** which are almost isostructural if we exclude their solvent content. However, based on the vicinity of the coordination geometries of the metal ions in **1**, **2** and **3**, it is reasonable to compare their respective photomagnetic properties.

The three compounds reported in this work show a significant photomagnetic response. Some of the characteristics of their photo-induced states are common with the other photomagnetic systems based on $[M(CN)_8]^{4-}$ anions. First, the photo-induced states are formed at 10 K with a blue light irradiation. Second, they also have a high thermal stability, and the recovering of the original magnetic properties (i.e., before irradiation) occurs around room temperature. Several mechanisms are proposed in the literature to explain the observed photomagnetic properties: metal-to-metal charge transfer (MMCT) or spin crossover (SCO) mechanisms. The metal-to-metal charge transfer is possible where the $[M(CN)_8]^{4-}$ units can be easily oxidized by the presence of reductive species, as, for example, Cu^{2+} ions. In this case, the presence of a metal-to-metal charge transfer transition (from

Mo^{4+}-Cu^{2+} to Mo^{5+}-Cu^{+}) in optical spectra appearing around 500 nm is the characteristic feature of this mechanism [22]. For compound **1** of this study, no MMCT is observed in its optical spectrum. Another indirect proof of the absence of MMCT mechanism for **1** is the comparison with the photomagnetic properties of **2**. In **2**, the Cu^{2+} ion has been substituted with Zn^{2+} ion which cannot easily form a Zn^{+} ion, thus excluding the MMCT mechanism.

The similarities of photomagnetic properties of **1** and **2** are nicely shown with the photomagnetic difference curves of **1** where the Cu^{2+} contributions are removed by considering the difference of χT or M before and after irradiation (Figure 6). The resulting χT vs. T and M vs. H plots display a strong similarity with the plots of **2** (Figure 4). This suggests that the photomagnetic properties of **1** and **2** come from the $[Mo(CN)_8]^{4-}$ anions. As mentioned in the introduction, this anion can display a SCO between a $S = 0$ state and a $S = 1$ state. Recently, we have investigated the photo-induced singlet-triplet trapping in the $K_4[Mo(CN)_8]\cdot 2H_2O$ that is accompanied by a breaking of one Mo-CN bond in the crystalline state [6]. To evaluate if this mechanism is active in **1** and **2**, we have analyzed the hypothesis of the photo-induced formation of the triplet state. Because the photomagnetic properties of **2** are not fully reversible, we only analyzed the magnetic data of **1** after the removal of the Cu^{2+} contributions (Figure 6). The triplet state was computed using the anisotropy parameters calculated in [6], namely $|D/k_B| = 20$ K and $g = 1.9$. To simulate correctly the properties of **1**, we also used a partial population of the triplet state at 75% ($p = 0.75$). Figure 6 shows the good reproducibility of the experimental data at a low temperature (T < 120 K), considering the triplet state. At higher temperatures, relaxation that is not considered in the theoretical model probably occurs, and leads to discrepancies with the experiment. This comparative analysis suggests the presence of a SCO mechanism centered on the $[Mo(CN)_8]^{4-}$ anions in **1** and **2**.

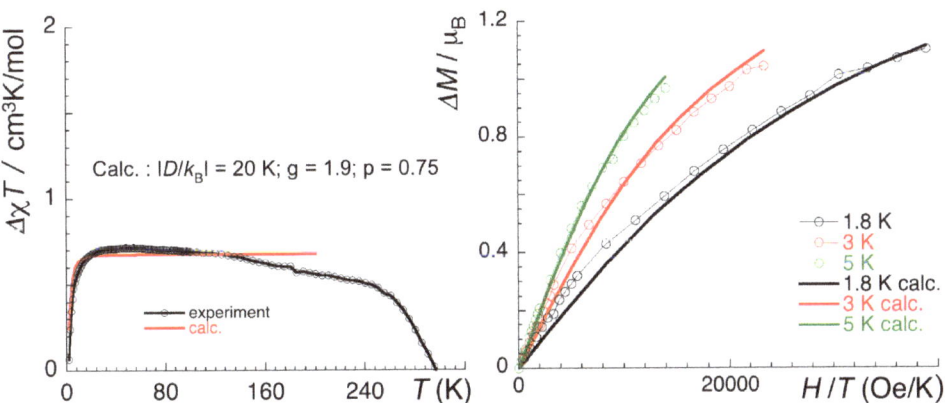

Figure 6. (**Left**) $\chi T = f(T)$ plot of **1** obtained with the difference of χT before and after irradiation (open dark points). In red, theoretical model for the triplet state (see text). (**Right**) $M = f(H/T)$ of the reduced magnetizations obtained with the difference of M before and after irradiation (open points) at different temperatures. In colored lines, theoretical models for the triplet state (see text).

The photomagnetic properties of the two Zn-based compounds **2** and **3** are similar but **3** displays a lower photoconversion than **2**, as shown by the lower observed values in the χT vs. T and M vs. H plots. By analogy with **2**, the observed behavior can be interpreted as a spin crossover from a low spin $S = 0$ to an high spin $S = 1$ for a $5d^2$ complex. Compared to other systems containing the $[W^{IV}(CN)_8]^{4-}$ units, the photoexcited state of **3** has a significant magnetic response, even at 200 K. This feature shows that the photo-induced state in **3** has a lifetime much higher that the lifetimes observed in 3d spin crossover metal ions exhibiting the LIESST phenomenon [23].

To conclude, a series of new photomagnetic compounds based on isolated octacyanometalates are reported in this work. Their photomagnetic properties have been analyzed as a photo-induced singlet-triplet crossover on the $[M(CN)_8]^{4-}$. Remarkably, we have shown that the incorporation of $[M^{IV}(CN)_8]^{4-}$ with bulky $[\{M(tren)\}_3(\mu\text{-tren})]^{6+}$ cations leads to a high thermal stability (above 200 K). This is quite an interesting result because the thermal stability of the photo-induced magnetic changes in the reference $K_4[Mo(CN)_8] \cdot 2H_2O$ compound is below 65 K [6]. To check if these photo-induced changes are also accompanied with a M-CN bond breaking, as already reported in the $K_4[Mo^{IV}(CN)_8] \cdot 2H_2O$ compound, other measurements are necessary such as photocrystallography at low temperature. These compounds are not the best candidates for that because of their huge solvent content. Therefore, the solvent composition of the compounds should be improved, for instance, by using organic cations to limit the presence of water during the crystallization and therefore most likely to increase the crystallinity and consequently the accuracy of the crystal structure determination. To further explain the structure and photomagnetic property relationship in ionic compounds built with $[M(CN)_8]^{4-}$ anions, the exploration of other type of cations to change the spatial arrangement of photomagnetic-active $[M^{IV}(CN)_8]^{4-}$ is still highly demanded.

4. Materials and Methods

4.1. General Remarks

$K_4[Mo^{IV}(CN)_8] \cdot 2H_2O$ and $K_4[W^{IV}(CN)_8] \cdot 2H_2O$ were synthesized successfully by following the procedures according to the literature [24].

4.2. Synthesis

4.2.1. Preparation of CutrenMo (1)

Compound **1** was prepared by a layering technique. A mixture of 3 mL solution $CuCl_2 \cdot 2H_2O$ (35.7 mg, 2.0 mmol) and 528 mg of tren ligand was diffused through 20 mL MeOH:H_2O (1:1) solution into the bottom 1 mL solution of $K_4[Mo^{IV}(CN)_8] \cdot 2H_2O$ (100 mg, 2.0 mmol). Green crystals of compound **1** would appear after one week of slow diffusion. Elemental analysis for **1** is as follows. Anal. Calcd for $[[\{Cu(tren)\}_3(\mu\text{-tren})]_{12}[Mo(CN)_8]_6 \cdot 45H_2O \cdot 2MeOH$, Cu12Mo6C146N112H386O47: C, 30.23%; H, 6.71%; N, 27.04%. Found: C, 31.08%; H, 6.29%; N, 27.70% FT-IR (cm^{-1}): 3263vs [ν(O-H), ν(N-H)]; 3147m, 2956w, 2923w, 2886w, 2825w [ν(C-H)]; 2091vs [ν(C≡N)]; 1591m [γ(O-H)]; 1471m, 1311w [ν(C-N), ν(C-C)]; 1101w, 1061ms, 997vs, 981m, 900w, 900w, 872w, 750w, 633s(br) [γ(N-H out-of-plane)].

4.2.2. Preparation of ZntrenMo (2)

In the first step, we mixed a 2 mL solution of $ZnCl_2$ (30.0 mg, 2.2 mmol) with 240 mg solution of tren ligand. The above mixture solution was added slowly to 1 mL solution of $K_4[Mo^{IV}(CN)_8] \cdot 2H_2O$ (50 mg, 1.0 mmol) avoid shaking. Then 1.5 mL of MeOH was slowly added, and yellow crystals of compound **2** would appear after one night. FT-IR (cm^{-1}): 3257vs [ν(O-H), ν(N-H)]; 3145m, 2964w, 2869w [ν(C-H)]; 2098vs [ν(C≡N)]; 1585m [γ(O-H)]; 1473m, 1322w [ν(C-N), ν(C-C)]; 1083w, 1054w, 1007ms, 989vs, 885m, 865w, 655s(br) [γ(N-H out-of-plane)].

4.2.3. Preparation of ZntrenW (3)

Similarly to compound **2**, we first mixed a 2 mL solution of $ZnCl_2$ (30.0 mg, 2.2 mmol) with 240 mg solution of tren ligand. The above mixture solution was added slowly to 1 mL solution of $K_4[W^{IV}(CN)_8] \cdot 2H_2O$ (58 mg, 1.0 mmol) avoid shaking. Then 1.5 mL of MeOH was slowly added, and yellow crystals of compound **3** would appear after one night. FT-IR (cm^{-1}): 3246vs [ν(O-H), ν(N-H)]; 3149m, 2960w, 2917w, 2894w, 2871w [ν(C-H)]; 2091vs [ν(C≡N)]; 1581m [γ(O-H)]; 1473m, 1322w [ν(C-N), ν(C-C)]; 1083w, 1056w, 1007ms, 983vs, 885m, 865w, 655s(br) [γ(N-H out-of-plane)].

4.3. Physical Measurements

4.3.1. Infrared Spectroscopy

The FT-IR spectra were recorded in the range of 650 cm^{-1}–4000 cm^{-1} on a Thermo-Fisher Nicolet^{TM} 6700 ATR (attenuated total reflection) spectrometer equipped with a Smart iTR diamond window on pure solid samples.

4.3.2. UV-Visible Spectroscopy

Solid-state UV-vis-NIR absorption spectra were recorded with a PerkinElmer Lambda 35 UV/vis spectrophotometer equipped with a PerkinElmer Labsphere on pure solid samples.

4.3.3. Magnetic Measurements

All magnetic properties were measured by a Quantum Design MPMS XL system in the range of temperatures of 1.8–300 K. Photomagnetic studies were conducted on a smaller sample (ca. 0.5 mg) sealed in a small PVE bag fixed with Scotch tape, blocked tightly between two transparent polypropylene films and mounted in the probe equipped with an optical fiber entry enabling the transmission of laser light of 405 nm line (P $\approx$ 3 mW/cm^2) into the sample space. We used for the three compounds the molecular weights shown in Table A1. To compare the magnetic data of the three compounds, the plots were obtained considering per M′$_2$Mo or M′$_2$W units. This means that the molecular weights have been divided by 6 for **1** in Figure 3, and by 3 for **2** and **3** in Figures 4 and 5. Diamagnetism of the sample holders and of the constituent atoms (Pascal's tables) was accounted for in all the obtained magnetic and photomagnetic data.

4.3.4. Powder X-ray Diffraction (PXRD)

PXRD was performed on a PANalytical X'PERT MDP-PRO diffractometer (Cu Kα radiation) equipped with a graphite monochromator using the θ-θ Bragg–Brentano geometry. The sample was deposited on a silicon holder for Bragg–Brentano geometry.

4.3.5. Elemental Analysis

Elemental analyses of C, H and N were carried out with a German Elementary Vario EL III instrument.

4.3.6. Single-Crystal X-ray Crystallography

Data collection and reduction for **1** and **2** were performed on a Microfocus rotating anode (Rigaku FRX) operating at 45 kV and 66 mA at the CuKα edge (λ = 1.54184 Å) with a partial chi goniometer. The X-ray source is equipped with high-flux Osmic Varimax mirrors and a Dectris Pilatus 300K detector. Data collection and reduction for **3** were performed on the Bruker Apex II instrument operating at 50 kV and 30 mA using molybdenum radiation Mo Kα [λ = 0.71073 Å]. The crystal structures were solved by direct methods using SHELXT and refined using a F2 full-matrix least-squares technique of SHELXL2014/7 [25] included in the OLEX-2 1.2 [26] software packages. The non-H atoms were refined anisotropically, adopting weighted full-matrix least squares on F2. CCDC 2083835, 2083836 and 2083837 contain the supplementary crystallographic data for compounds **1**, **2** and **3**, and additional crystallographic information is available in the Supporting Information. The structural data presented as figures were prepared with the use of the OLEX-2 software. Geometries of metal centers are estimated with the Continuous Shape Measures (CShM) analysis using of SHAPE v2.0 software [27].

Supplementary Materials: The following are available online at https://www.mdpi.com/article/10.3390/magnetochemistry7070097/s1, IR, UV-Vis spectra, additional crystallographic tables and figures, additional magnetic figures.

Author Contributions: The authors have contributed equally to the conceptualization and execution of this work and wrote the manuscript together. All authors have read and agreed to the published version of the manuscript.

Funding: The Chinese Scholarship Council is acknowledged for the phD funding of Xinghui Qi. Financial supports from the CNRS (Pessac), University of Bordeaux and the CPER are acknowledged.

Institutional Review Board Statement: Not applicable.

Informed Consent Statement: Not applicable.

Data Availability Statement: The data are available by corresponding authors.

Acknowledgments: We are grateful for the fruitful discussions with S. Pillet and E. E. Bendeif in Lorraine University.

Conflicts of Interest: The authors declare no conflict of interest.

Appendix A

Table A1. Crystal and experimental data for **1**, **2** and **3**.

Compound/CCDC Number	1/2083835	2/2083836	3/2083837
Formula	$Mo_6Cu_{12}C_{146}H_{386}N_{112}O_{47}$	$Mo_3Zn_6C_{72}H_{180}N_{56}O_{18}$	$W_3Zn_6C_{72}H_{178}N_{56}O_{17}$
$D_{calc.}$ / g cm^{-3}	1.424	1.406	1.537
μ mm^{-1}	4.121	4.036	3.760
Formula Weight	5801.61	2798.68	3044.39
T/K	130(2)	130(2)	150(2)
Crystal System	monoclinic	monoclinic	monoclinic
Space Group	$P2_1/c$	Cc	Cc
a/Å	30.4150(4)	32.1487(9)	31.8599(9)
b/Å	37.1820(5)	16.8699(3)	16.9574(3)
c/Å	22.5401(2)	25.4013(6)	25.3107(6)
β/°	98.0410(10)	106.487(3)	106.013(3)
V/Å^3	25239.8(5)	13209.8(6)	13143.8(6)
Z	4	4	4
Z'	1	1	1
Wavelength/Å	1.54184	1.54184	0.71073
Radiation type	CuK$_a$	CuK$_a$	MoK$_a$
Q_{min}/°	2.309	2.867	1.701
Q_{max}/°	73.591	74.083	26.733
Measured refl.	199397	46339	151852
Measured indep. refl.	49550	19947	27860
Observed indep. refl.	35018	19214	22758
R_{int}	0.0760	0.0775	0.0796
Parameters	2908	1450	1438
Restraints	0	2	20
Largest diff. peak (e-·A^{-3})	19.876	1.7437	1.938
Deepest diff. hole (e-·A^{-3})	−2.959	−1.561	−1.187
GooF (S)	1.764	1.042	1.018
wR_2 (all data)	0.4624	0.2071	0.0959
wR_2	0.4298	0.2049	0.0894
R_1 (all data)	0.2056	0.0791	0.0574
R_1	0.1830	0.0769	0.0397

Table A2. Continuous shape measurements (CShM) for metal ions in **1**, **2** and **3**.

[{Cu(tren)}$_3$(μ-tren)]$_{12}$[Mo(CN)$_8$]$_6$·45H$_2$O·2MeOH **1**						
Mo(CN)$_8$	Mo1	Mo2	Mo3	Mo4	Mo5	Mo6
SAPR	0.218	0.254	0.535	0.230	0.652	0.321
TDD	2.201	2.008	1.679	1.899	1.232	1.794
JBTPR	2.821	2.284	2.252	2.660	1.971	2.360
BTPR	2.220	1.709	1.649	2.077	1.384	1.738

Table A2. Cont.

[{Cu(tren)}$_3$(μ-tren)]$_{12}$[Mo(CN)$_8$]$_6$·45H$_2$O·2MeOH 1							
CuN$_5$	Cu1	Cu2	Cu3	Cu4	Cu5	Cu6
TBPY	0.420	0.377	0.904	0.540	0.368	0.443
SPY	4.269	4.425	3.226	4.175	4.805	4.970
JTBPY	3.967	3.731	3.928	3.661	3.794	3.911
CuN$_5$	Cu7	Cu8	Cu9	Cu10	Cu11	Cu12
TBPY	0.380	0.321	2.225	0.348	0.514	0.789
SPY	4.975	4.832	1.491	5.009	4.016	3.401
JTBPY	3.966	3.639	4.730	3.705	4.016	3.977

[{Zn(tren)}$_3$(μ-tren)]$_2$[Mo(CN)$_8$]$_3$·18H$_2$O 2							
Mo(CN)$_8$	Mo1	Mo2	Mo3			
SAPR	0.232	0.590	0.836			
TDD	2.106	1.567	1.372			
JBTPR	2.245	1.907	1.819			
BTPR	1.663	1.422	1.111			
ZnN$_5$	Zn1	Zn2	Zn3	Zn4	Zn5	Zn6
TBPY	0.817	0.773	1.076	0.771	0.814	0.760
SPY	5.582	4.987	5.149	5.715	4.698	5.558
JTBPY	2.594	2.368	2.152	2.654	2.510	2.548

[{Zn(tren)}$_3$(μ-tren)]$_2$[W(CN)$_8$]$_3$·17H$_2$O 3							
W(CN)$_8$	W1	W2	W3			
SAPR	0.969	0.295	0.532			
TDD	1.110	1.922	1.781			
JBTPR	1.868	2.104	1.934			
BTPR	1.151	1.540	1.355			
ZnN$_5$	Zn1	Zn2	Zn3	Zn4	Zn5	Zn6
TBPY	0.792	0.916	0.742	0.976	0.852	0.889
SPY	5.393	4.487	5.890	4.707	4.803	5.930
JTBPY	2.514	2.535	2.449	2.285	2.290	2.413

The numbers in the tables correspond to the S shape measures relative to the square antiprism (SAPR), triangular dodecahedron (TDD J84), Johnson elongated triangular bipyramid (JBTPR J14) and biaugmented trigonal prism (BTPR J50) for M(CN)$_8$ unit; Trigonal bipyramid (TBPY), Spherical square pyramid (SPY) and Johnson trigonal bipyramid (JTBPY J12) for M′N$_5$ unit. When the respective shape measure parameter equals zero, the real geometry coincides with the idealized one. For each site, the minimum calculated shape measure is given in violet.

References

1. Adamson, A.; Perumareddi, J.R. Photochemistry of aqueous octacyanomolybdate(IV) ion, Mo(CN)$_8^{-4}$. *Inorg. Chem.* **1964**, *4*, 247–248. [CrossRef]
2. Samotus, A.; Szklarzewicz, J. Photochemistry of transition metal octacyanides and related compounds. Past, present and future. *Coord. Chem. Rev.* **1993**, *125*, 63–74. [CrossRef]
3. Mitra, R.; Mohan, H. Tetracyano-oxo(hydroxo) complexes of Mo(IV): Their interconversion ans relation o the primary photoaquation product of the Mo(CN)$_8^{-4}$ ion. *J. Inorg. Nucl. Chem.* **1974**, *36*, 3739–3743. [CrossRef]
4. Sieklucka, B.; Podgajny, R.; Korzeniak, T.; Nowicka, B.; Pinkowicz, D.; Koziel, M. A decade of Octacyanides in Polynuclear Molecular materials. *Eur. J. Inorg. Chem.* **2011**, 305–326. [CrossRef]
5. Chorazy, S.; Zakrzewski, J.J.; Magott, M.; Korzeniak, T.; Nowicka, B.; Pinkowicz, D.; Podgajny, R.; Sieklucka, B. Octacyanidometallates for multifunctional molecule-based materials. *Chem. Soc. Rev.* **2020**, *49*, 5945–6001. [CrossRef] [PubMed]

6. Qi, X.; Pillet, S.; de Graaf, C.; Magott, M.; Bendeif, E.E.; Guionneau, P.; Rouzieres, M.; Marvaud, V.; Stefanczyk, O.; Pinkowicz, D.; et al. Photoinduced Mo-CN Bond Breakage in Octacyanomolybdate Leading to Spin Triplet Trapping. *Angew. Chem. Int. Ed. Engl.* **2020**, *59*, 3117–3121. [CrossRef] [PubMed]
7. Mobian, P.; Kern, J.-P.; Sauvage, J.-P. Light-Driven Machine Prototypes Based on Dissociative Excited States: Photoinduced Decoordination and Thermal Recoordination of a Ring in a Ruthenium(ii)-Containing [2]Catenane. *Angew. Chem. Int. Ed.* **2004**, *43*, 2392–2395. [CrossRef] [PubMed]
8. Soupart, A.; Alary, F.; Heully, J.-L.; Elliott, P.I.P.; Dixon, I.M. Recent progress in ligand release reaction mechanims: Theortical insights focusing on Ru(II) ^{3}MC states. *Coord. Chem. Rev.* **2020**, *408*, 231184. [CrossRef]
9. Chen, J.; Browne, W.R. Photochemistry of iron complexes. *Coord. Chem. Rev.* **2018**, *374*, 15–35. [CrossRef]
10. Venkataramani, S.; Jana, U.; Dommaschk, M.; Sonnichsen, F.D.; Tuzcek, F.; Herges, R. Magnetic bistability of molecules in homogeneous solution at room temperature. *Science* **2011**, *331*, 445–448. [CrossRef]
11. Su, S.-Q.; Wu, S.-Q.; Baker, M.L.; Bencok, P.; Azuma, N.; Miyazaki, Y.; Nakano, M.; Kang, S.; Shiota, Y.; Yoshizawa, K.; et al. Quenching and Restoration of Orbital Angular Momentum through a Dynamic Bond in a Cobalt(II) Complex. *J. Am. Chem. Soc.* **2020**, *26*, 11434–11441. [CrossRef] [PubMed]
12. Guionneau, P.; Le Gac, F.; Kaiba, A.; Costa, J.S.; Chasseau, D.; Létard, J.-F. A reversible metal–ligand bond break associated to a spin-crossover. *Chem. Commun.* **2007**, *3723*, 3723–3725. [CrossRef]
13. Aguila, D.; Dechambenoit, P.; Rouzières, M.; Mathonière, C.; Clérac, R. Direct crystallographic evidence of the reversible photo-formation and thermo-rupture of a coordination bond inducing spin-crossover phenomenon. *Chem. Comm.* **2017**, *53*, 11588–11591. [CrossRef]
14. Perumareddi, J.R.; Liehr, A.; Adamson, A. Ligand Field Theory of Transition Metal Cyanide Complexes. Part I. The Zero, One and Two Electron or Hole Configuration. *J. Am. Chem. Soc.* **1963**, *85*, 249–259. [CrossRef]
15. Duggan, M.; Ray, N.; Hataway, B.; Tomlison, G.; Brint, P.; Pelin, K. Crystal Structure and Electronic Properties of Ammine[tris(2-aminoethyl) amine]copper(II) Diperchlorate and Potassium Penta-amminecopper(II)Tris(hexaf luorophosphate). *J. Chem. Soc. Dalton Trans.* **1980**, 1342–1348. [CrossRef]
16. Shek, I.Y. A novel trinuclear copper(II) complex bridged by tren: [Cu$_3$(tren)$_4$][Pt(CN)$_4$]$_3$·2H$_2$O. *New J. Chem.* **1999**, *23*, 1049–1050. [CrossRef]
17. Cirera, J.; Ruiz, E.; Alvarez, S. Continuous Shape measures as a stereochemical tools in Organometallic Chemistry. *Organometallics* **2005**, *24*, 1556–1562. [CrossRef]
18. Korzeniak, T.; Mathonière, C.; Kaiba, A.; Guionneau, P.; Koziel, M.; Sieklucka, B. First example of photomagnetic effects in ionic pairs [Ni(bipy)$_3$]$_2$[Mo(CN)$_8$]·12H2O. *Inorg. Chim. Acta* **2008**, *361*, 3500–3504. [CrossRef]
19. Bridonneau, N.; Long, J.; Cantin, J.L.; Von Bardeleben, J.; Pillet, S.; Bendeif, E.E.; Aravena, D.; Ruiz, E.; Marvaud, V. First Evidence of Light-induced Spin Transition in Molybdenum(IV). *Chem. Commun.* **2015**, *51*, 8229–8232. [CrossRef]
20. Magott, M.; Stefanczyk, O.; Sieklucka, B.; Pinkowicz, D. Octacyanidotungstate(IV) Coordination Chains Demonstrate a Light-Induced Excited Spin State Trapping Behavior and Magnetic Exchange Photoswitching. *Angew. Chem. Int. Ed.* **2017**, *56*, 13283–13287. [CrossRef] [PubMed]
21. Magott, M.; Reczyński, M.; Gaweł, B.; Sieklucka, B.; Pinkowicz, D. A photomagnetic sponge: High-temperature light-induced ferrimagnet controlled by water sorption. *J. Am. Chem. Soc.* **2018**, *140*, 15876–15882. [CrossRef]
22. Ohkoshi, S.I.; Tokoro, H.; Hozumi, T.; Zhang, Y.; Hashimoto, K.; Mathonière, C.; Bord, I.; Rombaut, G.; Verelst, M.; Cartier Dit Moulin, C. Photoinduced magnetization in copper octacyanomolybdate. *J. Am. Chem. Soc.* **2006**, *128*, 270–277. [CrossRef]
23. Chastanet, G.; Desplanches, C.; Baldé, C.; Rosa, P.; Marchivie, M.; Guionneau, P.A. Critical review of the T(LIESST) temepeature in soun crrossover materails—What it is and what it is not. *Chem. Sq.* **2018**, *2*, 2. [CrossRef]
24. Szklarzewicz, J.; Matoga, D.; Lewiński, K. Photocatalytical Decomposition of Hydrazine in K$_4$[Mo(CN)$_8$] Solution: X-ray Crystal Structure of (PPh$_4$)$_2$[Mo(CN)$_4$O(NH$_3$)]·2H$_2$O. *Inorg. Chim. Acta* **2007**, *360*, 2002–2008. [CrossRef]
25. Sheldrick, G.M. Acta Crystallogr, Sect. A. *A Found Cryst.* **2008**, *64*, 112–122. [CrossRef] [PubMed]
26. Dolomanov, O.V.; Blake, A.J.; Champness, N.R.; Schroder, M. OLEX: New software for visualization and analysis of extended crystal structures. *J. Appl. Crystallogr.* **2003**, *36*, 1283–1284. [CrossRef]
27. Llunell, M.; Cirera, J.; Alemany, P.; Alvarez, S. *SHAPE v. 2.1.*; University of Barcelona: Barcelona, Spain, 2013.

Review

The Peter Day Series of Magnetic (Super)Conductors [†]

Samia Benmansour *[ ] and Carlos J. Gómez-García *[ ]

ICMol, Departamento de Química Inorgánica, Universidad de Valencia, C/Catedrático José Beltrán 2, 46980 Paterna, Spain
* Correspondence: sam.ben@uv.es (S.B.); carlos.gomez@uv.es (C.J.G.-G.);
  Tel.: +34-963544423 (S.B.); +34-963544423 (C.J.G.-G.); Fax: +34-963543273 (S.B.); +34-963543273 (C.J.G.-G.)
† In memory of Peter Day, an enthusiastic and clever chemist and a good friend.

Abstract: Here, we review the different series of (super)conducting and magnetic radical salts prepared with organic donors of the tetrathiafulvalene (TTF) family and oxalato-based metal complexes (ox = oxalate = $C_2O_4^{2-}$). Although most of these radical salts have been prepared with the donor bis(ethylenedithio)tetrathiafulvalene (BEDT-TTF = ET), we also include all the salts prepared with other TTF-type donors such as tetrathiafulvalene (TTF), tetramethyl-tetrathiafulvalene (TM-TTF), bis(ethylenediseleno)tetrathiafulvalene (BEST), bis(ethylenedithio)tetraselenafulvalene (BETS) and 4,5-bis((2S)-2-hydroxypropylthio)-4′,5′-(ethylenedithio)tetrathiafulvalene (DMPET). Most of the oxalate-based complexes are monomers of the type $[M^{III}(C_2O_4)_3]^{3-}$, $[Ge(C_2O_4)_3]^{2-}$ or $[Cu(C_2O_4)_2]^{2-}$, but we also include the reported salts with $[Fe_2(C_2O_4)_5]^{4-}$ dimers, $[M^{II}(H_2O)_2[M^{III}(C_2O_4)_3]_2]^{4-}$ trimers and homo- or heterometallic extended 2D layers such as $[M^{II}M^{III}(C_2O_4)_3]^{-}$ and $[M^{II}{}_2(C_2O_4)_3]^{2-}$. We will present the different structural families and their magnetic properties (such as diamagnetism, paramagnetism, antiferromagnetism, ferromagnetism and even long-range magnetic ordering) that coexist with interesting electrical properties (such as semiconductivity, metallic conductivity and even superconductivity). We will focus on the electrical and magnetic properties of the so-called Day series formulated as β″-(BEDT-TTF)$_4$[A^{I}M^{III}(C$_2$O$_4$)$_3$]·G, which represents the largest family of paramagnetic metals and superconductors reported to date, with more than fifty reported examples.

Keywords: oxalato; tris(oxalato) complexes; TTF; BEDT-TTF; radical salts; conducting; superconducting; metallic; conductivity; paramagnetism; ferromagnetism

Citation: Benmansour, S.; Gómez-García, C.J. The Peter Day Series of Magnetic (Super)Conductors. *Magnetochemistry* **2021**, *7*, 93. https://doi.org/10.3390/magnetochemistry7070093

Academic Editor: Marius Andruh

Received: 31 May 2021
Accepted: 23 June 2021
Published: 26 June 2021

Publisher's Note: MDPI stays neutral with regard to jurisdictional claims in published maps and institutional affiliations.

Copyright: © 2021 by the authors. Licensee MDPI, Basel, Switzerland. This article is an open access article distributed under the terms and conditions of the Creative Commons Attribution (CC BY) license (https://creativecommons.org/licenses/by/4.0/).

1. Introduction

Among the many legacies of Peter Day's work, besides the Robin and Day classification for mixed-valence compounds [1], one of the best known is the discovery of the superconducting paramagnetic radical salts prepared with bis(ethylenedithio)tetrathiafulvalene BEDT-TTF (ET, Scheme 1) and [M(ox)$_3$]$^{3-}$ anions (M = Fe, Cr, Ga, etc.) [2,3] The discovery of the first paramagnetic molecular superconductors boosted the research in the field and led to the preparation of around twenty-five molecular superconductors in the so-called Peter Day series. The first member of this series, initially formulated with a water molecule as β″-(BEDT-TTF)$_4$[(H$_2$O)Fe^{III}(C$_2$O$_4$)$_3$]·PhCN [2,3], and later with a H$_3$O$^+$ cation as β″-(BEDT-TTF)$_4$[(H$_3$O)Fe^{III}(C$_2$O$_4$)$_3$]·PhCN [4], showed a superconducting transition at 8.5(3) K, one of the highest T$_c$ observed in any molecular superconductor to date. As Peter Day noticed in his first report, this discovery paved the way to a synthetic strategy for obtaining further magnetic molecular superconductors [3]. Only two months later, P. Day's group published a deeper characterization of this compound and a second phase (*pseudo*-κ or κ′) obtained with the same donor (BEDT-TTF) and the same anion ([Fe(C$_2$O$_4$)$_3$]$^{3-}$), but with different A$^+$ cations: κ′-(BEDT-TTF)$_4$[AFe^{III}(C$_2$O$_4$)$_3$]·PhCN (A = K$^+$ and NH$_4^+$). This orthorhombic (*Pbcn*) pseudo-kappa or κ′ phase was also paramagnetic, but it was a semiconductor (as we will see below, all the known members of this κ′ phase are semiconductors) [2]. These two initial reports in 1995 initiated the search for novel molecular conductors and superconductors with interesting magnetic properties. Besides the more than one hundred metals, semiconductors and superconductors that will be presented in this review, this search also led to the synthesis of the first molecular superconducting antiferromagnets [5,6], metallic ferromagnets [7,8] or field-induced magnetic superconductors [9].

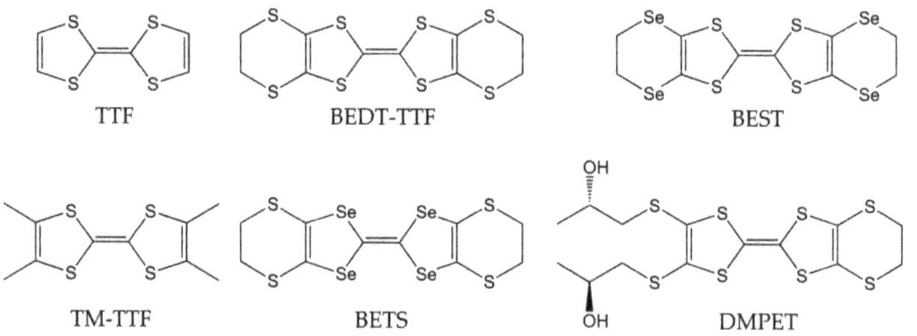

Scheme 1. Donors of the TTF family that have been combined with metal-oxalato complexes or lattices: tetrathiafulvalene (TTF), tetramethyl-tetrathiafulvalene (TMTTF), bis(ethylenedithio)tetrathiafulvalene (BEDT-TTF = ET), bis(ethylenediseleno)tetrathiafulvalene (BEDS-TTF = BEST), bis(ethylenedithio)tetraselenafulvalene (BEDT-TSF = BETS) and 4,5-bis((2S)-2-hydroxypropylthio)-4',5'-(ethylenedithio)tetrathiafulvalene (DMPET).

The synthesis of these radical salts was performed in H-shaped or U-shaped electrochemical cells containing two compartments separated by a glass frit. The anode compartment is filled with a solution of the TTF-type donor (Scheme 1) in solvents such as CH_2Cl_2 and $CHCl_2$-CH_2Cl. The cathode compartment is filled with a solution containing the desired oxalate-based magnetic anion dissolved in different solvents or mixtures of solvents. As we will see below, the choice of the solvents is crucial in many cases to determine the structure and properties of the radical salt obtained. Very often, a few drops of water were added to furnish the needed H_3O^+ cations. As we will see in Section 4.1, in some cases, a crown ether such as 18-crown-6 was added to the cathode to increase the solubility of the K^+ or NH_4^+ salts of the anions. Application of a very low constant DC current (in the 0.1–1 μA range) through platinum electrodes (usually with a diameter of 1 mm) results in a slow oxidation of the TTF-type donors in the presence of the magnetic oxalate-based anions that yields, after a few days or weeks, high-quality single crystals in most cases. The co-crystallization of neutral donor molecules together with the oxidized ones leads, in most cases, to lattices with a mixed-valence state in the donors and, therefore, to a combination of electrical and magnetic properties [2,4].

As it will be shown here, the continuous efforts of P. Day's group (and others) led to the synthesis and characterization of more than one hundred radical salts prepared with different TTF-type donors (Scheme 1) and metal-oxalato complexes (Figure 1). Although a revision with a part of the work conducted in this area was published by P. Day in 2004 [10], since then, the number of published salts has multiplied by a factor of four, passing from less than thirty to more than one hundred and twenty. Therefore, as a late homage to the legacy of Peter Day and his group in the area of molecular conductors, here, we will revise the work conducted in this field since the first report of a molecular paramagnetic superconductor by P. Day's group in 1995 [3].

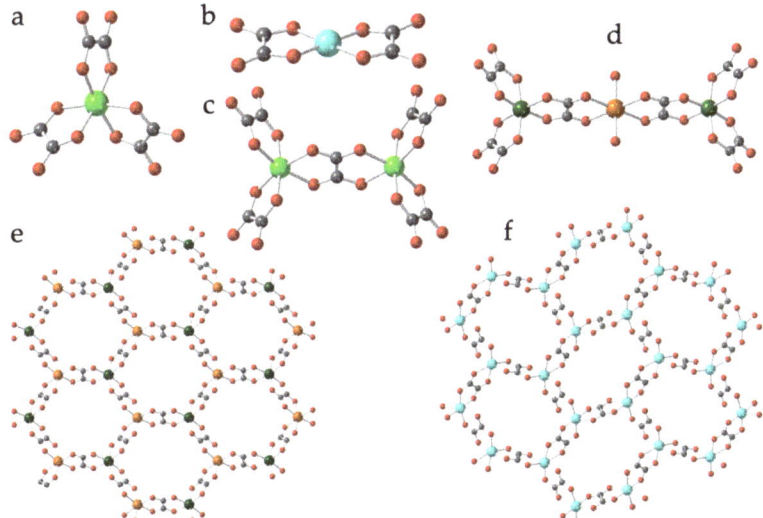

Figure 1. Metal-oxalato anions combined with TTF-type donors: (**a**) monomers [M^{III}(C$_2$O$_4$)$_3$]$^{3-}$ (M^{III} = Fe, Cr, Mn, Co, Al, Ga, Rh, Ru and Ge) and [Ge(C$_2$O$_4$)$_3$]$^{2-}$, (**b**) the monomer [Cu(C$_2$O$_4$)$_2$]$^{2-}$, (**c**) dimers [M^{III}$_2$(C$_2$O$_4$)$_5$]$^{4-}$ (M^{III} = Fe and Cr), (**d**) trimers {Mn(H$_2$O)$_2$[M^{III}(C$_2$O$_4$)$_3$]$_2$}$^{4-}$, (**e**) heterometallic layers [Mn^{II}M^{III}(C$_2$O$_4$)$_3$]$^-$ (M^{III} = Cr and Rh) and (**f**) homometallic layers [Cu^{II}$_2$(C$_2$O$_4$)$_3$]$^{2-}$. Color code: Fe = light green, Cr = dark green, Mn = orange Cu = blue, C = gray and O = red.

In this review, we present all the reported structures of radical salts with TTF-type donors and metal-oxalato complexes, divided into five different sections. In the first section, we present Peter Day's series with more than fifty reported monoclinic conductors and superconductors formulated as β"-(BEDT-TTF)$_4$[AM^{III}(C$_2$O$_4$)$_3$]·G with A$^+$ = H$_3$O$^+$, NH$_4$$^+$ and K$^+$; M^{III} = Fe, Cr, Ga, Rh and Ru; and G = C$_6$H$_5$N (py), PhNO$_2$, PhCN, 1,2-PhCl$_2$, PhI, PhBr, PhCl, PhF, dimethylformamide (dmf), 2-Clpy, 2-Brpy, 3-Clpy, 3-Brpy and CH$_2$Cl$_2$.

In the second section, we show all the reported BEDT-TTF radical salts with the orthorhombic *pseudo*-κ or κ' phase formulated as κ'-(BEDT-TTF)$_4$[AM^{III}(C$_2$O$_4$)$_3$]·G, with A$^+$ = H$_3$O$^+$, NH$_4$$^+$ and K$^+$; M^{III} = Fe, Cr, Co, Al, Mn, Rh and Ru; and G = PhCN and 1,2-PhCl$_2$.

In the third section, we present all the salts with BEDT-TTF and [M(C$_2$O$_4$)$_3$]$^{n-}$ anions with other crystallographic phases including those containing 18-crown-6, those with two different BEDT-TTF layers, those with 3:1 BEDT-TTF:[M(C$_2$O$_4$)$_3$]$^{3-}$ stoichiometry, those with other different and unusual packings and those with the monomeric dianions [Ge(C$_2$O$_4$)$_3$]$^{2-}$ and [Cu(C$_2$O$_4$)$_2$]$^{2-}$.

The fourth section shows all the reported salts with metal-oxalato complexes other than the monomeric complexes. These complexes include dimers of the type [Fe$_2$(C$_2$O$_4$)$_5$]$^{4-}$, trimers such as {M^{II}(H$_2$O)$_2$[M^{III}(C$_2$O$_4$)$_3$]$_2$}$^{4-}$, heterometallic M(II)/M(III) layers such as [Mn^{II}M^{III}(C$_2$O$_4$)$_3$]$^-$ (M^{III} = Cr and Rh) and homometallic M(II)/M(II) layers such as [Cu^{II}$_2$(C$_2$O$_4$)$_3$]$^{2-}$.

Finally, in the fifth section, we will include the radical salts prepared with metal-oxalate complexes and other TTF-type donors such as tetrathiafulvalene (TTF), tetramethyltetrathiafulvalene (TMTTF), bis(ethylenediseleno)tetrathiafulvalene (BEST), bis(ethylenedithio)tetraselenafulvalene (BETS) and 4,5-bis((2S)-2-hydroxypropylthio)-4',5'-(ethylenedithio)tetrathiafulvalene (DMPET) (Scheme 1).

2. The Superconducting Monoclinic β″ Phase

2.1. β″-BEDT-TTF Salts with the [Fe(C$_2$O$_4$)$_3$]$^{3-}$ Anion

As it can be seen in Table 1 (and 2), there are a total of 54 reported radical salts with BEDT-TTF and [M(C$_2$O$_4$)$_3$]$^{3-}$ anions showing the β″ packing mode. Among them, there are 35 salts with the [Fe(C$_2$O$_4$)$_3$]$^{3-}$ anion, formulated as β″-(BEDT-TTF)$_4$[AFe(C$_2$O$_4$)$_3$]·G with A$^+$ = H$_3$O$^+$ or NH$_4^+$ and G = pyridine (py), PhNO$_2$, PhCN, 1,2-PhCl$_2$, PhI, PhBr, PhCl, PhF, dmf, 2-Clpy, 2-Brpy, 3-Clpy and 3-Brpy (Table 1). All these isostructural salts crystallize in the monoclinic space group $C2/c$ and show alternating layers of BEDT-TTF donors and anionic layers in the ab plane (Figure 2).

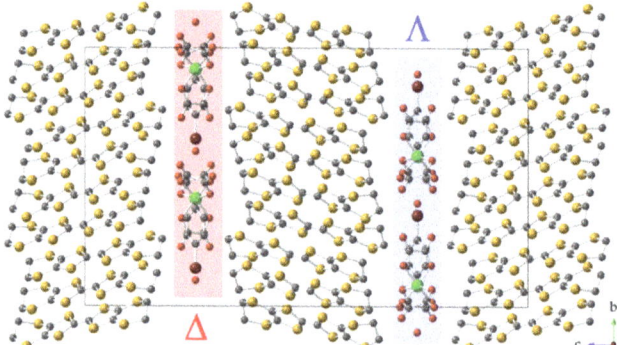

Figure 2. View of the alternating cationic/anionic layers at 292 K in compound β″-(BEDT-TTF)$_4$[(H$_3$O)Fe(C$_2$O$_4$)$_3$]·PhBr (15). Color code: Fe = light green, C = gray, O = red, S = yellow and Br = brown. H atoms are omitted for clarity.

The BEDT-TTF layer presents the so-called β″ packing motif where the donor molecules form parallel chains with the BEDT-TTF molecules tilted with respect to the chain direction (Figure 3a) [11]. The chains contain two independent BEDT-TTF molecules (A and B) following the sequence … AABB … Along the chain, the A-type molecules (in red in Figure 3b) show an eclipsed packing with their neighboring A-type molecule but are displaced half-rings along the long molecular axis with respect to the neighboring B-type molecules. In contrast, B-type molecules (in blue in Figure 3b) are displaced half-rings with both neighboring molecules (A and B). There are numerous S···S contacts between BEDT-TTF molecules of consecutive stacks that favor electron delocalization. The anionic layer contains the [Fe(C$_2$O$_4$)$_3$]$^{3-}$ anions, the A$^+$ cations and the solvent molecule (G). The [Fe(C$_2$O$_4$)$_3$]$^{3-}$ anions and the A$^+$ cations form a hexagonal honeycomb layer with the Fe ions and A cations located in the vertices of the hexagons and the solvent molecules in the center of the hexagonal cavities (Figure 3b).

There are two different anionic layers in the unit cell. Each layer contains a single enantiomer, resulting in an achiral salt with the layers following the sequence ···⊗-Λ-⊗-Λ.··· (Figure 2). The Fe-O bond distances are, as expected, very similar in all the structures.

In all these salts, the charge of the anionic [AFe(C$_2$O$_4$)$_3$]$^{2-}$ layers is balanced by four ET molecules. The charge estimated for each BEDT-TTF molecule from the bond distances in the BEDT-TTF molecules [12] (and from Raman spectroscopy) is +0.5, leading to a mixed-valence state with a typical $3/4$ filling of the four BEDT-TTF HOMO bands, resulting in a high electrical conductivity, as we will see below.

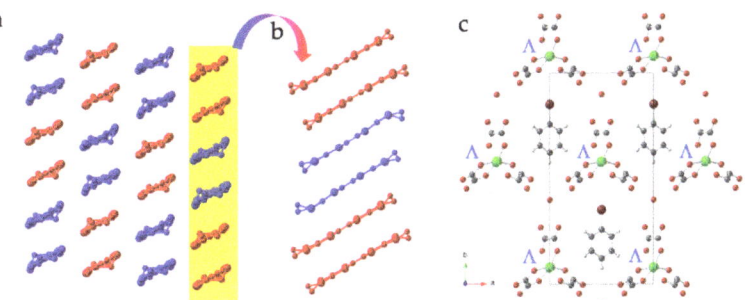

Figure 3. Structure of β''-(BEDT-TTF)$_4$[(H$_3$O)Fe(C$_2$O$_4$)$_3$]·PhBr (**15**): (**a**) View of the BEDT-TTF layer with the β'' packing mode. A and B molecules are represented in red and blue, respectively. (**b**) Side view of a BEDT-TTF chain. (**c**) View of one hexagonal anionic layer (with the Λ enantiomers) with the PhBr molecules in the hexagonal cavity. Color code: Fe = light green, C = gray, O = red, S = yellow, Br = brown and H = white.

Despite the large number of reported structures in the series β''-(BEDT-TTF)$_4$[AFe(C$_2$O$_4$)$_3$]·G (Table 1), there is only one reported example [13] with NH$_4^+$ (and one where the exact composition could not be determined from X-ray data) [14]. All the other solved structures contain the H$_3$O$^+$ cation. In contrast, there are many different solvents that can be found in the hexagonal cavities of the anionic layers. Thus, as it can be seen in Table 1, besides the PhCN molecule used in the first example (**35**) [3], there are also reported salts with PhNO$_2$ (**3, 4**), pyridine (py) (**1, 2**), PhF (**19, 20**), PhCl (**17, 18**), PhBr (**15, 16**), PhI (**14**), 1,2-PhCl$_2$ (**13**), 2-Clpy (**29, 30**), 3-Clpy (**33**), 2-Brpy (**31, 32**), 3-Brpy (**34**) and mixtures of PhCN and py (**5–12**) or PhCN with halobenzenes (**21–27**) and even with dimethylformamide (dmf) (**28**), the only non-aromatic solvent used in the β'' series with Fe(III) [13].

The key role of the solvent on the physical properties was soon noticed by P. Day's group after the synthesis of the py (**1, 2**) and PhNO$_2$ (**4**) derivatives with the same anion and the same structure [14,15]. Thus, the pyridine derivative showed a metal–insulator transition at T_{M-I} = 116 K, whereas the PhNO$_2$ derivative became a superconductor at T_c = 6.2 K, slightly below the PhCN derivative that showed a T_c of 8.5 K [3]. Further studies showed that the solvent molecules interact with the ethylene groups of the BEDT-TTF molecules of consecutive layers and play a key role in the order/disorder observed in these ethylene groups. The stronger the solvent–BEDT-TTF interaction, the better, since it facilitates the ordering of the ethylene groups and, therefore, favors the superconducting state. The lack of superconductivity in the salts with pyridine and the study by P. Day's group [16] showing a modulation of T_c with the PhCN/py ratio in the series of salts (ET)$_4$[(H$_3$O)Fe(C$_2$O$_4$)$_3$]·(PhCN)$_x$(py)$_{1-x}$ (**5–12**) further supported this conclusion. Detailed studies by E. B. Yagubskii et al. confirmed this effect with other different solvents and mixtures of them [17–19]. A final proof of this key role of the solvent is provided by the series of salts (ET)$_4$[(H$_3$O)Fe(C$_2$O$_4$)$_3$]·PhX with X = F, Cl, Br and I (**14–20**) [20]. In this series with halobenzene derivatives, the polarity and size of the halobenzene molecules determine the presence of superconducting transitions at 1.0 and 4.0 K for X = F and Br, respectively, but no superconducting transition is observed for X = Cl above 0.4 K, and a semiconducting behavior is observed for X = I. In this series, the differences were attributed to the combined effect of the size and the electronegativity of X. For X = F, the solvent–BEDT-TTF interaction is quite strong, despite a longer distance, given the high inductive effect of F. For X = Br, the larger size of X reduces the solvent–BEDT-TTF distance, increasing the interaction. For X = Cl, we are in an intermediate situation that does not favor the interaction (too far and not enough electronegativity). Finally, for X = I, the PhI molecule is too large and cannot fit in the hexagonal cavity and appears slightly out of the hexagonal cavity, with a tilted orientation that displaces the BEDT-TTF molecules, resulting in a loss of the metallic behavior. In fact, this compound is the only semiconducting salt reported in P. Day's series with Fe(III) (Table 1) [20].

A final interesting aspect of this series is the presence of a structural transition from the high-temperature monoclinic $C2/c$ space group to a triclinic P-1 one below ca. 200 K, only observed in the salts with G = PhF (**19**, **20**), PhCl (**17**, **18**), PhBr (**15**, **16**), 2-Clpy (**29**, **30**) and 2-Brpy (**31**, **32**). The first preliminary observation of this transition was reported in compound $(ET)_4[(H_3O)Fe(C_2O_4)_3]\cdot PhBr$ (**15**, **16**), where a change in the unit cell parameters to a lower symmetry phase was observed at ca. 180–200 K by single-crystal X-ray data and heat capacity measurements [21]. This result was later confirmed and studied in more detail in the series $(ET)_4[(H_3O)Fe(C_2O_4)_3]\cdot PhX$ (**15–20**) and $(ET)_4[(H_3O)Fe(C_2O_4)_3]\cdot (PhCN)_x(PhX)_{1-x}$, with X = F (**21**, **22**), Cl (**24**, **25**) and Br (**26**, **27**) [18,20], and in compounds $(ET)_4[(H_3O)Fe(C_2O_4)_3]\cdot 2$-Clpy (**29**, **30**) and $(ET)_4[(H_3O)Fe-(C_2O_4)_3]\cdot 2$-Brpy (**31**, **32**) [19]. The symmetry loss that appears due to the change in the space group from the monoclinic $C2/c$ space group to the triclinic P-1 one leads to the appearance of two independent BEDT-TTF layers in the unit cell (layers I and II, Figure 4), both keeping the β″ packing motif (Figure 4a,c). The number of independent BEDT-TTF molecules increases from two in the monoclinic $C2/c$ phase (A and B) to four in the triclinic P-1 phase (A-B in layer I and C-D in layer II).

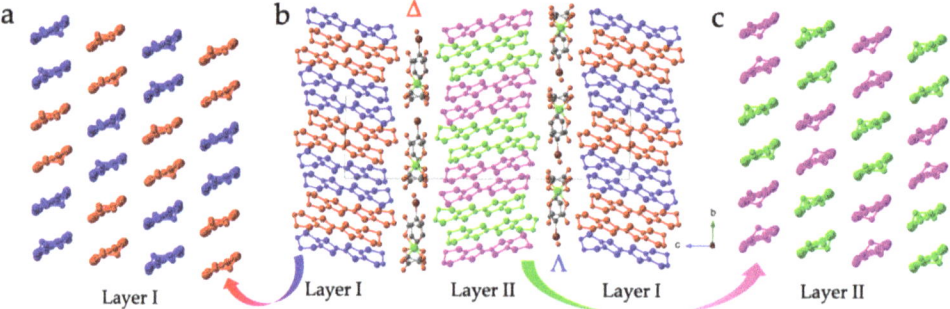

Figure 4. (**a**) View of layer I formed by A (red) and B (blue) BEDT-TTF molecules. (**b**) View of the structure of the triclinic P-1 phase of β″-$(ET)_4[(H_3O)Fe(C_2O_4)_3]\cdot PhBr$ (**16**) at 120 K showing the two independent BEDT-TTF layers (I and II). (**c**) View of layer II formed by C (green) and D (pink) BEDT-TTF molecules.

As in the monoclinic phase, the chains are also formed following the sequence ... AABB ... or ... CCDD ..., and the overlap between neighboring molecules follows the same scheme in both layers: A-A is eclipsed but A-B and B-B are shifted in layer I, and C-C is eclipsed but C-D and D-D are shifted in layer II.

Nevertheless, the most important changes in the triclinic P-1 phase are observed in the anionic layer, where the change in symmetry implies a decrease in four of the six sides of the hexagons that forces a change in the location and orientation of the solvent molecules. Thus, in the monoclinic $C2/c$ phase, the hexagons are planar, and the solvent molecule is located with the C-X bond in the hexagon plane pointing towards one Fe atom (Figure 5a). In contrast, in the triclinic P-1 phase, there is a reduction in the sides of the hexagons, the hexagons are not planar and the C-X bond is slightly out of the average plane and does not point to one Fe atom (Figure 5b).

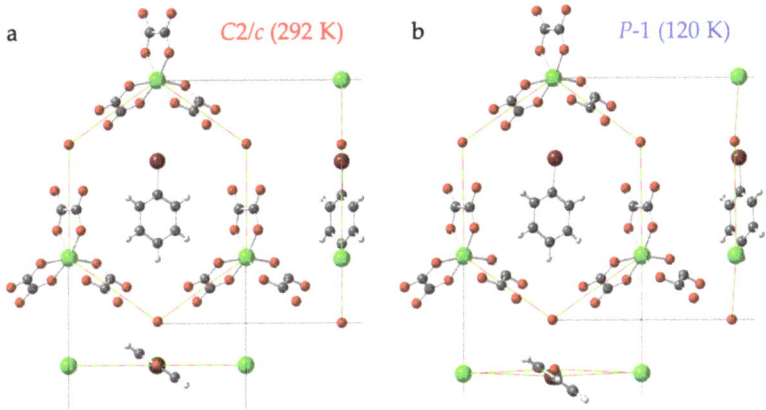

Figure 5. Front and side views of one hexagonal cavity in (**a**) the monoclinic $C2/c$ phase at 292 K and (**b**) the triclinic P-1 phase at 120 K in the radical salt β''-(BEDT-TTF)$_4$[(H$_3$O)Fe(C$_2$O$_4$)$_3$]·PhBr (**15**, **16**). Color code: Fe = light green, C = gray, O = red, Br = brown and H = white.

The most interesting aspect of the series β''-(ET)$_4$[AFe(C$_2$O$_4$)$_3$]·G (**1–35**) is their electrical properties. As mentioned above, compound (ET)$_4$[(H$_3$O)Fe(C$_2$O$_4$)$_3$]·PhCN (**35**) is the first molecular paramagnetic superconductor and shows an ordering temperature of 6.5–8.5 K (depending on the quality of the single crystals and on the exact synthetic conditions) [2–4,17]. There is a total of seventeen superconductors reported to date in the β''-(ET)$_4$[AFe(C$_2$O$_4$)$_3$]·G series, with T_c ranging from 1.0 to 8.5 K (Table 1). These superconductors include the derivatives with G = PhNO$_2$ (**4**) with T_c = 6.2 K [14]; G = (PhCN)$_x$(py)$_{1-x}$ (**5, 7–12**) with T_c = 3.9–7.3 K [16]; G = PhBr (**15**) with T_c = 4.0 K [21] (Figure 6); G = PhF (**19**) with an onset of T_c = 1.0 K [20] (Figure 6); G = (PhCN)$_{0.4}$(PhF)$_{0.6}$ (**21**) with T_c = 6.0 K [17]; G = (PhCN)$_{0.86}$(PhCl$_2$)$_{0.14}$ (**23**) with T_c = 7.2 K [17]; G = (PhCN)$_{0.35}$(PhCl)$_{0.65}$ (**24**) with T_c = 6.0 K [17]; G = (PhCN)$_{0.17}$(PhBr)$_{0.83}$ (**26**) with T_c = 4.2 K [17]; G = 2-ClPy (**29**) with T_c = 2.4–4.0 K [19]; G = 2-BrPy (**31**) with T_c = 4.3 K [19]; and G = PhCN (**35**) with T_c = 8.5 K [2–4]. Except for the salt with G = PhI (**14**) [20] (Figure 6), and for a salt with H$_2$O and G = PhNO$_2$ (**3**) [22], all the other reported salts in the β''-(ET)$_4$[(H$_3$O)Fe(C$_2$O$_4$)$_3$]·G series are metallic at room temperature and show metal–semiconductor or metal–insulator transitions at lower temperatures (Table 1).

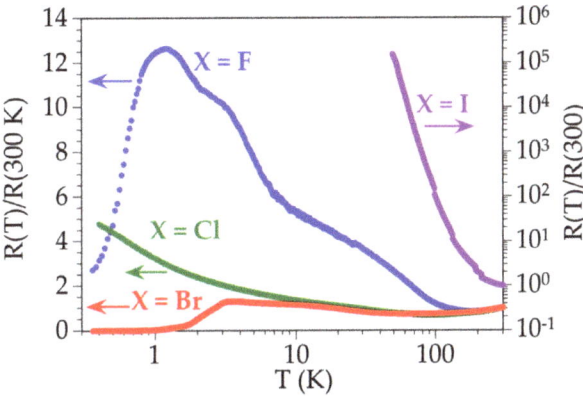

Figure 6. Electrical conductivity of the series β''-(BEDT-TTF)$_4$[(H$_3$O)Fe(C$_2$O$_4$)$_3$]·PhX with X = F (**19**), Cl (**17**), Br (**15**) and I (**14**).

All the superconductors in the β"-(ET)$_4$[(H$_3$O)Fe(C$_2$O$_4$)$_3$]·G series show abrupt transitions with a sharp decrease in the resistivity values. Most of the salts reach zero resistance around one kelvin below the onset of the superconducting transition (Figures 6 and 7). The band structure calculations show that the monoclinic β" phase is a 2D metal with stronger inter-chain than intra-chain interactions. These calculations also show the formation of $^3/_4$ filled bands with the Fermi level intersecting the two upper bands, leading to the observed metallic behavior [13].

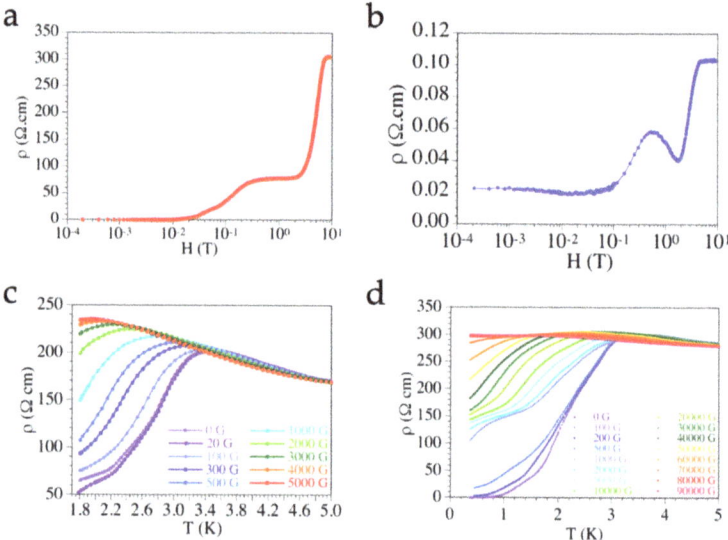

Figure 7. (**a**) Field dependence of the resistivity of compounds β"-(ET)$_4$[(H$_3$O)Fe(C$_2$O$_4$)$_3$]·PhBr (**15**) with the magnetic field applied perpendicular to the conducting layers. (**b**) Field dependence of the resistivity of compound β"-(ET)$_4$[(H$_3$O)Fe(C$_2$O$_4$)$_3$]·PhF (**19**) with the magnetic field applied parallel to the conducting layers. (**c**,**d**) Temperature dependence of the resistivity of compound β"-(ET)$_4$[(H$_3$O)Fe(C$_2$O$_4$)$_3$]·PhBr (**15**) with different applied DC fields parallel (**c**) and perpendicular (**d**) to the conducting layer.

Magnetoresistance measurements show that these salts are type II superconductors with low first critical fields of a few mT (H_{c1}, beginning of the penetration of the magnetic field) and very high second critical fields of several tesla (H_{c2}, complete suppression of the superconductivity) (Figure 7a,b). Furthermore, as expected for these quasi-2D superconductors, the effect of the magnetic field and the values of the critical fields are anisotropic and strongly depend on the direction of the applied DC field (Figure 7c,d).

The magnetic properties of these paramagnetic superconductors are those expected for isolated high-spin S = 5/2 [Fe(C$_2$O$_4$)$_3$]$^{3-}$ anions plus a temperature-independent paramagnetism arising from the conducting sublattice (Pauli paramagnetism, Figure 8a). At very low temperatures, there is a sharp decrease in the magnetic moment due to the presence of a zero-field splitting (ZFS) in the S = 5/2 ground spin state.

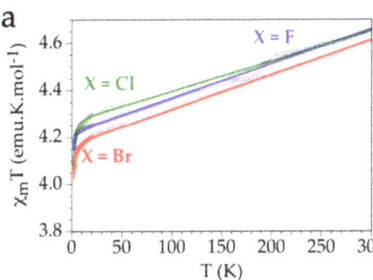
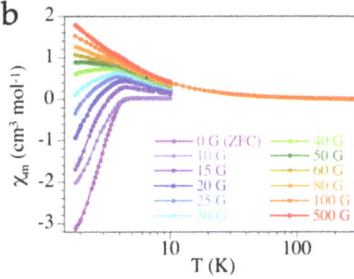

Figure 8. (a) Temperature dependence of the $\chi_m T$ product for the series β''-(BEDT-TTF)$_4$[(H$_3$O)Fe(C$_2$O$_4$)$_3$]·PhX with X = F (**19**), Cl (**17**) and Br (**15**). (b) Temperature dependence of the zero field and field-cooled susceptibility with different applied fields for compound β''-(BEDT-TTF)$_4$[(H$_3$O)Fe(C$_2$O$_4$)$_3$]·PhBr (**15**).

Additionally, when the samples are cooled under zero magnetic field (zero field cooling, ZFC), the superconducting transitions can be detected by the appearance of negative magnetization values (Meissner effect, Figure 8b) below the transition temperature. Heating the samples with increasing magnetic fields shows the progressive cancellation of the Meissner effect and an increase in the susceptibility values, typical of type II superconductivity (Figure 8b). For magnetic fields above a certain value, the susceptibility becomes positive at any temperature since the paramagnetic contribution of the [Fe(C$_2$O$_4$)$_3$]$^{3-}$ anion dominates the magnetic response of the sample (Figure 8b).

Magnetic measurements in the presence of an alternating magnetic field (AC susceptibility) further confirm the superconducting transitions in these radical salts and allow a precise estimation of T_c and of the critical fields. Thus, AC measurements show the presence of a negative in-phase signal (χ'_m) below T_c, very similar to the ZFC susceptibility. This in-phase signal also reduces its absolute value when a DC field is applied and becomes positive at any temperature above a critical DC field when the paramagnetic contribution of the [Fe(C$_2$O$_4$)$_3$]$^{3-}$ anion becomes dominant (Figure 9a). AC measurements also show and an out-of-phase signal (χ''_m) that becomes non-zero at T_c (Figure 9b). The application of a DC field also reduces the χ''_m signal, which eventually cancels above the critical field (Figure 9b).

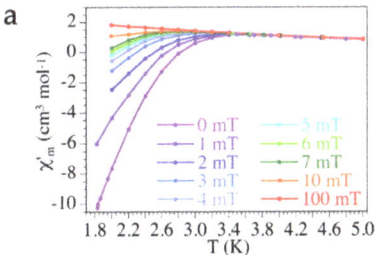
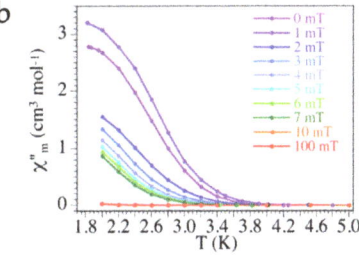

Figure 9. AC measurements of compound β''-(BEDT-TTF)$_4$[(H$_3$O)Fe(C$_2$O$_4$)$_3$]·PhBr (**15**): temperature dependence of (**a**) the in-phase (χ'_m) and (**b**) the out-of-phase (χ''_m) AC susceptibility with different applied DC fields.

Table 1. Fe-containing radical salts of P. Day's series: β''-(ET)$_4$[AFe(C$_2$O$_4$)$_3$]·G (1–35).

#	CCDC	Formula [a]	SG [b]	Elect. Prop	A$^+$	G	Ref.
1	BEMPEO	(ET)$_4$[(H$_3$O)Fe(ox)$_3$]·0.5 py	C2/c	M > 116 K	H$_3$O$^+$	Py	[15]
2	BEMQAL	(ET)$_4$[(H$_3$O)Fe(ox)$_3$]·py	C2/c	-	H$_3$O$^+$	py	[15]
3	COQNEB	(ET)$_4$[(H$_2$O)Fe(ox)$_3$]·PhNO$_2$	C2/c	σ = 10 S/cm Semi	-	PhNO$_2$	[22]
4	ECOPIV	(ET)$_4$[(H$_3$O/NH$_4$)Fe(ox)$_3$]·PhNO$_2$	C2/c	T$_c$ = 6.2 K	H$_3$O$^+$/NH$_4^+$	PhNO$_2$	[14]
5	KILFOB	(ET)$_4$[(H$_3$O)Fe(ox)$_3$]·(PhCN)$_{0.78}$(py)$_{0.22}$	C2/c	T$_c$ = 3.9 K	H$_3$O$^+$	PhCN/py	[16]
6	KILFOB01	(ET)$_4$[(H$_3$O)Fe(ox)$_3$]·(PhCN)$_{0.77}$(py)$_{0.23}$	C2/c	-	H$_3$O$^+$	PhCN/py	[16]
7	KILFOB02	(ET)$_4$[(H$_3$O)Fe(ox)$_3$]·(PhCN)$_{0.66}$(py)$_{0.34}$	C2/c	T$_c$ = 5.8 K	H$_3$O$^+$	PhCN/py	[16]
8	KILFOB03	(ET)$_4$[(H$_3$O)Fe(ox)$_3$]·(PhCN)$_{0.62}$(py)$_{0.38}$	C2/c	T$_c$ = 6.9 K	H$_3$O$^+$	PhCN/py	[16]
9	KILFOB04	(ET)$_4$[(H$_3$O)Fe(ox)$_3$]·(PhCN)$_{0.57}$(py)$_{0.43}$	C2/c	T$_c$ = 6.7 K	H$_3$O$^+$	PhCN/py	[16]
10	KILGOC	(ET)$_4$[(H$_3$O)Fe(ox)$_3$]·(PhCN)$_{0.46}$(py)$_{0.54}$	C2/c	T$_c$ = 5.9 K	H$_3$O$^+$	PhCN/py	[16]
11	KILGUI	(ET)$_4$[(H$_3$O)Fe(ox)$_3$]·(PhCN)$_{0.39}$(py)$_{0.61}$	C2/c	T$_c$ = 4.2 K	H$_3$O$^+$	PhCN/py	[16]
12	KILHAP	(ET)$_4$[(H$_3$O)Fe(ox)$_3$]·(PhCN)$_{0.10}$(py)$_{0.90}$	C2/c	T$_c$ = 7.3 K	H$_3$O$^+$	PhCN/py	[16]
13	PONMEL	(ET)$_4$[(H$_3$O)Fe(ox)$_3$]·1,2-PhCl$_2$	C2/c	M > 0.5 K	H$_3$O$^+$	1,2-PhCl$_2$	[23]
14	QAXSIT	(ET)$_4$[(H$_3$O)Fe(ox)$_3$]·PhI	C2/c	σ = 3.4 S/cm E$_a$ = 64 meV	H$_3$O$^+$	PhI	[20]
15	SAPWEM	(ET)$_4$[(H$_3$O)Fe(ox)$_3$]·PhBr	C2/c	T$_c$ = 4.0 K	H$_3$O$^+$	PhBr	[21]
16	SAPWEM02	(ET)$_4$[(H$_3$O)Fe(ox)$_3$]·PhBr	P-1	T$_c$ = 4.0 K	H$_3$O$^+$	PhBr	[20]
17	UJOXAT	(ET)$_4$[(H$_3$O)Fe(ox)$_3$]·PhCl	C2/c	M > 0.4 K	H$_3$O$^+$	PhCl	[17,20]
18	UJOXAT01	(ET)$_4$[(H$_3$O)Fe(ox)$_3$]·PhCl	P-1	M > 4.2 K	H$_3$O$^+$	PhCl	[18,20]
19	UJOXEX	(ET)$_4$[(H$_3$O)Fe(ox)$_3$]·PhF	C2/c	T$_c$ = 1.0 K	H$_3$O$^+$	PhF	[17,20]
20	UJOXEX01	(ET)$_4$[(H$_3$O)Fe(ox)$_3$]·PhF	P-1	T$_c$ = 1.0 K	H$_3$O$^+$	PhF	[18,20]
21	UJOXIB	(ET)$_4$[(H$_3$O)Fe(ox)$_3$]·(PhCN)$_{0.4}$(PhF)$_{0.6}$	C2/c	T$_c$ = 6.0 K	H$_3$O$^+$	PhCN/PhF	[17]
22	UJOXIB01	(ET)$_4$[(H$_3$O)Fe(ox)$_3$]·(PhCN)$_{0.4}$(PhF)$_{0.6}$	P-1	T$_c$ = 6.0 K	H$_3$O$^+$	PhCN/PhF	[18]
23	UJOXOH	(ET)$_4$[(H$_3$O)Fe(ox)$_3$]·(PhCN)$_{0.86}$(PhCl$_2$)$_{0.14}$	C2/c	T$_c$ = 7.2 K	H$_3$O$^+$	PhCN/PhCl$_2$	[17]

Table 1. Cont.

#	CCDC	Formula [a]	SG [b]	Elect. Prop	A$^+$	G	Ref.
24	UJOYAU	(ET)$_4$[(H$_3$O)Fe(ox)$_3$]·(PhCN)$_{0.35}$(PhCl)$_{0.65}$	C2/c	T$_c$ = 6.0 K	H$_3$O$^+$	PhCN/PhCl	[17]
25	UJOYAU02	(ET)$_4$[(H$_3$O)Fe(ox)$_3$]·(PhCN)$_{0.35}$(PhCl)$_{0.65}$	P-1	T$_c$ = 6.0 K	H$_3$O$^+$	PhCN/PhCl	[18]
26	UJOYEY	(ET)$_4$[(H$_3$O)Fe(ox)$_3$]·(PhCN)$_{0.17}$(PhBr)$_{0.83}$	C2/c	T$_c$ = 4.2 K	H$_3$O$^+$	PhCN/PhBr	[17]
27	UJOYEY01	(ET)$_4$[(H$_3$O)Fe(ox)$_3$]·(PhCN)$_{0.17}$(PhBr)$_{0.83}$	P-1	T$_c$ = 4.2 K	H$_3$O$^+$	PhCN/PhBr	[18]
28	UMACEQ	(ET)$_4$[(NH$_4$)Fe(ox)$_3$]·dmf	C2/c	M > 4.2 K	NH$_4^+$	dmf	[13]
29	YUYTUJ	(ET)$_4$[(H$_3$O)Fe(ox)$_3$]·2-Clpy	C2/c	T$_c$ = 2.4–4.0 K	H$_3$O$^+$	2-Clpy	[19]
30	YUYTUJ01	(ET)$_4$[(H$_3$O)Fe(ox)$_3$]·2-Clpy	P-1	T$_c$ = 2.4–4.0 K	H$_3$O$^+$	2-Clpy	[19]
31	YUYVEV	(ET)$_4$[(H$_3$O)Fe(ox)$_3$]·2-Brpy	C2/c	T$_c$ = 4.3 K	H$_3$O$^+$	2-Brpy	[19]
32	YUYVEV01	(ET)$_4$[(H$_3$O)Fe(ox)$_3$]·2-Brpy	P-1	T$_c$ = 4.3 K	H$_3$O$^+$	2-Brpy	[19]
33	YUYVOF	(ET)$_4$[(H$_3$O)Fe(ox)$_3$]·3-Clpy	C2/c	M > 0.5 K	H$_3$O$^+$	3-Clpy	[19]
34	YUYVUL	(ET)$_4$[(H$_3$O)Fe(ox)$_3$]·3-Brpy	C2/c	M > 0.5 K	H$_3$O$^+$	3-Brpy	[19]
35	ZIGYET	(ET)$_4$[(H$_3$O)Fe(ox)$_3$]·PhCN	C2/c	T$_c$ = 6.5–8.5 K	H$_3$O$^+$	PhCN	[2–4,17]

[a] ox = oxalate = C$_2$O$_4^{2-}$; py = pyridine; dmf = dimethylformamide. [b] SG = space group.

2.2. β″-BEDT-TTF Salts with Other $[M(C_2O_4)_3]^{3-}$ Anions (M ≠ Fe)

Soon after the discovery of superconductivity in compound β″-(ET)$_4$[(H$_3$O)Fe(C$_2$O$_4$)$_3$]·PhCN, P. Day's group explored the [Cr(C$_2$O$_4$)$_3$]$^{3-}$ anion and prepared the salts β″-(ET)$_4$[(H$_3$O)Cr(C$_2$O$_4$)$_3$]·G with G = PhCN (**47**) [24,25] and PhNO$_2$ (**40**) [14], which also present superconducting transitions at T$_c$ = 6.0 and 5.8 K, respectively. As it can be seen in Table 2, the [Cr(C$_2$O$_4$)$_3$]$^{3-}$ anion has also been combined in BEDT-TTF salts with other solvents such as PhBr, PhCl, CH$_2$Cl$_2$, dmf, 2-Clpy and 2-Brpy with A$^+$ cations such as H$_3$O$^+$, K$^+$, NH$_4^+$ and mixtures of them, giving rise to a superconducting salt with T$_c$ = 1.7 K for A$^+$/G = H$_3$O$^+$/PhBr (**53**) [21], two metallic salts that remain metallic down to low temperatures for A$^+$/G = (K$^+$/NH$_4^+$)/dmf (**50**) and K$^+$/dmf (**51**) [13] and three metallic salts that show metal–insulator transitions at low temperature for A$^+$/G = (K$^+$/H$_3$O$^+$)/2-Clpy (**36**), (K$^+$/H$_3$O$^+$)/2-Brpy (**37**) and H$_3$O$^+$/PhCl (**54**) [26,27]. In one case, with A$^+$/G = H$_3$O$^+$/CH$_2$Cl$_2$ (**48**), the obtained salt is a semiconductor (although not a classical one), most probably due to the lack of interactions between the BEDT-TTF and solvent molecules [28].

Parallel to the Cr-containing [Cr(C$_2$O$_4$)$_3$]$^{3-}$ anion, P. Day's group also checked the diamagnetic [Ga(C$_2$O$_4$)$_3$]$^{3-}$ anion and obtained one superconductor formulated as (ET)$_4$[(H$_3$O)Ga(C$_2$O$_4$)$_3$]·PhNO$_2$ (**46**) [29] with T$_c$ = 7.5 K, and also the pyridine derivative (ET)$_4$[(H$_3$O)Ga(C$_2$O$_4$)$_3$]·py (**45**), which might be a superconductor below 2 K, although no clear evidence was observed [29]. E. B. Yagubskii et al. also used the [Ga(C$_2$O$_4$)$_3$]$^{3-}$ anion to prepare the semiconducting salt (ET)$_4$[K$_{0.33}$(H$_3$O)$_{0.67}$Ga(C$_2$O$_4$)$_3$]·PhBr (**44**) [30], and the metallic salts (ET)$_4$[K$_{0.8}$(H$_3$O)$_{0.2}$Ga(C$_2$O$_4$)$_3$]·G with G = 2-Clpy (**38**) and 2-Brpy (**39**) [27]. Yagubskii's group also prepared the first example in this series with a 4d metal: (ET)$_4$[K$_{0.7}$(H$_3$O)$_{0.3}$Ru(C$_2$O$_4$)$_3$]·PhBr (**49**). This compound is a superconductor with T$_c$ ranging from 2.8 to 6.3 K, depending on the measured sample [31].

There is also a preliminary account of a salt with the anion [Mn(C$_2$O$_4$)$_3$]$^{3-}$ formulated as (ET)$_4$[(H$_3$O)Mn(C$_2$O$_4$)$_3$]·PhBr (**52**) that is also a superconductor with T$_c$ = 2.0 K [20].

Finally, L. Martin et al. prepared the only known examples in this series with the [Rh(C$_2$O$_4$)$_3$]$^{3-}$ anion: (BEDT-TTF)$_4$[ARh(C$_2$O$_4$)$_3$]·PhX with A/X = NH$_4^+$/Br (**41**), H$_3$O$^+$/F (**42**) and NH$_4^+$/Cl (**43**) [32]. The PhBr derivative is a superconductor with T$_c$ = 2.5 K, but the PhF and PhCl derivatives are metallic with metal–insulator transitions at 180 and 10 K, respectively (Table 2). As observed for the [Fe(C$_2$O$_4$)$_3$]$^{3-}$ anion, the adequate size of PhBr seems to be at the origin of the superconducting transition in the salt with [Rh(C$_2$O$_4$)$_3$]$^{3-}$.

A detailed revision of Tables 1 and 2 shows that among the β″-(ET)$_4$[AM(C$_2$O$_4$)$_3$]·G salts with pure solvents, the one that has produced more superconductors is bromobenzene. This solvent has originated, to date, five superconductors, with A$^+$/M^{III} = H$_3$O$^+$/Fe (**15**), NH$_4^+$/Rh (**41**), (H$_3$O$^+$/K$^+$)/Ru (**49**), H$_3$O$^+$/Mn (**52**) and H$_3$O$^+$/Cr (**53**). The other solvents that have given rise to more superconductors are PhNO$_2$, which has produced three superconductors, with A$^+$/M^{III} = (H$_3$O$^+$/NH$_4^+$)/Fe (**4**), (H$_3$O$^+$/NH$_4^+$)/Cr (**40**) and H$_3$O$^+$/Ga (**46**), and PhCN, with two superconductors, with A$^+$/M^{III} = H$_3$O$^+$/Fe (**35**) and H$_3$O$^+$/Cr (**47**). It is interesting to note that the size of the PhBr, PhCN and PhNO$_2$ molecules seems to fit very well in the hexagonal cavity, and their large size allows a close contact with the ethylene groups of the BEDT-TTF molecules. This size effect of the PhBr solvent is clearly seen in the halobenzenes, where PhCl is too small and does not give superconductivity and PhI is too big and does not fit in the hexagonal cavities [20].

Table 2. Radical salts of the series β''-(ET)$_4$[AM(C$_2$O$_4$)$_3$]·G with M $\neq$ Fe (**36-54**).

#	CCDC	Formula [a]	SG [b]	Elect. Prop.	A$^+$	M	G	Ref.
36	CIWMED	(ET)$_4$[K$_{0.8}$(H$_3$O)$_{0.2}$Cr(ox)$_3$]·2-Clpy	C2/c	T$_{MI}$ ≈ 10 K	K$^+$/H$_3$O$^+$	Cr	2-Clpy	[27]
37	CIWMIH	(ET)$_4$[K$_{0.8}$(H$_3$O)$_{0.2}$Cr(ox)$_3$]·2-Brpy	C2/c	T$_{MI}$ ≈ 10 K	K$^+$/H$_3$O$^+$	Cr	2-Brpy	[27]
38	CIWMON	(ET)$_4$[K$_{0.8}$(H$_3$O)$_{0.2}$Ga(ox)$_3$]·2-Clpy	C2/c	T$_{MI}$ ≈ 10 K	K$^+$/H$_3$O$^+$	Ga	2-Clpy	[27]
39	CIWMUT	(ET)$_4$[K$_{0.8}$(H$_3$O)$_{0.2}$Ga(ox)$_3$]·2-Brpy	C2/c	T$_{MI}$ ≈ 10 K	K$^+$/H$_3$O$^+$	Ga	2-Brpy	[27]
40	ECOPUH	(ET)$_4$[(H$_3$O/NH$_4$)Cr(ox)$_3$]·PhNO$_2$	C2/c	T$_c$ = 5.8 K	H$_3$O$^+$/NH$_4^+$	Cr	PhNO$_2$	[14]
41	FEBLIK	(ET)$_4$[(NH$_4$)Rh(ox)$_3$]·PhBr	C2/c	T$_c$ = 2.5 K	NH$_4^+$	Rh	PhBr	[32]
42	FECDAV	(ET)$_4$[(H$_3$O)Rh(ox)$_3$]·PhF	C2/c	T$_{MI}$ ≈ 180 K	H$_3$O$^+$	Rh	PhF	[32]
43	FECDID	(ET)$_4$[(NH$_4$)Rh(ox)$_3$]·PhCl	C2/c	T$_{MI}$ ≈ 10 K	NH$_4^+$	Rh	PhCl	[32]
44	HOBROH	(ET)$_4$[K$_{0.33}$(H$_3$O)$_{0.67}$Ga(ox)$_3$]·PhBr	C2/c	Semi	K$^+$/H$_3$O$^+$	Ga	PhBr	[30]
45	HUNQIQ	(ET)$_4$[(H$_3$O)Ga(ox)$_3$]·py	C2/c	T$_c$ ≈ 2 K ??	H$_3$O$^+$	Ga	py	[29]
46	HUNQUC	(ET)$_4$[(H$_3$O)Ga(ox)$_3$]·PhNO$_2$	C2/c	T$_c$ = 7.5 K	H$_3$O$^+$	Ga	PhNO$_2$	[29]
47	JUPGUW01	(ET)$_4$[(H$_3$O)Cr(ox)$_3$]·PhCN	C2/c	T$_c$ = 6.0 K	H$_3$O$^+$	Cr	PhCN	[4,24,25]
48	MEQZIR	(ET)$_4$[(H$_3$O)Cr(ox)$_3$]·CH$_2$Cl$_2$	C2/c	Semi	H$_3$O$^+$	Cr	CH$_2$Cl$_2$	[28]
49	UDETUU	(ET)$_4$[K$_{0.7}$(H$_3$O)$_{0.3}$Ru(ox)$_3$]·PhBr	C2/c	T$_c$ = 2.8–6.3 K	K$^+$/H$_3$O$^+$	Ru	PhBr	[31]
50	UMACAM	(ET)$_4$[(K/NH$_4$)Cr(ox)$_3$]·dmf	C2/c	M > 4.2	K$^+$/NH$_4^+$	Cr	dmf	[13]
51	UMACIU	(ET)$_4$[KCr(ox)$_3$]·dmf	C2/c	M > 4.2	K$^+$	Cr	dmf	[13]
52	-	(ET)$_4$[(H$_3$O)Mn(ox)$_3$]·PhBr	C2/c	T$_c$ = 2.0 K	H$_3$O$^+$	Mn	PhBr	[20]
53	-	(ET)$_4$[(H$_3$O)Cr(ox)$_3$]·PhBr	C2/c	T$_c$ = 1.7 K	H$_3$O$^+$	Cr	PhBr	[21,26]
54	-	(ET)$_4$[(H$_3$O)Cr(ox)$_3$]·PhCl	C2/c	M > 130 K $\sigma_{300} = 3 \times 10^{-3}$ S/cm	H$_3$O$^+$	Cr	PhCl	[26]

[a] ox = oxalate = C$_2$O$_4^{2-}$; py = pyridine; dmf = dimethylformamide. [b] SG = space group.

3. The Semiconducting *pseudo*-κ or κ' Phase

κ'-BEDT-TTF Radical Salts with $[M(C_2O_4)_3]^{3-}$ Anions

Besides the monoclinic $C2/c$ β'' salts described in the previous section, P. Day's group also obtained an orthorhombic *Pbcn* phase with the same $[Fe(C_2O_4)_3]^{3-}$ anion and the same PhCN solvent, but with $A^+ = NH_4^+$ and K^+ [2]. These orthorhombic salts show an original packing, called *pseudo*-κ or κ' phase, and present the same general formula as the β'' phase: κ'-(BEDT-TTF)$_4$[AM(C$_2$O$_4$)$_3$]·G (Table 3). The structure of the κ' phase also consists of alternating cationic layers containing the BEDT-TTF molecules and anionic layers containing the A^+ cations, the $[M(C_2O_4)_3]^{3-}$ anions and the solvent molecules (Figure 10).

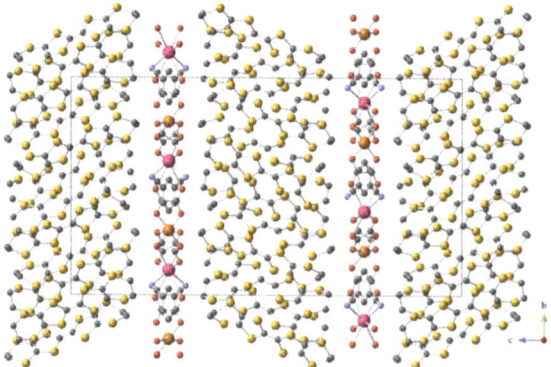

Figure 10. View of the alternating cationic/anionic layers in compound κ'-(BEDT-TTF)$_4$[(H$_3$O)Mn(C$_2$O$_4$)$_3$]·PhCN (**55**). Color code: Mn = orange, K = pink, C = gray, O = red, N = blue and S = yellow. H atoms are omitted for clarity.

The BEDT-TTF layer is also formed by two independent BEDT-TTF molecules noted as A and B. A-type molecules bear a charge of +1, whereas B-type molecules are neutral. The BEDT-TTF layer presents the so-called pseudo-κ or κ' packing motif formed by eclipsed A-A dimers surrounded by six B-type monomers in a distorted hexagonal arrangement (Figure 11a). As in the monoclinic β'' phase, the anionic layer contains the $[M(C_2O_4)_3]^{3-}$ anions and the A^+ cations forming a hexagonal lattice with the solvent molecules (G) inside the hexagonal cavities. The main difference is that now the anionic layers are equivalent since they contain both enantiomers in a 1:1 ratio arranged in alternating rows (Figure 11b). In both phases, the anionic layers are related by a C_2 axis. This different disposition of the ⊗ and Λ enantiomers in the anionic layers is at the origin of the different packing motifs of the BEDT-TTF molecules in both phases [4].

As already noted by Peter Day's group [4], a close look at the cation–anion interlayer interactions shows that the relative orientation of the BEDT-TTF molecules depends on the chirality of the closest $[M(C_2O_4)_3]^{3-}$ anions. Thus, in the monoclinic β'' phase, when the ethylene groups of the BEDT-TTF molecules interact with the terminal O atoms of ⊗-$[M(C_2O_4)_3]^{3-}$ anions, the BEDT-TTF chains run from bottom right to top left (Figure 12a), whereas, when they interact with Λ-$[M(C_2O_4)_3]^{3-}$ anions, the BEDT-TTF chains run from bottom left to top right (Figure 12b). In the orthorhombic κ' phase, the anionic layers contain both enantiomers arranged in rows located close to the BEDT-TTF dimers, and, accordingly, the orientation of these dimers follows the same trend: they run from bottom right to top left when they interact with ⊗-$[M(C_2O_4)_3]^{3-}$ anions (in green in Figure 12c) and from bottom left to top right when they interact with Λ-$[M(C_2O_4)_3]^{3-}$ anions (in orange in Figure 12c). The monomers in the κ' phase (in pink in Figure 12c) do not interact with the $[M(C_2O_4)_3]^{3-}$ anions and are packed following the BEDT-TTF dimers that they enclose.

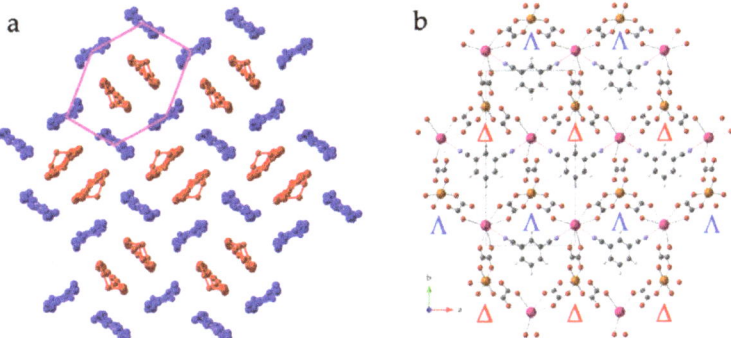

Figure 11. Structure of κ′-(BEDT-TTF)$_4$[KMn(C$_2$O$_4$)$_3$]·PhCN (**55**): (**a**) View of the BEDT-TTF layer with the κ′ packing mode (A and B molecules are drawn in red and blue, respectively). (**b**) View of one hexagonal anionic layer showing the Λ and ⊗ enantiomers arranged in rows and the PhCN molecules in the hexagonal cavities (the CN groups are disordered over two positions). Color code in (**b**): Mn = orange, K = pink, C = gray, O = red, N = blue and H = white.

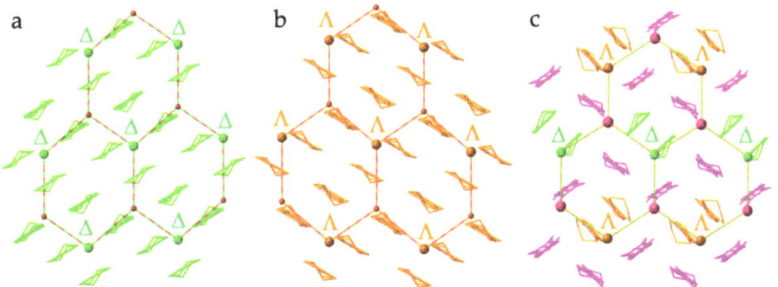

Figure 12. View of the relative orientation of the BEDT-TTF molecules with the chirality of the [M(C$_2$O$_4$)$_3$]$^{3-}$ anions closest to them (only the metal centers and the A$^+$ cations are shown): (**a**) ⊗-type layer in β″-(ET)$_4$[(H$_3$O)Fe(C$_2$O$_4$)$_3$]·PhBr (**15**). (**b**) Λ-type layer in β″-(ET)$_4$[(H$_3$O)Fe(C$_2$O$_4$)$_3$] PhBr (**15**). (**c**) Alternating ⊗-Λ layer in κ′-(ET)$_4$[KMn(C$_2$O$_4$)$_3$]·PhCN (**55**) showing the two different orientations of the BEDT-TTF dimers depending on the chirality (⊗ = green, Λ = orange) of the closest [M(C$_2$O$_4$)$_3$]$^{3-}$ anion.

The orthorhombic *pseudo*-κ phase has been observed in a total of nine salts to date (Table 3). Interestingly, in all these salts, the solvent is always PhCN (in one case with a small fraction of PhCl$_2$) [17]. Despite the low number of reported salts, this series presents a large variability in the anionic layer with three different A$^+$ cations (K$^+$, NH$_4^+$ or H$_3$O$^+$) and seven different metal ions (M = Mn, Rh, Cr, Co, Al, Ru and Fe).

Analysis of the bond distances in the BEDT-TTF molecules indicates that the molecules forming the dimers (A-type) are completely ionized, whereas the isolated BEDT-TTF molecules (B-type) are neutral. The presence of totally ionized (BEDT-TTF)$_2^{2+}$ dimers surrounded by neutral BEDT-TTF molecules precludes the electron delocalization in the BEDT-TTF layers, and, accordingly, all the reported κ′ salts are semiconductors (Table 3) with activation energies in the range 140 to 245 meV in all cases (Figure 13).

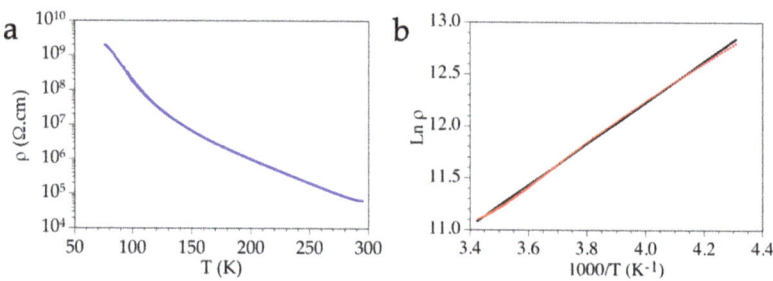

Figure 13. (a) Thermal dependence of the resistivity and (b) Arrhenius plot (Ln ρ vs. 1/T) for the salt κ'-(ET)$_4$[KMn(C$_2$O$_4$)$_3$]·PhCN (**55**) (red line is the fit to the Arrhenius equation for a classical semiconductor).

Band structure calculations indicate that there is a very strong intradimer (A-A) interaction corresponding to the overlap of the BEDT-TTF$^+$ molecules of the almost eclipsed face-to-face dimers. The transfer integrals are around ten times larger than those between the neutral isolated molecules and the dimers (A-B). The transfer integrals between the isolated neutral molecules (B-B) are very small except for one case [2]. The presence of (BEDT-TTF)$_2{}^{2+}$ dimers leads to the formation of a full band and an empty band, separated by a band gap, resulting in the observed semiconducting behavior in all these orthorhombic salts [4].

Table 3. Radical salts of the orthorhombic P. Day series κ'-(ET)$_4$[AM(C$_2$O$_4$)$_3$]·G (**55–63**).

#	CCDC	Formula [a]	SG [b]	Elect. Prop.	A$^+$	M	G	Ref.
55	CISMEZ	κ'-(ET)$_4$[KMn(ox)$_3$]·PhCN	*Pbcn*	$\sigma = 2 \times 10^{-5}$ S/cm $E_a = 180$ meV	K$^+$	Mn	PhCN	[33]
56	FECDEZ	κ'-(ET)$_4$[(NH$_4$)Rh(ox)$_3$]·PhCN	*Pbcn*	$E_a = 245$ meV	NH$_4{}^+$	Rh	PhCN	[32]
57	JUPGUW	κ'-(ET)$_4$[(H$_3$O)Cr(ox)$_3$]·PhCN	*Pbcn*	$E_a = 153$ meV	H$_3$O$^+$	Cr	PhCN	[4,25]
58	QIWMOY	κ'-(ET)$_4$[(NH$_4$)Co(ox)$_3$]·PhCN	*Pbcn*	$E_a = 225$ meV	NH$_4{}^+$	Co	PhCN	[4]
59	QIWMUE	κ'-(ET)$_4$[(NH$_4$)Al(ox)$_3$]·PhCN	*Pbcn*	$E_a = 222$ meV	NH$_4{}^+$	Al	PhCN	[4]
60	UDETOO	κ'-(ET)$_4$[K$_{0.8}$(H$_3$O)$_{0.2}$Ru(ox)$_3$]·PhCN	*Pbcn*	Semi	K$^+$/H$_3$O$^+$	Ru	PhCN	[31]
61	UJOXUN	κ'-(ET)$_4$[(H$_3$O)Fe(ox)$_3$]·(PhCN)$_{0.88}$(PhCl$_2$)$_{0.12}$	*Pbcn*	Semi	H$_3$O$^+$	Fe	PhCl$_2$/PhCN	[17]
62	ZIWNEY	κ'-(ET)$_4$[(NH$_4$)Fe(ox)$_3$]·PhCN	*Pbcn*	$\sigma = 10^{-4}$ S/cm $E_a = 140$ meV	NH$_4{}^+$	Fe	PhCN	[2,4]
63	ZIWNIC	κ'-(ET)$_4$[KFe(ox)$_3$]·PhCN	*Pbcn*	$\sigma = 10^{-4}$ S/cm $E_a = 140$ meV	K$^+$	Fe	PhCN	[2]

[a] ox = oxalate = C$_2$O$_4{}^{2-}$; [b] SG = space group.

4. Other Phases with BEDT-TTF and Oxalato Complexes

Besides the β'' and κ' phases, the intense research in the field has led to the synthesis of several other crystallographic phases with BEDT-TTF, different [M(C$_2$O$_4$)$_3$]$^{3-}$ anions (or the dianions [Ge(C$_2$O$_4$)$_3$]$^{2-}$ and [Cu(C$_2$O$_4$)$_2$]$^{2-}$), different solvents (including chiral ones) and even the inclusion of 18-crown-6 molecules. In this section, we will revise all these salts and classify them according to their composition and crystallographic phase.

4.1. BEDT-TTF Salts with [M(C$_2$O$_4$)$_3$]$^{3-}$ Anions and 18-Crown-6

The use of the crown ether 18-crown-6 in order to solubilize the precursor NH$_4{}^+$ salts of different [M(C$_2$O$_4$)$_3$]$^{3-}$ anions has led to the inclusion of the 18-crown-6 molecules into the crystal structure in a total of eight salts (Table 4). The first of these 18-crown-6-containing salts was also reported by P. Day's group in compound β''-(BEDT-TTF)$_4$[(H$_3$O)Cr(C$_2$O$_4$)$_3$]$_2$-

[(H$_3$O)$_2$(18-crown-6)]·5H$_2$O (**64**) [34]. The structure shows alternating cationic and anionic layers (Figure 14a). The anionic layers are formed by two layers of [M(C$_2$O$_4$)$_3$]$^{3-}$ anions separated by a layer with the 18-crown-6 molecules and the crystallization water molecules (Figure 14a). Each of the two [M(C$_2$O$_4$)$_3$]$^{3-}$ layers contains only one enantiomer (Figure 14b). The cationic layers are formed by two independent BEDT-TTF molecules with the β'' packing mode (Figure 14c) with chains of BEDT-TTF molecules following the sequence . . . AABB . . . with the same overlap mode (Figure 14d) as the superconducting monoclinic phase. This salt presents proton channels formed by the 18-crown-6 molecules containing H$_3$O$^+$ cations (Figure 14e) and presents a proton conductivity above 10^{-3} S/cm at room temperature. There is a Ga/NH$_4^+$ derivative with the same structure: β''-(ET)$_4$[(NH$_4$)Ga(C$_2$O$_4$)$_3$]$_2$[(NH$_4$)$_2$(18-crown-6)]·5H$_2$O (**68**), also reported by P. Day's group [35]. Interestingly, both salts are metallic at room temperature but show metal–insulator transitions at 190 and 240 K, respectively.

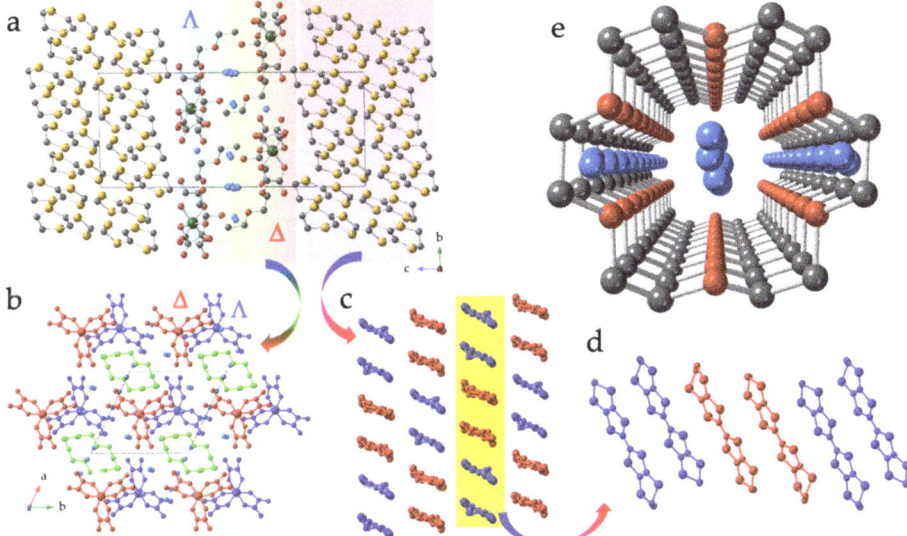

Figure 14. Structure of β''-(ET)$_4$[(H$_3$O)Cr(C$_2$O$_4$)$_3$]$_2$[(H$_3$O)$_2$(18-crown-6)]·5H$_2$O (**64**). (**a**) View of the alternating cationic and anionic layers parallel to the *ab* plane. (**b**) Top view of the anionic triple layer containing the ⊗-[Cr(C$_2$O$_4$)$_3$]$^{3-}$ anions (in red), the 18-crown-6 ether (in green) and the Λ-[Cr(C$_2$O$_4$)$_3$]$^{3-}$ anions (in blue). (**c**) View of the BEDT-TTF layers showing the β'' packing (A- and B-type molecules are drawn in red and blue, respectively). (**d**) Side view of one BEDT-TTF stack. (**e**) View of the proton channels formed by the 18-crown-6 molecules. Color code in (**a**,**e**): Cr = dark green, C = gray, O = red, O$_{water}$ = blue and S = yellow. H atoms are omitted for clarity.

A second phase obtained with 18-crown-6 is the series of salts formulated as β''-(ET)$_2$[(H$_2$O)(NH$_4$)$_2$M(C$_2$O$_4$)$_3$]·18-crown-6, with M = Ir (**65**) [36], Ru (**66**) [36], Cr (**67**) [37] and Rh (**69**) [38] (Table 3). The structure resembles that of the previously described compounds β''-(ET)$_4$[(A)M(C$_2$O$_4$)$_3$]$_2$[(A)$_2$(18-crown-6)]·5H$_2$O with A$^+$/M = H$_3$O$^+$/Cr (**64**) and NH$_4^+$/Ga (**68**). It also consists of alternating layers of BEDT-TTF molecules with the β'' packing motif and double anionic layers separated by the 18-crown-6 molecules, although, now, the 18-crown-6 molecules are inserted in the hexagonal cavities formed by the [M(C$_2$O$_4$)$_3$]$^{3-}$ anions.

The Cr and Rh salts are superconductors with T$_c$ = 4.0–4.9 K and 2.7 K, respectively [37,38]. This phase is the second one with a [M(C$_2$O$_4$)$_3$]$^{3-}$ complex to show superconductivity and is the phase with single-donor packing with the widest gap between conducting layers. The Ir and Ru salts are not superconductors due to the shorter interlayer M–M distance (worse 2D systems). The M–I transition in these salts may originate from changes in the anionic layer which induce changes in the BEDT-TTF layer, although Berezinski–Kosterlitz–Thouless (BKT) effects are not discarded [36]. The Ir salt is the first radical salt with a 5d tris(oxalato)metalate anion [36].

There are two additional radical salts with $[M(C_2O_4)_3]^{3-}$ anions containing 18-crown-6: α-(ET)$_{10}$(18-crown-6)$_6$K$_6$[Fe(C$_2$O$_4$)$_3$]$_4$·24H$_2$O (**70**) [39] and (ET)$_4$[Ga(C$_2$O$_4$)$_3$](18-crown-6)(H$_2$O)$_6$ (**71**). In compound **70**, the packing is original and consists of three types of alternating layers: BEDT-TTF layers (A-layers), layers containing [Fe(C$_2$O$_4$)$_3$]$^{3-}$ anions, water, 18-crown-6 and K$^+$ cations (B-layers) and layers with water, 18-crown-6 and K$^+$ cations (C-layers). These three layers alternate following the sequence . . . ABCBABCBA . . . This salt behaves as a classical semiconductor with an activation energy of 15 meV [39]. The last salt with 18-crown-6: (ET)$_4$[Ga(C$_2$O$_4$)$_3$](18-crown-6)(H$_2$O)$_6$ (**71**), has been deposited in the CCDC database but has not been published yet.

Table 4. Radical salts of BEDT-TTF with $[M(C_2O_4)_3]^{3-}$ anions containing 18-crown-6 (**64–71**).

#	CCDC	Formula [a]	SG [b]	Elect. Prop.	A$^+$	M	G	Ref.
64	ACAGUG	β''-(ET)$_4$[(H$_3$O)Cr(ox)$_3$]$_2$[(H$_3$O)$_2$(18-c-6)]·5H$_2$O	P-1	σ = 300 S/cm T_{MI} = 190 K	H$_3$O$^+$	Cr	H$_2$O/18-c-6	[34,35]
65	COLWUY	β''-(ET)$_2$[(H$_2$O)(NH$_4$)$_2$Ir(ox)$_3$]·18-c-6	P-1	$T_{MI} \approx$ 100 K	NH$_4^+$	Ir	H$_2$O/18-c-6	[36]
66	COLYOU	β''-(ET)$_2$[(H$_2$O)(NH$_4$)$_2$Ru(ox)$_3$]·18-c-6	P-1	$T_{MI} \approx$ 155 K	NH$_4^+$	Ru	H$_2$O/18-c-6	[36]
67	FENHEO	β''-(ET)$_2$[(H$_2$O)(NH$_4$)$_2$Cr(ox)$_3$]·18-c-6	P-1	T_c = 4.0–4.9 K	NH$_4^+$	Cr	H$_2$O/18-c-6	[37]
68	FEQQAU	β''-(ET)$_4$[(NH$_4$)Ga(ox)$_3$]$_2$[(NH$_4$)$_2$(18-c-6)]·5H$_2$O	P-1	σ = 200 S/cm T_{MI} = 240 K	NH$_4^+$	Ga	H$_2$O/18-c-6	[35]
69	KATLAV	β''-(ET)$_2$[(H$_2$O)(NH$_4$)$_2$Rh(ox)$_3$]·18-c-6	P-1	T_c = 2.7 K	NH$_4^+$	Rh	H$_2$O/18-c-6	[38]
70	NIHPEA	α-(ET)$_{10}$(18-c-6)$_6$K$_6$[Fe(ox)$_3$]$_4$·24H$_2$O	P2$_1$/c	E_a = 105 meV	K$^+$	Fe	H$_2$O/18-c-6	[39]
71	UJEYIR	(ET)$_4$[Ga(ox)$_3$](18-c-6)(H$_2$O)$_6$	P-1	-	-	Ga	H$_2$O/18-c-6	[c]

[a] ox = oxalate = C$_2$O$_4$$^{2-}$; 18-c-6 = 18-crown-6 = C$_{12}$H$_{24}$O$_6$; [b] SG = space group. [c] Unpublished results.

4.2. BEDT-TTF Salts with $[M(C_2O_4)_3]^{3-}$ Anions and Two Different Donor Layers

The use of larger non-planar solvent molecules and chiral ones has led to some nice examples of double-layered phases. These chiral and large non-planar solvent molecules do not fit in the hexagonal cavities of the anionic layers and are forced to cross the cavities, resulting in anionic layers with two different sides. This asymmetry of the two sides of the anionic layers induces two different packing modes in the BEDT-TTF molecules, resulting in salts with two different cationic layers. There are two different series of double-layered phases: the α,β'', and the α,κ'.

The α,β'' phase has been observed in the series α,β''-(ET)$_4$[(NH$_4$)M(C$_2$O$_4$)$_3$]·G with M/G = Ga/PhN(CH$_3$)CHO (**72**) [40], M/G = Ga/PhCH$_2$CN (**73**) [40], M/G = Fe/PhCOCH$_3$ (**74**) [40], M/G = Fe/(S)-PhC(OH)HCH$_3$ (**75**) [41] and M/G = Fe/(R/S)-PhC(OH)HCH$_3$ (**77**) [41] (Table 5). These salts are isostructural and crystallize in the triclinic P-1 space group, except the salt α,β''-(ET)$_4$[(NH$_4$)Fe(C$_2$O$_4$)$_3$]·(S)-PhC(OH)HCH$_3$ (**75**) that contains a single enantiomer (S) of the chiral solvent PhC(OH)HCH$_3$ and crystallizes in the P1 space group. The α,β'' salts show alternating cationic and anionic layers following the sequence . . . /β''/⊗/α/Λ/ . . . (Figure 15a). The anionic layers are very similar to those of the monoclinic β'' phase: they contain the $[M(C_2O_4)_3]^{3-}$ anions and the NH$_4^+$ cations forming a hexagonal honeycomb lattice with the solvent molecules located in the hexagonal cavities. There are two homochiral layers that alternate along the *c* axis (Figure 15b,c), each with a different [M(C$_2$O$_4$)$_3$]$^{3-}$ enantiomer and a slightly different orientation of the polar group of the solvent molecule. The main difference with the monoclinic β" phase is that, now, there are two different sides since the large solvent molecules cross the hexagonal cavities with the X groups pointing to one of the two sides to generate a corrugated side, next to the β" layer, and a smoother face, next to the α layer (Figure 15a). The polar groups of the solvent molecules point to the β" layer (purple circles in Figure 15a).

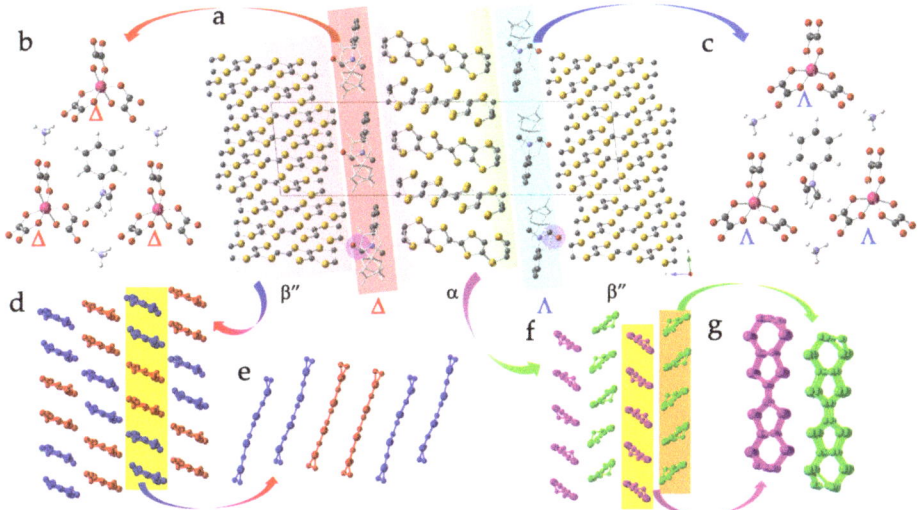

Figure 15. Structure of α,β"-(ET)$_4$[(NH$_4$)Ga(C$_2$O$_4$)$_3$]·PhN(CH$_3$)CHO (**72**): (**a**) View of the alternating cationic and anionic layers following the sequence . . . β"-⊗-α-Λ-β" . . . The purple circles mark the position of the CHO groups of the solvent molecules pointing to the β" layer. (**b**) View of the ⊗ layer. (**c**) View of the Λ layer. (**d**) View of the β" layer showing the two independent BEDT-TTF molecules (A and B) in red and blue, respectively. (**e**) Side view of one BEDT-TTF chain. (**f**) View of the α layer showing the two independent BEDT-TTF molecules (C and D) in pink and green, respectively. (**g**) Frontal view of the C- and D-type chains. Color code in (**a–c**): Ga = pink, C = gray, O = red, N = blue, H = white and S = yellow. Except in (**b,c**), H atoms are omitted for clarity.

There are two different cationic layers, with β" and α packing modes, alternating along the *c* axis. The β" layer (Figure 15d) is formed by two independent BEDT-TTF molecules (A and B) packing in chains with the sequence . . . AABB . . . , as in the monoclinic β" phase. The only difference is that, now, the overlap between the BEDT-TTF molecules in the chain is eclipsed for AA and AB but shifted for BB. This overlap mode generates groups of four eclipsed BEDT-TTF molecules (Figure 15e). In the α layer, the molecules of consecutive stacks are tilted in opposite directions (Figure 15f). There are also two independent molecules (C and D) that form two different stacks containing only C or only D molecules packed with an eclipsed overlap (Figure 15g).

Compounds **72**-**74**, also reported by P. Day's group [40], were the first examples of the α,β" phase (which is the third phase found in the ET$_4$ series, after the β" and κ' ones). All the reported α,β" salts show relatively high room temperature conductivities but are semiconductors, although band structure calculations indicate that both BEDT-TTF layers should be metallic [40].

The other double-layer phase is the α,κ' one, reported in two isostructural compounds with the same solvent molecule (1,2-dibromobenzene): α,κ'-(ET)$_4$[K$_{0.45}$(H$_3$O)$_{0.55}$Ga(C$_2$O$_4$)$_3$]·1,2-PhBr$_2$ (**76**) [30] and α,κ'-(ET)$_4$[(H$_3$O)Fe(C$_2$O$_4$)$_3$]·1,2-PhBr$_2$ (**78**) [42] (Table 5). The structure of these salts also consists of alternating cationic and anionic layers (Figure 16a). There are also two alternating cationic layers with two different packing motifs: κ' and α (Figure 16b,c). The κ' layers are formed by four independent BEDT-TTF molecules (A–D) forming two different face-to-face dimers (AA and CC) surrounded by isolated BEDT-TTF molecules (B and D) (Figure 16b). The α layers also contain four independent BEDT-TTF molecules (E–H) forming three different stacks containing only E, only H and alternating F/G molecules (Figure 16c). The BEDT-TTF molecules are packed in an eclipsed way in the three stacks. The anionic layers present the same hexagonal disposition observed in the β" and κ' phases and contain both enantiomers of the $[M(C_2O_4)_3]^{3-}$ anions (arranged in parallel rows, as observed in the κ' phase), together with A^+ cations (Figure 16d). The 1,2-PhBr$_2$ solvent molecules are located in the cavities with the two Br atoms pointing towards the α layer (Figure 16a). This asymmetry in both faces of the anionic layers induces the crystallization of two different layers, as observed in the α,β" phase. Both salts are metallic down to low temperatures, although they do not show superconductivity. This metallic behavior is attributed to the α layers that show a homogeneous charge distribution in the BEDT-TTF molecules, in contrast to the κ' layers that present the same charge localization observed in the other κ' phases: (BEDT-TTF)$_2^{2+}$ dimers surrounded by neutral BEDT-TTF molecules.

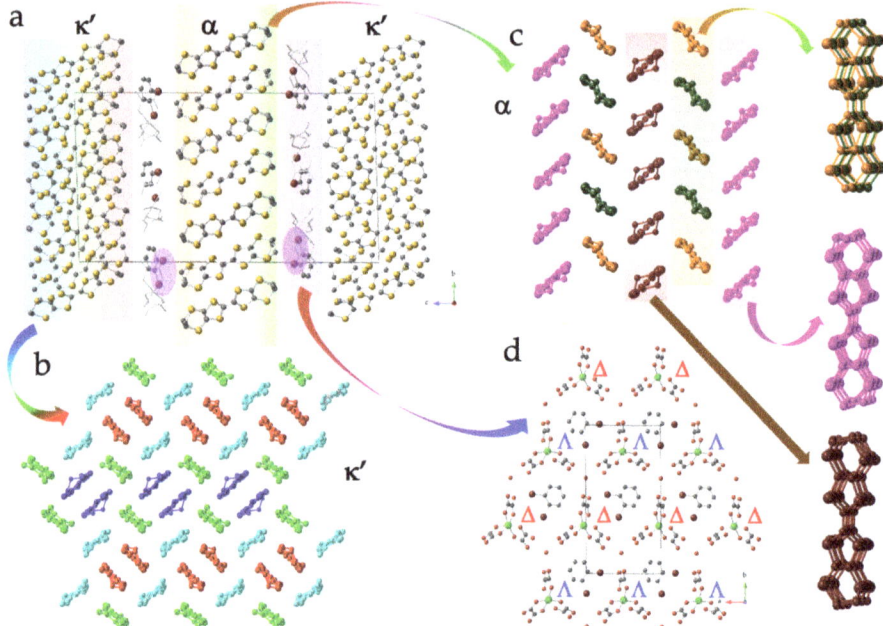

Figure 16. Structure of α,κ'-(ET)$_4$[(H$_3$O)Fe(C$_2$O$_4$)$_3$]·1,2-PhBr$_2$ (**78**): (**a**) View of the alternating cationic and anionic layers following the sequence κ'-⊗/Λ-α-⊗/Λ-κ'. The purple circles mark the position of the two Br atoms of the solvent molecules pointing to the α layer. (**b**) View of the κ' layer showing the four independent BEDT-TTF molecules in red, blue, green and light blue. (**c**) View of the α layer showing the four independent BEDT-TTF molecules in pink, orange, brown and dark green and the three eclipsed stacks. (**d**) View of the anionic layer with parallel rows of ⊗ and Λ enantiomers. Color code in (**a**,**d**): Fe = green, C = gray, O = red, Br = brown and S = yellow. H atoms are omitted for clarity.

Table 5. Radical salts of BEDT-TTF with [M(C₂O₄)₃]³⁻ anions with double-layer phases (72–78).

#	CCDC	Formula [a]	SG [b]	Elect. Prop.	A⁺	M	G	Ref.
72	AQUZUH	α,β''-(ET)₄[(NH₄)Ga(ox)₃]·PhN(CH₃)CHO	P-1	σ = 0.26–0.60 S/cm	NH₄⁺	Ga	PhN(CH₃)CHO	[40]
73	ARABAW	α,β''-(ET)₄[(NH₄)Ga(ox)₃]·PhCH₂CN	P-1	σ = 0.24–1.34 S/cm	NH₄⁺	Ga	PhCH₂CN	[40]
74	ARABEA	α,β''-(ET)₄[(NH₄)Fe(ox)₃]·PhCOCH₃	P-1	-	NH₄⁺	Fe	PhCOCH₃	[40]
75	CILDIL	α,β''-(ET)₄[(NH₄)Fe(ox)₃]·(S)-PhC(OH)HCH₃	P1	σ = 5.3 S/cm, T_{MI} = 170 K	NH₄⁺	Fe	(S)-PhC(OH)OCH₃	[41]
76	HOBRIB	α,κ'-(ET)₄K₀.₄₅(H₃O)₀.₅₅[Ga(ox)₃]·1,2-PhBr₂	P-1	Metal	K⁺/H₃O⁺	Ga	1,2-PhBr₂	[30]
77	NIPTEM	α,β''-(ET)₄[(NH₄)Fe(ox)₃]·(R/S)-PhC(OH)HCH₃	P-1	σ = 10.4 S/cm, T_{MI} = 150 K	NH₄⁺	Fe	(R/S)-PhC(OH)HCH₃	[41]
78	TANDIX	α,κ'-(ET)₄[(H₃O)Fe(ox)₃]·1,2-PhBr₂	P-1	M > 1.5 K	H₃O⁺	Fe	PhBr₂	[42]

[a] ox = oxalate = $C_2O_4^{2-}$; [b] SG = space group.

4.3. BEDT-TTF:[M(C$_2$O$_4$)$_3$]$^{3-}$ Phases with 3:1 Stoichiometry

Although most of the prepared radical salts of BEDT-TTF with [M(C$_2$O$_4$)$_3$]$^{3-}$ anions present a 4:1 stoichiometry, as observed in the β'', κ', α,β'' and α,κ' phases and even in some salts with 18-crown-6, there are also radical salts with a 3:1 stoichiometry and the general formula (ET)$_3$[AM(C$_2$O$_4$)$_3$]·G with A$^+$ = Na$^+$ and NH$_4$$^+$; M = Cr and Al and small solvent molecules in all cases (G = CH$_3$NO$_2$, CH$_2$Cl$_2$, CH$_3$CN, EtOH and dmf, Table 6). Interestingly, all these salts have been prepared with chiral solvents or mixtures of chiral and non-chiral solvents.

By using the chiral solvent (R)-(-)-carvone with small solvent molecules such as CH$_3$CN and CH$_3$NO$_2$, P. Day and L. Martin obtained a series of three isostructural chiral salts formulated as (BEDT-TTF)$_3$[NaM(C$_2$O$_4$)$_3$]·G with M/G = Al/CH$_3$NO$_2$ (**79**) [43], Cr/CH$_3$CN (**84**) [44] and Cr/CH$_3$NO$_2$ (**86**) [45] that crystallize in the monoclinic chiral $P2_1$ space group (Table 6). The structure of this 3:1 phase also consists of alternating BEDT-TTF and anionic layers (Figure 17b). The BEDT-TTF layers contain three independent BEDT-TTF molecules arranged in rows containing BC dimers alternating with A monomers tilted ca. 45° with respect to the dimers (Figure 17a). This disposition resembles the κ' phase, but now each dimer is surrounded by four monomers and four dimers (compared to six monomers in the κ' phase). The anionic layer contains Na$^+$ cations and a single enantiomer of the [M(C$_2$O$_4$)$_3$]$^{3-}$ anions, forming a hexagonal honeycomb lattice (Figure 17c). The small solvent molecules (CH$_3$CN or CH$_3$NO$_2$) are located in the center of the hexagons. It is to be noted that now the hexagons are smaller and, accordingly, the organic layers contain only three BEDT-TTF molecules per hexagonal cavity. The inhomogeneous charge distribution with (BEDT-TTF)$_2$$^{2+}$ dimers and neutral monomers and the lack of short intermolecular contacts explain the semiconductor behavior observed in these salts [43–45].

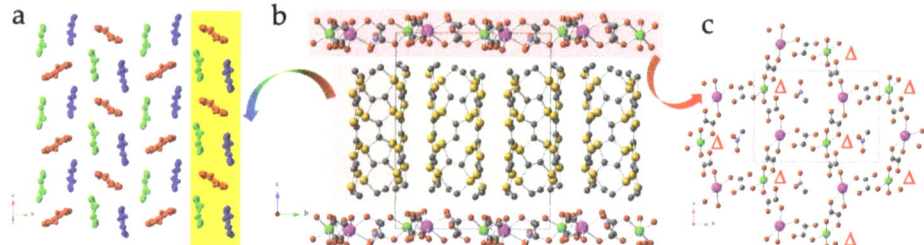

Figure 17. Structure of (ET)$_3$[NaAl(C$_2$O$_4$)$_3$]·CH$_3$NO$_2$ (**79**): (**a**) View of the BEDT-TTF layer showing the rows with alternating dimers and monomers. The three independent BEDT-TTF molecules are depicted in red, blue and green. (**b**) View of the alternating cationic and anionic layers. (**c**) View of the homochiral anionic layer. Color code in (**b**,**c**): Al = green, Na = pink, C = gray, O = red, N = blue and S = yellow. H atoms are omitted for clarity.

A very similar structure was found by L. Martin et al. in compounds (BEDT-TTF)$_3$[LiM(C$_2$O$_4$)$_3$]·EtOH with M = Cr (**81**) and Fe (**82**) [46]. Although these compounds were also prepared with a chiral solvent (Λ-carvone), there is a disorder between the Li$^+$ and M^{III} centers that leads to a non-chiral $P2_1/c$ space group. The BEDT-TTF and the anionic layers show the same structure as compounds **79**, **84** and **86** (Figure 17), although they are not, strictly speaking, isostructural since compounds **81** and **82** are not chiral.

L. Martin et al., using the same chiral solvent (R)-(-)-carvone, also obtained two isostructural chiral salts formulated as (BEDT-TTF)$_3${Na[⊗-Cr(C$_2$O$_4$)$_3$]$_{0.64}$[Λ-Cr(C$_2$O$_4$)$_3$]$_{0.36}$}·CH$_3$NO$_2$ (**85**) [45] and (BEDT-TTF)$_3$[(NH$_4$)$_{0.83}$Cr$_{1.17}$(C$_2$O$_4$)$_3$]·CH$_3$NO$_2$ (**80**) [43] that crystallize in the orthorhombic $P2_12_12_1$ chiral space group. The structure of these two salts also consists of alternating anionic and cationic layers, but now the organic layers present a different packing with double rows of face-to-face BEDT-TTF dimers (A-B) alternating with single rows of isolated BEDT-TTF molecules (C) tilted ca. 90° with respect to the dimers (Figure 18a). The charge distribution, determined from the bond distances in the BEDT-TTF molecules, indicates an inhomogeneous charge distribution with charges of 0.33, 0.63 and 0.88 for the three independent molecules. The anionic layer is similar to that observed in the $P2_1$ salts with the same stoichiometry (Figure 17c) and also contains a single enantiomer of the $[Cr(C_2O_4)_3]^{3-}$ anion. In compound **80**, there is an excess of the $[Cr(C_2O_4)_3]^{3-}$ anion that results in a different charge distribution in the BEDT-TTF molecules, although no electrical properties are reported. The isostructural NaCr derivative (**85**) [45] presents an excess of the ⊗ enantiomer, due to a disorder of the Cr and Na positions, and a slightly different charge distribution in the BEDT-TTF molecules since, now, the anionic layers have a charge of -2 per formula unit.

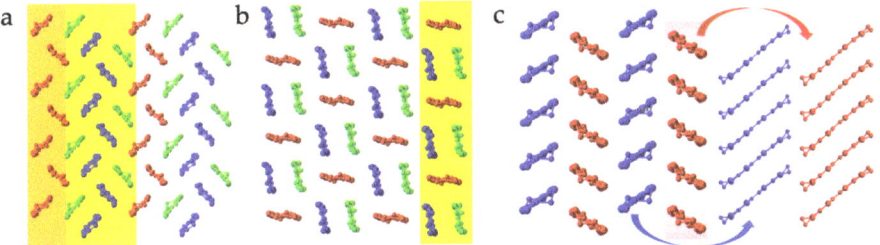

Figure 18. (**a**) Structure of the organic layer in $(BEDT-TTF)_3[(NH_4)_{0.83}Cr_{1.17}(C_2O_4)_3]\cdot CH_3NO_2$ (**80**) showing the double stacks with dimers (in green and blue) and monomers (in red). (**b**) Structure of the organic layer in $(BEDT-TTF)_3[NaCr(C_2O_4)_3]\cdot EtOH$ (**87**) showing the stacks with dimers (in green and blue) and monomers (in red). (**c**) Structure of the organic layer in θ-$(BEDT-TTF)_3[NaCr(C_2O_4)_3]\cdot dmf$ (**88**) showing the alternating stacks (in red and blue). H atoms are omitted for clarity.

L. Martin et al. also used the chiral solvent R-(-)-carvone with small solvent molecules such as CH_2Cl_2 and EtOH to prepare the isostructural salts $(ET)_3[Na(\otimes\text{-}Cr(C_2O_4)_3)_{0.56}(\Lambda\text{-}Cr(C_2O_4)_3)_{0.44}]\cdot CH_2Cl_2$ (**83**) [47] and $(ET)_3[NaCr(C_2O_4)_3]\cdot EtOH$ (**87**) [44] that crystallize in the triclinic $P1$ chiral space group. Salt **83** was the first chiral BEDT-TTF salt with $[M(C_2O_4)_3]^{3-}$ anions and also the first radical salt in this series with Na^+ ions in the anionic layer [47]. The structure of these two salts also consists of alternating anionic and cationic layers, but now the organic layers present a very slightly different packing to those of the $P2_1$ salts with similar compositions but different solvent molecules (**79**, **84** and **86**). In the $P1$ salts (**83**, **87**), the organic layers are also formed by rows containing alternating BEDT-TTF dimers and monomers, but now the monomers are tilted ca. 80° with respect to the dimers (compared to ca. 45° in the $P2_1$ salts, Figure 17b), and the offset inside the dimers is smaller in the $P1$ salts. Consecutive stacks are shifted to form a chessboard of monomers and dimers (Figure 18b).

Finally, when using dmf and the chiral solvent R-(-)-carvone, L. Martin et al. also obtained a 3:1 salt formulated as θ-$(BEDT-TTF)_3[NaCr(C_2O_4)_3]\cdot dmf$ (**88**) [44] that also crystallizes in the triclinic $P1$ chiral space group (as the CH_2Cl_2 and EtOH derivatives) but shows a completely different packing in the BEDT-TTF layer. Thus, salt **88** has only two independent BEDT-TTF molecules that pack in segregated parallel stacks with a tilt angle close to 116° between molecules of different stacks, giving rise to the so-called θ phase (Figure 18c). The two stacks show a similar shift of the BEDT-TTF molecules of half-rings along the stack. The anionic layer contains a single enantiomer of the $[Cr(C_2O_4)_3]^{3-}$ anions and Na^+ cations with the classical honeycomb structure [44].

Table 6. Radical salts of BEDT-TTF with $[M(C_2O_4)_3]^{3-}$ anions with 3:1 stoichiometry (**79–88**).

#	CCDC	Formula [a]	SG [b]	Elect. Prop.	A^+	M	G	Ref.
79	BOYTIU	$(ET)_3[NaAl(ox)_3] \cdot CH_3NO_2$	$P2_1$	$E_a = 140$ meV	Na^+	Al	CH_3NO_2	[43]
80	BOYTOA	$(ET)_3[(NH_4)_{0.83}Cr_{1.17}(ox)_3] \cdot CH_3NO_2$	$P2_12_12_1$	$E_a = 140$ meV	Na^+	Cr	CH_3NO_2	[43]
81	DUDWUW	$(ET)_3[LiCr(ox)_3] \cdot EtOH$	$P2_1/c$	$E_a = 179$ meV	Li^+	Cr	EtOH	[46]
82	DUDXAD	$(ET)_3[LiFe(ox)_3] \cdot EtOH$	$P2_1/c$	$E_a = 126$ meV	Li^+	Fe	EtOH	[46]
83	DUXNOA	$(ET)_3[Na(\Delta\text{-}Cr(ox)_3)_{0.56}(\Lambda\text{-}Cr(ox)_3)_{0.44}] \cdot CH_2Cl_2$	$P1$	$E_a = 69$ meV	Na^+	Cr	CH_2Cl_2	[47]
84	KOGMUQ01	$(ET)_3[NaCr(ox)_3] \cdot CH_3CN$	$P2_1$	$\sigma = 0.038$ S/cm $E_a = 172$ meV	Na^+	Cr	CH_3CN	[44]
85	XUNXOU	$(ET)_3[Na[\otimes\text{-}Cr(ox)_3]_{0.64}[\Lambda\text{-}Cr(ox)_3]_{0.36}] \cdot CH_3NO_2$	$P2_12_12_1$	$\sigma = 0.5$ S/cm $E_a = 80$ meV	Na^+	Cr	CH_3NO_2	[45]
86	XUNXOU01	$(ET)_3[NaCr(ox)_3] \cdot CH_3NO_2$	$P2_1$	$\sigma = 0.045$ S/cm $E_a = 79$ meV	Na^+	Cr	CH_3NO_2	[45]
87	YUCLOZ	$(ET)_3[NaCr(ox)_3] \cdot EtOH$	$P1$	Semi	Na^+	Cr	EtOH	[44]
88	YUCLUF	$\theta\text{-}(ET)_3[NaCr(ox)_3] \cdot dmf$	$P1$	Semi	Na^+	Cr	dmf	[44]

[a] ox = oxalate = $C_2O_4^{2-}$; [b] SG = space group.

The lack of a homogeneous charge distribution on the BEDT-TTF molecules and the absence of short intermolecular contacts in all these chiral salts preclude the existence of an electron delocalization, and, accordingly, all these 3:1 salts obtained with chiral solvents are semiconductors or insulators (Table 6).

4.4. Other Phases of BEDT-TTF Salts with $[M(C_2O_4)_3]^{3-}$ Anions

Besides all the above-described phases, the combination of BEDT-TTF donors with $[M(C_2O_4)_3]^{3-}$ anions has also led to other crystal phases with unusual stoichiometries and/or packings in the cationic and anionic layers (Table 7).

The first of these salts is α'''-(ET)$_9$Na$_{18}$[Fe(C$_2$O$_4$)$_3$]$_8$·24H$_2$O (**95**) [39], also obtained with the $[Cr(C_2O_4)_3]^{3-}$ anion in α'''-(ET)$_9$Na$_{18}$[Cr(C$_2$O$_4$)$_3$]$_8$·24H$_2$O (**91**) [47]. These salts present a very original structure with four different layers following the sequence ... ABCD ... alternating along the c direction (Figure 19b). A-type layers contain BEDT-TTF molecules with a very unusual α''' packing (Figure 19a), B and D layers contain Na$^+$ cations and $[Fe(C_2O_4)_3]^{3-}$ anions with one single enantiomer in each layer (Figure 19c) and C layers contain Na$^+$ cations and H$_2$O molecules in rows parallel to the b axis (Figure 19d) [39]. In the very unusual α''' packing, the BEDT-TTF molecules form parallel columns as in the α and β'' packings, but in one of every three columns, the BEDT-TTF molecules are tilted in the opposite direction (Figure 19a). There are, thus, two columns with one orientation (++) formed by three independent BEDT-TTF molecules (A–C) with the sequence ... ABC ..., and one with the opposite orientation (-) formed by two independent BEDT-TTF molecules (D and E) with the sequence ... DDE ... In both stacks (+ and -), the BEDT-TTF molecules are eclipsed. The α''' packing is, therefore, described as parallel columns following the sequence ... /+/+/-/+/+/-/ ... The anionic layer is also original since, now, the vertices of the hexagons contain alternating Fe(III) ions and (Na$^+$)$_2$ dimers and there is an additional $[Fe(C_2O_4)_3]^{3-}$ anion in the center of the hexagon with its oxalate ligands pointing towards the Na$^+$ dimers (Figure 19c). Interestingly, this unusual disposition has recently been found in a salt with the $[Fe(NA)_3]^{3-}$ anion (NA = nitranilato ligand = $C_6O_4(NO_2)_2)^{2-}$) [48], which is topologically identical to the oxalate ligand and also forms tris-chelato anionic complexes $[M^{III}(L)_3]^{3-}$ (L = anilato-type ligand) [49] and even $[A^+M^{III}(L_3)]^{2-}$ honeycomb layers [50], such as the ones here described for oxalate in previous sections. This original salt, **95**, has an inhomogeneous charge distribution in the five independent BEDT-TTF molecules, and, therefore, it is a semiconductor.

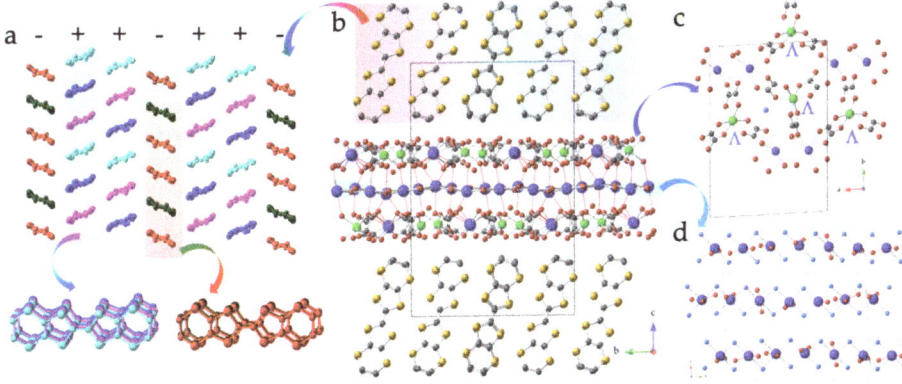

Figure 19. Structure of α'''-(ET)$_9$Na$_{18}$[Fe(C$_2$O$_4$)$_3$]$_8$·24H$_2$O (**95**): (**a**) View of the BEDT-TTF layer showing the two rows (++) with one orientation of the BEDT-TTF molecules (blue, pink and light blue molecules) and the row (-) with the opposite orientation (green and red molecules) and their eclipsed overlap. (**b**) View of the alternating cationic and anionic/Na$^+$/anionic layers. (**c**) View of one of the two layers containing the $[Fe(C_2O_4)_3]^{3-}$ anions and the Na$^+$ cations. (**d**) View of the layer with the Na$^+$ ions and the water molecules. Color code: Fe = green, Na = blue, C = gray, O = red, O$_{water}$ = light blue and S = yellow. H atoms are omitted for clarity.

Salts α-(BEDT-TTF)$_{12}$[Fe(C$_2$O$_4$)$_3$]$_2$·15H$_2$O (**93**) and α-(BEDT-TTF)$_{12}$[Fe(C$_2$O$_4$)$_3$]$_2$·16H$_2$O (**94**) are two solvates that, despite differing only in one water molecule, show different unit cell parameters and, therefore, are not isostructural [51]. The structure of both salts consists of cationic and anionic layers alternating along the *c* axis. Both salts show the α packing mode in the BEDT-TTF layer, formed by parallel stacks with an alternating orientation of the BEDT-TTF molecules. The only difference between both salts is that there are twelve independent BEDT-TTF molecules in **93**, compared to only three in salt **94**. The anionic layers in both compounds are formed by isolated [Fe(C$_2$O$_4$)$_3$]$^{3-}$ anions surrounded by water molecules connected through H bonds in a 2D lattice (Figure 20a). The only difference is the presence of an extra water molecule in **94** and a slight twist in one oxalato ligand. Although the average charge per BEDT-TTF molecule is +0.5, there is an inhomogeneous charge distribution that results in a semiconducting behavior in both salts [51].

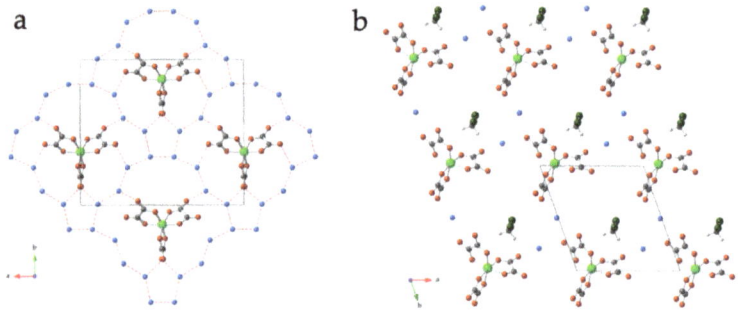

Figure 20. (**a**) View of the anionic layer in salt α-(BEDT-TTF)$_{12}$[Fe(C$_2$O$_4$)$_3$]$_2$·15H$_2$O (**93**) showing the [Fe(C$_2$O$_4$)$_3$]$^{3-}$ anions and the water molecules around the anions. Red dotted lines represent the H bonds between the water molecules. (**b**) View of the anionic layers in β″-(BEDT-TTF)$_5$[Fe(C$_2$O$_4$)$_3$]·2H$_2$O·CH$_2$Cl$_2$ (**96**). Color code: Fe = green, C = gray, O = red, O$_{water}$ = light blue and Cl = dark green.

When using (NEt$_4$)$_3$[Fe(C$_2$O$_4$)$_3$] as a precursor salt and CH$_2$Cl$_2$ as a solvent, the lack of small cations in the medium such as H$_3$O$^+$, Na$^+$, K$^+$ or NH$_4^+$ results in an original salt with no other cations in the structure. This salt, formulated as β″-(BEDT-TTF)$_5$[Fe(C$_2$O$_4$)$_3$]·2H$_2$O·CH$_2$Cl$_2$ (**96**) [52], shows cationic and anionic layers alternating along the *c* direction. The cationic layers of this 5:1 salt contain five independent BEDT-TTF molecules with the β″ packing mode, as in the superconducting 4:1 phase. The anionic layer is quite original since it contains isolated [Fe(C$_2$O$_4$)$_3$]$^{3-}$ anions packed in parallel rows, surrounded by the H$_2$O and CH$_2$Cl$_2$ molecules (Figure 20b). There are two different anionic layers with only one enantiomer each. The analysis of the bond distances shows an inhomogeneous charge distribution with four BEDT-TTF molecules with a charge of +0.5 and one BEDT-TTF with a charge of +1. This inhomogeneous charge distribution explains the semiconducting behavior of this salt [52].

Another original salt is α-(BEDT-TTF)$_6$[Fe(C$_2$O$_4$)$_3$] (**92**) [17], which shows a 6:1 stoichiometry with an α packing type in the BEDT-TTF layer. Unfortunately, the crystals are not stable, precluding the determination of the anionic layer. The use of a large solvent such as 1,2,4-trichlorobenzene may be at the origin of the unusual structure of this salt [17].

The use of Li$_3$[Fe(C$_2$O$_4$)$_3$] as a precursor salt, besides the above-mentioned 3:1 phases (ET)$_3$[LiM(C$_2$O$_4$)$_3$]·EtOH (**81** and **82**) obtained with a chiral solvent, also gave crystals of a 4:1 phase: η-(ET)$_4$[(H$_2$O)LiFe(C$_2$O$_4$)$_3$] (**90**), when no chiral solvent was used [46]. The structure of this salt consists of cationic and anionic layers alternating along the *b* axis. In this salt, the BEDT-TTF layer shows an original η packing mode (Figure 21a) formed by parallel stacks of eclipsed BEDT-TTF molecules (Figure 21b) tilted with respect to the stack direction (as in the β″ phase), but now the orientation of the molecules changes every two stacks, following the sequence: ... /+/+/−/−/ ... (Figure 21a). This original packing can be considered as a mixture of the β″ phase, where all the stacks have the same orientation: (+/+/+/+), and the α phase, where the stacks show alternating orientations: (+/−/+/−). The two stacks with a given orientation are equivalent and are formed by two independent BEDT-TTF molecules following the sequence ... AB ... (or ... CD ... in the stacks with the opposite orientation, Figure 21a). The BEDT-TTF molecules are eclipsed in both types of stacks (Figure 21b). The anionic layer is also very original: it shows a distorted hexagonal packing of the $[Fe(C_2O_4)_3]^{3-}$ anions and the Li^+ cations. This distortion is due to the presence of a water molecule in the layer connected to the Li^+ ions, resulting in elongated $Li(H_2O)^+$ entities that occupy alternating vertices of the hexagons (Figure 21c). The elongated hexagonal cavities are occupied by disordered CH_2Cl_2 solvent molecules. The BEDT-TTF molecules show an inhomogeneous charge distribution, and, accordingly, this original salt is a semiconductor (Table 7) [46].

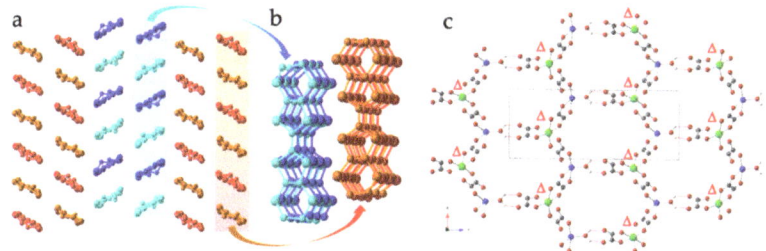

Figure 21. Structure of η-$(ET)_4[(H_2O)LiFe(C_2O_4)_3]$ (**90**). (**a**) View of the cationic layer showing the four independent BEDT-TTF molecules (as light blue, dark blue, red and orange). (**b**) View of the eclipsed packing of the BEDT-TTF molecules. (**c**) View of the anionic layer showing the elongated hexagonal cavities. Thin red lines represent the H bonds between the water molecule coordinated to the Li^+ ions and the terminal oxalate oxygen atoms. Color code in (**c**): Fe = green, Li = blue, C = gray and O = red. H atoms in (**a**) are omitted for clarity.

A final example of a radical salt with an unusual structure is a 5:1 phase formulated as α″-$(BEDT-TTF)_5[Ga(C_2O_4)_3]\cdot 3.4H_2O\cdot 0.6EtOH$ (**89**) [27]. This salt is the first orthorhombic *Pbca* BEDT-TTF salt with an oxalate complex. The structure consists of cationic and anionic layers alternating along the c axis (Figure 22c). The cationic layers show an α″ or η packing mode (Figure 22a), previously observed in η-$(BEDT-TTF)_4[(H_2O)LiFe(C_2O_4)_3]$ (**90**) [46], although in salt **89**, each BEDT-TTF stack contains five independent molecules with the sequence ... ABCDE ... , showing a dislocation every five molecules along the stack (Figure 22b), whereas in compound **90**, there are four BEDT-TTF independent molecules in two different stacks with the sequences ... AB ... and ... CD ... The anionic layers are formed by rows along the a direction with alternating ⊗ and Λ enantiomers of the $[Ga(C_2O_4)_3]^{3-}$ anions (Figure 22d).

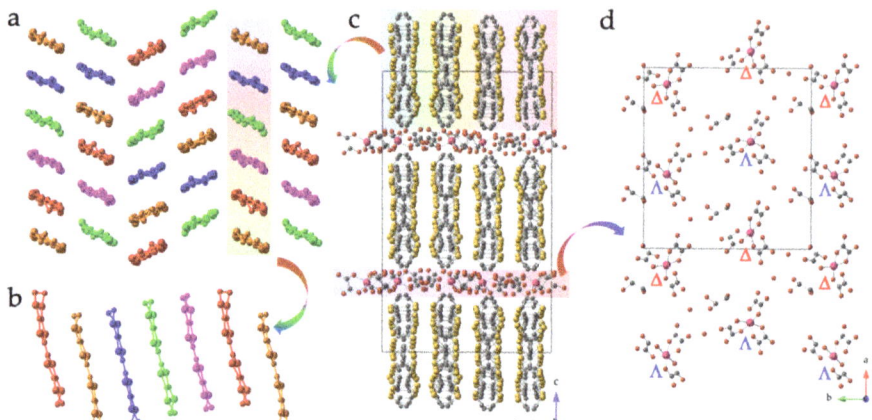

Figure 22. Structure of α''-(BEDT-TTF)$_5$[Ga(C$_2$O$_4$)$_3$]·3.4H$_2$O·0.6EtOH (**89**). (**a**) View of the cationic layer showing the five independent BEDT-TTF molecules (as blue, red, orange, green and pink). (**b**) Side view of the BEDT-TTF stack showing the pentameric repeating unit. (**c**) View of the alternating cationic and anionic layers along the *c* direction. (**d**) View of the anionic layer showing the rows along the *a* axis with alternating ⊗ and Λ [Ga(C$_2$O$_4$)$_3$]$^{3-}$ anions. Color code in (**b**): Ga = pink, C = gray and O = red. H atoms are omitted for clarity.

Table 7. Radical salts of BEDT-TTF with [M(C$_2$O$_4$)$_3$]$^{3-}$ anions showing other α and β phases (**89–96**).

#	CCDC	Formula [a]	SG [b]	Elect. Prop.	A$^+$	M	G	Ref.
89	CIWNAA	α''-(ET)$_5$[Ga(ox)$_3$]·3.4H$_2$O·0.6EtOH	*Pbca*	E_a = 71 meV	-	Ga	EtOH/H$_2$O	[27]
90	DUDWOQ	η-(ET)$_4$[(H$_2$O)LiFe(ox)$_3$]	*P2$_1$/c*	σ = 0.41 S/cm E_a = 80 meV	Li$^+$/H$_2$O	Fe	-	[46]
91	DUXNUG	α'''-(ET)$_9$Na$_{18}$[Cr(ox)$_3$]$_8$·24H$_2$O	*P-1*	E_a = 66 meV	Na$^+$	Cr	H$_2$O	[47]
92	IPEKAQ	α-(ET)$_6$[Fe(ox)$_3$]	*P2$_1$*	-	-	Fe	H$_2$O/EtOH ?	[17]
93	KIVKAC	α-(ET)$_{12}$[Fe(ox)$_3$]$_2$·15H$_2$O	*C2/c*	σ = 0.055 S/cm	H$_3$O$^+$	Fe	H$_2$O	[51]
94	KIVKEG	α-(ET)$_{12}$[Fe(ox)$_3$]$_2$·16H$_2$O	*C2/c*	σ = 0.111 S/cm	H$_3$O$^+$	Fe	H$_2$O	[51]
95	NIHPAW	α'''-(ET)$_9$Na$_{18}$[Fe(ox)$_3$]$_8$·24H$_2$O	*P-1*	E_a = 77 meV	Na$^+$	Fe	H$_2$O	[39]
96	OGUPAI	β''-(ET)$_5$[Fe(ox)$_3$](H$_2$O)$_2$CH$_2$Cl$_2$	*P-1*	σ = 4 S/cm E_a = 30 meV	-	Fe	CH$_2$Cl$_2$/H$_2$O	[52]

[a] ox = oxalate = C$_2$O$_4$$^{2-}$; [b] SG = space group.

4.5. BEDT-TTF Salts with [Ge(C$_2$O$_4$)$_3$]$^{2-}$ and [Cu(C$_2$O$_4$)$_2$]$^{2-}$ Dianions

Interestingly, besides all the above-mentioned M(III)-based [M(C$_2$O$_4$)$_3$]$^{3-}$ anions, there are also two oxalate-based dianions that have been combined with BEDT-TTF (Table 8): (i) the Ge(IV)-based [Ge(C$_2$O$_4$)$_3$]$^{2-}$ anion, which presents the same 3:1 stoichiometry and octahedral geometry as the previously used [M(C$_2$O$_4$)$_3$]$^{3-}$ anions, although with a −2 charge, and (ii) the Cu(II)-based [Cu(C$_2$O$_4$)$_2$]$^{2-}$ anion, also with a −2 charge, but with a 2:1 stoichiometry and a square planar geometry (Figure 1b). As we will show here, the change in the charge in the [Ge(C$_2$O$_4$)$_3$]$^{2-}$ anion leads to very important changes in the structure and properties of these radical salts.

The first salt with the dianion [Ge(C$_2$O$_4$)$_3$]$^{2-}$ was reported by P. Day's group and shows a 2:1 stoichiometry: (BEDT-TTF)$_2$[Ge(C$_2$O$_4$)$_3$]·PhCN (**97**) [53]. Surprisingly, this salt does not show alternating cationic and anionic layers but a chessboard arrangement of BEDT-TTF face-to-face dimers and [Ge(C$_2$O$_4$)$_3$]$^{2-}$ dianions interspersed with layers of

PhCN solvent molecules (Figure 23a,b). The presence of isolated (BEDT-TTF)$_2^{2+}$ dimers is at the origin of the semiconducting behavior shown by this salt [53].

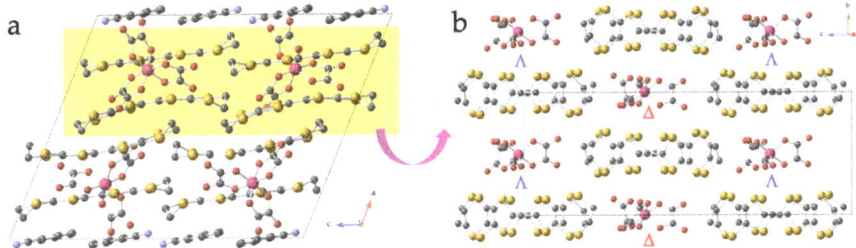

Figure 23. Structure of (BEDT-TTF)$_2$[Ge(C$_2$O$_4$)$_3$]·PhCN (**97**). (**a**) View of the *ac* plane. (**b**) Projection perpendicular to the *bc* plane showing the chessboard arrangement of the BEDT-TTF dimers and the [Ge(C$_2$O$_4$)$_3$]$^{2-}$ anions. Color code: Ge = pink, C = gray, O = red, N = blue and S = yellow. H atoms are omitted for clarity.

Salts (BEDT-TTF)$_5$[Ge(C$_2$O$_4$)$_3$]$_2$ (**98**) and (BEDT-TTF)$_7$[Ge(C$_2$O$_4$)$_3$]$_2$·0.87CH$_2$Cl$_2$·0.09H$_2$O (**99**) were prepared using the same conditions but changing the solvent (chiral R-(-)-carvone for **98** and CH$_2$Cl$_2$ for **99**) [54]. The change in the solvent leads to two very different structures and stoichiometries. Salt (BEDT-TTF)$_5$[Ge(C$_2$O$_4$)$_3$]$_2$ (**98**) contains mixed layers parallel to the *ac* plane with BEDT-TTF molecules and [Ge(C$_2$O$_4$)$_3$]$^{2-}$ anions (Figure 24a). In these layers, the BEDT-TTF molecules form diagonal stacks where the molecules are displaced along their long molecular axis. These stacks are separated by isolated [Ge(C$_2$O$_4$)$_3$]$^{2-}$ anions (Figure 24b) [54].

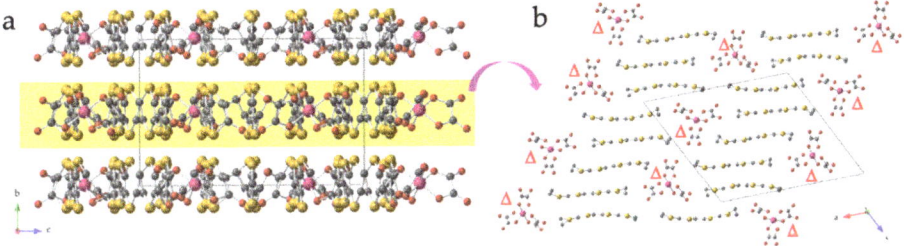

Figure 24. Structure of (BEDT-TTF)$_5$[Ge(C$_2$O$_4$)$_3$]$_2$ (**98**). (**a**) View of the mixed layers parallel to the *ac* plane. (**b**) Projection of the *ac* plane showing the stacks of BEDT-TTF molecules separated by rows of isolated [Ge(C$_2$O$_4$)$_3$]$^{2-}$ anions. Color code: Ge = pink, C = gray, O = red and S = yellow. H atoms are omitted for clarity.

In contrast, salt (BEDT-TTF)$_7$[Ge(C$_2$O$_4$)$_3$]$_2$·0.87CH$_2$Cl$_2$·0.09H$_2$O (**99**) shows alternating cationic and anionic layers parallel to the *bc* plane (Figure 25c), with the BEDT-TTF molecules packed with the α packing mode (Figure 25a) [54]. There are four independent BEDT-TTF molecules packed following the sequence ... ABCDCBA ... , with a dislocation in the stacks every seven molecules (Figure 25b). The anionic layer is very original since it shows [Ge(C$_2$O$_4$)$_3$]$^{2-}$ anions grouped in homochiral dimers with a water molecule in between them forming H bonds with both monomers (Figure 25c). Dimers with different chirality alternate along the *c* axis. Disordered CH$_2$Cl$_2$ molecules are located between the [Ge(C$_2$O$_4$)$_3$]$^{2-}$ dimers.

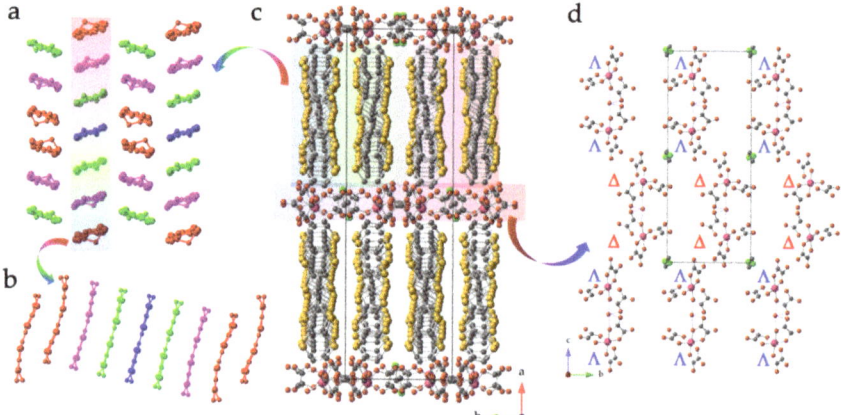

Figure 25. Structure of (BEDT-TTF)$_7$[Ge(C$_2$O$_4$)$_3$]$_2$·0.87CH$_2$Cl$_2$·0.09H$_2$O (**99**). (**a**) View of the cationic layer showing the four independent BEDT-TTF molecules (A–D, as red, pink, green and blue, respectively). (**b**) Side view of the chain. (**c**) View of the alternating cationic and anionic layers. (**d**) View of the anionic layer showing the [Ge(C$_2$O$_4$)$_3$]$^{2-}$ dimers H bonded to a water molecule (H bonds as thin blue lines). Color code in (**b**,**c**): Ge = pink, C = gray, O = red, Cl = green and S = yellow. H atoms are omitted for clarity.

Although salts **97-99** were prepared with the NH$_4^+$ salt of the [Ge(C$_2$O$_4$)$_3$]$^{2-}$ dianion, the NH$_4^+$ cation did not enter in the structure of the salts. Attempts to change the cation, using the Cs$^+$ salt, led to a new and original salt, although without the Cs$^+$ cation: (BEDT-TTF)$_4$[Ge(C$_2$O$_4$)$_3$]·0.5CH$_2$Cl$_2$ (**100**) [55]. This salt also presents alternating cationic and anionic layers (Figure 26b), both layers being original. There are four independent BEDT-TTF molecules (A–D) packed in parallel stacks following the sequence ... ABCD ... (Figure 26a), with a dislocation every four molecules (Figure 26b). Within each group of the four BEDT-TTF molecules, the two central ones (B and C) are completely ionized, whereas the external ones (A and D) are neutral, resulting in (BEDT-TTF)$_2^{2+}$ dimers surrounded by neutral monomers from the electronic point of view. This charge distribution results in a semiconducting behavior, as observed experimentally (Table 8) [55]. The anionic layer is also original. It contains dimers of [Ge(C$_2$O$_4$)$_3$]$^{2-}$ anions (with ⊗ and Λ chirality) with a CH$_2$Cl$_2$ molecule connecting both anions (Figure 26c).

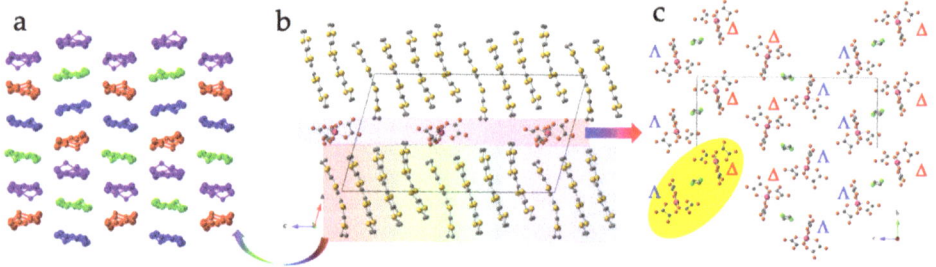

Figure 26. Structure of (BEDT-TTF)$_4$[Ge(C$_2$O$_4$)$_3$]·0.5CH$_2$Cl$_2$ (**100**). (**a**) View of the cationic layer showing the four independent BEDT-TTF molecules (as blue, green, purple and red). (**b**) View of the alternating cationic and anionic layers showing the dislocation in the BEDT-TTF stacks every four molecules. (**c**) View of the anionic layer showing the [Ge(C$_2$O$_4$)$_3$]$^{2-}$ dimers (highlighted in yellow) with a disordered CH$_2$Cl$_2$ molecule located between them. Color code in (**b**,**c**): Ge = pink, C = gray, O = red and S = yellow. H atoms are omitted for clarity.

The only salt reported with the $[Cu(C_2O_4)_2]^{2-}$ dianion: $(BEDT-TTF)_4[Cu(C_2O_4)_2]$ (**101**), is also the first radical salt prepared with any metal-oxalate complex. It shows alternating cationic and anionic layers parallel to the *ac* plane (Figure 27b). The BEDT-TTF layers are formed by two independent molecules with the β packing mode (Figure 27a). The $[Cu(C_2O_4)_2]^{2-}$ dianions are isolated and show a square planar geometry (Figure 27c). The salt is metallic down to 65 K, where it shows a metal–semiconductor transition with a low activation energy of 15 meV below 65 K (Table 8) [56,57].

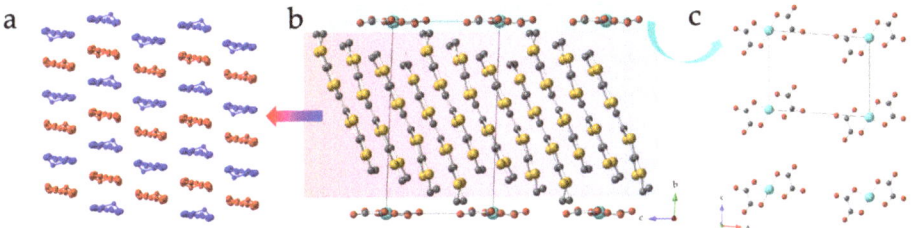

Figure 27. Structure of $(BEDT-TTF)_4[Cu(C_2O_4)_2]$ (**101**). (**a**) View of the cationic layer showing the β packing mode of the two independent BEDT-TTF molecules (red and blue). (**b**) View of the alternating cationic and anionic layers. (**c**) View of the anionic layer showing the isolated square planar $[Cu(C_2O_4)_2]^{2-}$ anions. Color code in (**b**,**c**): Cu = blue, C = gray, O = red and S = yellow. H atoms are omitted for clarity.

Table 8. Radical salts of BEDT-TTF with the $[Ge(C_2O_4)_3]^{2-}$ and $[Cu(C_2O_4)_2]^{2-}$ dianions (**97–101**).

#	CCDC	Formula [a]	SG [b]	Elect. Prop.	G	Anion	Ref.
97	MAJYUR	$(ET)_2[Ge(ox)_3] \cdot PhCN$	$P2_1/c$	$E_a = 127$ meV	PhCN	$[Ge(ox)_3]^{2-}$	[53]
98	MUVFUF	$(ET)_5[Ge(ox)_3]_2$	C2	$\sigma = 10^{-3}$ S/cm $E_a = 225$ meV	-	$[Ge(ox)_3]^{2-}$	[54]
99	MUVGAM	$(ET)_7[Ge(ox)_3]_2 \cdot 0.87CH_2Cl_2 \cdot 0.09H_2O$	C2/c	$\sigma = 1.75$ S/cm $E_a = 117\text{–}172$ meV	CH_2Cl_2/H_2O	$[Ge(ox)_3]^{2-}$	[54]
100	PADDOQ	$(ET)_4[Ge(ox)_3] \cdot 0.5CH_2Cl_2$	$P2_1/c$	$\sigma = 4.7 \times 10^{-3}$ S/cm $E_a = 224$ meV	CH_2Cl_2	$[Ge(ox)_3]^{2-}$	[55]
101	SOJLUY	$(ET)_4[Cu(ox)_2]$	P-1	$T_{M-I} = 65$ K $E_a = 15$ meV	-	$[Cu(ox)2]^{2-}$	[56,57]

[a] ox = oxalate = $C_2O_4^{2-}$; [b] SG = space group.

5. BEDT-TTF Salts with Oxalate Dimers and 2D Lattices

Besides monomeric tri-anions such as $[M(C_2O_4)_3]^{3-}$ (M = Fe, Cr, Co, Al, Ga, Mn, Ru, Rh, Ir, etc.) and dianions such as $[Ge(C_2O_4)_3]^{2-}$ and $[Cu(C_2O_4)_2]^{2-}$, there are a few reported radical salts of BEDT-TTF with the dimer $[Fe_2(C_2O_4)_5]^{4-}$ and even with extended homo- and heterometallic hexagonal honeycomb lattices such as $[MnCr(C_2O_4)_3]^-$, $[MnRh(C_2O_4)_3]^-$ and $[Cu_2(C_2O_4)_3]^{2-}$ (Table 9).

The only known salt with BEDT-TTF and a dimeric anion is: $(BEDT-TTF)_4[Fe_2(C_2O_4)_5]$ (**105**), also reported by P. Day's group [58]. In this radical salt, the dimeric anion $[Fe_2(C_2O_4)_5]^{4-}$ is formed in situ from the precursor $[Fe(C_2O_4)_3]^{3-}$ monomer in the electrochemical cell. The structure of this unusual salt consists of mixed layers parallel to the *ac* plane (Figure 28a) containing the $[Fe_2(C_2O_4)_5]^{4-}$ dimers interspersed with BEDT-TTF molecules that form stacks running parallel to the *a* direction (Figure 28b). The $[Fe_2(C_2O_4)_5]^{4-}$ dimers are formed by two $[Fe(C_2O_4)_3]^{3-}$ monomers with different chirality sharing an oxalate bridge (Figure 1c). There are two independent BEDT-TTF molecules, both with a +1 charge, in agreement with the 4:1 stoichiometry and the -4 charge of the anion, resulting in a semiconducting salt with a high activation energy and a low room temperature conductivity (Table 9) [58]. The oxalate-bridged Fe(III) dimer in this salt shows, as expected, a weak antiferromagnetic interaction with

$J = -3.44$ cm^{-1}, similar to that found in the TTF and TM-TTF salts with the same dimer that will be described in the next section [59,60].

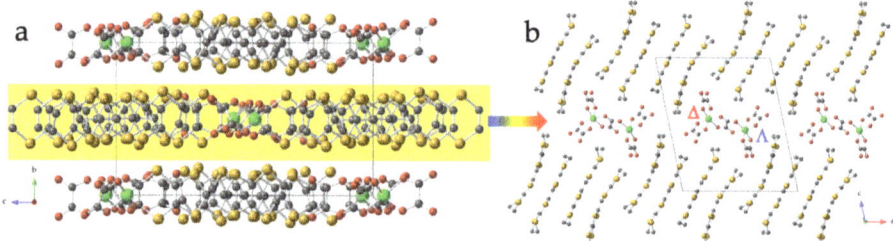

Figure 28. (a) View of the structure of (BEDT-TTF)$_4$[Fe$_2$(C$_2$O$_4$)$_5$] (**105**) showing the mixed layers parallel to the *ac* plane. (b) Top view of one mixed layer showing the stacks of the BEDT-TTF molecules and the [Fe$_2$(C$_2$O$_4$)$_5$]$^{4-}$ dimeric anion in between the stacks of BEDT-TTF molecules. Color code: Fe = green, C = gray, O = red and S = yellow. H atoms are omitted for clarity.

There are three closely related BEDT-TTF salts prepared with a heterometallic honeycomb oxalate-based layer: (ET)$_{2.53}$[MnCr(C$_2$O$_4$)$_3$]·CH$_2$Cl$_2$ (**103**), (ET)$_{2.53}$[MnRh(C$_2$O$_4$)$_3$]·CH$_2$Cl$_2$ (**104**) and (ET)$_3$[MnCr(C$_2$O$_4$)$_3$] (**106**) [7,8]. These three salts are isostructural and show alternating layers of BEDT-TTF molecules and anionic honeycomb layers (Figure 29b). The BEDT-TTF molecules are tilted around 45° with respect to the anionic layer and show a β packing mode (Figure 29a). The anionic layers show the classical honeycomb structure and contain Mn(II) and Cr(III) ions (or Rh(III) in **104**) with alternating chirality connected by oxalate bridges (Figure 29c).

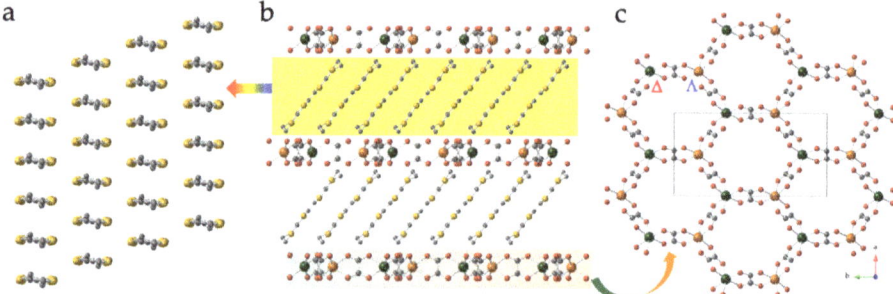

Figure 29. Structure of (BEDT-TTF)$_3$[MnCr(C$_2$O$_4$)$_3$] (**106**). (a) View of the BEDT-TTF layer showing the β packing mode. (b) View of the alternating cationic and anionic layers. (c) View of the honeycomb [MnCr(C$_2$O$_4$)$_3$]$^-$ anionic layer. Color code: Cr = dark green, Mn = orange, C = gray, O = red and S = yellow. H atoms are omitted for clarity.

Salt (BEDT-TTF)$_3$[MnCr(C$_2$O$_4$)$_3$] (**106**) was the first molecular compound showing metallic conductivity and a ferromagnetic long-range ordering, with an ordering temperature T_{curie} of ca. 5.5 K (Figure 30a) [7]. This salt presents magnetoresistance below ca. 10 K (Figure 30b), indicating that both sublattices are *quasi*-independent from the electronic point of view. Attempts to change the metal ions and the donor molecules and the use of other solvents led to different radical salts [61], including isostructural compounds (ET)$_{2.53}$[MnCr(C$_2$O$_4$)$_3$]·CH$_2$Cl$_2$ (**103**) and (ET)$_{2.53}$[MnRh(C$_2$O$_4$)$_3$]·CH$_2$Cl$_2$ (**104**) that are also metallic and show long-range magnetic ordering, although in compound **104**, there is a broad metal to semiconducting transition at around 100 K [8]. The use of this honeycomb layer with other TTF-type donors will be revised in the next section.

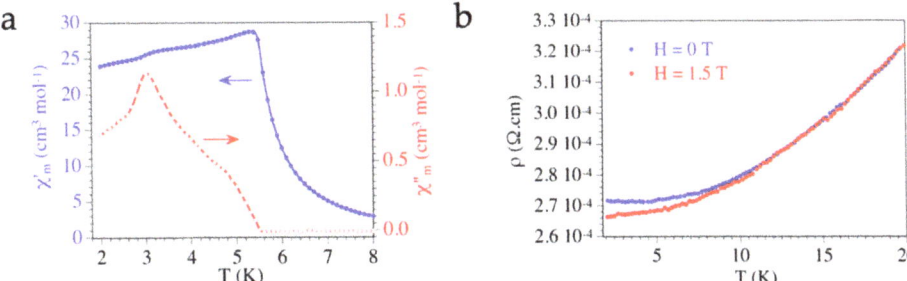

Figure 30. (a) Temperature dependencies of the in-phase (χ'_m) and the out-of-phase (χ''_m) AC susceptibility of (BEDT-TTF)$_3$[MnCr(C$_2$O$_4$)$_3$] (**106**) showing the long-range ordering at ca. 5.5 K. (b) Electrical conductivity of a single crystal of (BEDT-TTF)$_3$[MnCr(C$_2$O$_4$)$_3$] (**106**) with and without applied DC magnetic field.

A last example of extended lattices is provided by the series of salts formulated as (BEDT-TTF)$_3$[Cu$_2$(C$_2$O$_4$)$_3$]·G, with G = H$_2$O (**108**), CH$_2$Cl$_2$ (**102**) and CH$_3$OH (**107**) [62]. The structure of these isostructural salts shows alternating cationic and anionic layers parallel to the *ab* plane (Figure 31b). The BEDT-TTF layers show a θ^{21} packing mode, similar to that observed in α'''-(ET)$_9$Na$_{18}$[Fe(C$_2$O$_4$)$_3$]$_8$·24H$_2$O (**95**), formed by two stacks of BEDT-TTF tilted in one direction (+/+) and one stack tilted in the opposite direction (-), following the sequence (... /+/+/-/+/+/-/ ...) (Figure 31a). The anionic layer is also a honeycomb hexagonal lattice formed by Cu(II) ions in the vertices of the hexagons and oxalate bridges as the sides of the hexagons. The crystallization solvent molecules are located in the hexagonal cavities. As a result of the expected Jahn–Teller distortions in the Cu(II) ions, the hexagons are quite distorted (Figure 31c). The properties of the CH$_3$OH derivative (the only one reported to date) show that this radical salt is a semiconductor with an activation energy of 50 meV and presents a moderate antiferromagnetic Cu–Cu interaction through the oxalate bridge [62].

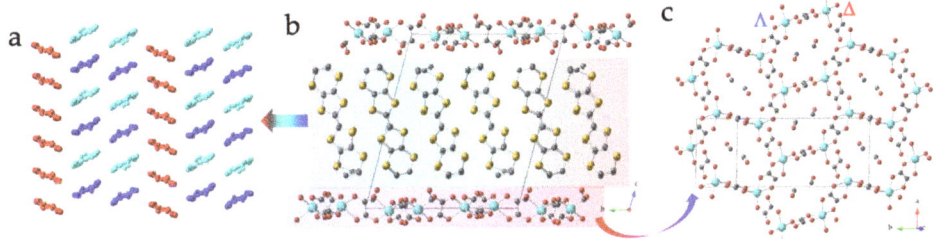

Figure 31. Structure of (BEDT-TTF)$_3$[Cu$_2$(C$_2$O$_4$)$_3$]·2CH$_3$OH (**107**). (a) View of the BEDT-TTF layer showing the θ^{21} packing mode with the three independent BEDT-TTF molecules (in red, blue and light blue). (b) View of the alternating cationic and anionic layers. (c) View of the distorted honeycomb [Cu$_2$(C$_2$O$_4$)$_3$]$^{2-}$ anionic layer with the CH$_3$OH molecules in the hexagonal cavities. Color code in (**b**,**c**): Cu = light blue, C = gray, O = red and S = yellow. H atoms are omitted for clarity.

The tuneability of these metallic magnets was evidenced by the synthesis of different derivatives of the metallic ferromagnet (BEDT-TTF)$_3$[MnCr(C$_2$O$_4$)$_3$] (**106**). One of these derivatives is the salt (BEDT-TTF)$_3$[CoCr(C$_2$O$_4$)$_3$]·CH$_2$Cl$_2$, which showed an ordering temperature of 9.2 K and a high electrical conductivity of 1 S/cm at room temperature, although no metallic behavior was observed, maybe because the measurements were performed on a pressed pellet as no big single crystals could be obtained [63]. The other derivatives, prepared with different TTF-type donors, will be revised in the next section.

Table 9. Radical salts of BEDT-TTF with metal-oxalate dimers and 2D lattices (**102–108**).

#	CCDC	Formula [a]	SG [b]	Elect. Prop.	Anion	G	Ref.
102	CEMMUF	(ET)$_3$[Cu$_2$(ox)$_3$]·CH$_2$Cl$_2$	P-1	-	[Cu$_2$(ox)$_3$]$^{2-}$	CH$_2$Cl$_2$	c
103	IPOZIY	(ET)$_{2.53}$[MnCr(ox)$_3$]·CH$_2$Cl$_2$	P-1	σ = 10 S/cm metal > 0.4 K	[MnCr(ox)$_3$]$^-$	CH$_2$Cl$_2$	[8]
104	IPOZOE	(ET)$_{2.53}$[MnRh(ox)$_3$]·CH$_2$Cl$_2$	P-1	σ = 13 S/cm metal > 100 K	[MnRh(ox)$_3$]$^-$	CH$_2$Cl$_2$	[8]
105	LOHWIO	(ET)$_4$[Fe$_2$(ox)$_5$]	P2$_1$/n	σ = 2 × 10^{-3} S/cm E$_a$ = 1200 meV	[Fe$_2$(ox)$_5$]$^{4-}$	-	[58]
106	NALVIG	(ET)$_3$[MnCr(ox)$_3$]	P-1	σ = 250 S/cm metal > 0.3 K	[MnCr(ox)$_3$]$^{2-}$	-	[7]
107	SAMMEA	(ET)$_3$[Cu$_2$(ox)$_3$]·2CH$_3$OH	P-1	σ = 4 S/cm E$_a$ = 50 meV	[Cu$_2$(ox)$_3$]$^{2-}$	CH$_3$OH	[62]
108	WUXWET	(ET)$_3$[Cu$_2$(ox)$_3$]·2H$_2$O	P-1	-	[Cu$_2$(ox)$_3$]$^{2-}$	H$_2$O	c

[a] ox = oxalate = C$_2$O$_4$$^{2-}$; [b] SG = space group. [c] Unpublished results.

6. Radical Salts of Metal-Oxalate Anions with Other TTF-Type Donor Molecules

Although BEDT-TTF is, by far, the most used donor molecule with metal-oxalato complexes, with more than one hundred reported salts (see Tables 1–9), other TTF-type donors have been combined with metal-oxalato complexes and lattices (Scheme 1). Here, we will revise all these salts, prepared with donors such as tetrathiafulvalene (TTF), tetramethyl-tetrathiafulvalene (TM-TTF), bis(ethylenediseleno)tetrathiafulvalene (BEDS-TTF = BEST), bis(ethylenedithio)tetraselenafulvalene (BEDT-TSF = BETS) and 4,5-bis((2S)-2-hydroxypropylthio)-4′,5′-(ethylenedithio)tetrathiafulvalene (DMPET).

6.1. TTF and TM-TTF Salts with Oxalate Complexes

There are only two radical salts with [M(C$_2$O$_4$)$_3$]$^{3-}$ anions and the donor TTF: (TTF)$_7$-[Fe(C$_2$O$_4$)$_3$]$_2$·4H$_2$O (**114**) [59,60] and (TTF)$_3$[Ru(C$_2$O$_4$)$_3$]·0.5EtOH·4H$_2$O (**118**) [64]. Salt **114** is an unusual 7:2 salt that shows corrugated mixed layers containing TTF molecules and the [Fe(C$_2$O$_4$)$_3$]$^{3-}$ anions with the water molecules located between the layers (Figure 32a) [60]. The top view of these layers shows chains of TTF molecules, running parallel to the *a* axis (perpendicular to the layer), surrounded by TTF dimers and [Fe(C$_2$O$_4$)$_3$]$^{3-}$ anions (Figure 32b). The TTF dimers contain two independent TTF molecules (C and D), and the TTF chains are formed by two different independent TTF molecules (A and B) following the sequence ... ABB ... (Figure 32c). The [Fe(C$_2$O$_4$)$_3$]$^{3-}$ anions form homochiral rows running along the *b* axis, with opposite chirality in alternating rows, meaning that two of the four [Fe(C$_2$O$_4$)$_3$]$^{3-}$ anions surrounding the TTF chains are ⊗ enantiomers and the other two are Λ (Figure 32b).

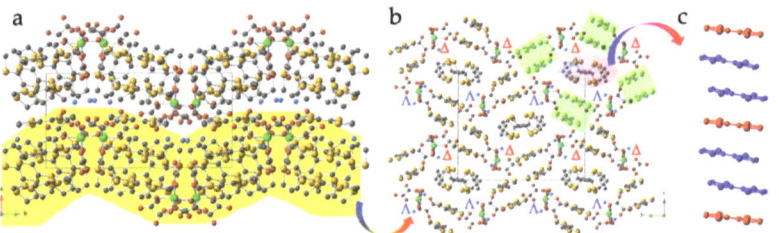

Figure 32. Structure of (TTF)$_7$[Fe(C$_2$O$_4$)$_3$]$_2$·4H$_2$O (**114**): (**a**) View of the corrugated mixed layers. (**b**) Top view (down the *a* direction) of the mixed layers showing the TTF dimers (green rectangles) and the TTF chains. (**c**) Side view of the TTF chains showing the ... ABB ... sequence. Color code in (**a**,**b**): Fe = green, C = gray, O = red, O$_{water}$ = blue and S = yellow. H atoms are omitted for clarity.

The other salt with TTF and a monomeric $[M(C_2O_4)_3]^{3-}$ anion, $(TTF)_3[Ru(C_2O_4)_3] \cdot 0.5$EtOH$\cdot 4H_2O$ (**118**), was the first radical salt prepared with a 4d $[M(C_2O_4)_3]^{3-}$ anion and shows a very original structure in both sublattices [64]. Its structure consists of alternating cationic and anionic layers parallel to the *ab* plane (Figure 33a). There are two different anionic layers each containing a single enantiomer. The anionic layers contain the $[M(C_2O_4)_3]^{3-}$ anions and the solvent molecules with a hexagonal arrangement (Figure 33b), although different to the one observed in the BEDT-TTF salts of previous sections since, now, there is no extra A^+ cation orienting the terminal O atoms of the oxalate ligands. The TTF layers are also original: they contain three independent TTF molecules (A–C) arranged in AA dimers and chains following the sequence . . . BBCC . . . (Figure 33c). The TTF molecules of consecutive layers run in opposite directions (Figure 33d). This salt shows a paramagnetic behavior with contributions from the $[Ru(C_2O_4)_3]^{3-}$ $S = \frac{1}{2}$ anions and from the TTF molecules that show a charge localization, in agreement with the semiconducting behavior observed in this salt [64].

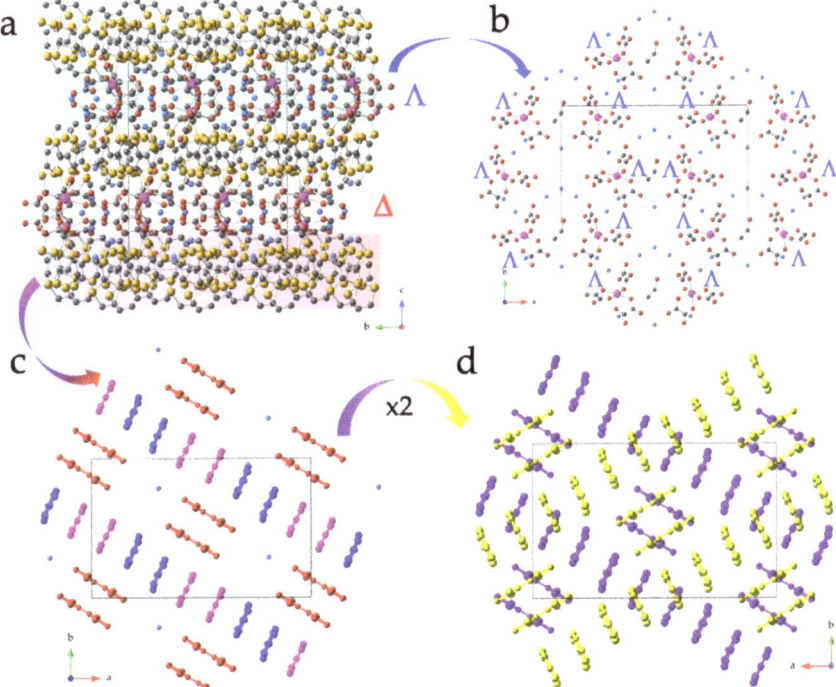

Figure 33. Structure of $(TTF)_3[Ru(C_2O_4)_3] \cdot 0.5$EtOH$\cdot 4H_2O$ (**118**): (**a**) View of the alternating cationic and anionic layers parallel to the *ab* plane. (**b**) View of the Λ anionic layer. (**c**) View of the cationic layer showing the AA dimers (in red) and the . . . BBCC . . . chains (in blue and pink). (**d**) View of two consecutive TTF layers (in yellow and purple) showing the opposite orientations of the chains and the dimers. Color code in (**a**,**b**): Ru = pink, C = gray, O = red, O$_{water}$ = blue and S = yellow. H atoms are omitted for clarity.

There are two other salts with oxalate complexes and TTF, although they do not contain monomeric $[M(C_2O_4)_3]^{3-}$ anions but an $[Fe_2(C_2O_4)_5]^{4-}$ dimer in $(TTF)_5[Fe_2(C_2O_4)_5] \cdot 2$Ph-CH$_3 \cdot 2H_2O$ (**115**) [59,60], or a $\{Mn(H_2O)_2[Cr(C_2O_4)_3]_2\}^{4-}$ trimer in $(TTF)_4\{Mn(H_2O)_2[Cr(C_2O_4)_3]_2\} \cdot 14H_2O$ (**120**) [65,66]. Interestingly, salts **115** and **120** are the first examples of any salt with the previously unknown $[Fe_2(C_2O_4)_5]^{4-}$ dimer and $\{Mn(H_2O)_2[Cr(C_2O_4)_3]_2\}^{4-}$ trimer, respectively. In both cases, the anions are assembled during the electrochemical synthesis of the radical salts from the corresponding $[M(C_2O_4)_3]^{3-}$ anions (and Mn^{2+} cations in salt **120**).

Salt **115** shows alternating cationic and anionic layers parallel to the *ab* plane (Figure 34b). The cationic layers contain two independent TTF molecules (A and B). A-type TTF molecules form chains running parallel to the *b* axis, whereas B-type molecules are monomers located between the chains. The packing of the TTF molecules is unique because not only the planes of the A- and B-type molecules are orthogonal but also their long molecular axis (Figure 34a,b). The $[Fe_2(C_2O_4)_5]^{4-}$ dimer is formed by the fusion of one $\otimes$ and one Λ monomer through a bis-bidentate oxalate ligand (Figure 1c). These dimers are packed in rows along the *b* axis with $PhCH_3$ molecules located between the rows (Figure 34c). The electrical properties show that this salt is a semiconductor (Table 10), in agreement with the presence of neutral isolated TTF molecules. The magnetic properties show that the $[Fe_2(C_2O_4)_5]^{4-}$ dimer presents a weak antiferromagnetic Fe–Fe coupling with J = -3.57 cm^{-1}, through the oxalate bridge, as observed in the other salts prepared with this dimer [58,60].

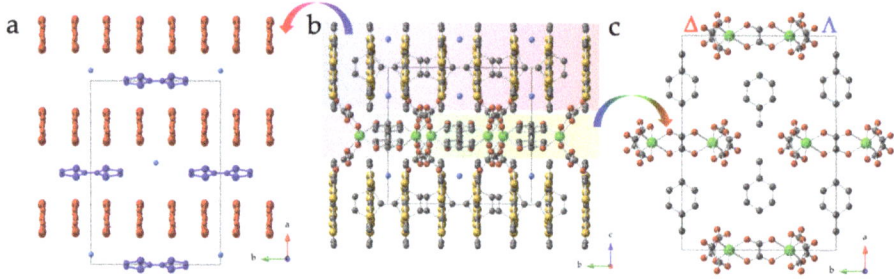

Figure 34. Structure of $(TTF)_4[Fe_2(C_2O_4)_5]\cdot 2PhCH_3\cdot 2H_2O$ (**115**): (**a**) View of the cationic layer showing the chains along the *b* axis containing the A-type molecules (in red) and the isolated orthogonal monomers (in blue). (**b**) View of the alternating cationic and anionic layers parallel to the *ab* plane. (**c**) View of the anionic layer. Color code in (**b**,**c**): Fe = green, C = gray, O = red, O_{water} = blue and S = yellow. H atoms are omitted for clarity.

The last salt with TTF and an oxalate complex is $(TTF)_4\{Mn(H_2O)_2[Cr(C_2O_4)_3]_2\}\cdot 14H_2O$ (**120**) [65,66]. This salt was the first one prepared with TTF and any oxalate complex and contains the then-unknown $\{Mn(H_2O)_2[Cr(C_2O_4)_3]_2\}^{4-}$ trimer. The structure shows alternating cationic and anionic layers parallel to the *ab* plane (Figure 35b). There are three independent TTF molecules forming orthogonal AA and BC dimers (similar to the κ-phase in BEDT-TTF) [67], although now the dimers form stacks separated by water molecules (Figure 35a). The anionic layer contains linear trimeric $\{Mn(H_2O)_2[Cr(C_2O_4)_3]_2\}^{4-}$ anions formed by the fusion of two $[Cr(C_2O_4)_3]^{3-}$ monomers with different chirality with a Mn(II) ion coordinated by two equatorial bidentate oxalato ligands and two axial H_2O molecules (Figure 1d). These trimers are packed in a rhombic disposition with strong H bonds between the two coordinated water molecules and the terminal O atoms of the oxalato ligands of neighboring trimers (Figure 35c).

The magnetic properties show the presence of a weak ferromagnetic Cr–Mn coupling with J = 1.08 cm^{-1}, although the salt is a semiconductor since the TTF molecules are completely oxidized and there are no short intermolecular contacts (Table 10) [65,66]. An additional interest of this salt is the possibility to change the anion by simply changing Cr(III) to Fe(III) and Mn(II) to other metal ions such as Fe(II), Co(II), Ni(II), Cu(II) and Zn(II). In this way, it was possible to prepare the series of salts $(TTF)_4\{M^{II}(H_2O)_2[M^{III}(C_2O_4)_3]_2\}\cdot nH_2O$ with M^{III}/M^{II} = Cr/Mn, Cr/Fe, Cr/Co, Cr/Ni, Cr/Cu, Cr/Zn, Fe/Mn, Fe/Fe, Fe/Co, Fe/Ni and Fe/Zn [66]. The use of different metal ions in these series allowed a modulation of the magnetic coupling that is weak ferromagnetic for the Cr(III) derivatives and weak antiferromagnetic for the Fe(III) ones [66].

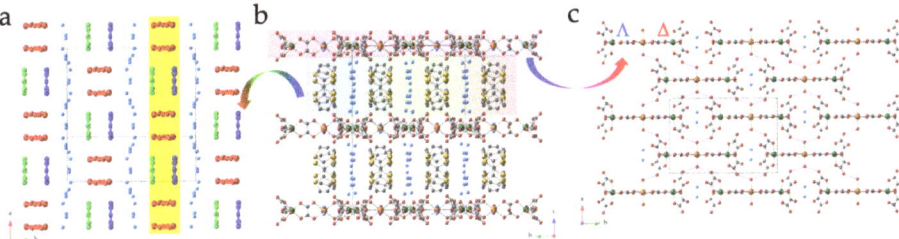

Figure 35. Structure of (TTF)$_4${Mn(H$_2$O)$_2$[Cr(C$_2$O$_4$)$_3$]$_2$}·14H$_2$O (**120**): (**a**) View of the cationic layer showing the orthogonal AA (in red) and BC (green and blue) dimers arranged in chains separated by water molecules. (**b**) View of the alternating cationic and anionic layers parallel to the *ab* plane. (**c**) View of the anionic layer. Thin blue lines are the H bonds between the anions. Color code in (**a**,**c**): Cr = dark green, Mn = orange, C = gray, O = red, O$_{water}$ = blue and S = yellow. H atoms are omitted for clarity.

The then-unknown [Fe$_2$(C$_2$O$_4$)$_5$]$^{4-}$ dimer was also obtained with the donor tetramethyl-tetrathiafulvalene (TM-TTF) and published with the TTF salt described above [60]. This salt, formulated as (TM-TTF)$_4$[Fe$_2$(C$_2$O$_4$)$_5$]·PhCN·4H$_2$O (**116**), shows mixed layers of TM-TTF molecules and [Fe$_2$(C$_2$O$_4$)$_5$]$^{4-}$ anions with crystallization PhCN and water molecules (Figure 36a). The layers contain stacks of TM-TTF separated by rows with the [Fe$_2$(C$_2$O$_4$)$_5$]$^{4-}$ anions and crystallization solvent molecules (Figure 36b). The TM-TTF stacks are formed by two independent TM-TTF molecules following the sequence ... AABB ... Although the molecular planes of the donor molecules are parallel, the long axis of the molecules is slightly tilted (Figure 36c). As the TM-TTF molecules are completely oxidized, the salt is a semiconductor with a high activation energy (Table 10). As observed in the TTF and BEDT-TTF salts of the same [Fe$_2$(C$_2$O$_4$)$_5$]$^{4-}$ anion, in compound **116**, this anion shows a weak antiferromagnetic Fe–Fe coupling with J = −3.69 cm^{-1}, through the oxalate bridge, similar to those of the other salts with this dimer [58,60].

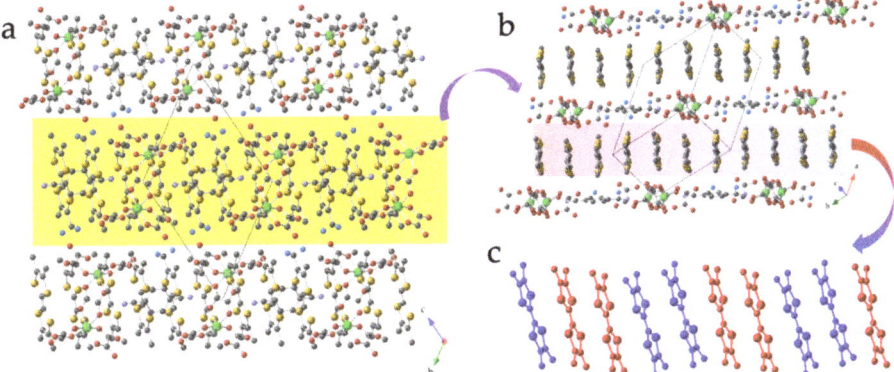

Figure 36. Structure of (TM-TTF)$_4$[Fe(C$_2$O$_4$)$_3$]$_2$·PhCN·4H$_2$O (**116**): (**a**) Side view of the mixed cationic and anionic layers. (**b**) Top view of the mixed layer showing the TM-TTF chains separated by [Fe(C$_2$O$_4$)$_5$]$^{4-}$ anions and water molecules. (**c**) View of the TM-TTF chains formed by A (red) and B (blue) TM-TTF molecules. Color code in (**a**,**b**): Fe = green, C = gray, O = red, O$_{water}$ = blue and S = yellow. H atoms are omitted for clarity.

6.2. Salts with Se-Containing Donors (BEST and BETS)

There are two selenium-containing derivatives of BEDT-TTF that have also been used with oxalate complexes: bis(ethylenediseleno)tetrathiafulvalene (BEDS-TTF = BEST), which contains four Se atoms in the outer rings, and bis(ethylenedithio)tetraselenafulvalene

(BEDT-TSF = BETS), which contains four Se atoms in the inner rings (Scheme 1). With the donor BEST, a total of five salts have been reported [68]. There are two isostructural salts with benzoic acid and water as crystallization solvents: (BEST)$_4$[M(C$_2$O$_4$)$_3$]·PhCOOH·H$_2$O, with M = Cr (**109**) and Fe (**110**), and a couple of isostructural salts with the same stoichiometry but different solvents formulated as (BEST)$_4$[M(C$_2$O$_4$)$_3$]·1.5H$_2$O, with M = Cr (**111**) and Fe (**113**). The fifth salt is a 9:2 salt formulated as (BEST)$_9$[Fe(C$_2$O$_4$)$_3$]$_2$·7H$_2$O (**112**) [68].

The structure of salts (BEST)$_4$[M(C$_2$O$_4$)$_3$]·PhCOOH·H$_2$O, with M = Cr (**109**) and Fe (**110**) consists of alternating cationic and anionic layers parallel to the *ab* plane (Figure 37b). The cationic layer is formed by four independent BEST molecules (A–D) packed with the β packing mode (Figure 37a) in parallel stacks, with a step every four BEST molecules (Figure 37b), following the sequence ... ABCDDCBA ... (Figure 37a). The anionic layers show an original arrangement with pairs of anions with different chirality separated by PhCOOH molecules, also arranged in pairs, and water molecules forming moderate H bonds (Figure 37c). A- and C-type BEST molecules are completely ionized, whereas B and D molecules bear a charge of +1/2. The presence of totally ionized BEST molecules is responsible of the semiconducting behavior of these two salts (Table 10) [68]. The magnetic properties show that there is no noticeable contribution from the organic layers, and the salts behave as isolated S = 3/2 or S = 5/2 ions for **109** and **110**, respectively.

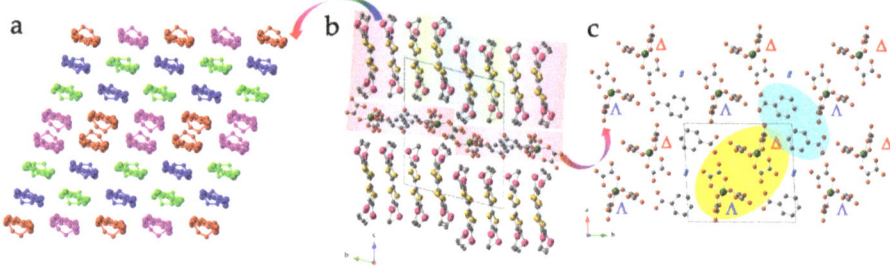

Figure 37. Structure of (BEST)$_4$[Cr(C$_2$O$_4$)$_3$]·PhCOOH·H$_2$O (**109**): (**a**) View of the cationic layer showing the β packing formed by four independent BEST molecules (in red, blue, green and pink). (**b**) View of the alternating cationic and anionic layers parallel to the *ab* plane. (**c**) View of the anionic layer. Color code in (**b**,**c**): Cr = dark green, C = gray, O = red, O$_{water}$ = blue, Se = pink and S = yellow. H atoms are omitted for clarity.

Salts (BEST)$_4$[M(C$_2$O$_4$)$_3$]·1.5H$_2$O, with M = Cr (**111**) and Fe (**113**), are isostructural and represent a pair of solvates of **109** and **110**, respectively, since they show the same stoichiometry and composition and only differ in the crystallization solvent molecules (PhCOOH and H$_2$O in **109** and **110**, compared to only H$_2$O in **111** and **113**) [68]. The structure of salts **111** and **113** also consists of alternating cationic and anionic layers parallel to the *ab* plane (Figure 38b). The cationic layer contains two independent BEST molecules (A and B) packed in parallel stacks following the sequence ... ABAB ... , also with the β packing mode (Figure 38a). The anions appear with a disorder since they are located close to an inversion center. When only one of the two possible locations is considered, the anions show zigzag chains with alternating chirality along the *c* axis (Figure 38c). The A-type BEST molecules are completely ionized, whereas B molecules bear a charge of +1/2, in agreement with the stoichiometry and the anionic charge. The presence of alternating totally ionized BEST molecules results in a semiconducting behavior for these two salts (Table 10). The magnetic properties are, as expected, similar to those of salts **109** and **110** since the Cr(III) S = 3/2 and Fe(III) S = 5/2 ions are isolated from the magnetic point of view and there is no noticeable contribution from the organic layers [68].

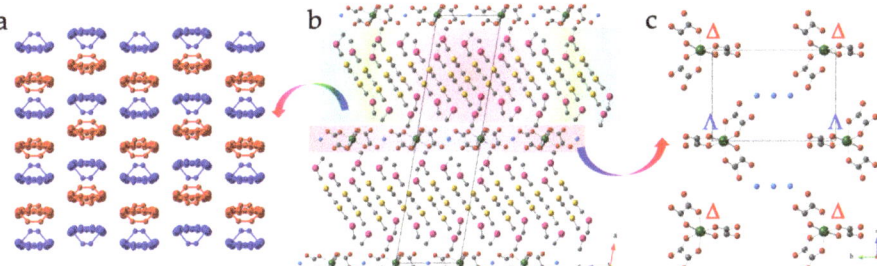

Figure 38. Structure of (BEST)$_4$[Cr(C$_2$O$_4$)$_3$]·1.5H$_2$O (**111**): (**a**) View of the cationic layer showing the β packing formed by two independent BEST molecules (in red and blue). (**b**) View of the alternating cationic and anionic layers parallel to the *ac* plane. (**c**) View of the anionic layer showing only one of the two orientations of the [Cr(C$_2$O$_4$)$_3$]$^{3-}$ anions. Color code in (**b**,**c**): Cr = dark green, C = gray, O = red, O$_{water}$ = blue, Se = pink and S = yellow. H atoms are omitted for clarity.

The last salt containing the donor BEST is a very original 9:2 phase, formulated as (BEST)$_9$[Fe(C$_2$O$_4$)$_3$]$_2$·7H$_2$O (**112**) [68]. This salt also presents alternating cationic and anionic layers parallel to the *ab* plane (Figure 39b), and the cationic layer also shows the β packing mode (Figure 39a). The main difference with the other BEST layers is the presence of five independent BEST molecules (A–E) that form two different chains (I and II) with steps every three BEST molecules (Figure 39b). Chains of type I are formed by D and E molecules packed following the sequence ... DED ..., whereas chains of type II contain three independent molecules (A–C) packed following the sequence ... ABC ... There are two chains of type II and one of type I alternating in the direction perpendicular to the stacks following the sequence ... /I/II/II/ ... (Figure 39a). The [Fe(C$_2$O$_4$)$_3$]$^{3-}$ anions are located to form dimers with the opposite chirality, and the crystallization water molecules are located in between the anions forming several H bonds with the anions and with the water molecules (Figure 39c). The inhomogeneous charge distribution of the five BEST-TTF molecules is at the origin of the semiconducting behavior observed in this salt (Table 10). As in the previous BEST salts, the magnetic properties correspond to isolated [Fe(C$_2$O$_4$)$_3$]$^{3-}$ anions with no noticeable contribution from the organic layers [68].

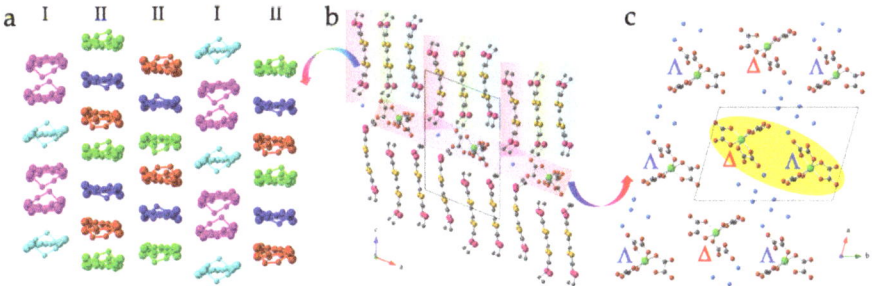

Figure 39. Structure of (BEST)$_9$[Fe(C$_2$O$_4$)$_3$]$_2$·7H$_2$O (**112**): (**a**) View of the cationic layer showing the β packing and the two different stacks (I and II) with the five independent BEST molecules (A–E in red, dark blue, green, pink and light blue, respectively). (**b**) View of the alternating cationic and anionic layers parallel to the *ab* plane. (**c**) View of the anionic layer showing the [Fe(C$_2$O$_4$)$_3$]$^{3-}$ anions and the crystallization water molecules. Color code in (**b**,**c**): Fe = green, C = gray, O = red, O$_{water}$ = blue, Se = pink and S = yellow. H atoms are omitted for clarity.

The other Se-containing donor, bis(ethylenedithio)tetraselenafulvalene (BETS), has never been combined with monomeric [M(C$_2$O$_4$)$_3$]$^{3-}$ anions but only with two extended honeycomb lattices, in compounds (BETS)$_3$[Cu$_2$(C$_2$O$_4$)$_3$]·2CH$_3$OH (**117**) [69] and (BETS)$_3$-[MnCr(C$_2$O$_4$)$_3$]·CH$_2$Cl$_2$ (**119**) [70].

Compound (BETS)$_3$[Cu$_2$(C$_2$O$_4$)$_3$]·2CH$_3$OH (**117**) contains alternating cationic and anionic layers parallel to the *ab* plane (Figure 40b) [69]. The organic layer presents the θ^{21} phase with two stacks tilted in one direction and one stack in the opposite direction (Figure 40a), also observed in the BEDT-TTF derivatives with the same [Cu$_2$(C$_2$O$_4$)$_3$]$^{2-}$ layer (compound **107**, see above). The anionic sublattice in **117** is a hexagonal honeycomb layer that shows important distortion due to the Jahn–Teller effect on the Cu(II) ions (Figure 40c). The electrical properties show that the BETS salt is a much better electrical conductor than the BEDT-TTF one with a higher room temperature conductivity and a metallic behavior down to 180 K (the BEDT-TTF salt is semiconducting). The improved electrical properties are attributed to the enhanced intermolecular interactions when S is substituted by Se. The Cu···Cu interaction is also antiferromagnetic, as in the BEDT-TTF derivative [69].

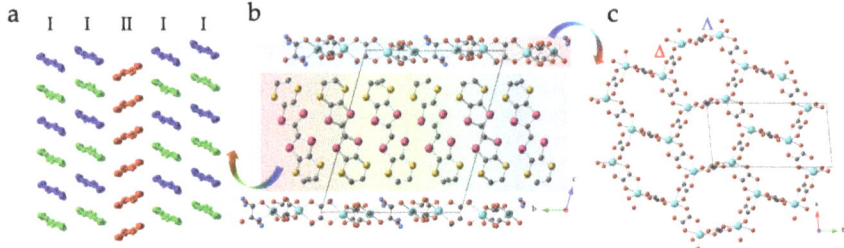

Figure 40. Structure of (BETS)$_3$[Cu$_2$(C$_2$O$_4$)$_3$]·2CH$_3$OH (**117**): (**a**) View of the cationic layer showing the θ^{21} packing and the two different stacks (I and II) with the three independent BEST molecules (A–C in red, dark blue and green, respectively). (**b**) View of the alternating cationic and anionic layers parallel to the *ab* plane. (**c**) View of the anionic layer showing the [Cu$_2$(C$_2$O$_4$)$_3$]$^{3-}$ honeycomb lattice. Color code in (**b**,**c**): Cu = light blue, C = gray, O = red, O$_{water}$ = blue, Se = pink and S = yellow. H atoms are omitted for clarity.

The other salt prepared with BETS is (BETS)$_3$[MnCr(C$_2$O$_4$)$_3$]·CH$_2$Cl$_2$ (**119**) [70]. This salt is the first one prepared with the BETS donor and an oxalate complex and also shows alternating cationic and anionic layers parallel to the *ab* plane. The cationic layer shows the α packing mode, where consecutive stacks are tilted in opposite directions (Figure 41a). The anionic layer is identical to the one observed in the BEDT-TTF derivative: it shows the classical honeycomb structure with alternating Mn(II) and Cr(III) centers connected through oxalato bridges. This salt is metallic down to 150 K (Figure 41b) and shows the expected ferromagnetic long-range order below 5.3 K (Figure 41c). In this case, the change of S to Se in the donor molecule did not improve the electrical properties [70].

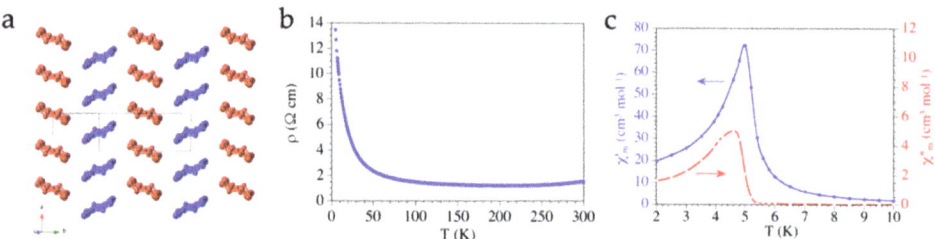

Figure 41. (**a**) View of the cationic layer in (BETS)$_3$[MnCr(C$_2$O$_4$)$_3$]·CH$_2$Cl$_2$ (**119**) showing the α packing with the two different stacks formed by the A- (in red) and B-type (in blue) BETS molecules. H atoms are omitted for clarity. (**b**) Thermal variation in the electrical resistivity of compound **119**. (**c**) Thermal variation in the in-phase (χ'_m) and out-of-phase (χ''_m) susceptibilities in compound **119**.

Finally, although its structure could not be determined, the [MnCr(C$_2$O$_4$)$_3$]$^-$ lattice has also been combined with BEST to obtain a radical salt with a formula of (BEST)$_3$[MnCr(C$_2$O$_4$)$_3$] that shows a low room temperature conductivity of 10^{-6} S/cm and a ferromagnetic order at T$_c$ = 5.6 K [61]. The change of Mn(II) to Co(II) in this honeycomb lattice leads to the isostructural lattice [CoCr(C$_2$O$_4$)$_3$]$^-$ with a higher ordering temperature that has also been combined with the donors BEST and BETS to prepare salts (BEST)$_3$[CoCr(C$_2$O$_4$)$_3$] and (BETS)$_3$[CoCr(C$_2$O$_4$)$_3$]. These two salts show ferromagnetic ordering temperatures of 10.8 and 9.2 K, respectively, and show room temperature conductivities of 10^{-6} S/cm for the BEST salt and 2.3 S/cm for the BETS one [61].

6.3. Salts with Other Donors

Finally, there is one reported salt with the [Fe$_2$(C$_2$O$_4$)$_5$]$^{4-}$ dimeric anion and the chiral donor 4,5-bis((2S)-2-hydroxypropylthio)-4′,5′-(ethylenedithio)tetrathiafulvalene (DMPET-TTF, Scheme 1): (DMPET)$_4$[Fe$_2$(C$_2$O$_4$)$_5$] (**121**) [71]. The structure of this compound shows alternating cationic and anionic layers parallel to the *ac* plane (Figure 42b), although the large size of the two 2-hydroxypropylthio groups precludes the formation of parallel donor stacks. Thus, the cationic layers are formed by single stacks of the DMPET-TTF molecules running along the *c* axis (Figure 42a). The anionic layers contain well-isolated [Fe$_2$(C$_2$O$_4$)$_5$]$^{4-}$ anions as a result of the large size of the DMPET molecules (Figure 42c). No physical properties are reported for this salt [71].

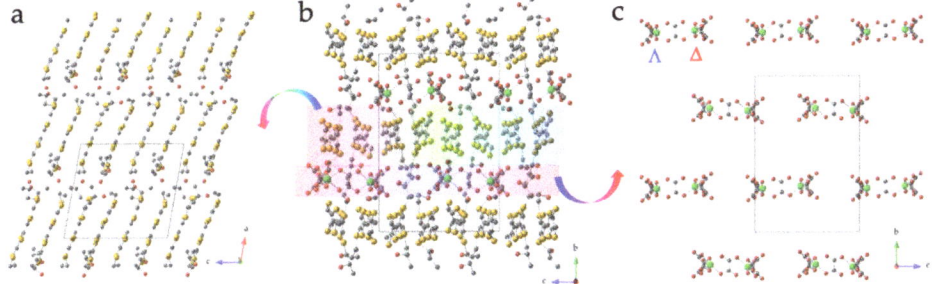

Figure 42. Structure of (DMPET-TTF)$_4$[Fe$_2$(C$_2$O$_4$)$_5$] (**121**): (**a**) View of the cationic layer showing the individual chains of DMPET-TTF molecules along the c direction. (**b**) View of the alternating cationic and anionic layers parallel to the *ac* plane. (**c**) View of the anionic layer showing the [Fe$_2$(C$_2$O$_4$)$_5$]$^{4-}$ anions. Color code: Fe = green, C = gray, O = red and S = yellow. H atoms are omitted for clarity.

Finally, although there is no structural report, the honeycomb lattice [MnCr(C$_2$O$_4$)$_3$]$^-$ has also been combined with other donors such as bis(ethylenethio)tetrathiafulvalene (BET), bis(methylenedithio)tetrathiafulvalene (BMDT-TTF), bis(ethylenedioxo)tetrathiafulvalene (BEDO-TTF) and bis(ethylenedithio)trithiaselenafulvalene (ET-1Se), whereas the isostructural [CoCr(C$_2$O$_4$)$_3$]$^-$ lattice was combined with the donor BET. The ferromagnetic ordering temperatures are in the range 5.0–5.6 K for the MnCr lattices and 13.0 K for the CoCr one. The room temperature conductivities, measured on pressed pellets, are quite high, in the range 0.1–21 S/cm [61].

Table 10. Radical salts of metal-oxalate anions with other TTF-type donor molecules (109–121).

#	CCDC	Formula [a]	SG [b]	Elect. Prop.	Donor [c]	G	Ref
109	CEWMEX	(BEST)$_4$[Cr(ox)$_3$]·PhCOOH·H$_2$O	P-1	σ = 1.5 S/cm E_a = 49 meV	BEST	PhCOOHH$_2$O	[68]
110	CEWMIB	(BEST)$_4$[Fe(ox)$_3$]·PhCOOH·H$_2$O	P-1	σ = 6.4 S/cm E_a = 54 meV	BEST	PhCOOHH$_2$O	[68]
111	CEWMOH	(BEST)$_4$[Cr(ox)$_3$]·1.5H$_2$O	$C2/m$	σ = 8.5 S/cm E_a = 62 meV	BEST	H$_2$O	[68]
112	CEWMUN	(BEST)$_9$[Fe(ox)$_3$]$_2$·7H$_2$O	P-1	σ = 2.4 S/cm E_a = 44 meV	BEST	H$_2$O	[68]
113	CEWNAU	(BEST)$_4$[Fe(ox)$_3$]·1.5H$_2$O	$C2/m$	σ = 14.0 S/cm E_a = 60 meV	BEST	H$_2$O	[68]
114	DIQFOY	(TTF)$_7$[Fe(ox)$_3$]$_2$·4H$_2$O	$P2_1/c$	σ = 10^{-4} S/cm E_a = 279 meV	TTF	H$_2$O	[59,60]
115	DIQFUE	(TTF)$_5$[Fe$_2$(ox)$_5$]·2PhCH$_3$·2H$_2$O	$C2/m$	σ = 1.8 × 10^{-6} S/cm	TTF	PhCH$_3$H$_2$O	[59,60]
116	DIQGAL	(TMTTF)$_4$[Fe$_2$(ox)$_5$]·PhCN·4H$_2$O	P-1	σ = 2.2 × 10^{-3} S/cm E_a = 290 meV	TMTTF	PhCNH$_2$O	[59,60]
117	NIDDIP	(BETS)$_3$[Cu$_2$(ox)$_3$]·2CH$_3$OH	P-1	M > 180 K	BETS	CH$_3$OH	[69]
118	OLABAE	(TTF)$_3$[Ru(ox)$_3$]·0.5EtOH·4H$_2$O	$C2/c$	σ = 1.5 × 10^{-4} S/cm E_a = 61 meV	TTF	EtOHH$_2$O	[64]
119	RUDNOT02	(BETS)$_3$[MnCr(ox)$_3$]·CH$_2$Cl$_2$	P-1	M > 150 K	BETS	CH$_2$Cl$_2$	[70]
120	TUHDOP	(TTF)$_4$[Mn(H$_2$O)$_2$[Cr(ox)$_3$]$_2$]·14H$_2$O	$C2/c$	σ = 2 × 10^{-4} S/cm E_a = 200 meV	TTF	H$_2$O	[65,66]
121	VIPYUQ	(DMPET)$_4$[Fe$_2$(ox)$_5$]	$P2_1$	-	DMPET	-	[71]

[a] ox = oxalate = C$_2$O$_4$$^{2-}$; [b] SG = space group; [c] BEST = bis(ethylenediseleno)tetrathiafulvalene; TTF = tetrathiafulvalene; TMTTF = tetramethyl-tetrathiafulvalene; BETS = bis(ethylenedithio)tetraselenafulvalene; DMPET = 4,5-bis((2S)-2-hydroxypropylthio)-4′,5′-(ethylenedithio)tetrathiafulvalene.

7. Conclusions

The seminal work of Peter Day's group in 1995 [2,3] with the synthesis of the first paramagnetic superconductors in the family of salts β''-(BEDT-TTF)$_4$[AM(C$_2$O$_4$)$_3$]·G (A = monocation, M = trivalent metal ion and G = solvent) is at the origin of the largest series of molecular metals and superconductors prepared to date. This series constitutes a paradigmatic example of the tuneability of molecular materials, as shown by the large number of related compounds prepared by changing: (i) the A$^+$ cation with other monocations (H$_3$O$^+$, NH$_4^+$, K$^+$, Na$^+$ and Li$^+$) or even with dications (Mn^{2+}, Co^{2+} and Cu^{2+}); (ii) the TTF-type donors (TTF, TM-TTF, BEST, BETS, BET, BEDO, BMDT-TTF, etc.); (iii) the M(III) ion (Fe, Cr, Ga, Mn, Rh, Ru, Al, Co and Ir), or even M(IV) as Ge(IV) and M(II) as Cu(II); and (iv) the solvent molecule (PhCN, PhNO2, PhF, PhCl, PhBr, PhI, PhCOOH, H$_2$O, CH$_2$Cl$_2$, CH$_3$OH, py, dmf, Cl-Py, Br-py, etc.). These relatively easy-to-perform modifications have led to the synthesis of more than one hundred and twenty radical salts with oxalate complexes combining electrical properties (semiconductors, metals and superconductors) with magnetic properties (paramagnetism, ferro-, ferri- and antiferromagnetic couplings and even long-range magnetic ordering). These series constitute, by far, the largest family of multifunctional molecular materials prepared to date.

No doubt, the research on this type of salt was boosted by the preparation by P. Day's group of the first molecular paramagnetic superconductor in 1995 [2,3]. Since then, several groups have prepared and characterized many different salts to try to understand the key aspects of these magnetic superconductors and to improve the magnetic and/or the electrical properties. Furthermore, this search has allowed the synthesis of magnetic conductors with other properties such as chirality or proton conductivity.

As a homage to the legacy of the late Peter Day, we have shown, here, the so-called Day series of radical salts formulated as β''-(BEDT-TTF)$_4$[AM(C$_2$O$_4$)$_3$]·G, with more than fifty reported structures to date, and of the closely related series prepared with other oxalate complexes and different donors. Many of these related series were also initiated by Peter Day's group, and still, twenty-five years later, most of the most active researchers in this field were part of his group in the past.

Author Contributions: Both authors contributed equally to the manuscript. Both authors have read and agreed to the published version of the manuscript.

Funding: This research was founded by the Spanish MINECO (project CTQ2017-87201-P AEI/FEDER, UE) and the Generalidad Valenciana (project PrometeoIII/2019/076).

Acknowledgments: This research was funded by the Spanish MINECO (project CTQ2017-87201-P AEI/FEDER, UE) and the Generalidad Valenciana (project PrometeoIII/2019/076), who we thank for the financial support. We also thank all the co-authors that appear in our contributions in this field.

Conflicts of Interest: The authors declare no conflict of interest.

References

1. Robin, M.B.; Day, P. Mixed Valence Chemistry-A Survey and Classification. *Adv. Inorg. Chem. Radiochem.* **1968**, *10*, 247–422.
2. Kurmoo, M.; Graham, A.W.; Day, P.; Coles, S.J.; Hursthouse, M.B.; Caulfield, J.L.; Singleton, J.; Pratt, F.L.; Hayes, W. Superconducting and Semiconducting Magnetic Charge Transfer Salts: (BEDT-TTF)$_4$AFe(C$_2$O$_4$)$_3$·C$_6$H$_5$CN (A = H$_2$O, K, NH$_4$). *J. Am. Chem. Soc.* **1995**, *117*, 12209–12217. [CrossRef]
3. Graham, A.W.; Kurmoo, M.; Day, P. β''-(BEDT-TTF)$_4$[(H$_2$O)Fe(C$_2$O$_4$)$_3$]·PhCN: The First Molecular Superconductor Containing Paramagnetic Metal Ions. *J. Chem. Soc. Chem. Commun.* **1995**, 2061–2062. [CrossRef]
4. Martin, L.; Turner, S.S.; Day, P.; Guionneau, P.; Howard, J.A.K.; Hibbs, D.E.; Light, M.E.; Hursthouse, M.B.; Uruichi, M.; Yakushi, K. Crystal Chemistry and Physical Properties of Superconducting and Semiconducting Charge Transfer Salts of the Type (BEDT-TTF)$_4$[A^{I}M^{III}(C$_2$O$_4$)$_3$]·PhCN (A^I = H$_3$O, NH$_4$, K; M^{III} = Cr, Fe, Co, Al; BEDT-TTF = Bis(Ethylenedithio)Tetrathiafulvalene). *Inorg. Chem.* **2001**, *40*, 1363–1371. [CrossRef]
5. Ojima, E.; Fujiwara, H.; Kato, K.; Kobayashi, H.; Tanaka, H.; Kobayashi, A.; Tokumoto, M.; Cassoux, P. Antiferromagnetic Organic Metal Exhibiting Superconducting Transition, κ-(BETS)$_2$FeBr$_4$ [BETS = Bis(Ethylenedithio)Tetraselenafulvalene]. *J. Am. Chem. Soc.* **1999**, *121*, 5581–5582. [CrossRef]
6. Kobayashi, H.; Fujiwara, E.; Fujiwara, H.; Tanaka, H.; Otsuka, T.; Kobayashi, A.; Tokumoto, M.; Cassoux, P. Antiferromagnetic Organic Superconductors, BETS$_2$FeX$_4$ (X = Br, Cl). *Mol. Cryst. Liq. Cryst.* **2002**, *380*, 139–144. [CrossRef]

7. Coronado, E.; Galán-Mascarós, J.R.; Gómez-García, C.J.; Laukhin, V. Coexistence of Ferromagnetism and Metallic Conductivity in a Molecule-Based Layered Compound. *Nature* **2000**, *408*, 447–449. [CrossRef]
8. Coronado, E.; Galán-Mascarós, J.R.; Gómez-García, C.J.; Martínez-Ferrero, E.; van Smaalen, S. Incommensurate Nature of the Multilayered Molecular Ferromagnetic Metals Based on Bis(Ethylenedithio)Tetrathiafulvalene and Bimetallic Oxalate Complexes. *Inorg. Chem.* **2004**, *43*, 4808–4810. [CrossRef]
9. Uji, S.; Shinagawa, H.; Terashima, T.; Yakabe, T.; Terai, Y.; Tokumoto, M.; Kobayashi, A.; Tanaka, H.; Kobayashi, H. Magnetic-Field-Induced Superconductivity in a Two-Dimensional Organic Conductor. *Nature* **2001**, *410*, 908–910. [CrossRef]
10. Coronado, E.; Day, P. Magnetic Molecular Conductors. *Chem. Rev.* **2004**, *104*, 5419–5448. [CrossRef]
11. Mori, T. Structural Genealogy of BEDT-TTF-Based Organic Conductors I. Parallel Molecules: Beta and Beta Phases. *Bull. Chem. Soc. Jpn.* **1998**, *71*, 2509–2526. [CrossRef]
12. Guionneau, P.; Kepert, C.J.; Bravic, G.; Chasseau, D.; Truter, M.R.; Kurmoo, M.; Day, P. Determining the Charge Distribution in BEDT-TTF Salts. *Synth. Met.* **1997**, *86*, 1973–1974. [CrossRef]
13. Prokhorova, T.G.; Khasanov, S.S.; Zorina, L.V.; Buravov, L.I.; Tkacheva, V.A.; Baskakov, A.A.; Morgunov, R.B.; Gener, M.; Canadell, E.; Shibaeva, R.P.; et al. Molecular Metals Based on BEDT-TTF Radical Cation Salts with Magnetic Metal Oxalates as Counterions: β''-(BEDT-TTF)$_4$A[M(C$_2$O$_4$)$_3$]·DMF (A = NH$_4^+$, K$^+$; M = Cr^{III}, Fe^{III}). *Adv. Funct. Mater.* **2003**, *13*, 403–411. [CrossRef]
14. Rashid, S.; Turner, S.S.; Day, P.; Howard, J.A.K.; Guionneau, P.; McInnes, E.J.L.; Mabbs, F.E.; Clark, R.J.H.; Firth, S.; Biggs, T. New Superconducting Charge-Transfer Salts (BEDT-TTF)$_4$[A·M(C$_2$O$_4$)$_3$]·C$_6$H$_5$NO$_2$ (A = H$_3$O Or NH$_4$, M = Cr Or Fe, BEDT-TTF = Bis(Ethylenedithio)Tetrathiafulvalene). *J. Mater. Chem.* **2001**, *11*, 2095–2101. [CrossRef]
15. Turner, S.S.; Day, P.; Malik, K.M.A.; Hursthouse, M.B.; Teat, S.J.; MacLean, E.J.; Martin, L.; French, S.A. Effect of Included Solvent Molecules on the Physical Properties of the Paramagnetic Charge Transfer Salts β''-(BEDT-TTF)$_4$[(H$_3$O)Fe(C$_2$O$_4$)$_3$]·solvent (BEDT-TTF = Bis(Ethylenedithio)Tetrathiafulvalene). *Inorg. Chem.* **1999**, *38*, 3543–3549. [CrossRef]
16. Akutsu-Sato, A.; Akutsu, H.; Yamada, J.; Nakatsuji, S.; Turner, S.S.; Day, P. Suppression of Superconductivity in a Molecular Charge Transfer Salt by Changing Guest Molecule: β''-(BEDT-TTF)$_4$[(H$_3$O)Fe(C$_2$O$_4$)$_3$](C$_6$H$_5$CN)$_x$(C$_5$H$_5$N)$_{1-x}$. *J. Mater. Chem.* **2007**, *17*, 2497–2499. [CrossRef]
17. Prokhorova, T.G.; Buravov, L.I.; Yagubskii, E.B.; Zorina, L.V.; Khasanov, S.S.; Simonov, S.V.; Shibaeva, R.P.; Korobenko, A.V.; Zverev, V.N. Effect of Electrocrystallization Medium on Quality, Structural Features, and Conducting Properties of Single Crystals of the (BEDT-TTF)$_4$A^I[Fe^{III}(C$_2$O$_4$)$_3$]·G. *CrystEngComm* **2011**, *13*, 537–545. [CrossRef]
18. Zorina, L.V.; Khasanov, S.S.; Simonov, S.V.; Shibaeva, R.P.; Bulanchuk, P.O.; Zverev, V.N.; Canadell, E.; Prokhorova, T.G.; Yagubskii, E.B. Structural Phase Transition in the β''-(BEDT-TTF)$_4$H$_3$O[Fe(C$_2$O$_4$)$_3$]·G Crystals (Where G is a Guest Solvent Molecule). *CrystEngComm* **2012**, *14*, 460–465. [CrossRef]
19. Prokhorova, T.G.; Buravov, L.I.; Yagubskii, E.B.; Zorina, L.V.; Simonov, S.V.; Zverev, V.N.; Shibaeva, R.P.; Canadell, E. Effect of Halopyridine Guest Molecules on the Structure and Superconducting Properties of β''-[Bis(Ethylenedithio)tetrathiafulvalene]$_4$(H$_3$O)[Fe(C$_2$O$_4$)$_3$]·Guest Crystals. *Eur. J. Inorg. Chem.* **2015**, *2015*, 5611–5620. [CrossRef]
20. Coronado, E.; Curreli, S.; Giménez-Saiz, C.; Gómez-García, C.J. The Series of Molecular Conductors and Superconductors ET$_4$[AFe(C$_2$O$_4$)$_3$]·PhX (ET = Bis(Ethylenedithio)Tetrathiafulvalene; (C$_2$O$_4$)$^{2-}$ = Oxalate; A$^+$ = H$_3$O$^+$, K$^+$; X = F, Cl, Br, and I): Influence of the Halobenzene Guest Molecules on the Crystal Structure and Superconducting Properties. *Inorg. Chem.* **2012**, *51*, 1111–1126. [CrossRef]
21. Coronado, E.; Curreli, S.; Giménez-Saiz, C.; Gómez-García, C.J. A Novel Paramagnetic Molecular Superconductor Formed by Bis(Ethylenedithio)Tetrathiafulvalene, Tris(Oxalato) Ferrate(III) Anions and Bromobenzene as Guest Molecule: ET$_4$[(H$_3$O)Fe(C$_2$O$_4$)$_3$]·C$_6$H$_5$Br. *J. Mater. Chem.* **2005**, *15*, 1429–1436. [CrossRef]
22. Sun, S.Q.; Wu, P.J.; Zhang, Q.C.; Zhu, D.B. The New Semiconducting Magnetic Charge Transfer Salt (BEDT-TTF)$_4$·H$_2$O·Fe(C$_2$O$_4$)$_3$·C$_6$H$_5$-NO$_2$: Crystal Structure and Physical Properties. *Mol. Cryst. Liq. Cryst.* **1998**, *319*, 259–269. [CrossRef]
23. Zorina, L.; Prokhorova, T.; Simonov, S.; Khasanov, S.; Shibaeva, R.; Manakov, A.; Zverev, V.; Buravov, L.; Yagubskii, E. Structure and Magnetotransport Properties of the New Quasi-Two-Dimensional Molecular Metal β''-(BEDT-TTF)$_4$H$_3$O[Fe(C$_2$O$_4$)$_3$]·C$_6$H$_4$Cl$_2$. *J. Exp. Theor. Phys.* **2008**, *106*, 347–354. [CrossRef]
24. Martin, L.; Turner, S.S.; Day, P.; Mabbs, F.E.; McInnes, E.J.L. New Molecular Superconductor Containing Paramagnetic Chromium(III) Ions. *Chem. Commun.* **1997**, 1367–1368. [CrossRef]
25. Martin, L.; Turner, S.S.; Day, P.; Malik, K.M.A.; Coles, S.J.; Hursthouse, M.B. Polymorphism Based on Molecular Stereoisomerism in Tris(Oxalato) Cr(III) Salts of BEDT-TTF [Bis(Ethylenedithio)Tetrathiafulvalene]. *Chem. Commun.* **1999**, 513–514. [CrossRef]
26. Coronado, E.; Curreli, S.; Giménez-Saiz, C.; Gómez-García, C.J. New Magnetic Conductors and Superconductors Based on BEDT-TTF and BEDS-TTF. *Synth. Met.* **2005**, *154*, 245–248. [CrossRef]
27. Prokhorova, T.G.; Yagubskii, E.B.; Zorina, L.V.; Simonov, S.V.; Zverev, V.N.; Shibaeva, R.P.; Buravov, L.I. Specific Structural Disorder in an Anion Layer and its Influence on Conducting Properties of New Crystals of the (BEDT-TTF)$_4$A$^+$[M^{3+}(Ox)$_3$]G Family, Where G is 2-Halopyridine; M is Cr, Ga; A$^+$ is [K$_{0.8}$(H$_3$O)$_{0.2}$]$^+$. *Crystals* **2018**, *8*, 92. [CrossRef]
28. Rashid, S.; Turner, S.S.; Le Pevelen, D.; Day, P.; Light, M.E.; Hursthouse, M.B.; Firth, S.; Clark, R.J.H. β''-(BEDT-TTF)$_4$[(H$_3$O)Cr(C$_2$O$_4$)$_3$]CH$_2$Cl$_2$: Effect of Included Solvent on the Structure and Properties of a Conducting Molecular Charge-Transfer Salt. *Inorg. Chem.* **2001**, *40*, 5304–5306. [CrossRef]

29. Akutsu, H.; Akutsu-Sato, A.; Turner, S.S.; Le Pevelen, D.; Day, P.; Laukhin, V.; Klehe, A.; Singleton, J.; Tocher, D.A.; Probert, M.R.; et al. Effect of Included Guest Molecules on the Normal State Conductivity and Superconductivity of β′-(ET)$_4$[(H$_3$O)Ga(C$_2$O$_4$)$_3$]·G (G = Pyridine, Nitrobenzene). *J. Am. Chem. Soc.* **2002**, *124*, 12430–12431. [CrossRef]
30. Prokhorova, T.G.; Buravov, L.I.; Yagubskii, E.B.; Zorina, L.V.; Simonov, S.V.; Shibaeva, R.P.; Zverev, V.N. Metallic Bi- and Monolayered Radical Cation Salts Based on Bis(Ethylenedithio) tetrathiafulvalene (BEDT-TTF) with the Tris(Oxalato)Gallate Anion. *Eur. J. Inorg. Chem.* **2014**, 3933–3940. [CrossRef]
31. Prokhorova, T.G.; Zorina, L.V.; Simonov, S.V.; Zverev, V.N.; Canadell, E.; Shibaeva, R.P.; Yagubskii, E.B. The First Molecular Superconductor Based on BEDT-TTF Radical Cation Salt with Paramagnetic Tris(Oxalato)Ruthenate Anion. *Crystengcomm* **2013**, *15*, 7048–7055. [CrossRef]
32. Martin, L.; Morritt, A.L.; Lopez, J.R.; Nakazawa, Y.; Akutsu, H.; Imajo, S.; Ihara, Y.; Zhang, B.; Zhang, Y.; Guo, Y. Molecular Conductors from Bis(Ethylenedithio)Tetrathiafulvalene with Tris(Oxalato)Rhodate. *Dalton Trans.* **2017**, *46*, 9542–9548. [CrossRef] [PubMed]
33. Benmansour, S.; Sánchez-Máñez, Y.; Gómez-García, C.J. Mn-Containing Paramagnetic Conductors with Bis(Ethylenedithio)Tetrathiafulvalene (BEDT-TTF). *Magnetochemistry* **2017**, *3*, 7. [CrossRef]
34. Rashid, S.; Turner, S.S.; Day, P.; Light, M.E.; Hursthouse, M.B.; Firth, S.; Clark, R.J.H. The First Molecular Charge Transfer Salt Containing Proton Channels. *Chem. Commun.* **2001**, 1462–1463. [CrossRef]
35. Akutsu-Sato, A.; Akutsu, H.; Turner, S.S.; Day, P.; Probert, M.R.; Howard, J.A.K.; Akutagawa, T.; Takeda, S.; Nakamura, T.; Mori, T. The First Proton-Conducting Metallic Ion-Radical Salts. *Angew. Chem. Int. Ed.* **2005**, *44*, 292–295. [CrossRef]
36. Morritt, A.L.; Lopez, J.R.; Blundell, T.J.; Canadell, E.; Akutsu, H.; Nakazawa, Y.; Imajo, S.; Martin, L. 2D Molecular Superconductor to Insulator Transition in the β″-(BEDT-TTF)$_2$[(H$_2$O)(NH$_4$)$_2$M(C$_2$O$_4$)$_3$]·18-Crown-6 Series (M = Rh, Cr, Ru, Ir). *Inorg. Chem.* **2019**, *58*, 10656–10664. [CrossRef]
37. Martin, L.; Lopez, J.R.; Akutsu, H.; Nakazawa, Y.; Imajo, S. Bulk Kosterlitz–Thouless Type Molecular Superconductor β″-(BEDT-TTF)$_2$[(H$_2$O)(NH$_4$)$_2$Cr(C$_2$O$_4$)$_3$]·18-Crown-6. *Inorg. Chem.* **2017**, *56*, 14045–14052. [CrossRef]
38. Martin, L.; Morritt, A.L.; Lopez, J.R.; Akutsu, H.; Nakazawa, Y.; Imajo, S.; Ihara, Y. Ambient-Pressure Molecular Superconductor with a Superlattice Containing Layers of Tris(Oxalato)Rhodate Enantiomers and 18-Crown-6. *Inorg. Chem.* **2017**, *56*, 717–720. [CrossRef]
39. Martin, L.; Day, P.; Clegg, W.; Harrington, R.W.; Horton, P.N.; Bingham, A.; Hursthouse, M.B.; McMillan, P.; Firth, S. Multi-Layered Molecular Charge-Transfer Salts Containing Alkali Metal Ions. *J. Mater. Chem.* **2007**, *17*, 3324–3329. [CrossRef]
40. Akutsu, H.; Akutsu-Sato, A.; Turner, S.S.; Day, P.; Canadell, E.; Firth, S.; Clark, R.J.H.; Yamada, J.; Nakatsuji, S. Superstructures of Donor Packing Arrangements in a Series of Molecular Charge Transfer Salts. *Chem. Commun.* **2004**, *10*, 18–19. [CrossRef]
41. Martin, L.; Day, P.; Akutsu, H.; Yamada, J.; Nakatsuji, S.; Clegg, W.; Harrington, R.W.; Horton, P.N.; Hursthouse, M.B.; McMillan, P.; et al. Metallic Molecular Crystals Containing Chiral or Racemic Guest Molecules. *CrystEngComm* **2007**, *9*, 865–867. [CrossRef]
42. Zorina, L.V.; Khasanov, S.S.; Simonov, S.V.; Shibaeva, R.P.; Zverev, V.N.; Canadell, E.; Prokhorova, T.G.; Yagubskii, E.B. Coexistence of Two Donor Packing Motifs in the Stable Molecular Metal α-Pseudo-κ-(BEDT-TTF)$_4$(H$_3$O)[Fe(C$_2$O$_4$)$_3$]·C$_6$H$_4$Br$_2$. *CrystEngComm* **2011**, *13*, 2430–2438. [CrossRef]
43. Martin, L.; Akutsu, H.; Horton, P.N.; Hursthouse, M.B.; Harrington, R.W.; Clegg, W. Chiral Radical-Cation Salts of BEDT-TTF Containing a Single Enantiomer of Tris(Oxalato)Aluminate(III) and -chromate(III). *Eur. J. Inorg. Chem.* **2015**, 1865–1870. [CrossRef]
44. Martin, L.; Akutsu, H.; Horton, P.N.; Hursthouse, M.B. Chirality in Charge-Transfer Salts of BEDT-TTF of Tris(Oxalato)Chromate(III). *CrystEngComm* **2015**, *17*, 2783–2790. [CrossRef]
45. Martin, L.; Day, P.; Horton, P.; Nakatsuji, S.; Yamada, J.; Akutsu, H. Chiral Conducting Salts of BEDT-TTF Containing a Single Enantiomer of Tris(Oxalato)Chromate(III) Crystallised from a Chiral Solvent. *J. Mater. Chem.* **2010**, *20*, 2738–2742. [CrossRef]
46. Martin, L.; Engelkamp, H.; Akutsu, H.; Nakatsuji, S.; Yamada, J.; Horton, P.; Hursthouse, M.B. Radical-Cation Salts of BEDT-TTF with Lithium Tris(Oxalato)Metallate(Iii). *Dalton Trans.* **2015**, *44*, 6219–6223. [CrossRef]
47. Martin, L.; Day, P.; Nakatsuji, S.; Yamada, J.; Akutsu, H.; Horton, P. A Molecular Charge Transfer Salt of BEDT-TTF Containing a Single Enantiomer of Tris(Oxalato)Chromate(III) Crystallised from a Chiral Solvent. *CrystEngComm* **2010**, *12*, 1369–1372. [CrossRef]
48. Benmansour, S.; Gómez-García, C.J. A Heterobimetallic Anionic 3,6-Connected 2D Coordination Polymer Based on Nitranilate as Ligand. *Polymers* **2016**, *8*, 89. [CrossRef]
49. Benmansour, S.; Gómez-Claramunt, P.; Vallés-García, C.; Mínguez Espallargas, G.; Gómez-García, C.J. Key Role of the Cation in the Crystallization of Chiral Tris(Anilato)Metalate Magnetic Anions. *Cryst. Growth Des.* **2016**, *16*, 518–526. [CrossRef]
50. Benmansour, S.; Vallés-García, C.; Gómez-Claramunt, P.; Mínguez Espallargas, G.; Gómez-García, C.J. 2D and 3D Anilato-Based Heterometallic M(I)M(III) Lattices: The Missing Link. *Inorg. Chem.* **2015**, *54*, 5410–5418. [CrossRef]
51. Martin, L.; Day, P.; Barnett, S.A.; Tocher, D.A.; Horton, P.N.; Hursthouse, M.B. Magnetic Molecular Charge-Transfer Salts Containing Layers of Water and Tris(Oxalato)Ferrate(III) Anions. *CrystEngComm* **2008**, *10*, 192–196. [CrossRef]
52. Zhang, B.; Zhang, Y.; Liu, F.; Guo, Y. Synthesis, Crystal Structure, and Characterization of Charge-Transfer Salt: (BEDT-TTF)$_5$[Fe(C$_2$O$_4$)$_3$]·(H$_2$O)$_2$·CH$_2$Cl$_2$ (BEDT-TTF = Bis(Ethylenedithio) Tetrathiafulvalene). *CrystEngComm.* **2009**, *11*, 2523–2528. [CrossRef]

53. Martin, L.; Turner, S.S.; Day, P.; Guionneau, P.; Howard, J.A.K.; Uruichi, M.; Yakushi, K. Synthesis, Crystal Structure and Properties of the Semiconducting Molecular Charge-Transfer Salt (BEDT-TTF)$_2$Ge(C$_2$O$_4$)$_3$·PhCN [BEDT-TTF = Bis(Ethylenedithio)Tetrathiafulvalene]. *J. Mater. Chem.* **1999**, *9*, 2731–2736. [CrossRef]
54. Martin, L.; Day, P.; Nakatsuji, S.; Yamada, J.; Akutsu, H.; Horton, P.N. BEDT-TTF Tris(Oxalato)Germanate(IV) Salts with Novel Donor Packing Motifs. *Bull. Chem. Soc. Jpn.* **2010**, *83*, 419–423. [CrossRef]
55. Lopez, J.R.; Akutsu, H.; Martin, L. Radical-Cation Salt with Novel BEDT-TTF Packing Motif Containing Tris(Oxalato)Germanate(IV). *Synth. Met.* **2015**, *209*, 188–191. [CrossRef]
56. Wang, P.; Bandow, S.; Maruyama, Y.; Wang, X.; Zhu, D. Physical and Structural Properties of a New Organic Conductor (BEDT-TTF)$_4$Cu(C$_2$O$_4$)$_2$. *Synth. Met.* **1991**, *44*, 147–157. [CrossRef]
57. Qian, M.; Rudert, R.; Luger, P.; Ge, C.; Wang, X. Structure of the 1:4 Complex of Bis[1,2-Oxalato(2-)] Copper(II) and Bis(Ethylenedithio)Tetrathiafulvalene (BEDT-TTF). *Acta Cryst. C* **1991**, *47*, 2358–2362. [CrossRef]
58. Rashid, S.; Turner, S.S.; Day, P.; Light, M.E.; Hursthouse, M.B. Molecular Charge-Transfer Salt of BEDT-TTF [Bis(Ethylenedithio)Tetrathiafulvalene] with the Oxalate-Bridged Dimeric Anion [Fe$_2$(C$_2$O$_4$)$_5$]$^{4-}$. *Inorg. Chem.* **2000**, *39*, 2426–2428. [CrossRef]
59. Clemente-Leon, M.; Coronado, E.; Galán-Mascarós, J.R.; Giménez-Saiz, C.; Gómez-García, C.J.; Fabre, J.M. Molecular Conductors Based upon TTF-Type Donors and Octahedral Magnetic Complexes. *Synth. Met.* **1999**, *103*, 2279–2282. [CrossRef]
60. Coronado, E.; Galán-Mascarós, J.R.; Gómez-García, C.J. Charge Transfer Salts of Tetrathiafulvalene Derivatives with Magnetic Iron(III) Oxalate Complexes: [TTF]$_7$[Fe(Ox)$_3$]$_2$·4H$_2$O, [TTF]$_5$[Fe$_2$(Ox)$_5$]·2PhMe·2H$_2$O and [TMTTF]$_4$[Fe$_2$(Ox)$_5$]·PhCN ·4H$_2$O (TMTTF = Tetramethyltetrathiafulvalene). *J. Chem. Soc. Dalton Trans.* **2000**, 205–210. [CrossRef]
61. Alberola, A.; Coronado, E.; Galán-Mascarós, J.R.; Giménez-Saiz, C.; Gómez-García, C.J.; Martínez-Ferrero, E.; Murcia-Martínez, A. Multifunctionality in Hybrid Molecular Materials: Design of Ferromagnetic Molecular Metals. *Synth. Met.* **2003**, *135*, 687–689. [CrossRef]
62. Zhang, B.; Zhang, Y.; Zhu, D. (BEDT-TTF)$_3$Cu$_2$(C$_2$O$_4$)$_3$(CH$_3$OH)$_2$: An Organic-Inorganic Hybrid Antiferromagnetic Semiconductor. *Chem. Commun.* **2012**, *48*, 197–199. [CrossRef] [PubMed]
63. Alberola, A.; Coronado, E.; Galán-Mascarós, J.R.; Giménez-Saiz, C.; Gómez-García, C.J.; Romero, F.M. Multifunctionality in Hybrid Molecular Materials: Design of Ferromagnetic Molecular Metals and Hybrid Magnets. *Synth. Met.* **2003**, *133*, 509–513. [CrossRef]
64. Coronado, E.; Galán-Mascarós, J.R.; Giménez-Saiz, C.; Gómez-García, C.J.; Martínez-Agudo, J.M.; Martinez-Ferrero, E. Magnetic Properties of Hybrid Molecular Materials Based on Oxalato Complexes. *Polyhedron* **2003**, *22*, 2381–2386. [CrossRef]
65. Coronado, E.; Galán-Mascarós, J.R.; Giménez-Saiz, C.; Gómez-García, C.J.; Ruiz-Pérez, C.; Triki, S. Hybrid Molecular Materials Formed by Alternating Layers of Bimetallic Oxalate Complexes and Tetrathiafulvalene Molecules: Synthesis, Structure, and Magnetic Properties of TTF$_4${Mn(H$_2$O)$_2$[Cr(Ox)$_3$]$_2$}·14H$_2$O. *Adv. Mater.* **1996**, *8*, 737–740. [CrossRef]
66. Coronado, E.; Galán-Mascarós, J.R.; Giménez-Saiz, C.; Gómez-García, C.J.; Ruiz-Perez, C. Hybrid organic/inorganic Molecular Materials Formed by Tetrathiafulvalene Radicals and Magnetic Trimeric Clusters of Dimetallic Oxalate-Bridged Complexes: The Series (TTF)$_4${M^{II}(H$_2$O)$_2$[M^{III}(Ox)$_3$]$_2$}·nH$_2$O (M^{II} = Mn, Fe, Co, Ni, Cu and Zn; M^{III} = Cr and Fe; Ox = C$_2$O$_4{}^{2-}$). *Eur. J. Inorg. Chem.* **2003**, 2290–2298. [CrossRef]
67. Mori, T.; Mori, H.; Tanaka, S. Structural Genealogy of BEDT-TTF-Based Organic Conductors—II. Inclined Molecules: Theta, Alpha, and Chi Phases. *Bull. Chem. Soc. Jpn.* **1999**, *72*, 179–197. [CrossRef]
68. Coronado, E.; Curreli, S.; Giménez-Saiz, C.; Gómez-García, C.J.; Alberola, A. Radical Salts of Bis(Ethylenediseleno)Tetrathiafulvalene with Paramagnetic Tris(Oxalato)Metalate Anions. *Inorg. Chem.* **2006**, *45*, 10815–10824. [CrossRef]
69. Zhang, B.; Zhang, Y.; Wang, Z.; Gao, S.; Guo, Y.; Liu, F.; Zhu, D. BETS$_3$[Cu$_2$(C$_2$O$_4$)$_3$](CH$_3$OH)$_2$: An Organic-Inorganic Hybrid Antiferromagnetic Metal (BETS = Bisethylene(Tetraselenfulvalene)). *CrystEngComm* **2013**, *15*, 3529–3535. [CrossRef]
70. Alberola, A.; Coronado, E.; Galán-Mascarós, J.R.; Giménez-Saiz, C.; Gómez-García, C.J. A Molecular Metal Ferromagnet from the Organic Donor Bis(Ethylenedithio)Tetraselenafulvalene and Bimetallic Oxalate Complexes. *J. Am. Chem. Soc.* **2003**, *125*, 10774–10775. [CrossRef]
71. Awheda, I.; Krivickas, S.J.; Yang, S.; Martin, L.; Guziak, M.A.; Brooks, A.C.; Pelletier, F.; Le Kerneau, M.; Day, P.; Horton, P.N.; et al. Synthesis of New Chiral Organosulfur Donors with Hydrogen Bonding Functionality and their First Charge Transfer Salts. *Tetrahedron* **2013**, *69*, 8738–8750. [CrossRef]

Article

Structures and Properties of New Organic Molecule-Based Metals, (D)₂BrC₂H₄SO₃ [D = BEDT-TTF and BETS]

Hiroki Akutsu [1,*], Yuta Koyama [1], Scott S. Turner [2] and Yasuhiro Nakazawa [1]

[1] Department of Chemistry, Graduate School of Science, Osaka University, 1-1 Machikaneyama, Toyonaka, Osaka 560-0043, Japan; u293605g@alumni.osaka-u.ac.jp (Y.K.); nakazawa@chem.sci.osaka-u.ac.jp (Y.N.)
[2] Department of Chemistry, University of Surrey, Guildford GU2 7XH, UK; s.s.turner@surrey.ac.uk
* Correspondence: akutsu@chem.sci.osaka-u.ac.jp; Tel.: +81-6-6850-5399

Abstract: An organic anion, 2-bromoethanesulfonate ($BrC_2H_4SO_3^-$), provides one bis(ethylenedithio) tetrathiafulvalene (BEDT-TTF) and two bis(ethylenedithio)tetraselenafulvalene (BETS) salts, the compositions of which are β"-β"-(BEDT-TTF)₂BrC₂H₄SO₃ (**1**), β"-β"-(BETS)₂BrC₂H₄SO₃ (**2**), and θ-(BETS)₂BrC₂H₄SO₃ (**3**), respectively. Compound **1** shows a metal–insulator transition at around 70 K. Compound **2** is isomorphous to **1**, and **3** is polymorphic with **2**. Compounds **2** and **3** show metallic behavior at least down to 4.2 K. The pressure dependence of the electrical resistivity of **1** is also reported.

Keywords: organic conductors; organic anions; electrocrystallization; crystal structure; band structure; electrical resistivity; magnetic susceptibility

Citation: Akutsu, H.; Koyama, Y.; Turner, S.S.; Nakazawa, Y. Structures and Properties of New Organic Molecule-Based Metals, (D)₂BrC₂H₄SO₃ [D = BEDT-TTF and BETS]. *Magnetochemistry* **2021**, *7*, 91. https://doi.org/10.3390/magnetochemistry7070091

Academic Editor: Manuel Almeida

Received: 15 May 2021
Accepted: 18 June 2021
Published: 23 June 2021

Publisher's Note: MDPI stays neutral with regard to jurisdictional claims in published maps and institutional affiliations.

Copyright: © 2021 by the authors. Licensee MDPI, Basel, Switzerland. This article is an open access article distributed under the terms and conditions of the Creative Commons Attribution (CC BY) license (https://creativecommons.org/licenses/by/4.0/).

1. Introduction

Over the past half century, numerous organic conductors have been prepared [1], particularly electron donor–anion type conductors, based on TTF (tetrathiafulvalene), TMTTF (tetramethyltetrathiafulvalene), TMTSF (tetramethyltetraselenafulvalene), BEDT-TTF (bis(ethylenedithio)tetrathiafulvalene), BETS (bis(ethylenedithio)tetraselenafulvalene), etc., with a wide variety of counterions, which are commonly inorganic, such as BF_4^-, PF_6^-, ClO_4^-, Cl^-, Br^-, I^-, I_3^-, AuI_2^-, etc. Organic anions, which are less common than inorganic ions, have also been used. The first radical cation salts with organic anions were reported by D. R. Rosseinsky et al. in 1979 [2], where $CH_3CO_2^-$, maleate, fumarate, and $p\text{-MeC}_6H_4SO_2^-$ were used. Each electrocrystallization with TTF gave powder, microcrystalline, or blocklets. However, the crystal structures were not determined. According to a review about the organic conducting salts with organic and organometallic anions [3], in 1983, the crystal structures of a $CF_3SO_3^-$ salt of TMTSF was reported [4]. However, $CF_3SO_3^-$ is not clearly organic because the anion includes no hydrogens. In 1985, the crystal structures of TMTTF and TTMTTF (tetramethylthiotetrathiafulvalene) salts with hexacyanobutadiene (HGBD) were reported [5]. However, HGBD was used not as an anion but as an electron acceptor because the crystals were obtained by mixing hot acetonitrile solutions of the donor and the acceptor, namely HGBD. This indicates that these are donor–acceptor complexes. The crystal structure of the first organic conducting salt with an organic anion reported in 1988 by Peter Day's group [6] is (BEDT-TTF)₂($p\text{-CH}_3C_6H_4SO_3$). After the discovery, sulfonates were widely used as counterions of organic conductors [3,7–10] because sulfonates have relatively low pK_a (<1), which indicates that the bare monoanion state ($-SO_3^-$) is far more stable than the protonated state ($-SO_3H$). By contrast carboxylates ($-CO_2^-$) are not useful as counterions because of the relatively high pK_a (>3), where the protonated state ($-CO_2H$) is much more stable than the bare monoanion state ($-CO_2^-$). We have already reported several organic conducting salts with sulfonates [11–41], which are relatively large and anisotropic and several of which have provided unique salts having polar counterion layers [20,28,29,35,36,38,41]. Schematic diagrams of the crystal structures of the salts are shown in Figure S1. Here we report new BEDT-TTF and BETS salts of a relatively small and anisotropic sulfonate, $BrC_2H_4SO_3^-$.

2. Results and Discussion

A conventional constant-current electrocrystallization in a mixed solvent of PhCl (18 mL) and EtOH (2 mL) with BEDT-TTF (10 mg), $BrC_2H_4SO_3Na$ (44 mg), and 18-crown-6 ether (67 mg) gave black blocks and thick needles. X-ray analyses indicated that the blocks and needles had the same cell parameters. The resulting data were solved as β"-β"-(BEDT-TTF)$_2$(BrC$_2$H$_4$SO$_3$) (**1**). Using BETS (5 mg) instead of BEDT-TTF afforded dark green plates (**3**) with a small number of black blocks and needles (**2**). X-ray analyses indicated that **2** was isomorphous to **1**, and the composition of the major product **3** was θ-(BETS)$_2$(BrC$_2$H$_4$SO$_3$), which is a polymorph of **2** and the cell parameters are different from those of **2**. The crystallographic data of **1**, **2**, and **3** are shown in Table 1. Using o-C$_6$H$_4$Cl$_2$ (18 mL) instead of PhCl in the electrocrystallization of **1** yielded black thin plates (**4**), X-ray analysis of which indicated that the composition was (BEDT-TTF)$_3$(Br$_3$)$_5$, the structure and properties of which have already been reported [42]. This indicates that the $BrC_2H_4SO_3^-$ anion decomposed during the electrocrystallization process. Indeed, both the cell parameters and the crystal structure (Figure S2) were the same as reported in ref. [42]. However, the electrical resistivity measurement indicated that the salt **4** was metallic down to 4.2 K (Figure S3), whereas ref. [42] reported a broad metal–insulator (MI) transition at 120 K. The difference may be caused by a difference in the direction of the resistivity measurements and/or crystal conditions.

Table 1. Crystallographic data of **1**, **2**, and **3**.

Compound	1	2	3	3	3
Formula	C$_{22}$H$_{20}$O$_3$S$_{17}$Br	C$_{22}$H$_{20}$O$_3$S$_9$Se$_8$Br	C$_{22}$H$_{20}$O$_3$S$_9$Se$_8$Br	C$_{22}$H$_{20}$O$_3$S$_9$Se$_8$Br	C$_{22}$H$_{20}$O$_3$S$_9$Se$_8$Br
Fw [1]	957.32	1332.52	1267.52	1267.52	1267.52
Space Group	$P\bar{1}$	$P\bar{1}$	$C2/c$	$C2/c$	$C2/c$
a (Å)	5.8671(2)	5.93509(18)	35.850(2)	35.686(2)	35.633(2)
b (Å)	8.7790(2)	8.8364(2)	5.1587(3)	5.1193(3)	5.0986(4)
c (Å)	33.3201(7)	34.0428(9)	9.9900(6)	9.9154(7)	9.9154(7)
α (°)	89.076(6)	88.939(6)	90.0	90.0	90.0
β (°)	85.469(6)	85.697(6)	93.346(7)	92.695(7)	92.673(7)
γ (°)	75.793(5)	76.167(5)	90.0	90.0	90.0
V (Å^3)	1658.53(9)	1728.70(9)	1844.38(19)	1809.43(19)	1798.1(2)
Z	2	2	2	2	2
T (K)	150	150	290	150	110
d_{calc} (g·cm^{-1})	1.917	2.560	2.399	2.446	2.461
μ (cm^{-1}) [2]	23.434	101.896	95.505	97.349	97.963
F(000) [3]	966	1254	1254	1254	1254
2θ range (°)	4–55	4–55	4–55	4–55	4–55
Total ref.	16,127	16,694	8164	7936	8018
Unique ref.	7559	7871	2110	2081	2063
R_{int}	0.0325	0.0672	0.0733	0.0391	0.0431
Parameters	407	395	118	118	118
R_1 ($I > 2\sigma(I)$)	0.032	0.049	0.063	0.051	0.059
wR_2 (all data)	0.089	0.158	0.205	0.138	0.166
S [4]	1.009	1.042	1.060	1.119	1.088
$\Delta\rho_{max}$ (e Å^{-3})	1.49	1.94	1.18	2.18	2.20
$\Delta\rho_{min}$ (e Å^{-3})	−0.72	−1.73	−0.62	−1.92	−2.06
CCDC number	2,083,397	2,083,398	2,088,144	2,083,399	2,088,145

[1] Formula weight, [2] linear absorption coefficient, [3] total number of electrons in the unit cell, [4] goodness of fitness.

2.1. Crystal Structures

2.1.1. Crystal Structure of β″-β″-(BEDT-TTF)$_2$(BrC$_2$H$_4$SO$_3$) (1)

Figure 1a shows the crystal structure of **1**. Two BEDT-TTF molecules (A and B) and one BrC$_2$H$_4$SO$_3^-$ anion are crystallographically-independent. The unit cell has two independent donor layers, one of which consists of only the donor A (A layer) and the other of which consists of only the donor B (B layer). One of the two ethylene groups of A is disordered and refined over two positions, the refined occupancies of which are found to be 0.734 and 0.266. Both donor layers have quite similar β″-type arrangements, as shown in Figure 1b,c. Each donor layer alternates with an anionic layer. Figure 1d shows short contacts between S atoms of BEDT-TTF molecules and O atoms of sulfo (–SO$_3^-$) groups. In our previous reports, concerning sulfonate salts that are in the charge-ordered state, the donor having the shortest S···O contact with a sulfonate has the largest positive charge, perhaps because of the largest Madelung potential [20,31,35]. However, in salt **1**, both A and B molecules have short S···O contacts, the distances of which are almost the same. This suggests that no charge disproportionation occurs. The molecular charges of A and B were estimated from bond lengths according to the literature method [43]. The calculated charges, normalized by the total formula charge (and non-normalized charge), are +0.492 (+0.567) and +0.508 (+0.586) for A and B, respectively. Both values are close to +0.5, suggesting no charge disproportionation. Figure 1e shows the structure of the anionic layer. All anions in the layer orient along the same direction (//b), which confers a dipole moment on the anionic layer. However, an inversion center provides the other anionic layer in the unit cell with the opposite dipole moment. Therefore, no net dipole moment exists. According to our classification suggested in ref. [44], the crystal has a Type III dipole arrangement (Figure S1) [44,45]. In addition, the dipole moment of the BrC$_2$H$_4$SO$_3^-$ anion was calculated by MOPAC7 [46] using the geometry observed in **1** to be 9.9 D.

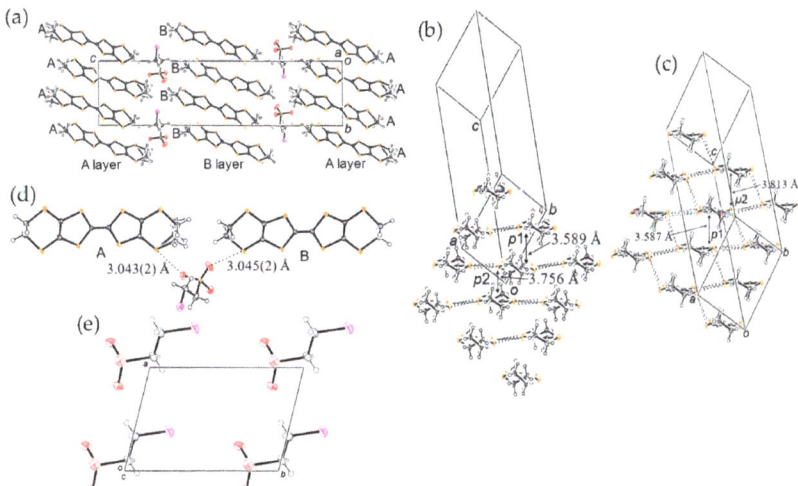

Figure 1. (a) Crystal structure, donor arrangements of (b) A and (c) B layers with two crystallographically-independent plane–plane distances, where the dashed lines indicate S···S contacts shorter than the van der Waals distance of 3.70 Å, (d) short contacts between donors and anions, and (e) structure of the anion layer of **1**.

2.1.2. Crystal Structure of β″-β″-(BETS)$_2$(BrC$_2$H$_4$SO$_3$) (2)

As shown in Table 1, the cell parameters of **2** are almost the same as **1**, suggesting that both are isomorphous. Therefore, the crystal packing structure of **2** is not shown. There are two independent donors (A and B) and one anion in the asymmetric unit. Similar to **1**, one of the two ethylene groups of A is disordered and refined over two positions. The occupancies were found to be 0.799 and 0.201. Figure 2a,b shows the donor arrangements of both A and B layers, which are almost the same as **1**. However, a larger number of short contacts were observed in A and B layers compared to **1**, suggesting that **2** has stronger donor–donor interactions than **1**. Figure 2c shows short contacts between S atoms of BETS molecules and O atoms of sulfo (–SO$_3^-$) groups. An A molecule has a S···O contact that is 0.032 Å shorter than that of B. However, the difference is one order smaller than those of charge-ordered salts [20,31,35]. The relatively small difference suggests that the short anon–donor interactions provide no charge-ordered state. In addition, this cannot be confirmed since there are no previous studies on the relationship between bond lengths in BETS and the molecular charges. The molecular arrangement of the anionic layers of **2** is almost the same as **1**, and we do not show the structure. All anions in the layer orient along the same direction (//b) and therefore the crystal also has a Type III dipole arrangement [44,45]. In addition, the dipole moment of the BrC$_2$H$_4$SO$_3^-$ anion was calculated by MOPAC7 [46] using the geometry observed in **2** to be 9.7 D. In addition, a CCDC search indicated that the structures of 72 BETS salts have already been reported, which consist of 21 κ–, 18 θ–, 7 λ–, 7 α–, 4 β–, and 15 other miscellaneous types of salts. However, no β″-type salts have been reported yet, indicating that **2** is the first BETS-based salt having a β″-type donor arrangement.

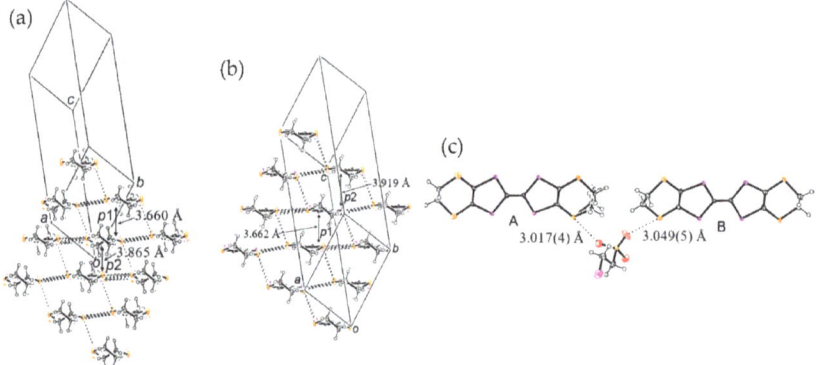

Figure 2. Donor arrangements of (**a**) A and (**b**) B layers of **2** with two crystallographically-independent plane–plane distances, where the dashed lines indicate short Se···Se (<3.80 Å), Se···S (<3.75 Å), and S···S contacts (<3.70 Å). (**c**) Short contacts between donors and anions of **2**.

2.1.3. Crystal Structure of θ-(BETS)$_2$(BrC$_2$H$_4$SO$_3$) (**3**)

The thin plate crystals were the main products under the electrocrystallization condition, as mentioned previously. Figure 3a shows the molecular structure of the asymmetric unit of **3**. There are a half of BETS and a quarter of anion in the asymmetric unit. The donor is not disordered and located about the center of symmetry, whereas the anion is heavily disordered. The asymmetric part of the anion consists of –SO$_3^-$, each occupancy of which is 0.25, and –CH$_2$CH$_2$Br disordered over two positions (–C61–C71–Br11 and –C62–C72–Br12), where each occupancy is 0.125. Thus, the anion is disordered over eight positions. Relatively large residual densities were observed due to the severe disorder. The asymmetric part of the anion is located about the center of symmetry, so that the anion layer is non-polar. The crystal structure is shown in Figure 3b. One donor layer and one anion layer are crystallographically-independent. Figure 3c shows the packing arrangement of the donor layer, which has a so called θ-type packing motif. The dihedral angle (θ) shown in Figure 3c of 98.1° was observed. A literature source [47] shows a phase diagram of the θ-type BEDT-TTF-based salts as a function of the dihedral angle (θ), where 98.1° is located in a superconducting phase. However, the salt **3** is BETS-based, not BEDT-TTF-based. BETS usually provides much stronger intermolecular interactions, which give much larger transfer integrals. Therefore, the interaction in **3** is stronger than that of the BEDT-TTF-based salt, which suggests that **3** is a stable metal [48]. The occupancy of the $BrC_2H_4SO_3^-$ anion of 0.25 suggests that there are four possible positions, one of which is occupied by an anion and the other three of which are vacant in the actual crystal. The anion is located about the center of symmetry, indicating that there are two possible positions on the inversion center. The length of the b axis of 5.1193(3) Å is too short for the anions to occupy each unit cell, suggesting that the anions exist every two-unit cells along the b axis. If anions exist in every two-unit cells regularly, which makes the length of the b axis double, the crystal usually provides satellite reflections and/or diffuse streaks on the X-ray photographs. No satellite reflections and/or diffuse streaks were observed in all 44 measured photographs. The lack of any superstructures suggests that the disorder is not so simple that each anion chain has unique periodicity. Furthermore, each donor, which also can form a superstructure of a $2k_F$ (four-fold) or $4k_F$ (two-fold) charge–density wave, gathers not to form any superstructures but to form a uniform stack. This again suggests that the salt is a stable metal.

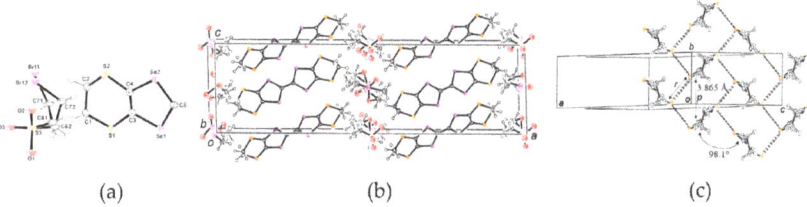

(a)   (b)   (c)

Figure 3. (**a**) Molecular structure of the asymmetric unit, (**b**) crystal structure, and (**c**) packing motif of the donor layer with one independent plane–plane distance and dihedral angle of **3** at 150 K.

2.2. Electrical Resistivity

Temperature dependences of electrical resistivities of **1**, **2**, and **3** are shown in Figure 4. The BEDT-TTF salt **1** shows a metal–insulator (MI) transition at around 70 K, and then the resistivity gradually increases, and the resistivity at 4.2 K is 20 times larger than that at 70 K. β″-salts can be classified into several groups according to structural features [49]. According to the classification, **1** belongs to the β″$_{211}$-type. Most β″$_{211}$-salts are stable metals apart from the two salts, $(BEDT-TTF)_2Br_2SeCN$ and $(BEDT-TTF)_2Cl_2SeCN$, which show relatively sharp MI transitions at 200 K. The isomorphous BETS-based salt **2** shows metallic behavior from room temperature down to 4.2 K. Salt **3** also shows metallic behavior across the whole temperature range.

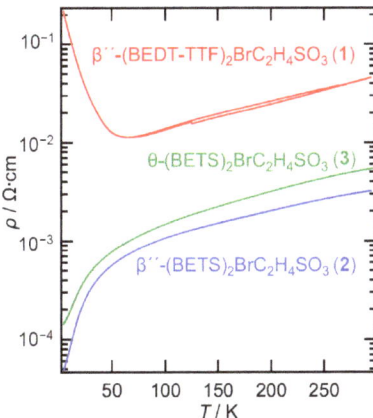

Figure 4. Temperature-dependent electrical resistivities of **1**, **2**, and **3**.

2.3. Magnetic Susceptibility

Temperature dependences of magnetic susceptibility of **1** and **3** are shown in Figure 5a,b, respectively. We have not yet obtained a sufficient quantity of **2** for SQUID measurement, and therefore the susceptibility data of **2** is not available. The magnetic susceptibility of **1** decreases monotonically from room temperature (RT) from 6 to 2×10^{-4} emu mol^{-1}, which are in the range of normal Pauli paramagnetism of organic BEDT-TTF-based metals [50], and then shows a more rapid decrease in susceptibility at the same temperature as was observed (70 K) for the metal–insulator (MI) transition. The susceptibility then becomes almost zero at the lowest measured temperature. The result indicates a non-magnetic ground state for **1**. Most β″-salts have charge-ordered ground states, suggesting that the ground state of **1** has a non-magnetic 0011 type of charge-ordered pattern [1]. We will discuss further this in the context of the electronic structure of **1** later. The susceptibility of **3** (Figure 5b) is almost constant, $\approx 2.5 \times 10^{-4}$ emu mol^{-1}, from RT to 175 K. On further lowering the temperature, the susceptibility decreases monotonically down to 2 K. The non-zero susceptibility, $\approx 1.2 \times 10^{-4}$ emu mol^{-1}, at the lowest temperature suggests that the sample has an itinerant nature down to 2 K. In addition, the ρ-T plot of **3** (Figure 5c) has an anomaly at 175 K, at which temperature the susceptibility starts decreasing, suggesting that there is a transition between two metallic phases at around 175 K. The temperature dependence of magnetic susceptibility of θ-(BETS)$_2$Cu$_2$Cl$_6$ [51], which is also a stable metal with θ = 100°, is similar to that of **3**. The broad decrease was observed from 2.3×10^{-4} above 120 K, where a structural phase transition ($Pbcn$ to $P2_12_12$) is observed, to 1.0×10^{-4} emu mol^{-1} below 70 K. X-ray analyses of **3** at 110 and 290 K were also performed (Table 1) to confirm whether there is a phase transition at around 175 K. However, cell parameters at 110 and 290 K are almost the same, and no structural phase transition was observed. The origin is not clear at present.

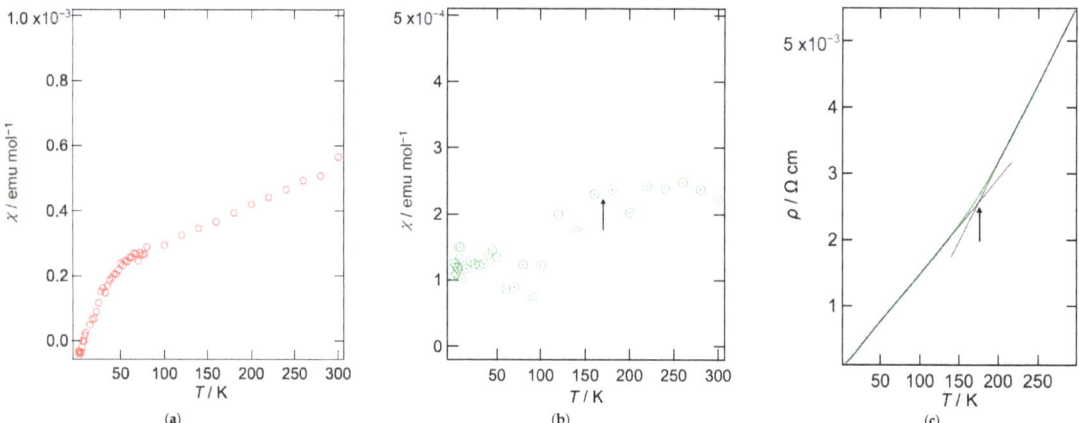

Figure 5. Temperature dependences of magnetic susceptibilities of (**a**) **1** and (**b**) **3**, where 0.46 and 0.30% of Curie tails have been subtracted, respectively. (**c**) Resistivity–temperature plots of **3**.

2.4. Band Structure Calculations

Band structures of **1**, **2**, and **3** were calculated using the Mori's band structure calculation software package [52]. The resultant overlap integrals are shown in Table 2. Figure 6 shows a schematic diagram of the donor layers with the directions and labelling of donor–donor interactions. The $p1$ values of both A and B layers in **1** is one and two orders larger than those of $p2$, respectively, indicating strong dimerization along the stacking direction in each layer, which is also confirmed by the intermolecular spacing; the plane–plane distances of $p1$ are 0.167 (A layer) and 0.226 Å (B layer) shorter than those of $p2$ (see

Figure 1b,c). The side-by-side interactions ($r1$, $r2$, and s) were larger than $p1$, suggesting that the salt is metallic along the side-by-side directions. Band dispersions and Fermi surfaces of **1** are shown in Figure 7a. There are quasi-1D electron sheets and hole pockets in the first Brillouin zone. The Fermi surface is open along the stacking direction ($//b$), suggesting that the metallic tendency along the side-by-side directions is stronger than that along the stacking direction. As previously described, the salt shows a MI transition, for which we speculate that a tetramerization along the side-by-side direction occurs (perhaps in the s direction because s values of A and B are the largest). The tetramerization will make the salt a band insulator to give a diamagnetic ground state, which is consistent with the magnetic behavior shown in Figure 5a. In addition, since the A and B layers of **1** have almost the same donor arrangements and overlap integrals (Table 2), both band dispersions and Fermi surfaces are also quite similar.

Table 2. Overlap integrals ($\times 10^{-3}$) of **1** and **2** at 150 K.

Salts	1		2	
Layers	A	B	A	B
$p1$	−5.03	−4.88	−6.43	−5.97
$p2$	−0.36	−0.06	+2.53	+3.50
$r1$	−9.80	−9.89	−17.73	−17.27
$r2$	−10.74	−9.60	−18.70	−16.87
s	−14.50	−14.98	−26.02	−26.96

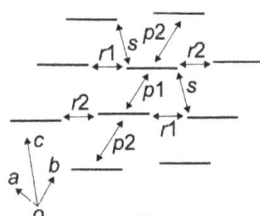

Figure 6. Schematic diagram of the structure of the donor layer with directions of interactions of **1** and **2**.

Similarly, the band dispersions and Fermi surfaces of A and B layers of **2** are also similar (Figure 7b) because, again, both layers have almost the same donor arrangements. Since **1** and **2** are isomorphous, their electronic structures are almost the same as compared in Figure 7a,b. However, **2** does not show a MI transition. The most significant difference between **1** and **2** is a degree of dimerization along the b axis. The values of $|p1|/|p2|$ of **1** are 14.0 and 81.3 for A and B layers, respectively, but the similar values for **2** are only 2.5 and 1.7 for A and B layers, respectively. The considerably smaller values for **2** suggest that the support for dimerization in **2** is far weaker than in **1**, which is not reflected in the intermolecular spacing. The plane–plane distances of $p1$ are 0.205 (A layer) and 0.257 Å (B layer) shorter than those of $p2$ (see Figure 2a,b). The differences in **2** are rather larger than those in **1**. The weak dimerization in **2** makes carriers less correlated and stabilizes the metallic state. In addition, the band width of the conduction bands of **2** are 1.77 eV, which is 1.9 times larger than that of the 0.93 eV of **1**.

The donor arrangement of **3** is simple so that there are only two crystallographically-independent overlap integrals, one of which (p) lies along the stacking direction, and the other (r) lies along the side-by-side direction, as shown in Figure 3c. Values of $p = 3.26$ and $r = 22.52 \times 10^{-3}$ were calculated. Figure 8 shows the band dispersions and Fermi surfaces of **3**. There is a simple and large 2D Fermi surface in the first Brillouin zone, and the band width of 1.80 eV was observed.

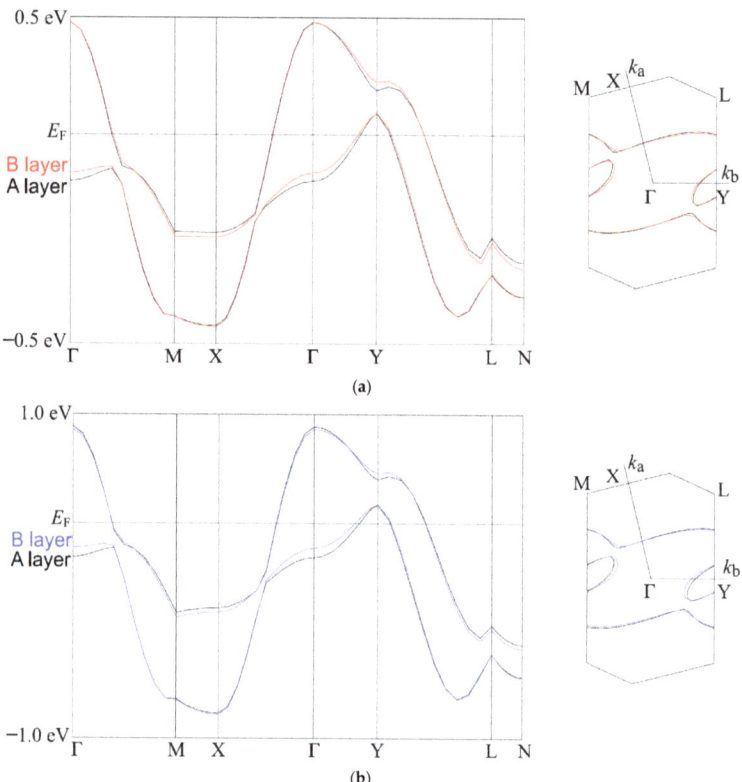

Figure 7. Band dispersions (left) and Fermi surfaces (right) of **1** (**a**) and **2** (**b**) at 150 K.

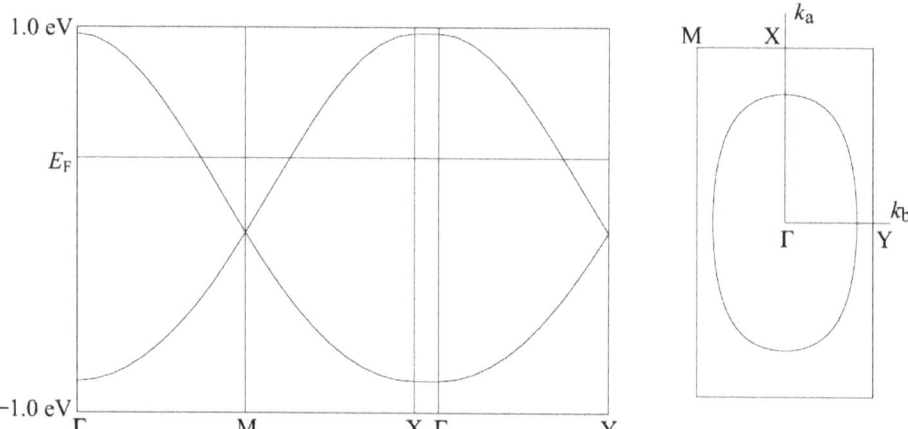

Figure 8. Band dispersions (**left**) and Fermi surfaces (**right**) of **3** at 150 K.

2.5. Electrical Resistivity under Pressure

For compound **1**, the electrical resistivity under static pressure up to 0.5 GPa (5.0 kbar) was measured. This was achieved using a clamp-type pressure cell from RT to 4.2 K (Figure 9). Applying 1.0 kbar of pressure made the MI transition sharper and moved the transition temperature (T_{MI}) 10 K higher to 80 K. By contrast the 2.0 kbar curve reveals an upturn at 55 K, which is 15 K lower than that at 1 bar, but the transition is still sharper than that at 1 bar. The T_{MI} at 3.0 kbar is 51 K, which is only 4 K smaller than that at 2.0 kbar; however, the resistivity at 4.2 K ($\rho_{4.2\,K}$) is more than two orders of magnitude smaller than that at 2 kbar. In fact, $\rho_{4.2\,K}$ decreases with increasing pressure from 3 to 5 kbar. The upturn for the MI transition almost disappears at 4.5 kbar, but we have not yet observed superconductivity at the measured pressures and down to 4.2 K.

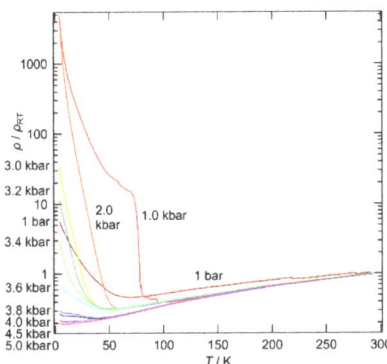

Figure 9. Electrical resistivity under pressure of **1**.

3. Materials and Methods

BEDT-TTF, purchased from Tokyo Chemical Industry Co. Ltd., Tokyo, Japan, and BrC$_2$H$_4$SO$_3$Na, purchased from FUJIFILM Wako Pure Chemical Corporation (Chuo-Ku Osaka, Japan), were used without purification. 18-Crown-6 ether was used after recrystallization from acetonitrile. PhCl, purchased from Kishida Chemicals, was distilled from P$_2$O$_5$. EtOH, special grade, purchased from Kishida Chemicals, and o-C$_6$H$_4$Cl$_2$, HPLC grade, purchased from FUJIFILM Wako Pure Chemical Corporation, were used without further purification. Electrocrystallization was performed using a conventional H-shaped cell with Pt wire (1 mm φ) electrodes, between which 0.9 μA for 20 days and 0.2 μA for 2 months were applied for BEDT-TTF and BETS salts, respectively.

Single crystal X-ray measurements were performed at 150 K for **1** and **2** and at 110, 150 and 290 K for **3** with a Rigaku Rapid II imaging plate system with MicroMax-007 HF/VariMax rotating-anode X-ray generator with confocal monochromated MoKα radiation. The crystallographic data of **1**, **2**, and **3** are listed in Table 1. The structures of **1**, **2**, and **3** were solved by SHELXT [53], and each structure refinement was completed with SHELXL [53] software. The intensities were corrected for absorption with ABSCOR (Higashi, T. (1995); ABSCOR. Rigaku Corporation, Tokyo, Japan). All non-hydrogen atoms were refined anisotropically. Hydrogen atom positions were calculated geometrically and refined using the riding model. The disordered ethylene moieties in **1** and **2** were constrained displacement parameters (EADP in SHELX). The anion in **3** was severely disordered and refined with bond distance and angle restraints (DFIX and DANG in SHELX) and constrained displacement parameters (EADP in SHELX). In the circumstance, relatively large residual densities were observed. In addition, **1** and **2** salts had relatively weaker reflections with $h + l$ = odd than those with $h + l$ = even, suggesting the possibility that the cell volumes become half. Actually, **1** and **2** can be solved using the cell parameters with the half volumes, **1'**: $P\bar{1}$, a = 5.8667(2), b = 8.7788(2), c = 16.6858(5) Å, α = 88.451(6), β = 84.438(6),

$\gamma = 75.791(5)°$, $V = 829.14(5)$ Å^3, $R(I > 2\sigma(I)) = 3.0\%$, wR(all data) = 8.1%, $R_{int} = 3.2\%$, and **2'**: $P\bar{1}$, $a = 5.9343(2)$, $b = 8.8354(3)$, $c = 17.0567(5)$ Å, $\alpha = 88.674(6)$, $\beta = 84.312(6)$, $\gamma = 76.164(5)°$, $V = 864.09(5)$ Å^3, $R(I > 2\sigma(I)) = 3.9\%$, wR(all data) = 11.6%, $R_{int} = 6.4\%$. Each asymmetric unit has one donor and a half of anion, the latter of which is located about a center of symmetry and which is disordered where Br atoms and crystallographically-independent SO$_3$ group are overlapped. Only one donor layer is crystallographically-independent in **1'** and **2'**. However, **1** and **2** have many reflections with $h + l$ = odd, which were significantly observed ($I \gg 3\sigma(I)$). For example, the intensity (I), sigma (σ), and $I/\sigma(I)$ of ($-2\ 2\ -7$) of 236.10, 3.04, and 77.66, respectively, for **1** and 123.45, 2.20, and 56.11, respectively, for **2** were observed. Moreover, **1'** and **2'** have the disorder of the anion, which does not exist in **1** and **2**. Therefore, we decided that **1** and **2** are correct. The relatively low R values of **1'** and **2'** indicate that each salt has pseudosymmetry because the structures of each two independent donor layers are quite similar in **1** and **2**. Electrical AC resistivity measurements from 4.2 to 300 K were measured by the conventional four-probe method using a HUSO HECS 994C1 four channel resistivity meter with cooling and heating rates of ≈0.5 K/min. Each resistivity value was recorded after averaging for 10 s. Temperature-dependent magnetic susceptibility of a polycrystalline sample from 2 to 300 K was measured using a Quantum Design MPMS-2S SQUID magnetometer. The magnetic susceptibility data of **1** and **3** were corrected for a contribution from an aluminum foil sample holder, and the diamagnetic contributions of the samples were estimated from Pascal's constant (-4.444 and -5.084×10^{-4} emu mol^{-1} for **1** and **3**, respectively). The contributions of the Al foil sample holders were of the same orders as those of conducting electrons. Due to the relatively low contribution of the conducting electrons, less than 20% of absolute error was sometimes observed.

4. Conclusions

We prepared three new BEDT-TTF- and BETS-based organic conductors with an organic sulfonate anion, bromoethanesulfonate, β''-β''-(BEDT-TTF)$_2$BrC$_2$H$_4$SO$_3$ (**1**), β''-β''-(BETS)$_2$BrC$_2$H$_4$SO$_3$ (**2**), and θ-(BETS)$_2$BrC$_2$H$_4$SO$_3$ (**3**). Salt **1** shows a MI transition at around 70 K. The isomorphous **2** does not show a MI transition and is metallic down to at least 4.2 K. Compound **3** is also a stable metal down to at least 4.2 K. The dihedral angle of **3** of 98.1° is located in the superconducting phase of the phase diagram of the θ-type BEDT-TTF salts. This suggests that if we can obtain the isomorphous BEDT-TTF-based salt, it may show superconductivity (a resistivity drop and a Meissner effect). Preparation of the θ-type BEDT-TTF salt is now in progress.

Supplementary Materials: The following are available online at https://www.mdpi.com/article/10.3390/magnetochemistry7070091/s1, Figure S1. Schematic diagrams of the crystal structures of Type I-IV salts where the electrical di-poles of the counterions are indicated by arrows, electrically conducting layers are shown as green squares, and counterion layers are shown as blue rectangles, Figure S2. Crystal structure of **4**, Figure S3. Temperature-dependent electrical resistivity of **4**.

Author Contributions: Conceptualization, H.A.; methodology, H.A.; validation, H.A., Y.K., and Y.N.; formal analysis, H.A. and Y.K; investigation, H.A. and Y.K.; resources, H.A. and Y.N.; data curation, H.A.; writing—original draft preparation, H.A.; writing—review and editing, S.S.T.; visualization, H.A.; supervision, H.A., S.S.T., and Y.N.; project administration, Y.N.; funding acquisition, H.A. All authors have read and agreed to the published version of the manuscript.

Funding: This research was partially funded by JSPS KAKENHI, grant number 17K05751.

Institutional Review Board Statement: Not applicable.

Informed Consent Statement: Not applicable.

Data Availability Statement: Data available on request from the corresponding author. Data available includes raw resistivity and SQUID magnetization data. The crystallographic data has been submitted to the Cambridge Crystallographic Data Centre. The deposition references are shown in Table 1. Available online: http://www.ccdc.cam.ac.uk/conts/retrieving.html (accessed on 15 March 2021).

Acknowledgments: The authors thank Keigo Furuta at Yokohama National University for providing the BETS molecule.

Conflicts of Interest: The authors declare no conflict of interest.

References

1. Mori, T. *Electronic Properties of Organic Conductors*; Springer: Tokyo, Japan, 2016.
2. Kaythirgamanan, P.; Mucklejohn, S.A.; Rosseinsky, D.R. Electrocrystallisation of Conductive Nonstoicheiornetric Adducts of Tetrathiafulvalene with Inorganic or Organic Anions, and of Similar Adducts of Tetracyanoquinodirnethane. *J. Chem. Soc. Chem. Commun.* **1979**, *2*, 86–87. [CrossRef]
3. Geiser, U.; Schlueter, J.A. Conducting Organic Radical Cation Salts with Organic and Organometallic Anions. *Chem. Rev.* **2004**, *104*, 5203–5241. [CrossRef] [PubMed]
4. Soling, H.; Rindorf, G.; Thorup, N. Di(4,4′,5,5′-tetramethyl-$\Delta^{2,2'}$-bi-1,3-diselenolyliden)ium trifluoromethanesulfonate, $C_{21}H_{24}F_3O_3SSe_8$, (TMTSF)$_2$CF$_3$SO$_3$. *Acta Cryst.* **1983**, *C39*, 490–491.
5. Katayama, C.; Honda, M.; Kumagai, H.; Tanaka, J.; Saito, G.; Inokuchi, H. Crystal Structures of Complexes between Hexacyanobutadiene and Tetramethyltetrathiafulvalene and Tetramethylthiotetrathiafulvalene. *Bull. Chem. Soc. Jpn.* **1985**, *58*, 2272–2278. [CrossRef]
6. Chasseau, D.; Watkin, D.; Rosseinsky, M.J.; Kurmoo, M.; Talham, D.R.; Day, P. Syntheses, crystal structures and physical properties of conducting salts (BEDT-TTF)$_2$X (X = CF$_3$SO$_3$ or *p*-CH$_3$C$_6$H$_4$SO$_3$). *Synth. Met.* **1988**, *24*, 117–125. [CrossRef]
7. Brezgunova, M.; Shin, K.-S.; Auban-Senzier, P.; Jeannin, O.; Fourmigué, M. Combining halogen bonding and chirality in a two-dimensional organic metal (EDT-TTF-I$_2$)$_2$(*D*-camphorsulfonate)·H$_2$O. *Chem. Commun.* **2010**, *46*, 3926–3928. [CrossRef]
8. Lakhdar, Y.; El-Ghayoury, A.; Zorina, L.; Mercier, N.; Allain, M.; Mézière, C.; Auban-Senzier, P.; Batail, P.; Giffard, M. Acentric Polymeric Chains in Radical Cation Salts of Tetrathiafulvalene Derivatives with the p-Carboxybenzenesulfonate Anion. *Eur. J. Inorg. Chem.* **2010**, *2010*, 3338–3342. [CrossRef]
9. Shin, K.-S.; Brezgunova, M.; Jeannin, O.; Roisnel, T.; Camerel, F.; Auban-Senzier, P.; Fourmigué, M. Strong Iodine···Oxygen Interactions in Molecular Conductors Incorporating Sulfonate Anions. *Cryst. Growth Des.* **2011**, *11*, 5337–5345. [CrossRef]
10. Camerel, F.; Helloco, G.L.; Guizouarn, T.; Jeannin, O.; Fourmigué, M.; Frąckowiak, A.; Olejniczak, I.; Świetlik, R.; Marino, A.; Collet, E.; et al. Correlation between Metal–Insulator Transition and Hydrogen-Bonding Network in the Organic Metal δ-(BEDT-TTF)$_4$[2,6-Anthracene-bis(sulfonate)]·(H$_2$O)$_4$. *Cryst. Growth Des.* **2013**, *13*, 5135–5145. [CrossRef]
11. Akutsu, H.; Yamada, J.; Nakatsuji, S. A New Organic Anion Consisting of the TEMPO Radical for Organic Charge-Transfer Salts: 2,2,6,6 Tetramethylpiperidinyloxy-4-sulfamate (TEMPO-NHSO$_3^-$). *Chem. Lett.* **2001**, *30*, 208–209. [CrossRef]
12. Akutsu, H.; Yamada, J.; Nakatsuji, S. Preparation and characterization of novel organic radical anions for organic conductors: TEMPO–NHSO$_3^-$ and TEMPO-OSO$_3^-$. *Synth. Met.* **2001**, *120*, 871–872. [CrossRef]
13. Akutsu, H.; Yamada, J.; Nakatsuji, S. New BEDT-TTF-based Organic Conductor Including an Organic Anion Derived from the TEMPO Radical, α-(BEDT-TTF)$_3$(TEMPO–NHCOCH$_2$SO$_3$)$_2$·6H$_2$O. *Chem. Lett.* **2003**, *32*, 1118–1119. [CrossRef]
14. Furuta, K.; Akutsu, H.; Yamada, J.; Nakatsuji, S. A Novel BEDT-TTF-based Organic Conducting Salt with a Ferrocene-containing Dianion, α-(BEDT-TTF)$_4$(Fe(Cp–CONHCH$_2$SO$_3$)$_2$)·4H$_2$O. *Chem. Lett.* **2004**, *33*, 1214–1215. [CrossRef]
15. Furuta, K.; Akutsu, H.; Yamada, J.; Nakatsuji, S. New organic functional anions: Ferrocenyl-(CONHCH$_2$SO$_3^-$)$_n$ (n = 1–2) and their TTF salts. *Synth. Met.* **2005**, *152*, 381–384. [CrossRef]
16. Akutsu, H.; Yamada, J.; Nakatsuji, S. Novel organic magnetic conductors based on organochalcogen donors and an organic magnetic anion, TEMPO-NHCOCH$_2$SO$_3^-$. *Synth. Met.* **2005**, *152*, 377–380. [CrossRef]
17. Akutsu, H.; Masaki, K.; Mori, K.; Yamada, J.; Nakatsuji, S. New organic free radical anions TEMPO-A-CO-(*o*-, *m*-, *p*-)C$_6$H$_4$SO$_3^-$ (A = NH, NCH$_3$, O) and their TTF and/or BEDT-TTF salts. *Polyhedron* **2005**, *24*, 2126–2132. [CrossRef]
18. Yamashita, A.; Akutsu, H.; Yamada, J.; Nakatsuji, S. New organic magnetic anions TEMPO–CONA(CH$_2$)$_n$SO$_3^-$ (n = 0–3 for A = H, n = 2 for A = CH$_3$) and their TTF, TMTSF and/or BEDT-TTF salts. *Polyhedron* **2005**, *16*, 2796–2802. [CrossRef]
19. Furuta, K.; Akutsu, H.; Yamada, J.; Nakatsuji, S.; Turner, S.S. The first organic molecule-based metal containing ferrocene. *J. Mater. Chem.* **2006**, *16*, 1504–1506. [CrossRef]
20. Akutsu, H.; Yamada, J.; Nakatsuji, S.; Turner, S.S. A novel BEDT-TTF-based purely organic magnetic conductor, α-(BEDT-TTF)$_2$(TEMPO-N(CH$_3$)COCH$_2$SO$_3$)·3H$_2$O. *Solid State Commun.* **2006**, *140*, 256–260. [CrossRef]
21. Furuta, K.; Akutsu, H.; Yamada, J.; Nakatsuji, S.; Turner, S.S. The first metallic salt containing ferrocene, β″-(BEDT-TTF)$_4$(Fe(Cp-CONHCH$_2$SO$_3$)$_2$)·2H$_2$O. In *Multifunctional Conducting Molecular Materials*; Saito, G., Wudl, F., Haddon, R.C., Tanigaki, K., Enoki, T., Katz, H.E., Maesato, M., Eds.; RSC Publishing: Dorchester, UK, 2007; pp. 147–150.
22. Akutsu, H.; Yamada, J.; Nakatsuji, S.; Turner, S.S. An anionic weak acceptor 2-aminomethylsulfo-3,5,6-trichloro-1, 4-benzoquinone and its BEDT-TTF-based charge-transfer salts. *Solid State Commun.* **2007**, *144*, 144–147. [CrossRef]

23. Akutsu, H.; Ohnishi, R.; Yamada, J.; Nakatsuji, S.; Turner, S.S. Novel Bis(ethylenedithio)tetrathiafulvalene-Based Organic Conductor with 1,1-Ferrocenedisulfonate. *Inorg. Chem.* **2007**, *46*, 8472–8474. [CrossRef]
24. Akutsu, H.; Sato, K.; Yamashita, S.; Yamada, J.; Nakatsuji, S.; Turner, S.S. The first organic paramagnetic metal containing the aminoxyl radical. *J. Mater. Chem.* **2008**, *18*, 3313–3315. [CrossRef]
25. Akutsu, H.; Yamashita, S.; Yamada, J.; Nakatsuji, S.; Turner, S.S. Novel Purely Organic Conductor with an Aminoxyl Radical, α-(BEDT-TTF)$_2$(PO–CONHCH$_2$SO$_3$)·2H$_2$O (PO = 2,2,5,5-Tetramethyl-3-pyrrolin-1-oxyl Free Radical). *Chem. Lett.* **2008**, *37*, 882–883. [CrossRef]
26. Akutsu, H.; Yamada, J.; Nakatsuji, S.; Turner, S.S. A new anionic acceptor, 2-sulfo-3,5,6-trichloro-1,4-benzoquinone and its charge-transfer salts. *CrystEngComm* **2009**, *11*, 2588–2592. [CrossRef]
27. Akutsu, H.; Sasai, T.; Yamada, J.; Nakatsuji, S.; Turner, S.S. New anionic acceptors Br$_2$XQNHCH$_2$SO$_3^-$ [X = Br, Br$_y$Cl$_{1-y}$ ($y \approx 0.5$), and Cl; Q=1,4-benzoquinone] and their charge-transfer salts. *Physica B* **2010**, *405*, S2–S5. [CrossRef]
28. Akutsu, H.; Yamashita, S.; Yamada, J.; Nakatsuji, S.; Hosokoshi, Y.; Turner, S.S. A Purely Organic Paramagnetic Metal, κ-β''-(BEDT-TTF)$_2$(PO–CONHC$_2$H$_4$SO$_3$), Where PO = 2,2,5,5-Tetramethyl-3-pyrrolin-1-oxyl Free Radical. *Chem. Mater.* **2011**, *23*, 762–764. [CrossRef]
29. Akutsu, H.; Kawamura, A.; Yamada, J.; Nakatsuji, S.; Turner, S.S. Anion polarity-induced dual oxidation states in a dual-layered purely organic paramagnetic charge-transfer salt, (TTF)$_3$(PO–CON(CH$_3$)C$_2$H$_4$SO$_3$)$_2$, where PO = 2,2,5,5-tetramethyl-3-pyrrolin-1-oxyl free radical. *CrystEngComm* **2011**, *13*, 5281–5284. [CrossRef]
30. Akutsu, H.; Maruyama, Y.; Yamada, J.; Nakatsuji, S.; Turner, S.S. A new BEDT-TTF-based organic metal with an anionic weak acceptor 2-sulfo-1,4-benzoquinone. *Synth. Met.* **2011**, *161*, 2339–2343. [CrossRef]
31. Akutsu, H.; Yamada, J.; Nakatsuji, S.; Turner, S.S. A New BEDT-TTF-Based Organic Charge Transfer Salt with a New Anionic Strong Acceptor, N,N'-Disulfo-1,4-benzoquinonediimine. *Crystals* **2012**, *2*, 182–192. [CrossRef]
32. Kanbayashi, N.; Akutsu, H.; Yamada, J.; Nakatsuji, S.; Turner, S.S. A new ferrocene-containing charge-transfer salt, (TTF)$_2$[Fe(C$_5$H$_4$-CH(CH$_3$)NHCOCH$_2$SO$_3$)$_2$]. *Inorg. Chem. Commun.* **2012**, *21*, 122–124. [CrossRef]
33. Akutsu, H.; Yamada, J.; Nakatsuji, S.; Turner, S.S. Structures and properties of a BEDT-TTF-based organic charge transfer salt and the zwitterion of ferrocenesulfonate. *Dalton Trans.* **2013**, *42*, 16351–16354. [CrossRef] [PubMed]
34. Akutsu, H.; Hashimoto, R.; Yamada, J.; Nakatsuji, S.; Nakazawa, Y.; Turner, S.S. Structures and properties of new ferrocene-based paramagnetic anion octamethylferrocenedisulfonate and its TTF salt. *Inorg. Chem. Commun.* **2015**, *61*, 41–47. [CrossRef]
35. Akutsu, H.; Ishihara, K.; Yamada, J.; Nakatsuji, S.; Turner, S.S.; Nakazawa, Y. A strongly polarized organic conductor. *CrystEngComm* **2016**, *18*, 8151–8154. [CrossRef]
36. Akutsu, H.; Ishihara, K.; Ito, S.; Nishiyama, F.; Yamada, J.; Nakatsuji, S.; Turner, S.S.; Nakazawa, Y. Anion Polarity-Induced Self-doping in Purely Organic Paramagnetic Conductor, α'-α'-(BEDT-TTF)$_2$(PO-CONH-m-C$_6$H$_4$SO$_3$)·H$_2$O where BEDT-TTF is Bis(ethylenedithio)tetrathiafulvalene and PO is 2,2,5,5-Tetramethyl-3-pyrrolin-1-oxyl Free Radical. *Polyhedron* **2017**, *136*, 23–29. [CrossRef]
37. Akutsu, H.; Hashimoto, R.; Yamada, J.; Nakatsuji, S.; Turner, S.S.; Nakazawa, Y. Structure and Properties of a BEDT-TTF-Based Organic Conductor with a Ferrocene-Based Magnetic Anion Octamethylferrocenedisulfonate. *Eur. J. Inorg. Chem.* **2018**, 3249–3252. [CrossRef]
38. Ito, H.; Edagawa, Y.; Pu, J.; Akutsu, H.; Suda, M.; Yamamoto, H.M.; Kawasugi, Y.; Haruki, R.; Kumai, R.; Takenobu, T. Electrolyte-Gating-Induced Metal-Like Conduction in Nonstoichiometric Organic Crystalline Semiconductors under Simultaneous Bandwidth Control. *Phys. Status Solidi* **2019**, *13*, 1900162–1900175. [CrossRef]
39. Akutsu, H.; Koyama, Y.; Turner, S.S.; Furuta, K.; Nakazawa, Y. Structures and Properties of New Organic Conductors: BEDT-TTF, BEST and BETS Salts of the HOC$_2$H$_4$SO$_3^-$ Anion. *Crystals* **2020**, *10*, 775. [CrossRef]
40. Akutsu, H.; Kohno, A.; Turner, S.S.; Nakazawa, Y. Structure and Properties of a New Purely Organic Magnetic Conductor, δ'-(BEDT-TTF)$_2$(PO-CONHCH(cyclopropyl)SO$_3$)·1.7H$_2$O. *Chem. Lett.* **2020**, *49*, 1345–1348. [CrossRef]
41. Akutsu, H.; Kohno, A.; Turner, S.S.; Yamashita, S.; Nakazawa, Y. Different electronic states in the isomorphous chiral vs. racemic organic conducting salts, β''-(BEDT-TTF)$_2$(S- and rac-PROXYL-CONHCH$_2$SO$_3$). *Mat. Adv.* **2020**, *1*, 3171–3175. [CrossRef]
42. Minemawari, H.; Naito, T.; Inabe, T. (ET)$_3$(Br$_3$)$_5$: A Metallic Conductor with an Unusually High Oxidation State of ET (ET = Bis(ethylenedithio)tetrathiafulvalene). *Chem. Lett.* **2007**, *36*, 74–75. [CrossRef]
43. Guionneau, P.; Kepert, C.J.; Bravic, G.; Chasseau, D.; Truter, R.M.; Kurmoo, M.; Day, P. Determine the charge distribution in BEDT-TTF salts. *Synth. Met.* **1997**, *86*, 1973–1974. [CrossRef]
44. Akutsu, H.; Turner, S.S.; Nakazawa, Y. New Dmit-Based Organic Magnetic Conductors (PO-CONH-C$_2$H$_4$N(CH$_3$)$_3$)[M(dmit)$_2$]$_2$ (M = Ni, Pd) Including an Organic Cation Derived from a 2,2,5,5-Tetramethyl-3-pyrrolin-1-oxyl (PO) Radical. *Magnetochemisry* **2017**, *3*, 11. [CrossRef]
45. Akutsu, H.; Ito, S.; Kadoya, T.; Yamada, J.; Nakatsuji, S.; Turner, S.S.; Nakazawa, Y. A new Ni(dmit)$_2$-based organic magnetic charge-transfer salt, (m-PO-CONH-N-methylpyridinium)[Ni(dmit)$_2$]·CH$_3$CN. *Inorg. Chim. Acta* **2018**, *482*, 654–658. [CrossRef]
46. Stewart, J.J.P. *MOPAC7*; Stewart Computational Chemistry: Colorado Springs, CO, USA, 1993.
47. Mori, H.; Tanaka, S.; Mori, T. Systematic study of the electronic state in θ-type BEDT-TTF organic conductors by changing the electronic correlation. *Phys. Rev. B* **1998**, *57*, 12023–12029. [CrossRef]
48. Mori, H.; Sakurai, N.; Tanaka, S.; Moriyama, H.; Mori, T.; Kobayashi, H.; Kobayashi, A. Control of electronic state by dihedral angle in θ-type bis(ethylenedithio)tetraselenafulvalene salts. *Chem. Matter.* **2000**, *12*, 2984–2987. [CrossRef]

49. Mori, T. Structural Genealogy of BEDT-TTF-Based Organic Conductors I. Parallel Molecules: β and β'' Phases. *Bull. Chem. Soc. Jpn.* **1998**, *71*, 2509–2526. [CrossRef]
50. Williams, J.M.; Ferraro, J.R.; Thorn, R.J.; Carlson, K.D.; Geiser, U.; Wang, H.H.; Kini, A.M.; Whangbo, M.H. *Organic Superconductors: Synthesis, Structure, Properties, and Theory*; Prentice Hall: Englewood Cliffs, NJ, USA, 1992.
51. Kobayashi, A.; Sato, A.; Arai, E.; Kobayashi, H.; Faulmann, C.; Kushch, N.; Cassoux, P. A stable molecular metal with a binuclear magnetic anion, θ-(BETS)$_2$Cu$_2$Cl$_6$. *Solid State Commun.* **1997**, *103*, 371–374. [CrossRef]
52. Mori, T.; Kobayashi, A.; Sasaki, Y.; Kobayashi, H.; Saito, G.; Inokuchi, H. The Intermolecular Interaction of Tetrathiafulvalene and Bis(ethylenedithio)tetrathiafulvalene in Organic Metals. Calculation of Orbital Overlaps and Models of Energy-band Structures. *Bull. Chem. Soc. Jpn.* **1984**, *57*, 627–633. [CrossRef]
53. Sheldrick, G.M. Crystal structure refinement with SHELXL. *Acta Cryst. C* **2015**, *71*, 3–8. [CrossRef]

Article

First Molecular Superconductor with the Tris(Oxalato)Aluminate Anion, β"-(BEDT-TTF)$_4$(H$_3$O)Al(C$_2$O$_4$)$_3$·C$_6$H$_5$Br, and Isostructural Tris(Oxalato)Cobaltate and Tris(Oxalato)Ruthenate Radical Cation Salts

Toby James Blundell [1], Michael Brannan [1], Joey Mburu-Newman [1], Hiroki Akutsu [2], Yasuhiro Nakazawa [2], Shusaku Imajo [3] and Lee Martin [1,*]

[1] School of Science and Technology, Nottingham Trent University, Clifton Lane, Clifton, Nottingham NG11 8NS, UK; toby.blundell@ntu.ac.uk (T.J.B.); michael.brannan2015@my.ntu.ac.uk (M.B.); joey.mburu-newman2015@my.ntu.ac.uk (J.M.-N.)
[2] Department of Chemistry, Graduate School of Science, Osaka University, 1-1 Machikaneyama-cho, Toyonaka, Osaka 560-0043, Japan; akutsu@chem.sci.osaka-u.ac.jp (H.A.); nakazawa@chem.sci.osaka-u.ac.jp (Y.N.)
[3] The Institute for Solid State Physics, The University of Tokyo, Kashiwa, Chiba 277-8581, Japan; imajo@issp.u-tokyo.ac.jp
* Correspondence: lee.martin@ntu.ac.uk

Citation: Blundell, T.J.; Brannan, M.; Mburu-Newman, J.; Akutsu, H.; Nakazawa, Y.; Imajo, S.; Martin, L. First Molecular Superconductor with the Tris(Oxalato)Aluminate Anion, β"-(BEDT-TTF)$_4$(H$_3$O)Al(C$_2$O$_4$)$_3$·C$_6$H$_5$Br, and Isostructural Tris(Oxalato)Cobaltate and Tris(Oxalato)Ruthenate Radical Cation Salts. *Magnetochemistry* **2021**, *7*, 90. https://doi.org/10.3390/magnetochemistry7070090

Academic Editor: Marius Andruh

Received: 21 May 2021
Accepted: 18 June 2021
Published: 22 June 2021

Publisher's Note: MDPI stays neutral with regard to jurisdictional claims in published maps and institutional affiliations.

Copyright: © 2021 by the authors. Licensee MDPI, Basel, Switzerland. This article is an open access article distributed under the terms and conditions of the Creative Commons Attribution (CC BY) license (https://creativecommons.org/licenses/by/4.0/).

Abstract: Peter Day's research group reported the first molecular superconductor containing paramagnetic metal ions in 1995, β"-(BEDT-TTF)$_4$(H$_3$O)Fe(C$_2$O$_4$)$_3$·C$_6$H$_5$CN. Subsequent research has produced a multitude of BEDT-TTF-tris(oxalato)metallate salts with a variety of structures and properties, including 32 superconductors to date. We present here the synthesis, crystal structure, and conducting properties of the newest additions to the Day series including the first superconductor incorporating the diamagnetic tris(oxalato)aluminate anion, β"-(BEDT-TTF)$_4$(H$_3$O)Al(C$_2$O$_4$)$_3$·C$_6$H$_5$Br, which has a superconducting T$_c$ of ~2.5 K. β"-(BEDT-TTF)$_4$(H$_3$O)Co(C$_2$O$_4$)$_3$·C$_6$H$_5$Br represents the first example of a β" phase for the tris(oxalato)cobaltate anion, but this salt does not show superconductivity.

Keywords: molecular conductor; superconductor; metal; semiconductor; BEDT-TTF; tris(oxalato)metallate

1. Introduction

The first paramagnetic superconductor, β"-(BEDT-TTF)$_4$(H$_3$O)Fe(C$_2$O$_4$)$_3$·C$_6$H$_5$CN, was discovered in 1995 by the group of Professor Peter Day at the Royal Institution of Great Britain [1]. The ability of tris(oxalato)metallate(III) anions, M(C$_2$O$_4$)$_3^{3-}$, to bridge through oxalate ions with monocations or metal(II) ions and form 2D sheets opened the door to a huge variety of structures and properties in radical cation salts with BEDT-TTF [2]. This family of salts includes not only paramagnetic superconductors, but also a ferromagnetic metal [3], antiferromagnetic semiconductor [4], and proton conductor [5,6].

Most of the reported salts in the BEDT-TTF-tris(oxalato)metallate family are 4:1 salts having the formula (BEDT-TTF)$_4$(**A**)**M**(C$_2$O$_4$)$_3$·**G**. The lattice consists of cation layers of BEDT-TTF alternating with anion layers where hydrogen bonding between the terminal ethylene groups of BEDT-TTF and the anion layer determine the donor molecule packing arrangement. The anion layers are built up of **M** and **A** bridged by oxalate ligands to form a honeycomb with guest molecules, **G**, contained within the hexagons.

The most widely studied 4:1 salts in this "Day series" are the β" salts, which crystallise in the monoclinic *C*2/*c* space group, of which 32 are superconductors [1,2,7–30]. The counter cation (**A** = H$_3$O$^+$/K$^+$/NH$_4^+$/Rb$^+$) and the tris(oxalato)metallate metal centre can be changed (**M** = Fe [1,7–19,25], Cr [20–25], Co [25], Al [25], Mn [17,26], Ga [24,27,28], Ru [29], Rh [30]), which has a small effect on the electrical properties of the material owing to the change in size of **A** and **M**. For example, β"-(BEDT-TTF)$_4$(**A**)**M**(C$_2$O$_4$)$_3$·**G**, where **G** = benzonitrile, sees a reduction of superconducting T_c to 5.5–6.0 K when **M** = Cr, compared to when **M** = Fe 7.0–8.5 K. A more marked effect on the electrical properties and the superconducting T_c [31] is observed when changing the guest molecule, **G**—the solvent used for the electrocrystallization. Changing **G** from benzonitrile to different sized and shaped guest molecules can alter the conducting properties from superconducting to metallic or semiconducting [2,7–30]. The highest superconducting T_c values are obtained when longer guest solvent molecules are used, which increase the b axis length the furthest, e.g., **G** = benzonitrile, nitrobenzene [31].

When **G** = benzonitrile, crystals of an additional 4:1 orthorhombic phase are also obtained when **M** = Fe or Cr. Crystals of this 4:1 orthorhombic phase are the only phase obtained with **G** = benzonitrile when **M** = Co [25], Al [25], or Rh [30], and the β″ phase has not been reported. This semiconducting phase crystallises in the orthorhombic space group $Pbcn$ with a pseudo-κ donor packing. (BEDT-TTF$^+$)$_2$ dimers are surrounded by neutral BEDT-TTF^0 monomers. The –C≡N group of benzonitrile is disordered over two positions directed towards **A**, rather than along the b axis towards **M**, as seen in the β″ salts. The chirality of the tris(oxalato)metallates in the anion layers differs between the β″ and pseudo-κ salts despite both having an overall racemic lattice. In the β″ salts, each anion layer contains only a single enantiomer of $M(C_2O_4)_3^{3-}$ with alternating layers being of the opposing enantiomer. However, in the pseudo-κ salts, each anion layer is identical with alternating rows of Δ or Λ enantiomers.

When **G** is too large to fit inside the honeycomb cavity of the anion layer, a 4:1 triclinic phase is obtained. The guest molecule in these salts protrudes on one side of the anion layer and not on the other. The two different faces of the anion layer then lead to two different packing modes of the donor layer within the same crystal, e.g., both α and β″ donor packing (**G** = PhCH$_2$CN, PhN(CH$_3$)CHO, PhCOCH$_3$, or PhCH(OH)CH$_3$) [32] or α and pseudo-κ (**G** =1,2-Br$_2$Ph) [33]. An α-β″ salt has also been obtained with the inclusion of a chiral guest molecule (**G** = sec-phenethyl alcohol, PhCH(OH)CH$_3$) in both the chiral S form and the racemic R/S form. A small difference in the metal–insulator transition temperature is observed between the racemic and the chiral salts owing to the disorder, which is found only in the racemate [34].

While the 4:1 salts make up the majority of BEDT-TTF-tris(oxalate)metallate salts, some semiconducting 3:1 salts have been obtained when using smaller guest molecules (**G** = DMF, acetonitrile, dichloromethane, nitromethane; cation = Li$^+$, Na$^+$, NH$_4^+$; metal = Fe [35], Cr [36–39], Al [39]). A 2:1 salt has also been reported in which an 18-crown-6 molecule is the guest in the honeycomb cavity, β″-(BEDT-TTF)$_2$[(H$_2$O)(NH$_4$)$_2$M(C$_2$O$_4$)$_3$]·18-crown-6 (**M** = Cr, Rh, Ru, Ir). Both the Cr and Rh salts show a bulk Berezinskii–Kosterlitz–Thouless superconducting transition [40–42]. Changing the counter cation **A** has produced several salts where the packing of the anion layer differs from the aforementioned honeycomb packing arrangement giving salts β′-(BEDT-TTF)$_5$[Fe(C$_2$O$_4$)$_3$]·(H$_2$O)$_2$·CH$_2$Cl$_2$ [43] (**A** = tetraethylammonium), η-(BEDT-TTF)$_4$(H$_2$O)LiFe(C$_2$O$_4$)$_3$ [35] (**A** = lithium), α‴-(BEDT-TTF)$_9$[Fe(C$_2$O$_4$)$_3$]$_8$Na$_{18}$(H$_2$O)$_{24}$ [44,45] (**A** = sodium, α-(BEDT-TTF)$_{10}$(18-crown-6)$_6$K$_6$[Fe(C$_2$O$_4$)$_3$]$_4$(H$_2$O)$_{24}$ [45] (**A** = potassium), and α-(BEDT-TTF)$_{12}$[Fe(C$_2$O$_4$)$_3$]$_2$·(H$_2$O)$_n$ [46] (**A** = potassium or caesium, n = 15 or 16). Changing the **M**(III) to Ge(IV) produces very different structures in the semiconductors (BEDT-TTF)$_2$[Ge(C$_2$O$_4$)$_3$]·PhCN [47], (BEDT-TTF)$_5$[Ge(C$_2$O$_4$)$_3$]$_2$ [48], (BEDT-TTF)$_7$[Ge(C$_2$O$_4$)$_3$]$_2$(CH$_2$Cl$_2$)$_{0.87}$(H$_2$O)$_{0.09}$ [48], and (BEDT-TTF)$_4$Ge(C$_2$O$_4$)$_3$·(CH$_2$Cl$_2$)$_{0.50}$ [49].

We report here the synthesis, crystal structures, and conducting properties of the first superconductor incorporating the tris(oxalato)aluminate anion, β″-(BEDT-TTF)$_4$(H$_3$O)Al(C$_2$O$_4$)$_3$·C$_6$H$_5$Br (Tc ~ 2.5 K), the first example of a β″ phase for the tris(oxalato)cobaltate anion (**G** = PhBr), and two new β″ salts from tris(oxalato)ruthenate (**G** = PhCl or PhF). Resistivity is also presented for β″-(BEDT-TTF)$_4$(H$_3$O)Ru(C$_2$O$_4$)$_3$·PhBr, which shows a superconducting T_c of 2.8 K, though the crystals were very thin and not suitable for a publishable X-ray dataset.

2. Results and Discussion

All salts β″-(BEDT-TTF)$_4$(H$_3$O)M(C$_2$O$_4$)$_3$·G (M-G = Al-PhBr, Co-PhBr, Ru-PhCl, Ru-PhF) are isostructural with the previously reported Day series β″-(BEDT-TTF)$_4$[(A)M(C$_2$O$_4$)$_3$]·G. Ru-PhBr is also isostructural, though the crystals were very thin and not suitable for a publishable X-ray dataset. They crystallise in the monoclinic space group C2/c. The asymmetric unit contains two crystallographically independent BEDT-TTF molecules, half an M(C$_2$O$_4$)$^{3-}$ molecule, half a guest halobenzene molecule, and half a H$_3$O$^+$ molecule (Table 1). The long-range structure consists of ordered alternating layers of BEDT-TTF donor molecules and M(C$_2$O$_4$)$^{3-}$ anions (Figure 1). The two crystallographically independent donor BEDT-TTF molecules form two-dimensional stacks along the *a/b* crystallographic axis in a β″ arrangement (Figure 2). A number of predominately side-to-side sulphur-sulphur interactions below the sum of the van der Waals radii are present (Table 2). The estimated charge on BEDT-TTF cations can be calculated via the method of Guionneau et al. [50] from the central C=C and C–S bond lengths of the TTF core and results in a charge of approximately +0.5 for each BEDT-TTF molecule, as expected (Table 3).

Table 1. Crystal data for β″-(BEDT-TTF)$_4$(H$_3$O)M(C$_2$O$_4$)$_3$·G salts.

Salt	Al-PhBr 290 K	Al-PhBr 110 K	Co-PhBr 150 K	Ru-PhF 293 K	Ru-PhCl 298 K
Formula	C$_{52}$H$_{40}$AlBrO$_{13}$S$_{32}$	C$_{52}$H$_{40}$AlBrO$_{13}$S$_{32}$	C$_{52}$H$_{40}$BrCoO$_{13}$S$_{32}$	C$_{52}$H$_{40}$FO$_{13}$RuS$_{32}$	C$_{52}$H$_{40}$O$_{13}$S$_{32}$ClRu
Fw (g mol^{-1})	2005.65	2005.65	2037.60	2018.83	2035.28
Crystal System	monoclinic	monoclinic	monoclinic	monoclinic	monoclinic
Space group	C2/c	C2/c	C2/c	C2/c	C2/c
Z	4	4	4	4	4
T (K)	290 (2)	110 (2)	150 (2)	293 (2)	293 (2)
a (Å)	10.2851 (2)	10.2520 (3)	10.2306 (3)	10.32786 (19)	10.32017 (19)
b (Å)	19.9472 (4)	19.7919 (7)	19.7508 (5)	19.9521 (4)	20.0264 (4)
c (Å)	35.597 (3)	35.4275 (11)	35.2520 (9)	34.9966 (6)	35.161 (3)
α (°)	90	90	90	90	90
β (°)	93.399 (7)	93.843 (7)	93.938 (7)	93.010 (7)	93.586 (7)
γ (°)	90	90	90	90	90
Volume (Å^3)	7290.1 (6)	7172.3 (4)	7106.3 (3)	7201.5 (2)	7252.6 (6)
Density (g cm^{-3})	1.827	1.857	1.905	1.862	1.864
μ (mm^{-1})	1.553	1.578	1.806	1.209	1.235
R_1	0.0547	0.0688	0.0431	0.0460	0.0442
wR (all data)	0.1401	0.1526	0.0914	0.318	0.1089

Table 2. S...S close contacts below the van der Waals distance in β″-(BEDT-TTF)$_4$(H$_3$O)M(C$_2$O$_4$)$_3$·G salts.

Contact (Å)	Al-PhBr 290 K	Al-PhBr 110 K	Co-PhBr 150 K	Ru-PhF 293 K	Ru-PhCl 298 K
S1...S7	3.4069 (13)	3.3795 (19)	3.3683 (11)	3.4015 (14)	3.4283 (11)
S3...S7	3.5046 (13)	3.458 (2)	3.4544 (11)	3.5369 (15)	3.5290 (11)
S2...S9	3.3138 (14)	3.2838 (19)	3.2864 (11)	3.3744 (15)	3.3528 (11)
S2...S11	3.3720 (13)	3.340 (2)	3.3413 (11)	3.3954 (15)	3.3842 (11)
S6...S15	3.5231 (14)	3.463 (2)	3.4475 (12)	3.5236 (17)	3.5190 (13)
S8...S15	3.5704 (15)	3.502 (2)	3.4904 (13)	3.6182 (17)	3.5869 (12)
S8...S10	3.5889 (14)	3.550 (2)	3.5551 (13)	3.6169 (16)	3.6031 (13)

Table 3. Average bond lengths (Å) in BEDT-TTF molecules with approximation of the charge on the molecules. δ = (b + c) − (a + d), Q = 6.347 − 7.463δ [50].

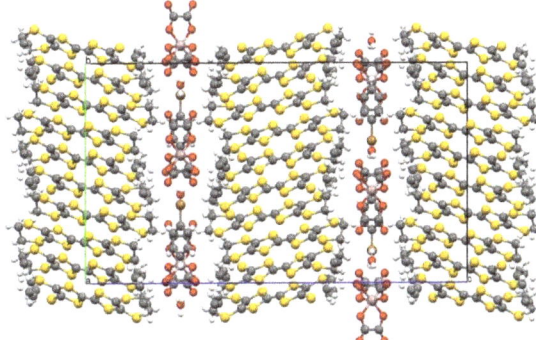

Salt	Donor	a	b	c	d	δ	Q
Al-PhBr 290 K	A	1.376	1.73925	1.7515	1.352	0.763	0.65
	B	1.373	1.74025	1.753	1.347	0.773	0.58
Al-PhBr 110 K	A	1.364	1.74475	1.75675	1.346	0.792	0.44
	B	1.376	1.7435	1.75725	1.3455	0.779	0.53
Co-PhBr 150 K	A	1.367	1.73775	1.75275	1.345	0.779	0.54
	B	1.369	1.738	1.7515	1.346	0.775	0.57
Ru-PhF 293 K	A	1.36	1.7385	1.74475	1.349	0.774	0.57
	B	1.367	1.73325	1.74525	1.354	0.758	0.69
Ru-PhCl 298 K	A	1.366	1.737	1.7475	1.349	0.770	0.60
	B	1.366	1.73575	1.74625	1.349	0.767	0.62

Figure 1. Layered structure of **Al-PhBr**. The other salts reported in this paper are isostructural. Carbon atoms are grey, hydrogen atoms white, oxygen atoms red, sulphur atoms yellow, aluminium atoms pink, and bromine atoms brown. The b axis is shown in green, and the c axis is shown in blue.

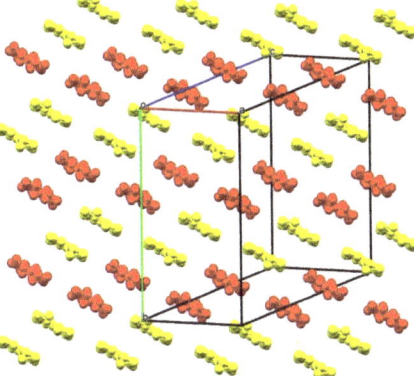

Figure 2. β″ BEDT-TTF layer packing in **Al-PhBr**. The other salts reported in this paper are isostructural. The two crystallographically independent BEDT-TTF molecules are shown in different colours. Hydrogens have been removed for clarity. The a axis is shown in red, the b axis in green, and the c axis in blue.

The anion layer consists of a honeycomb arrangement of $M(C_2O_4)^{3-}$, perpendicular to the long axis of the BEDT-TTF molecules, resulting in a hexagonal cavity that is occupied by the guest halobenzene molecule. Each anion layer contains a single enantiomer of the tris(oxalato)metallate ion with adjacent layers containing the alternate enantiomer, which gives an overall racemic lattice. The hexagonal cavity and the orientation of the guest halobenzene molecule within it are shown in Figure 3 and Table 4. Distances a, b, h, and w represent the dimensions of the hexagonal cavity. The latter two are the height and width of the cavity, respectively, and δ is the angle of the benzene ring plane relative to the plane of the hexagonal cavity (measured as the least-squares plane of the three metal atoms making up three corners of the hexagon). For **M** = Rh, we see a reduction in height (h) of the hexagonal cavity going from **G** = PhCl to the smaller PhF, accompanied by a reduction in the length of the b axis of the unit cell. For salt **Al-PhBr**, we observed a T_c of ~2.5 K (Figure 4), which is similar to previously published salts of β''-(BEDT-TTF)$_4$[(H$_3$O)M(C$_2$O$_4$)$_3$]·**G**, where **G** = bromobenzene. When applying a magnetic field along the c^* axis, the critical field of the superconductivity at 0.7 K is about 0.2 T. This is comparable to other salts in the Day series, for example: the Fe-DMF salt has a T_c of 2.0 K, and H_{c2} in a perpendicular field is ~0.1 T [51]. Higher T_c salts in the Day series have higher H_{c2} values (2–5 T) [52], and these quasi-2D superconductors are strongly anisotropic [53].

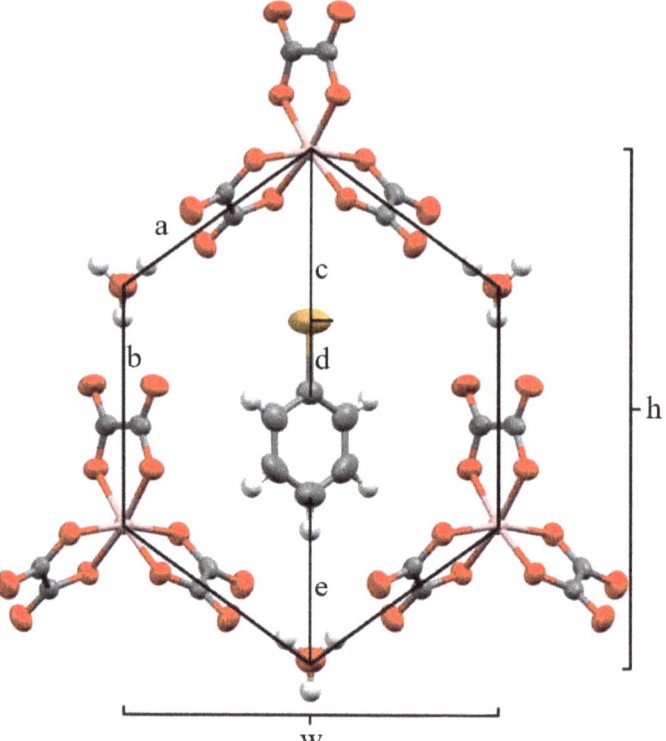

Figure 3. Honeycomb cavity in the anion layer of β''-(BEDT-TTF)$_4$(H$_3$O)M(C$_2$O$_4$)$_3$·**G** salts with measurement parameters labelled (a–e, w = width, h = height). Carbon atoms are grey, hydrogen atoms white, and oxygen atoms red. This image shows salt **Al-PhBr**, where aluminium atoms are pink and bromine atoms brown.

Table 4. Honeycomb cavity measurements in the anion layer (see Figure 3) of β''-(BEDT-TTF)$_4$(H$_3$O)M(C$_2$O$_4$)$_3 \cdot$ **G** salts.

Salt Temp.	Al-PhBr 290 K	Al-PhBr 110 K	Co-PhBr 150 K	Ru-PhF 293 K	Ru-PhCl 298 K
Distances (Å)					
a	6.269 (3)	6.255 (6)	6.249 (3)	6.292 (3)	6.312 (2)
b	6.387 (5)	6.111 (10)	6.286 (6)	6.380 (5)	6.377 (4)
c	4.5360 (16)	4.487 (3)	4.5212 (10)	4.804 (13)	4.331 (16)
d	1.894 (6)	1.905 (10)	1.902 (6)	1.377 (14)	1.737 (6)
e	4.401 (9)	4.338 (19)	4.296 (10)	4.786 (13)	4.543 (9)
h	13.560 (5)	13.481 (10)	13.464 (6)	13.572 (5)	13.649 (4)
w	10.2851 (2)	10.2520 (3)	10.2306 (3)	10.32786 (19)	10.32017 (19)
O4-cation	3.066 (5)	3.004 (11)	2.985 (6)	2.956 (6)	2.962 (5)
O6-cation	2.857 (3)	2.851 (6)	2.842 (3)	2.846 (6)	2.831 (3)
O1-cation	3.083 (5)	3.073 (9)	3.112 (5)	2.941 (5)	2.996 (4)
Angles (°)					
δ	33.522 (3)	33.378 (3)	33.60 (13)	33.74 (19)	32.677 (3)

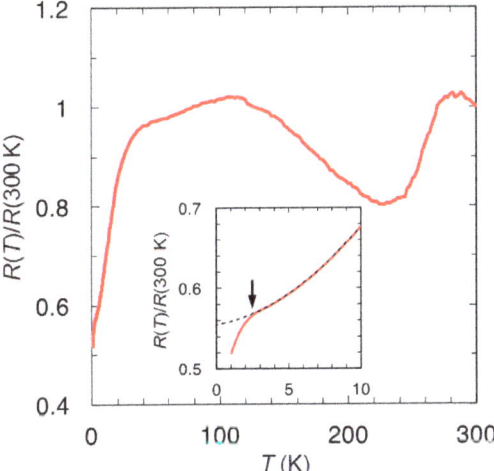

Figure 4. Electrical resistivity for **Al-PhBr**.

The Al^{3+} ion of tris(oxalato)aluminate is smaller than previous examples, where **M** = Fe [13,17], Ga [28], Rh [30], and Ru [29] (T$_c$ = ~3.8, ~3.0, ~2.9, ~2.8 K, respectively, for **G** = bromobenzene), and the T$_c$ is smaller for **M** = Al at ~2.5 K. A comparison of the b axis length of these bromobenzene salts at room temperature showed that the **M** = Fe salt has the longest at 20.0546(15) Å and also the highest ~3.8 K [13,17]; **M** = Rh has an intermediate b axis of 20.0458(4) Å and a T$_c$ of ~2.9 K [30]; while **M** = Al has the shortest b axis of 19.9472(4) Å and the lowest T$_c$ at ~2.5 K. A direct comparison with the **M** = Ga [28] and Ru [29] salts cannot be made owing to **A** = K$_x$(H$_3$O)$_{1-x}$ rather than H$_3$O for these salts. Salts with **M** = Cr [23] and Mn [17] have been reported with T$_c$s of 1.5 K and 2.0 K, respectively, but crystal structures are not published for the comparison of the b axes. Our crystals of **Co-PhBr** did not show superconductivity (Figure 5), with the b axis of this salt being much shorter than all other PhBr salts at 19.7508(5) Å.

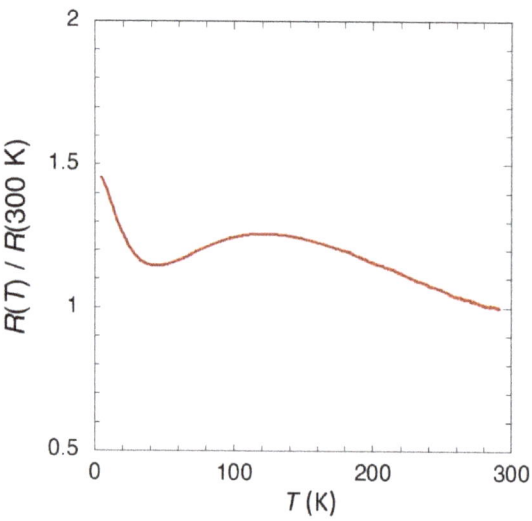

Figure 5. Electrical resistivity for **Co-PhBr**.

There are thirty-two superconductors to date having the formula β''-(BEDT-TTF)$_4$[(**A**)**M**(C$_2$O$_4$)$_3$]·**G** (**M** = Fe, Cr, Ga, Rh, Ru, Mn, **G** = guest molecule, **A** = H$_3$O$^+$/K$^+$/NH$_4^+$) [1,2,7–30]. There is negligible π-d interaction in β''-(BEDT-TTF)$_4$[(**A**)**M**(C$_2$O$_4$)$_3$]·**G** salts because the M^{3+} ions are located in the centre of the anion layer, distant from the BEDT-TTF layer (Figure 1). This is confirmed by the similar T$_c$s that are observed for isostructural salts with the same **A** and **G**, but which differ only in the presence of paramagnetic Fe^{3+} (S = 5/2) or non-magnetic Ga^{3+} [53,54]. A much more marked effect on the value of T$_c$ is observed when changing the guest molecule, **G**. Changing **M** and **G** leads to a change in the length of the unit cell dimensions. A correlation between the b axis length and superconducting T$_c$ has been observed through structural analysis [31]. The effect of chemical pressure through changing **G** and **M** is mainly attributed to the guest molecule, **G**, which is oriented with the R-group oriented in the b direction (Figure 3). The longest molecules, benzonitrile and nitrobenzene, have the highest T$_c$s observed in the family, and the relationship between T$_c$ and the guest molecule size can be observed in the series of salts with halobenzene guest molecules [31]. Only the higher T$_c$ salts in this family show insulating behaviour just above T$_c$ owing to charge disproportionation in these salts [55–58].

Figure 6 shows the resistivity of β''-(BEDT-TTF)$_4$(H$_3$O)Ru(C$_2$O$_4$)$_3$·**G**, where **G** = PhBr, PhCl, or PhF. **Ru-PhBr** for **A** = K$_x$(H$_3$O)$_{1-x}$ has previously been studied by Prokhorova et al. [29] with a sample-dependent T$_c$ in the range 2.8–6.3 K. Resistivity measurements on our crystals of β''-(BEDT-TTF)$_4$(H$_3$O)Ru(C$_2$O$_4$)$_3$·PhBr gave a T$_c$ of 2.8 K, which was as expected based on the b axis length [31]. Upon reducing the size of **G** from PhBr to PhCl or PhF, no superconductivity was observed. Both **Ru-PhCl** and **Ru-PhF** showed semiconducting behaviour (Figure 6). Both **Ru-PhCl** and **Ru-PhF** had shorter b axis lengths compared to the Ru-PhBr salt. However, the b axis lengths in semiconducting **Ru-PhCl** and **Ru-PhF** were longer than that in superconducting **Al-PhBr** (Table 3). This indicates that other factors, such as the shape and the electric dipole of the guest molecule, may have minor influences even though the b axis length predominantly affects the electronic state, including the T$_c$ [31].

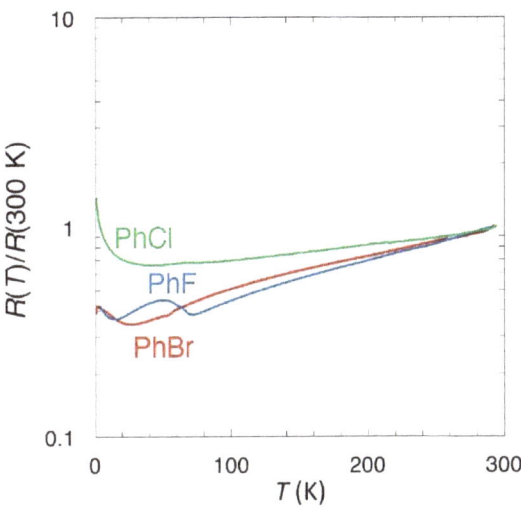

Figure 6. Electrical resistivity for β"-(BEDT-TTF)$_4$(H$_3$O)Ru(C$_2$O$_4$)$_3$·**G** where **G** = PhBr, PhCl, or PhF.

3. Materials and Methods

Bromobenzene, chlorobenzene, fluorobenzene, ethanol, and 18-crown-6 were purchased from Sigma Aldrich and used as received. BEDT-TTF was purchased from Sigma Aldrich (Gillingham, Dorset, UK) and recrystallised from chloroform.

3.1. Synthesis

Ammonium tris(oxalato)aluminate and tri(oxalato)cobaltate were synthesised by the method of Bailar and Jones [59]. Ammonium tris(oxalato)ruthenate was synthesised by the method of Kaziro et al [60].

Al-PhBr: One-hundred milligrams of ammonium tris(oxalato)aluminate and 200 mg of 18-crown-6 ether were dissolved in 10 mL 1,2,4-trichlorobenzene, 10 mL bromobenzene, and 2 mL ethanol. The solution was then filtered into the cathodic side of the H-cell, while 20 mg of BEDT-TTF was added to the anodic side of the H-cell. The level of solvent was allowed to equilibrate in the cell, and a platinum electrode was added to each side. A constant current of 0.8 µA was applied across the H-cell which gave small black crystals of **Al-PhBr** which were collected after 28 days.

Co-PhBr: One-hundred milligrams of ammonium tris(oxalato)cobaltate and 200 mg of 18-crown-6 ether were dissolved in 10 mL 1,2,4-trichlorobenzene, 10 mL bromobenzene, and 2 mL ethanol. Ten milligrams of BEDT-TTF were added to the anodic side of the H-cell. A constant current of 0.6 µA was applied across the H-cell which gave tiny black crystals of **Co-PhBr** which were collected after 14 days.

Ru-PhF: One-hundred milligrams of ammonium tris(oxalato)ruthenate and 200 mg of 18-crown-6 ether were dissolved in 10 mL 1,2,4-trichlorobenzene, 10 mL fluorobenzene, and 2 mL ethanol. Ten milligrams of BEDT-TTF were added to the anodic side of the H-cell. A constant current of 1.0 µA was applied across the H-cell which gave black block crystals of **Ru-PhF** which were collected after 28 days.

Ru-PhCl: One-hundred milligrams of ammonium tris(oxalato)ruthenate and 200 mg of 18-crown-6 ether were dissolved in 10 mL 1,2,4-trichlorobenzene, 10 mL chlorobenzene, and 2 mL ethanol. Ten milligrams of BEDT-TTF were added to the anodic side of the H-cell. A constant current of 1.0 µA was applied across the H-cell which gave black block crystals of **Ru-PhCl** which were collected after 28 days.

Ru-PhBr: One-hundred milligrams of ammonium tris(oxalato)ruthenate and 200 mg of 18-crown-6 ether were dissolved in 10 mL 1,2,4-trichlorobenzene, 10 mL chlorobenzene, and 2 mL ethanol. Ten milligrams of BEDT-TTF were added to the anodic side of the H-cell.

A constant current of 1.0 µA was applied across the H-cell which gave thin needle crystals of **Ru-PhBr** which were collected after 28 days. The crystals were very thin and not suitable for a publishable X-ray dataset.

3.2. Single-Crystal X-ray Crystallography

Data were collected using a RigakuRapid II (Tokyo, Japan) imaging plate system with the MicroMax-007 HF/VariMax rotating-anode X-ray generator and confocal monochromated Mo-Kα radiation.

3.3. Conducting Properties

Out-of-plane electrical resistance was measured using the standard four-terminal AC method with the current along the c^* axis. Four gold wires were attached using carbon paint on both plane surfaces of single crystals.

4. Conclusions

We reported the synthesis and characterization of β''-(BEDT-TTF)$_4$(H$_3$O)Al(C$_2$O$_4$)$_3$·C$_6$H$_5$Br (**Al-PhBr**), which represents the first superconductor in the Day series to contain the tris(oxalato)aluminate anion. This salt (**M** = Al) is isostructural with bromobenzene salts where **M** = Fe, Ga, Rh, Ru, Mn, Cr. A relationship between the b axis length and superconducting T_c has previously been observed in the Day series [31]. The b axis length of these bromobenzene salts at room temperature showed that the **M** = Fe salt had the longest b axis and also the highest T_c of ~3.8 K, while **M** = Al had the shortest b axis and the lowest T_c of ~2.5 K. We also reported the isostructural **M** = Co salt (**Co-PhBr**), which did not show superconductivity. The b axis of this salt was much shorter than all other bromobenzene salts. Isostructural salts **Ru-PhCl** and **Ru-PhF** were presented in which the b axes were longer than that observed in superconducting **Al-PhBr**, but these two ruthenium salts did not show superconductivity. This indicates that even though the b axis length predominantly affected the electronic state, including the T_c, other factors may also be at work, such as the shape and the electric dipole of the guest molecules, which may have minor influences on the electronic states.

Author Contributions: Synthesis, L.M., M.B. and J.M.-N.; X-ray crystallography, T.J.B., H.A. and Y.N.; conductivity measurements S.I., H.A. and Y.N.; writing—original draft preparation, L.M. and T.J.B.; project administration, L.M.; funding acquisition, L.M. All authors have read and agreed to the published version of the manuscript.

Funding: L.M. and T.J.B. would like to thank The Leverhulme Trust for financial support (RPG-2019-242).

Institutional Review Board Statement: Not applicable.

Informed Consent Statement: Not applicable.

Data Availability Statement: CCDC 2084688-2084692 contains supplementary X-ray crystallographic data for **Al-PhBr** (100 K), **Al-PhBr** (298 K), **Co-PhBr**, **Ru-PhCl**, and **Ru-PhF**, respectively. https://www.ccdc.cam.ac.uk/structures/.

Acknowledgments: This article is dedicated to Peter Day, who was an inspirational chemist and a dear friend to those authors of this paper who had the pleasure of working with him.

Conflicts of Interest: The authors declare no conflict of interest.

References

1. Kurmoo, M.; Graham, A.W.; Day, P.; Coles, S.J.; Hursthouse, M.B.; Caulfield, J.L.; Singleton, J.; Pratt, F.L.; Hayes, W.; Ducasse, L.; et al. Superconducting and Semiconducting Magnetic Charge Transfer Salts: (BEDT-TTF)4AFe(C$_2$O$_4$)$_3$·C$_6$H$_5$CN (A = H$_2$O, K, NH4). *J. Am. Chem. Soc.* **1995**, *117*, 12209–12217. [CrossRef]
2. Martin, L. Molecular conductors of BEDT-TTF with tris(oxalato)metallate anions. *Coord. Chem. Rev.* **2018**, *376*, 277–291. [CrossRef]
3. Coronado, E.; Galán-Mascarós, J.R.; Gómez-García, C.J.; Laukhin, V. Coexistence of ferromagnetism and metallic conductivity in a molecule-based layered compound. *Nat. Cell Biol.* **2000**, *408*, 447–449. [CrossRef] [PubMed]

4. Zhang, B.; Zhang, Y.; Zhu, D. (BEDT-TTF)$_3$Cu$_2$(C$_2$O$_4$)$_3$(CH$_3$OH)$_2$: An organic–inorganic hybrid antiferromagnetic semiconductor. *Chem. Commun.* **2011**, *48*, 197–199. [CrossRef] [PubMed]
5. Rashid, S.; Turner, S.S.; Day, P.; Light, M.E.; Hursthouse, M.B.; Firth, S.; Clark, R.J.H. The first molecular charge transfer salt containing proton channels. *Chem. Commun.* **2001**, *16*, 1462–1463. [CrossRef]
6. Akutsu-Sato, A.; Akutsu, H.; Turner, S.S.; Day, P.; Probert, M.R.; Howard, J.A.K.; Akutagawa, T.; Takeda, S.; Nakamura, T.; Mori, T. The First Proton-Conducting Metallic Ion-Radical Salts. *Angew. Chem. Int. Ed.* **2004**, *44*, 292–295. [CrossRef] [PubMed]
7. Sun, S.; Wu, P.; Zhang, Q.; Zhu, D. The New Semiconducting Magnetic Charge Transfer Salt (BEDT-TTF)4 • H$_2$O • Fe(C$_2$O$_4$)$_3$ • C$_6$H$_5$NO$_2$: Crystal Structure and Physical Properties. *Mol. Cryst. Liq. Cryst. Sci. Technol. Sect. A Mol. Cryst. Liq. Cryst.* **1998**, *319*, 259–269. [CrossRef]
8. Sun, S.; Wu, P.; Zhang, Q.; Zhu, D. The new semiconducting magnetic charge transfer salt (BEDT-TTF)$_4$·H$_2$O·Fe(C$_2$O$_4$)$_3$·C$_6$H$_5$NO$_2$: Crystal structure and physical properties. *Synth. Met.* **1998**, *94*, 161–166. [CrossRef]
9. Turner, S.S.; Day, P.; Malik, K.M.A.; Hursthouse, M.B.; Teat, S.J.; MacLean, E.J.; Martin, L.; French, S.A. Effect of Included Solvent Molecules on the Physical Properties of the Paramagnetic Charge Transfer Salts β″-(bedt-ttf)$_4$[(H$_3$O)Fe(C$_2$O$_4$)$_3$]·Solvent (bedttf = Bis(ethylenedithio)tetrathiafulvalene). *Inorg. Chem.* **1999**, *38*, 3543–3549. [CrossRef]
10. Rashid, S.; Turner, S.S.; Day, P.; Howard, J.A.K.; Guionneau, P.; McInnes, E.J.L.; Mabbs, F.E.; Clark, R.J.H.; Firth, S.; Biggs, T. New superconducting charge-transfer salts (BEDT-TTF)$_4$[AM(C$_2$O$_4$)$_3$]·C$_6$H$_5$NO$_2$ (A = H$_3$O or NH$_4$, M = Cr or Fe, BEDT-TTF = bis(ethylenedithio)tetrathiafulvalene). *J. Mater. Chem.* **2001**, *11*, 2095–2101. [CrossRef]
11. Prokhorova, T.; Khasanov, S.; Zorina, L.; Buravov, L.; Tkacheva, V.; Baskakov, A.; Morgunov, R.; Gener-Moret, M.; Canadell, E.; Shibaeva, R.; et al. Molecular Metals Based on BEDT-TTF Radical Cation Salts with Magnetic Metal Oxalates as Counterions: β″-(BEDT-TTF)$_4$A[M(C2O4)$_3$]·DMF (A = NH$_4^+$, K$^+$; M = CrIII, FeIII). *Adv. Funct. Mater.* **2003**, *13*, 403–411. [CrossRef]
12. Akutsu-Sato, A.; Kobayashi, A.; Mori, T.; Akutsu, H.; Yamada, J.; Nakatsuji, S.; Turner, S.; Day, P.; Tocher, D.; Light, M.; et al. Structures and Physical Properties of New β′-BEDT-TTF Tris-Oxalatometallate (III) Salts Containing Chlorobenzene and Halomethane Guest Molecules. *Synth. Met.* **2005**, *152*, 373–376. [CrossRef]
13. Coronado, E.; Curreli, S.; Giménez-Saiz, C.; Gómez-García, C.J. A novel paramagnetic molecular superconductor formed by bis(ethylenedithio)tetrathiafulvalene, tris(oxalato)ferrate(iii) anions and bromobenzene as guest molecule: ET$_4$[(H$_3$O)Fe(C$_2$O$_4$)$_3$]·C$_6$H$_5$Br. *J. Mater. Chem.* **2005**, *15*, 1429–1436. [CrossRef]
14. Akutsu-Sato, A.; Turner, S.S.; Akutsu, H.; Yamada, J.; Nakatsuji, S.; Day, P. Suppression of superconductivity in a molecular charge transfer salt by changing guest molecule: -(BEDT-TTF)$_4$[(H$_3$O)Fe(C$_2$O$_4$)$_3$](C$_6$H$_5$CN)x(C$_5$H$_5$N)1x. *J. Mater. Chem.* **2007**, *17*, 2497–2499. [CrossRef]
15. Zorina, L.V.; Prokhorova, T.G.; Simonov, S.V.; Khasanov, S.S.; Shibaeva, R.P.; Manakov, A.I.; Zverev, V.N.; Buravov, L.I.; Yagubskiĭ, É.B. Structure and magnetotransport properties of the new quasi-two-dimensional molecular metal β″-(BEDT-TTF)$_4$H$_3$O[Fe(C$_2$O$_4$)$_3$]·C$_6$H$_4$Cl$_2$. *J. Exp. Theor. Phys.* **2008**, *106*, 347–354. [CrossRef]
16. Prokhorova, T.G.; Korobenko, A.V.; Buravov, L.I.; Yagubskii, E.B.; Zorina, L.V.; Khasanov, S.S.; Simonov, S.; Shibaeva, R.P.; Zverev, V.N. Effect of electrocrystallization medium on quality, structural features, and conducting properties of single crystals of the (BEDT-TTF)$_4$AI[FeIII(C$_2$O$_4$)$_3$]G family. *CrystEngComm* **2011**, *13*, 537–545. [CrossRef]
17. Coronado, E.; Curreli, S.; Giménez-Saiz, C.; Gómez-García, C.J. The Series of Molecular Conductors and Superconductors ET$_4$[AFe(C$_2$O$_4$)$_3$]PhX (ET = bis(ethylenedithio)tetrathiafulvalene; (C$_2$O$_4$)$^{2-}$ = oxalate; A$^+$ = H$_3$O$^+$, K$^+$; X = F, Cl, Br, and I)· Influence of the Halobenzene Guest Molecules on the Crystal Structure and Superconducting Properties. *Inorg. Chem.* **2011**, *51*, 1111–1126. [CrossRef] [PubMed]
18. Zorina, L.V.; Khasanov, S.S.; Simonov, S.V.; Shibaeva, R.P.; Bulanchuk, P.O.; Zverev, V.N.; Canadell, E.; Prokhorova, T.G.; Yagubskii, E.B. Structural phase transition in the β″-(BEDT-TTF)$_4$H$_3$O[Fe(C$_2$O$_4$)$_3$]G crystals (where G is a guest solvent molecule). *CrystEngComm* **2011**, *14*, 460–465. [CrossRef]
19. Prokhorova, T.G.; Buravov, L.I.; Yagubskii, E.B.; Zorina, L.V.; Simonov, S.; Zverev, V.N.; Shibaeva, R.P.; Canadell, E. Effect of Halopyridine Guest Molecules on the Structure and Superconducting Properties of β″-[Bis(ethylenedithio)tetrathiafulvalene]$_4$(H$_3$O)[Fe(C$_2$O$_4$)$_3$]·Guest Crystals. *Eur. J. Inorg. Chem.* **2015**, *2015*, 5611–5620. [CrossRef]
20. Martin, L.; Turner, S.S.; Day, P.; Mabbs, F.E.; McInnes, E.J.L. New molecular superconductor containing paramagnetic chromium(iii) ions. *Chem. Commun.* **1997**, 1367–1368. [CrossRef]
21. Martin, L.; Turner, S.S.; Day, P.; Malik, K.M.A.; Coles, S.J.; Hursthouse, M.B. Polymorphism based on molecular stereoisomerism in tris(oxalato) Cr(III) salts of bedt-ttf [bis(ethylenedithio)tetrathiafulvalene]. *Chem. Commun.* **1999**, 513–514. [CrossRef]
22. Rashid, S.; Turner, S.S.; Le Pevelen, D.; Day, P.; Light, M.E.; Hursthouse, M.B.; Firth, S.; Clark, R.J.H. β″-(BEDT-TTF)$_4$[(H$_3$O)Cr(C$_2$O$_4$)$_3$]CH$_2$Cl$_2$: Effect of Included Solvent on the Structure and Properties of a Conducting Molecular Charge-Transfer Salt. *Inorg. Chem.* **2001**, *40*, 5304–5306. [CrossRef] [PubMed]
23. Coronado, E.; Curreli, S.; Giménez-Saiz, C.; Gómez-García, C. New magnetic conductors and superconductors based on BEDT-TTF and BEDS-TTF. *Synth. Met.* **2005**, *154*, 245–248. [CrossRef]
24. Prokhorova, T.G.; Yagubskii, E.B.; Zorina, L.V.; Simonov, S.V.; Zverev, V.N.; Shibaeva, R.P.; Buravov, L.I. Specific Structural Disorder in an Anion Layer and Its Influence on Conducting Properties of New Crystals of the (BEDT-TTF)$_4$A$^+$[M^{3+}(ox)$_3$]G Family, Where G Is 2-Halopyridine; M Is Cr, Ga; A$^+$ Is [K$_{0.8}$(H$_3$O)$_{0.2}$]$^+$. *Crystals* **2018**, *8*, 92. [CrossRef]
25. Martin, L.; Turner, S.S.; Day, P.; Guionneau, P.; Howard, J.A.K.; Hibbs, D.E.; Light, M.E.; Hursthouse, M.B.; Uruichi, M.; Yakushi, K. Crystal Chemistry and Physical Properties of Superconducting and Semiconducting Charge Transfer Salts of the Type (BEDT-

TTF)$_4$[AlMIII(C$_2$O$_4$)$_3$]PhCN (AI = H$_3$O, NH$_4$, K.; MIII = Cr, Fe, Co, Al; BEDT-TTF = Bis(ethylenedithio)tetrathiafulvalene). *Inorg. Chem.* **2001**, *40*, 1363–1371. [CrossRef]

26. Benmansour, S.; Sánchez-Máez, Y.; Gómez-García, C.J.; Sánchez-Máñez, Y. Mn-Containing Paramagnetic Conductors with Bis(ethylenedithio)tetrathiafulvalene (BEDT-TTF). *Magnetochemistry* **2017**, *3*, 7. [CrossRef]
27. Akutsu, H.; Akutsu-Sato, A.; Turner, S.S.; Le Pevelen, D.; Day, P.; Laukhin, V.; Klehe, A.-K.; Singleton, J.; Tocher, D.; Probert, M.R.; et al. Effect of Included Guest Molecules on the Normal State Conductivity and Superconductivity of β"-(ET)$_4$[(H$_3$O)Ga(C$_2$O$_4$)$_3$]G (G = Pyridine, Nitrobenzene). *J. Am. Chem. Soc.* **2002**, *124*, 12430–12431. [CrossRef] [PubMed]
28. Prokhorova, T.G.; Buravov, L.I.; Yagubskii, E.B.; Zorina, L.V.; Simonov, S.; Shibaeva, R.P.; Zverev, V.N. Metallic Bi- and Monolayered Radical Cation Salts Based on Bis(ethylenedithio)tetrathiafulvalene (BEDT-TTF) with the Tris(oxalato)gallate Anion. *Eur. J. Inorg. Chem.* **2014**, *2014*, 3933–3940. [CrossRef]
29. Prokhorova, T.G.; Zorina, L.V.; Simonov, S.V.; Zverev, V.N.; Canadell, E.; Shibaeva, R.P.; Yagubskii, E.B. The first molecular superconductor based on BEDT-TTF radical cation salt with paramagnetic tris(oxalato)ruthenate anion. *CrystEngComm* **2013**, *15*, 7048. [CrossRef]
30. Martin, L.; Morritt, A.L.; Lopez, J.R.; Nakazawa, Y.; Akutsu, H.; Imajo, S.; Ihara, Y.; Zhang, B.; Zhang, Y.; Guo, Y. Molecular conductors from bis(ethylenedithio)tetrathiafulvalene with tris(oxalato)rhodate. *Dalton Trans.* **2017**, *46*, 9542–9548. [CrossRef]
31. Imajo, S.; Akutsu, H.; Akutsu-Sato, A.; Morritt, A.L.; Martin, L.; Nakazawa, Y. Effects of Electron Correlations and Chemical Pressures on Superconductivity of β"-Type Organic Compounds. *Phys. Rev. Res.* **2019**, *1*, 33184. [CrossRef]
32. Akutsu, H.; Akutsu-Sato, A.; Turner, S.S.; Day, P.; Canadell, E.; Firth, S.; Clark, R.J.H.; Yamada, J.-I.; Nakatsuji, S. Superstructures of donor packing arrangements in a series of molecular charge transfer salts. *Chem. Commun.* **2004**, 18–19. [CrossRef] [PubMed]
33. Zorina, L.V.; Khasanov, S.S.; Simonov, S.V.; Shibaeva, R.P.; Zverev, V.N.; Canadell, E.; Prokhorova, T.G.; Yagubskii, E.B. Coexistence of two donor packing motifs in the stable molecular metal α-'pseudo-κ'-(BEDT-TTF)$_4$(H$_3$O)[Fe(C$_2$O$_4$)$_3$]·C$_6$H$_4$Br$_2$. *CrystEngComm* **2011**, *13*, 2430–2438. [CrossRef]
34. Martin, L.; Day, P.; Akutsu, H.; Yamada, J.-I.; Nakatsuji, S.; Clegg, W.; Harrington, R.W.; Horton, P.N.; Hursthouse, M.B.; McMillan, P.; et al. Metallic molecular crystals containing chiral or racemic guest molecules. *CrystEngComm* **2007**, *9*, 865–867. [CrossRef]
35. Martin, L.; Engelkamp, H.; Akutsu, H.; Nakatsuji, S.; Yamada, J.; Horton, P.; Hursthouse, M.B. Radical-cation salts of BEDT-TTF with lithium tris(oxalato)metallate(iii). *Dalton Trans.* **2015**, *44*, 6219–6223. [CrossRef] [PubMed]
36. Martin, L.; Day, P.; Horton, P.; Nakatsuji, S.; Yamada, J.; Akutsu, H.; Horton, P. Chiral conducting salts of BEDT-TTF containing a single enantiomer of tris(oxalato)chromate(III) crystallised from a chiral solvent. *J. Mater. Chem.* **2010**, *20*, 2738–2742. [CrossRef]
37. Martin, L.; Day, P.; Nakatsuji, S.; Yamada, J.; Akutsu, H.; Horton, P. A molecular charge transfer salt of BEDT-TTF containing a single enantiomer of tris(oxalato)chromate(III) crystallised from a chiral solvent. *CrystEngComm* **2009**, *12*, 1369–1372. [CrossRef]
38. Martin, L.; Akutsu, H.; Horton, P.N.; Hursthouse, M.B. Chirality in charge-transfer salts of BEDT-TTF of tris(oxalato)chromate(III). *CrystEngComm* **2015**, *17*, 2783–2790. [CrossRef]
39. Martin, L.; Akutsu, H.; Horton, P.N.; Hursthouse, M.B.; Harrington, R.W.; Clegg, W. Chiral Radical-Cation Salts of BEDT-TTF Containing a Single Enantiomer of Tris(oxalato)aluminate(III) and -chromate(III). *Eur. J. Inorg. Chem.* **2015**, *2015*, 1865–1870. [CrossRef]
40. Martin, L.; Morritt, A.L.; Lopez, J.R.; Akutsu, H.; Nakazawa, Y.; Imajo, S.; Ihara, Y. Ambient-pressure molecular superconductor with a superlattice containing layers of tris(oxalato)rhodate enantiomers and 18-crown-6. *Inorg. Chem.* **2017**, *56*, 717–720. [CrossRef]
41. Martin, L.; Lopez, J.R.; Akutsu, H.; Nakazawa, Y.; Imajo, S. Bulk Kosterlitz–Thouless Type Molecular Superconductor β"-(BEDT-TTF)$_2$[(H$_2$O)(NH$_4$)$_2$Cr(C$_2$O$_4$)$_3$]$_{18}$-crown-6. *Inorg. Chem.* **2017**, *56*, 14045–14052. [CrossRef] [PubMed]
42. Morritt, A.L.; Lopez, J.R.; Blundell, T.; Canadell, E.; Akutsu, H.; Nakazawa, Y.; Imajo, S.; Martin, L. 2D Molecular Superconductor to Insulator Transition in the β"-(BEDT-TTF)$_2$[(H$_2$O)(NH$_4$)$_2$M(C$_2$O$_4$)$_3$]$_{18}$-crown-6 Series (M = Rh, Cr, Ru, Ir). *Inorg. Chem.* **2019**, *58*, 10656–10664. [CrossRef]
43. Zhang, B.; Zhang, Y.; Liu, F.; Guo, Y. Synthesis, crystal structure, and characterization of charge-transfer salt: (BEDT-TTF)$_5$[Fe(C$_2$O$_4$)$_3$]·(H$_2$O)$_2$·CH$_2$Cl$_2$ (BEDT-TTF = bis(ethylenedithio)tetrathiafulvalene). *CrystEngComm* **2009**, *11*, 2523–2528. [CrossRef]
44. Martin, L.; Day, P.; Horton, P.; Bingham, A.; Hursthouse, M.B. The first molecular charge transfer salt containing layers of an alkali metal. *J. Low Temp. Phys.* **2006**, *142*, 417–420. [CrossRef]
45. Martin, L.; Day, P.; Clegg, W.; Harrington, R.W.; Horton, P.N.; Bingham, A.; Hursthouse, M.B.; McMillan, P.; Firth, S. Multi-layered molecular charge-transfer salts containing alkali metal ions. *J. Mater. Chem.* **2007**, *17*, 3324–3329. [CrossRef]
46. Martin, L.; Day, P.; Barnett, S.A.; Tocher, D.A.; Horton, P.N.; Hursthouse, M.B. Magnetic molecular charge-transfer salts containing layers of water and tris(oxalato)ferrate(iii) anions. *CrystEngComm* **2008**, *10*, 192–196. [CrossRef]
47. Martin, L.; Turner, S.S.; Day, P.; Guionneau, P.; Howard, J.A.K.; Uruichi, M.; Yakushi, K. Synthesis, crystal structure and properties of the semiconducting molecular charge-transfer salt (bedtttf)$_2$Ge(C$_2$O$_4$)$_3$ PhCN [bedt-ttfbis(ethylenedithio)tetrathiafulvalene]. *J. Mater. Chem.* **1999**, *9*, 2731–2736. [CrossRef]
48. Martin, L.; Day, P.; Nakatsuji, S.; Yamada, J.-I.; Akutsu, H.; Horton, P.N. BEDT-TTF Tris(oxalato)germanate(IV) Salts with Novel Donor Packing Motifs. *Bull. Chem. Soc. Jpn.* **2010**, *83*, 419–423. [CrossRef]
49. Lopez, J.R.; Akutsu, H.; Martin, L. Radical-cation salt with novel BEDT-TTF packing motif containing tris(oxalato)germanate(IV). *Synth. Met.* **2015**, *209*, 188–191. [CrossRef]

50. Guionneau, P.; Kepert, C.; Bravic, G.; Chasseau, D.; Truter, M.; Kurmoo, M.; Day, P. Determining the charge distribution in BEDT-TTF salts. *Synth. Met.* **1997**, *86*, 1973–1974. [CrossRef]
51. Audouard, A.; Laukhin, V.N.; Brossard, L.; Prokhorova, T.G.; Yagubskii, E.B.; Canadell, E. Combination frequencies of magnetic oscillations in $\beta''-(BEDT-TTF)_4(NH_4)[Fe(C_2O_4)_3]\cdot DMF$. *Phys. Rev. B* **2004**, *69*. [CrossRef]
52. Uji, S.; Iida, Y.; Sugiura, S.; Isono, T.; Sugii, K.; Kikugawa, N.; Terashima, T.; Yasuzuka, S.; Akutsu, H.; Nakazawa, Y.; et al. Fulde-Ferrell-Larkin-Ovchinnikov superconductivity in the layered organic superconductor $\beta''-(BEDT-TTF)_4[(H_3O)Ga(C_2O_4)_3]C_6H_5NO_2$. *Phys. Rev. B* **2018**, *97*, 144505. [CrossRef]
53. Bangura, A.F.; Coldea, A.I.; Singleton, J.; Ardavan, A.; Akutsu-Sato, A.; Akutsu, H.; Turner, S.S.; Day, P.; Yamamoto, T.; Yakushi, K. Robust superconducting state in the low-quasiparticle-density organic metals $\beta''-(BEDT-TTF)_4[(H_3O)M(C_2O_4)_3]Y$: Superconductivity due to proximity to a charge-ordered state. *Phys. Rev. B* **2005**, *72*, 014543. [CrossRef]
54. Coldea, A.I.; Bangura, A.F.; Singleton, J.; Ardavan, A.; Akutsu-Sato, A.; Akutsu, H.; Turner, S.S.; Day, P. Fermi-surface topology and the effects of intrinsic disorder in a class of charge-transfer salts containing magnetic ions: $\beta''-(BEDT-TTF)_4[(H_3O)M(C_2O_4)_3]Y$ (M = Ga, Cr, Fe; Y = C5H5N). *Phys. Rev. B* **2004**, *69*, 085112. [CrossRef]
55. Ihara, Y.; Seki, H.; Kawamoto, A. ^{13}C NMR Study of Superconductivity near Charge Instability Realized in β''-(BEDT-TTF)$_4$[(H$_3$O)Ga(C$_2$O$_4$)$_3$]C$_6$H$_5$NO$_2$. *J. Phys. Soc. Jpn.* **2013**, *82*, 83701. [CrossRef]
56. Ihara, Y.; Jeong, M.; Mayaffre, H.; Berthier, C.; Horvatić, M.; Seki, H.; Kawamoto, A. C ^{13}NMR study of the charge-ordered state near the superconducting transition in the organic superconductor $\beta''-(BEDT-TTF)_4(H_3O)Ga(C_2O_4)_3\cdot C_6H_5NO_2$. *Phys. Rev. B* **2014**, *90*, 121106. [CrossRef]
57. Ihara, Y.; Moribe, K.; Fukuoka, S.; Kawamoto, A. Microscopic coexistence of superconductivity and charge order in the organic superconductor $\beta''-(BEDT-TTF)_4(H_3O)Ga(C_2O_4)_3\cdot C_6H_5NO_2$. *Phys. Rev. B* **2019**, *100*, 060505. [CrossRef]
58. Imajo, S.; Akutsu, H.; Kurihara, R.; Yajima, T.; Kohama, Y.; Tokunaga, M.; Kindo, K.; Nakazawa, Y. Anisotropic Fully Gapped Superconductivity Possibly Mediated by Charge Fluctuations in a Nondimeric Organic Complex. *Phys. Rev. Lett.* **2020**, *125*, 177002. [CrossRef]
59. Bailar, J.C., Jr.; Jones, E.M.; Booth, H.S.; Grennert, M. Trioxalato Salts (Trioxalatoaluminiate, -Ferriate, -Chromiate, and -Cobaltiate). *Inorg. Synth.* **1939**, *1*, 35–38.
60. Kaziro, R.; Hambley, T.W.; Binstead, R.A.; Beattie, J.K. Potassium tris(oxalato)ruthenate (III). *Inorg. Chim. Acta* **1989**, *164*, 85–91. [CrossRef]

Article

Chiral Radical Cation Salts of Me-EDT-TTF and DM-EDT-TTF with Octahedral, Linear and Tetrahedral Monoanions [†]

Nabil Mroweh [1], Alexandra Bogdan [1], Flavia Pop [1], Pascale Auban-Senzier [2], Nicolas Vanthuyne [3], Elsa B. Lopes [4], Manuel Almeida [4] and Narcis Avarvari [1,*]

1. Univ Angers, CNRS, MOLTECH-Anjou, SFR MATRIX, F-49000 Angers, France; nabil.mroweh@cea.fr (N.M.); alexandra.bogdan@etud.univ-angers.fr (A.B.); flavia.pop@univ-angers.fr (F.P.)
2. Laboratoire de Physique des Solides, Université Paris-Saclay CNRS UMR 8502, Bât. 510, 91405 Orsay, France; pascale.senzier@universite-paris-saclay.fr
3. Aix Marseille Université, CNRS, Centrale Marseille, iSm2, 13013 Marseille, France; nicolas.vanthuyne@univ-amu.fr
4. Centro de Ciencias e Tecnologias Nucleares (C2TN) and Departmento de Engenharia e Ciencias Nucleares(DECN), Instituto Superior Técnico (IST), Universidade de Lisboa, E.N. 10, 2695-066 Bobadela LRS, Portugal; eblopes@ctn.tecnico.ulisboa.pt (E.B.L.); malmeida@ctn.tecnico.ulisboa.pt (M.A.)
* Correspondence: narcis.avarvari@univ-angers.fr; Tel.: +33-2-4173-5084
† Dedicated to the memory of Professor Peter Day.

Abstract: Methyl-ethylenedithio-tetrathiafulvalene (Me-EDT-TTF) (**1**) and dimethyl-ethylenedithio-tetrathiafulvalene (DM-EDT-TTF) (**2**) are valuable precursors for chiral molecular conductors, which are generally obtained by electrocrystallization in the presence of various counter-ions. The number of the stereogenic centers, their relative location on the molecule, the nature of the counter-ion and the electrocrystallization conditions play a paramount role in the crystal structures and conducting properties of the resulting materials. Here, we report the preparation and detailed structural characterization of the following series of radical cation salts: (i) mixed valence (**1**)$_2$AsF$_6$ as racemic, and (S) and (R) enantiomers; (ii) [(S)-**1**]AsF$_6$·C$_4$H$_8$O and [(R)-**1**]AsF$_6$·C$_4$H$_8$O where a strong dimerization of the donors is observed; (iii) (**1**)I$_3$ and (**2**)I$_3$ as racemic and enantiopure forms and (iv) [(meso)-**2**]PF$_6$ and [(meso)-**2**]XO$_4$ (X = Cl, Re), based on the new donor (meso)-**2**. In the latter, the two methyl substituents necessarily adopt axial and equatorial conformations, thus leading to a completely different packing of the donors when compared to the chiral form (S,S)/(R,R) of **2** in its radical cation salts. Single crystal resistivity measurements, complemented by thermoelectric power measurements in the case of (**1**)$_2$AsF$_6$, suggest quasi-metallic conductivity for the latter in the high temperature regime, with σ$_{RT}$ ≈ 1–10 S cm^{-1}, while semiconducting behavior is observed for the (meso)-**2** based salts.

Keywords: organic conductors; chirality; tetrathiafulvalene; EDT-TTF; crystal structures; electrical conductivity

1. Introduction

Substitution of a hydrogen atom at one or both carbon atoms of the ethylene bridge of the ethylenedithio-tetrathiafulvalene (EDT-TTF) precursor generates one or two stereogenic centers, respectively, as in methyl-EDT-TTF (Me-EDT-TTF) **1** and dimethyl-EDT-TTF (DM-EDT-TTF) **2**, thus providing chiral tetrathiafulvalenes [1,2] (Scheme 1). These precursors proved to be highly valuable since they afforded by electrocrystallization a series of enantiopure and racemic radical cation salts showing chirality and anion-dependent crystalline packing and conducting properties. For example, donor **1**, which has been only recently described [3], provided the complete series of mixed-valence salts (**1**)$_2$PF$_6$, with metal-like conductivity [3], while with the perchlorate anion only the enantiopure salts, formulated as [(S)-**1**]$_2$ClO$_4$ and [(R)-**1**]$_2$ClO$_4$, showed metallic conductivity in the high-temperature regime, the racemic form [(rac)-**1**]ClO$_4$ being a very poor conductor because of the formation of strong heterochiral dimers by the radical cations [4]. On the other hand, the dimethylated donor **2**, containing two stereogenic centers (*S*,*S*) or (*R*,*R*) afforded semiconducting 4:2 enantiopure salts and metallic 2:1 racemic salt with PF_6^- [5], whereas the use of ClO_4^- proved to be of huge importance across the preparation of the enantiomorphic salts [(*S*,*S*)-**2**]$_2$ClO$_4$ and [(*R*,*R*)-**2**]$_2$ClO$_4$ which allowed the first observation of the electrical magnetochiral anisotropy (eMChA) effect [6,7] in a bulk crystalline chiral conductor [8]. Thus, the addition of a second stereogenic center on the EDT-TTF platform has a paramount importance for the resulting crystalline materials. Interestingly, the use of **2** in combination with AsF_6^- or SbF_6^-, having larger volumes than PF_6^-, favored the formation of metallic enantiopure salts (**2**)$_2$XF$_6$ (X = As, Sb) as a consequence of the interplay between the anion size, its propensity to engage in intermolecular hydrogen bonding and chirality of the donor [9], with the much larger double octahedral shaped fluorinated dianion $[Ta_2F_{10}O]^{2-}$ semiconducting salts of **2** with a 3:1 stoichiometry having been obtained [10]. In order to develop the families of chiral conducting materials further, a first objective of the present work is directed to the association of the mono-substituted donor **1** with the larger anion AsF_6^-. Secondly, when considering the important role played by the linear monoanion I_3^- in the field of molecular conductors through the synthesis of the first ambient pressure organic superconductor (BEDT-TTF)$_2$I$_3$ [11–13], its use in radical cation salts with the donors **1** and **2** has been investigated and described herein. Note that tetramethylbis(ethylenedithio)-tetrathiafulvalene (TM-BEDT-TTF), the first reported enantiopure TTF derivative [14], provided 1:1 semiconducting radical cation salts with I_3^- [15], besides other conducting materials [16–18], thus demonstrating the drastic consequences of the introduction of substituents on the ethylene carbon atoms on the stoichiometry and properties of the resulting materials. Finally, another interesting comparison can be made between the two diastereomers of **2**, i.e., the (*S*,*S*)/(*R*,*R*) pair and the *meso* form (Scheme 1). Indeed, radical cation salts based on the related donor dimethyl-bis(ethylenedithio)-tetrathiafulvalene (DM-BEDT-TTF) [19,20], which also presents a similar pair of diastereomers (*S*,*S*)/(*R*,*R*) and *meso*, show striking differences between them. For example, with the PF_6^- anion, semiconducting orthorhombic [(*R*,*R*)-DM-BEDT-TTF]$_2$PF$_6$ and monoclinic [(*rac*)-DM-BEDT-TTF]$_2$PF$_6$ salts [21] have been described, while the (*meso*)-DM-BEDT-TTF form provided the triclinic β-[(*meso*)-DM-BEDT-TTF]$_2$PF$_6$ salt which showed a superconducting transition with T_c ~ 4.3 K under 4.0 kbar [22,23]. Thus, the third objective of the present work is to introduce the hitherto unknown (*meso*)-**2** donor and radical cation salts containing, especially, PF_6^- and ClO_4^- anions, to be compared with the corresponding materials with (*S*,*S*)/(*R*,*R*)-**2** [5,8].

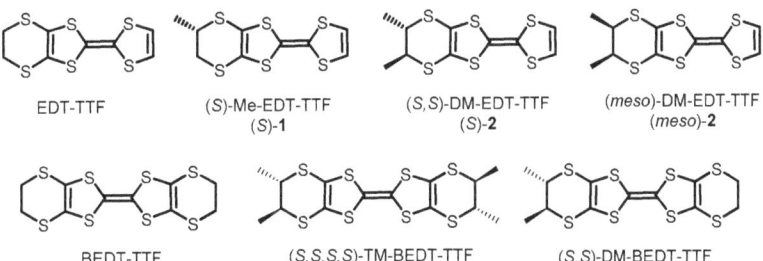

Scheme 1. EDT-TTF, Me-EDT-TTF (**1**), DM-EDT-TTF (**2**), BEDT-TTF, TM-BEDT-TTF and DM-BEDT-TTF donors. Specifically, **1** and **2** were used in the present study. Only the (*S*) enantiomers for **1** and **2** together with the (*meso*) form of **2** are represented.

We describe herein a series of radical cation salts of Me-EDT-TTF **1** with the AsF_6^- and I_3^- anions, the complete series of radical cation salts of (*S*,*S*)/(*R*,*R*)-**2** with I_3^- and, finally, the synthesis and structural characterization of the new donor (*meso*)-**2** together with its radical cation salts with PF_6^- and XO_4^- (X = Cl, Re) anions.

2. Results and Discussion

2.1. Radical Cation Salts of Me-EDT-TTF (1) with the AsF_6^- Anion

As outlined in the Introduction, donor **1** provided a complete series of radical cation salts (**1**)$_2$PF$_6$ for which the enantiopure compounds crystallized in the triclinic space group *P*1, with two independent donor molecules and one anion in the unit cell, while the racemic salt crystallized in the triclinic space group *P*–1 with one independent donor in the asymmetric unit and the anion located on an inversion center [3]. We describe here the analogous complete series of chiral radical cation salts of Me-EDT-TTF **1** with AsF_6^- and compare their structural features and conducting properties with the previously reported PF_6^- counterparts. Donor **1**, prepared according to our published procedure [3], afforded enantiopure and racemic salts by electrocrystallization in THF in the presence of ((*n*-Bu)$_4$N)AsF$_6$. Single crystals were obtained in the form of black plates following experimental conditions identical with those employed for the PF$_6$ series. Note that, with donor **1**, the use of tetrahydrofuran (THF) as an electrocrystallization solvent generally requires working temperatures of 2–3 °C in order to favor crystallization of radical cation salts.

[(*rac*)-**1**]$_2$AsF$_6$ is isostructural with the previously described [(*rac*)-**1**]$_2$PF$_6$ metallic salt [3], but also with the racemic [(*rac*)-**2**]$_2$PF$_6$ [5]. It crystallizes in the triclinic centrosymmetric space group *P*–1 with one independent donor molecule and half of anion, located on an inversion center, in the asymmetric unit (Figure 1a, Table S1). The ethylene bridge C7–C8 is disordered over two positions A and B with s.o.f. values of 0.62 and 0.38, respectively, whereas the methyl group C9 is not disordered and adopts an equatorial conformation, thus leading to the presence of both enantiomers (*S*) and (*R*) on the same crystallographic site. The four equatorial fluorine atoms are disordered on two positions each. The enantiopure salts [(*S*)-**1**]$_2$AsF$_6$ and [(*R*)-**1**]$_2$AsF$_6$ crystallize in the non-centrosymmetric triclinic space group *P*1 with two independent donor molecules and one anion in the asymmetric unit (Figure 1b for the (*S*) enantiomer). Since they are isostructural, only the structure of [(*S*)-**1**]$_2$AsF$_6$ will be detailed.

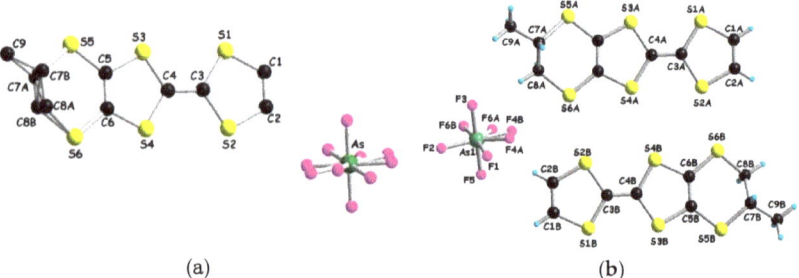

(a)   (b)

Figure 1. (**a**) Molecular structure of [(*rac*)-**1**]$_2$AsF$_6$. H atoms have been omitted for clarity. Equatorial F atoms are disordered on two positions each (s.o.f. 0.66 and 0.34). The ethylene bridge is disordered over two positions A (s.o.f. 0.62) and B (s.o.f. 0.38); (**b**) Molecular structure of [(*S*)-**1**]$_2$AsF$_6$. F4 and F6 atoms are disordered (s.o.f. 0.57 and 0.43).

Contrary to the racemic form, the donor molecules of the enantiopure salts have no occupational disorder; the methyl substituent adopts equatorial conformation in both independent molecules, and two of the fluorine atoms were modelled over two positions.

The lengths of the central C = C bond together with the internal C–S bonds, shown in Table 1 for [(*rac*)-**1**]$_2$AsF$_6$ and [(*S*)-**1**]$_2$AsF$_6$, are in agreement with a +0.5 oxidation state of the donor and are comparable with the values measured for the [(*rac*)-**1**]$_2$PF$_6$ salt [3].

Table 1. Selected C = C and C–S internal bond distances for [(rac)-1]$_2$AsF$_6$ and [(S)-1]$_2$AsF$_6$.

		Bond lengths (Å)		
		[(S)-1]$_2$AsF$_6$		[(rac)-1]AsF$_6$
A	C3A–C4A	1.367(9)	C3–C4	1.363(5)
	S1A–C3A	1.714(7)	S1–C3	1.739(4)
	S2A–C3A	1.759(7)	S2–C3	1.739(3)
	S3A–C4A	1.744(7)	S3–C4	1.740(3)
	S4A–C4A	1.736(7)	S4–C4	1.733(3)
B	C3B–C4B	1.374(10)		
	S1B–C3B	1.737(7)		
	S2B–C3B	1.753(7)		
	S3B–C4B	1.733(7)		
	S4B–C4B	1.726(7)		

A classical organic-inorganic segregation occurs in the packing of both enantiopure and racemic salts, with the donors adopting a β-type organization in parallel columns, with short intrastack and interstack S···S distances (Figure 2a and Figure S1). As in the case of the PF$_6^-$ homologous series, the donors engage in a complex set of intermolecular CH···F hydrogen bond interactions with the fluorine atoms of the anion (Figure 2b and Figure S2). When comparing the structure of the two series, the CH···F distances, such as those for CH$_{vinyl}$ and CH$_{Me}$, are either equal or slightly smaller for AsF$_6^-$ than those with PF$_6^-$, as a consequence of the longer As–F bond lengths (Tables S2 and S3). All fluorine atoms are involved in such hydrogen-bonding interactions.

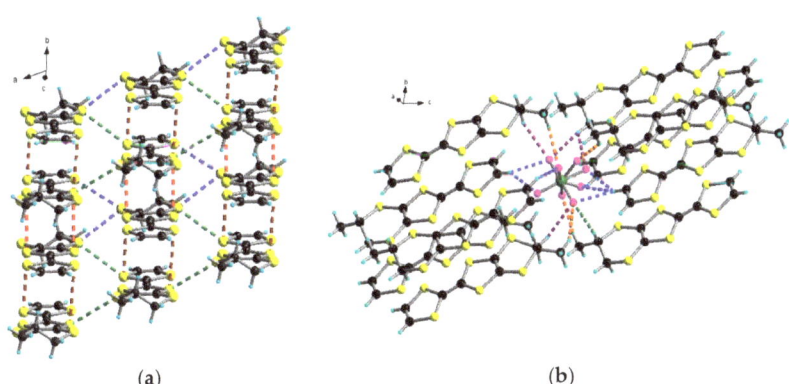

Figure 2. (a) Layer of donors in the packing of [(S)-1]$_2$AsF$_6$ with an emphasis on the S···S short contacts: red dotted lines (3.55–3.59 Å), brown dotted lines (3.62 Å), blue dotted lines (3.60–3.63 Å) and green dotted lines (3.68–3.70 Å); (b) C–H···F short contacts: blue dotted lines for CH$_{vinyl}$ (2.40–2.44–2.53 Å), orange dotted lines for Me (2.42–2.55 Å), violet dotted lines for CH$_2$ (2.49–2.66 Å) and green dotted lines for CH$_{Me}$ (2.53 Å).

Thus, this (1)$_2$AsF$_6$ series presents the same structural characteristics as the (1)$_2$PF$_6$ series and also as the isostructural one of donor 2 with AsF$_6^-$, namely (2)$_2$AsF$_6$ [9]. Since the latter two show metal-like behavior, it is reasonable to envisage similar conducting properties for the former. Although crystals of the series (1)$_2$AsF$_6$ were rather small and brittle, single crystal resistivity measurements could be performed for [(rac)-1]$_2$AsF$_6$ and [(R)-1]$_2$AsF$_6$. The metallic behavior in the high-temperature regime could not be detected, as in the case of two point contact measurements in the (**1**)$_2$PF$_6$ series, in spite of rather high values of the room temperature conductivities ($\sigma_{RT} \approx$ 1–10 S cm^{-1}) and very small activation energies (9–12 meV) (Figure 3a). Most likely, the presence of structural disorder on the donors and the anions prevents the observation of a metal-like conductivity. However, thermoelectric power measurements suggest metallic behavior in the high-temperature regime according to the small positive values of the Seebeck coefficient decreasing towards zero upon cooling (Figure 3b).

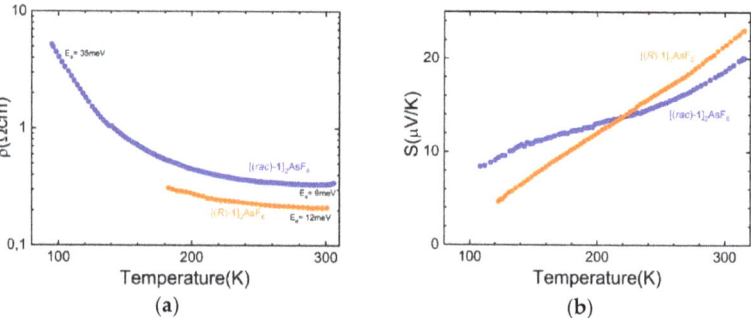

Figure 3. (**a**) Temperature dependence of the electrical resistivity ρ for single crystals of [(rac)-**1**]$_2$AsF$_6$ (blue curve) and [(R)-**1**]$_2$AsF$_6$ (orange curve) measured using four-in-line contacts; (**b**) Temperature dependence of the thermoelectric power for a single crystal of [(rac)-**1**]$_2$AsF$_6$ (blue curve) and [(R)-**1**]$_2$AsF$_6$ (orange curve).

These values of conductivity are comparable to those in the PF$_6^-$ series; moreover, no striking difference has been observed between the racemic and enantiopure materials. This last feature is not really surprising in view of the similar packing of the donors, although one might argue that the enantiopure salts could show in principle higher conductivity since there is no structural disorder on the donors. However, the anion is disordered in all the salts of the series; therefore, this feature can hamper such fine observations.

Surprisingly, in a second set of experiments realized with the enantiopure donor and [(n-Bu)$_4$N]AsF$_6$ in separate compartments, totally different 1:1 salts crystallized in the anodic compartment as black prismatic blocks. Once again, a working temperature of 2–3 °C was imposed in order to obtain crystalline materials. Accordingly, [(S)-**1**]AsF$_6$·C$_4$H$_8$O and [(R)-**1**]AsF$_6$·C$_4$H$_8$O radical cation salts are isostructural and crystallize in the triclinic system non-centrosymmetric system P1, with two independent donors A and B, two anions and two tetrahydrofuran (THF) molecules in the unit cell (Figure 4 and Figure S3 for the (R) enantiomer, Table S4).

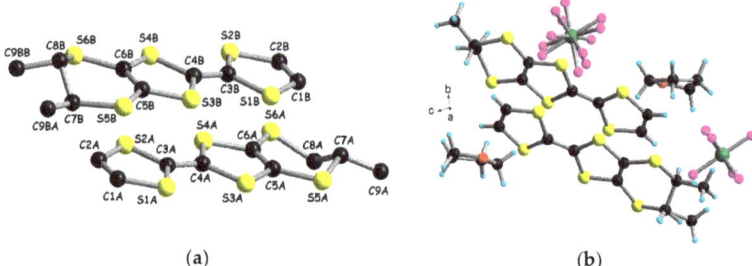

Figure 4. (**a**) The two independent donor molecules in the crystal structure of [(S)-**1**]AsF$_6$·C$_4$H$_8$O. C9B atom is disordered over two positions A (s.o.f. 0.69) and B (s.o.f. 0.31). H atoms have been omitted for clarity; (**b**) Molecular structure of [(S)-**1**]AsF$_6$·C$_4$H$_8$O. F1B-F6B atoms are disordered over two positions A (s.o.f. 0.57) and B (s.o.f. 0.43).

In one of the two donors, the methyl substituent C9B, located in an equatorial position, is disordered over two positions (C9BA and C9BB), which is different to the previous mixed-valence salts. Moreover, all the six fluorine atoms of one anion are disordered on two positions each. The stoichiometry of the two compounds indicates that both independent donors are in a radical cation state, in agreement with the values of the central C = C and internal C–S bond lengths (Table 2). The crystallization of these 1:1 phases is very likely favored by the much lower concentration of available AsF_6^- counter ion in the anodic compartment, leading to the complete oxidation of the donors before crystallization.

Table 2. Selected C = C and C–S internal bond distances for $[(S)\text{-}1]_2AsF_6$ and $[(S)\text{-}1]_2AsF_6$.

		Bond Lengths (Å)		
		$[(S)\text{-}1]AsF_6\cdot C_4H_8O$		$[(R)\text{-}1]AsF_6\cdot C_4H_8O$
A	C3A–C4A	1.396(13)	C3A–C4A	1.4214(13)
	S1A–C3A	1.737(9)	S1A–C3A	1.734(12)
	S2A–C3A	1.710(10)	S2A–C3A	1.7068(98)
	S3A–C4A	1.705(9)	S3A–C4A	1.6963(91)
	S4A–C4A	1.727(9)	S4A–C4A	1.7142(12)
B	C3B–C4B	1.389(13)	C3B–C4B	1.3577(13)
	S1B–C3B	1.729(10)	S1B–C3B	1.7306(89)
	S2B–C3B	1.710(10)	S2B–C3B	1.7154(12)
	S3B–C4B	1.705(10)	S3B–C4B	1.7128(12)
	S4B–C4B	1.735(9)	S4B–C4B	1.7485(88)

As is often observed in the case of salts based on fully oxidized donors, the radical cations form strong eclipsed dimers in the packing, with very short S···S intradimer distances of 3.33–3.37 Å and much longer lateral S···S interdimer distances (3.59–3.77 Å) (Figure 5a). The dimers are separated along the stacking direction by THF molecules (Figures S4 and S5 for the (R) enantiomer). The fluorine atoms are engaged in hydrogen bonding with protons of the donors but also with a CH_2 group of THF, thus providing a certain stability of the crystals against the desolvation (Figure 5b and Figure S6 for the (R) enantiomer, Tables S5 and S6). When considering this strong dimerization of the donors, it can be safely concluded that the two salts are insulators.

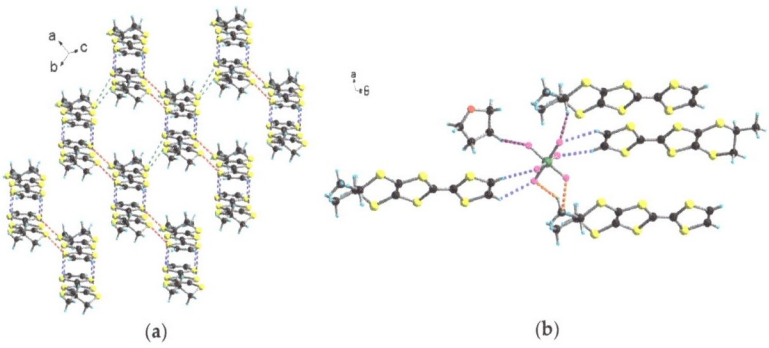

Figure 5. (a) Packing of the donors in the structure of $[(S)\text{-}1]AsF_6\cdot C_4H_8O$ with emphasis on the S···S short contacts: blue dotted lines for (3.33–3.37 Å), red dotted lines for (3.59–3.66 Å), green dotted lines for (3.75–3.77 Å); (b) View of the structure of $[(S)\text{-}1]AsF_6\cdot C_4H_8O$ with an emphasis on the C–H···F short contacts: blue dotted lines for CH_{vinyl} (2.41–2.64 Å), orange dotted lines for Me (2.59–2.62 Å), violet dotted lines for CH_2 and CH (2.64–2.65 Å).

2.2. Radical Cation Salts of Me-EDT-TTF (1) and DM-EDT-TTF (2) with the I_3^- Anion

As mentioned above, TM-BEDT-TTF afforded 1:1 radical cation salts with the tris(iodide) anion I_3^- [15]. On the other hand, BEDT-TTF provided the ambient pressure superconducting phase (BEDT-TTF)$_2$I$_3$ [11–13], while a 1:1 salt was described with EDT-TTF [24]. Here, we present our results on the complete series of radical cation salts of chiral EDT-TTF derived donors 1 and 2 with the I_3^- anion. The crystalline salts, collected as brown plates for 1 and black prisms for 2, were prepared by electrocrystallization of the respective donors in acetonitrile at 20 °C in the presence of [(n-Bu)$_4$N]I$_3$.

The racemic and enantiopure salts (1)I$_3$ are isostructural, with [(rac)-1]I$_3$ crystallizing in the centrosymmetric triclinic P–1 space group, while the enantiopure [(S)-1]I$_3$ and [(R)-1]I$_3$ are in the non-centrosymmetric triclinic space group P1 (Table S7). The asymmetric unit consists of two independent donor molecules and two anions for the latter (Figure 6a for [(S)-1]I$_3$) and one donor and two half anions for the former. In both independent molecules of [(S)-1]I$_3$, the methyl substituent (C9A and C9B) adopts an axial position. The central C = C bond lengths values, together with the internal C–S bond lengths (Table S8), are indicative of the +1 oxidation state, in agreement with the 1:1 stoichiometry.

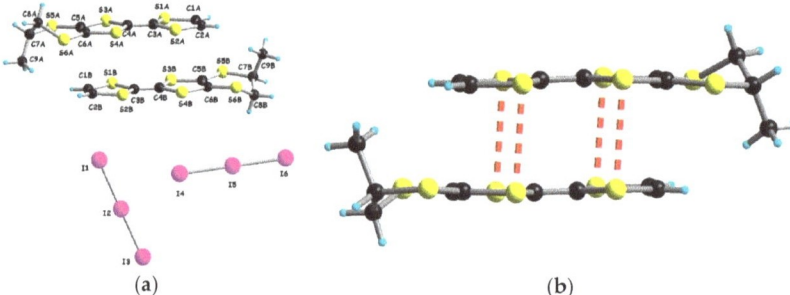

Figure 6. (a) Molecular structure of [(S)-1]I$_3$ together with the atom numbering scheme; (b) Dimer motif in the structure of [(S)-1]I$_3$ with emphasis on the S···S short contacts: red dotted lines for 3.34, 3.37, 3.38 and 3.42 Å.

The donors A and B are strongly dimerized as attested by the eclipsed arrangement and the very short intradimer S···S distances ranging between 3.34 and 3.42 Å (Figure 6b for [(S)-1]I$_3$). The packing of the donors is strongly reminiscent of the one observed in the structure of (EDT-TTF)I$_3$ [24]. Accordingly, one I$_3$ anion separates the layers of donors, and the other I$_3$ anion is embedded in the layer, with its axis parallel to the long axis of the donors, thus leading to only lateral overlap between the dimers (Figure 7 for [(R)-1]I$_3$).

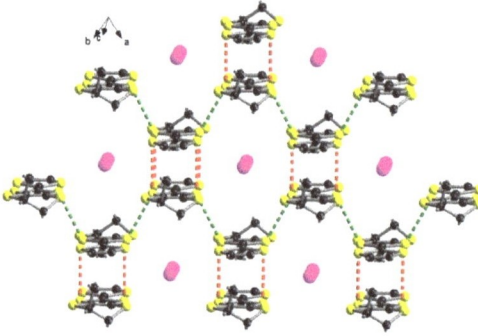

Figure 7. Packing of donors in the structure of [(R)-1]I$_3$ with emphasis on the S···S contacts shorter than 3.6 Å: red dotted lines for 3.33, 3.35 and 3.39 Å; green dotted lines for 3.45 and 3.48 Å.

Although the interdimer S···S distances for these lateral donor···donor contacts are as short as 3.45–3.48 Å (Figure 7), one can expect very weak conductivity for these radical cation salts of 1:1 composition when considering the strong dimerization of the donors.

Thus, within this series of (**1**)I$_3$, the presence of a methyl substituent on the ethylene bridge of EDT-TTF as in donor **1** does not have a strong impact on the packing, which is similar to the one observed in the parent (EDT-TTF)I$_3$ salt. What would be the consequence of introducing a second methyl substituent as in donor **2**? Accordingly, we have proceeded to electrocrystallization experiments of **2** in the presence of [(n-Bu)$_4$N]I$_3$ in similar conditions as for **1**. A complete series of isostructural crystalline radical cation salts formulated as (**2**)I$_3$ has been obtained as well, with the racemic compound having crystallized in the triclinic space group $P\bar{1}$ and the enantiopure counterparts in triclinic $P1$ (Table S9). The asymmetric unit of the former contains one independent donor and one anion, while those of the latter contain two independent donors and two anions. In contrast to the previous series based on donor **1**, now both tris(iodide) ions are aligned with the long axis of the donors, and the methyl groups are located in equatorial positions (eq, eq) (Figure 8).

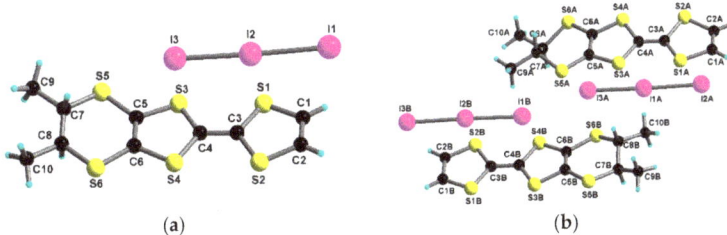

Figure 8. (**a**) Molecular structure of [(*rac*)-**2**]I$_3$ with together with the atom numbering scheme; (**b**) Molecular structure of [(*R,R*)-**2**]I$_3$ together with the atom numbering scheme.

As it was recently evidenced by DFT calculations, the difference in energy between the axial and equatorial conformers of DM-EDT-TTF amounts only to ≈3 kcal mol^{-1} [25], and even less for the monomethylated precursors [26], the occurrence of one form or the other in the solid state being determined mainly by the packing and establishment of intermolecular interactions. The analysis of the central C = C and internal C–S bonds lengths values allows the assignment of the oxidation state +1 to the donors, in agreement with the 1:1 stoichiometry. Since in this case both tris(iodide) anions are aligned with the long axis of the donors, the *c* direction, along which no donor···donor interaction takes place (Figures S7 and S8), mixed donor/anion layers are formed in the *ab* plane, with strongly dimerized donors interacting laterally (Figure 9a for (*rac*) and Figure 9b for (*R,R*)).

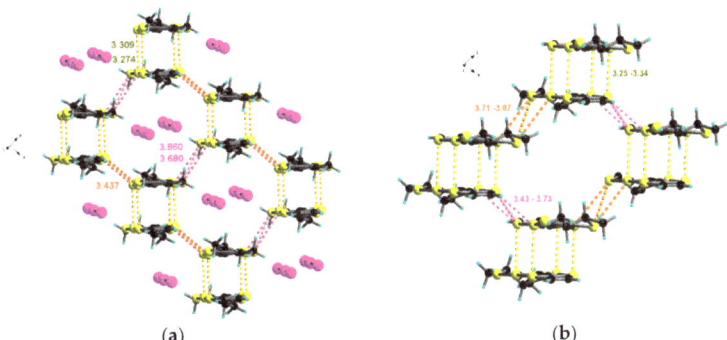

Figure 9. (**a**) Packing of the donors and highlight of the S···S short contacts for [(*rac*)-**2**]I$_3$; (**b**) Packing of the donors and highlight of the S···S short contacts for [(*R,R*)-**2**]I$_3$. The anions are not shown.

The intradimer S···S distances of 3.27–3.31 Å for [(*rac*)-**2**]I$_3$ and 3.25–3.34 Å for [(*R,R*)-**2**]I$_3$ are even shorter than in the previous series, probably as a consequence of the (eq, eq) conformation of the methyl groups, while the interdimer ones are slightly longer (Figure 9 and Figure S9). Once again, due to the strong dimerization, the conductivity of these materials is expected to be poor.

2.3. Synthesis and Structure of (meso)-DM-EDT-TTF

The fascinating properties shown by the radical cation salts of DM-EDT-TTF (**2**) in its chiral version, that is (*rac*), (*S,S*) and (*R,R*) forms, with different anions (ClO$_4$ [8], PF$_6$ [5], AsF$_6$ [9], SbF$_6$ [9], Ta$_2$F$_{10}$O [10]) prompted us to investigate the hitherto unknown (*meso*) form of DM-EDT-TTF ((*meso*)-**2** in Scheme 1). For its preparation, we have envisaged a three-step procedure, starting with a double nucleophilic substitution of (*meso*)-2,3-dibromobutane with dmit^{2-}, generated from the protected precursor **3** in basic conditions, to form (*meso*)-DM-DDDT **4**. Then, a classical phosphite-mediated cross-coupling reaction of **4** with the dicarboxylate dithiolene **5** afforded the TTF (*meso*)-**6**, which was subsequently heated in DMF in the presence of lithium bromide to yield (*meso*)-**2** upon a double decarboxylation (Scheme 2).

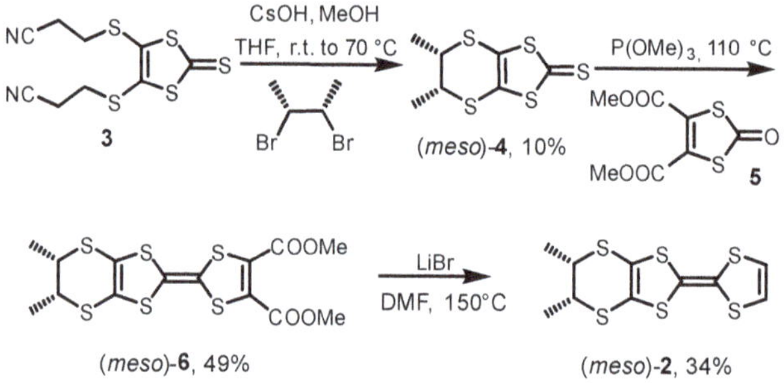

Scheme 2. Synthesis of (*meso*)-DM-EDT-TTF donor ((*meso*)-**2**).

The neutral donor (*meso*)-**2** crystallizes in the monoclinic centrosymmetric space group $P2_1/n$, with one independent molecule in the unit cell (Table S11). The dithiin six-membered ring shows a sofa-type conformation (dihedral angles C5–C6–S6–C7 and C6–C5–S5–C8 measure 5.05° and 23.2°, respectively), with axial (C10) and equatorial (C9) orientations of the methyl substituents (Figure 10) and boat-like conformation of the TTF unit. Central C=C and internal C–S bond distances have typical values for neutral donors (Table S12). In the packing, the donors organize in orthogonal dyads (Figure S10), with intermolecular S···S contacts of 3.54–3.87 Å.

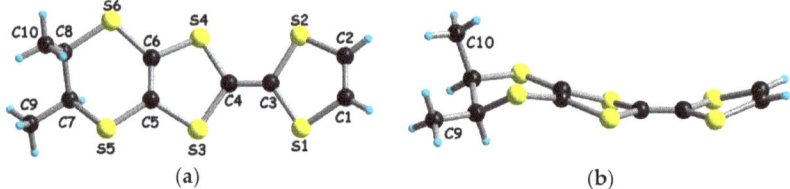

Figure 10. (a) Molecular structure of (*meso*)-**2**; (b) Lateral view.

2.4. Radical Cation Salts of (meso)-DM-EDT-TTF

Our prime objective was to obtain radical cation salts of (*meso*)-**2** with PF$_6^-$ and ClO$_4^-$ anions in order to compare them with those obtained with donors **1** and **2** (chiral form), previously reported. Accordingly, electrocrystallization of (*meso*)-**2** in the presence of [(n-Bu)$_4$N]PF$_6$ and [(n-Bu)$_4$N]ClO$_4$ provided 1:1 radical cation salts [(*meso*)-**2**]PF$_6$ and [(*meso*)-**2**]ClO$_4$, respectively, as black crystalline needles. Additionally, knowing that ClO$_4^-$ and ReO$_4^-$ anions usually afford isostructural salts, the use of [(n-Bu)$_4$N]ReO$_4$ leads to the formation of crystalline [(*meso*)-**2**]ReO$_4$, isostructural with its perchlorate congener (Table S11). A working temperature of 3 °C had to be imposed in the case of ClO$_4$ and ReO$_4$ salts, very likely because of the use of THF as co-solvent.

[(*meso*)-**2**]PF$_6$ crystallizes in the monoclinic centrosymmetric space group $P2_1/n$, with one independent donor molecule and one anion in the unit cell (Figure 11a). While in the chiral form of **2**, the methyl substituents can adopt either axial (ax) or equatorial (eq) conformations, the most common being the equatorial one in radical cation salts, thus maximizing the overlap between the donors, in (*meso*)-**2** the conformation is necessarily (ax, eq). In the packing, an organic-inorganic segregation takes place along the c direction, with the donors forming strong centrosymmetric dimers, as attested by the short intermolecular S···S contacts of 3.35–3.37 Å (Figure 11b). The dimers interact only laterally in the ab plane, yet with much longer S···S contacts (3.64–3.81 Å). A network of hydrogen bonding exists between fluorine atoms and the different hydrogen atoms of the donor: H$_{vinyl}$···F (2.32–2.36 Å) and H$_{Me}$···F (2.53–2.75 Å), H$_{CH}$···F (2.45–2.65 Å) (Figure S11).

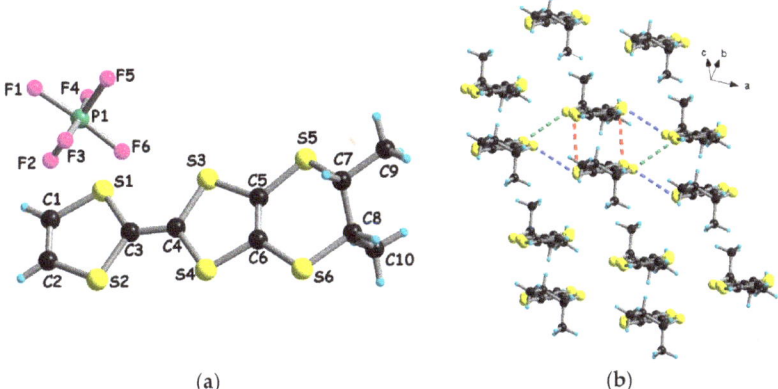

Figure 11. (a) Molecular structure of [(*meso*)-**2**]PF$_6$; (b) Packing diagram for [(*meso*)-**2**]PF$_6$ with S···S short contacts highlighted. Red dotted lines (3.35–3.37 Å), blue dotted lines (3.75–3.81 Å), green dotted lines (3.64–3.68 Å). The anions are not shown.

As mentioned above, [(*meso*)-**2**]ClO$_4$ and [(*meso*)-**2**]ReO$_4$ are isostructural, crystallizing in the triclinic centrosymmetric space group $P-1$, with one independent donor and one anion in the unit cell (Figures S12 and S13, Table S11). The donors organize in dyads, through short S···S interactions (3.34–3.41 Å), further interacting laterally (Figure 12a), while donors-anions' segregation establishes along the c direction (Figure 12b and Figure S14). The lateral S···S contacts are on average shorter in the ClO$_4$ and ReO$_4$ salts than in the PF$_6$ one.

A strong dimerization of radical cations, with little interdimer interactions, is generally detrimental for good electron transport properties. Indeed, single crystal resistivity measurements for [(*meso*)-**2**]PF$_6$ and [(*meso*)-**2**]ReO$_4$ show semiconducting behavior with room temperature conductivities σ_{RT} of 1.4 10^{-5} S cm^{-1} for the former and 1.6 10^{-4} S cm^{-1} for the latter, the activation energies E_a being around 300–340 meV (Figure 13). The origin of the higher conductivity of the ReO$_4$ salt compared to the PF$_6$ salt may be in the stronger lateral interdimer interactions observed in the former.

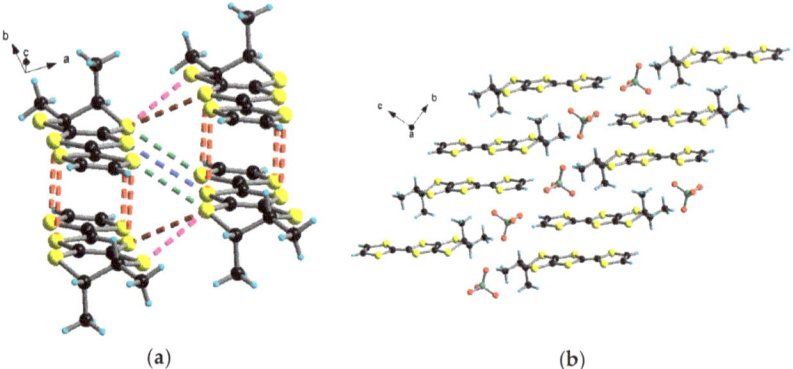

(a) (b)

Figure 12. (**a**) Packing diagram for [(*meso*)-**2**]ClO$_4$ in the *ab* plane with an emphasis on short S···S interactions. Red dotted lines (3.34–3.41 Å), green dotted lines (3.49 Å), blue dotted lines (3.79 Å), magenta lines (3.48 Å), brown lines (3.65 Å); (**b**) Packing diagram in the *bc* plane showing the donors-anions' segregation.

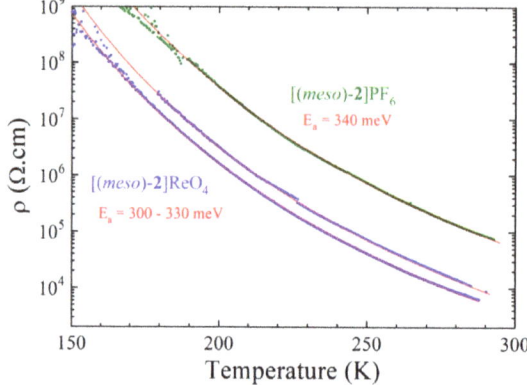

Figure 13. Temperature dependence of the electrical resistivity ρ for one single crystal of [(*meso*)-**2**]PF$_6$ (green curve) and two single crystals of [(*meso*)-**2**]ReO$_4$ (blue curves) measured using two contacts. The red lines are the fit to the activation law $\rho = \rho_0 \exp(E_a/T)$ in the 160–300 K temperature range giving the activation energy E_a.

Across the preparation and conducting properties of these radical cation salts based on (*meso*)-**2**, it is clear that the mutual orientation of the methyl substituents on the ethylene bridge plays a paramount role on the donor:anion stoichiometry and the packing of the donors. Indeed, in (*meso*)-**2**, one of the methyl substituents is necessarily axial, thus precluding strong axial overlap between donors, at the difference with (*S,S*)- and (*R,R*)-**2** where both methyl groups are equatorial in the most conducting salts.

3. Conclusions

In the continuation of our research lines on chiral materials and, more specifically, on chiral molecular conductors, we reported here a new series of radical cation salts based on the chiral donors Me-EDT-TTF (**1**) and DM-EDT-TTF (**2**) as racemic and enantiopure forms, obtained by electrocrystallization. The former provided mixed valence salts (**1**)$_2$AsF$_6$ with metal-like behavior in the high-temperature regime, which are isostructural with the previously described PF$_6$ counterpart, but also with the (**2**)$_2$AsF$_6$ series. Additionally, 1:1 enantiopure salts formulated as [(*S*)-**1**]AsF$_6$·C$_4$H$_8$O and [(*R*)-**1**]AsF$_6$·C$_4$H$_8$O have been obtained in slightly different conditions. With the tris(iodide) anion I$_3^-$, both donors afforded 1:1 salts, i.e., (**1**)I$_3$ and (**2**)I$_3$, as racemic and enantiopure forms, yet the disposition of the anions with respect to the donors is drastically different in the two series. Finally, the synthesis, characterization and single crystal X-ray structure of the new donor (*meso*)-**2** are described, together with its poorly semiconducting 1:1 radical cation salts with the PF$_6^-$, ClO$_4^-$ and ReO$_4^-$ anions. The striking difference between the salts formed by (*meso*)-**2** and those resulting from the chiral form (*S,S*)/(*R,R*)-**2** with the same anions is very likely due to the location of the methyl substituents which is necessarily axial, equatorial in the former, while in the latter they can adopt an equatorial, equatorial conformation, thus maximizing the packing. Throughout these different series of radical cation salts, the importance of the number of stereogenic centers, of their mutual arrangement in the case of (*meso*)-**2** and (*S,S*)/(*R,R*)-**2** and of the nature of the counter-ion on the donor:anion stoichiometry, packing of the donors and, ultimately, electron transport properties, are highlighted. Future work will be devoted to conductivity measurements under high pressures, and the use of these chiral precursors in electrocrystallization in the presence of magnetic anions. This last direction is particularly interesting when considering the possibility of combining chirality with conducting and magnetic properties [17,27], since in magnetic conductors, the existence of delocalized π-electrons and localized d-electrons may lead to exotic phenomena such as magnetic-field-induced superconductors, magnetoresistance effects and magnetic-field-switchable conductors, with applications in molecular electronics and spintronics [28–32].

4. Materials and Methods

All commercially available reagents and solvents were used as received unless otherwise noted. Dry tetrahydrofuran was directly used from the purification machines. Chloroform as a solvent for synthesis was passed through a short column of basic alumina prior to use. Chromatography purifications were performed on silica gel, and thin layer chromatograhy (TLC) was carried out using aluminum sheets precoated with silica gel. NMR spectra were acquired on a Bruker Avance DRX 300 spectrometer operating at 300 MHz for ^{1}H at room temperature in CDCl$_3$ solutions. ^{1}H NMR spectra were referenced to the residual protonated solvent (^{1}H). All chemical shifts are expressed in parts per million (ppm) downfield from external tetramethylsilane (TMS) using the solvent residual signal as an internal standard, and the coupling constant values (J) are reported in Hertz (Hz). The following abbreviations have been used: s, singlet; d, doublet; m, multiplet. Mass spectrometry MALDI–TOF MS spectra were recorded on a Bruker Biflex-IIITM apparatus equipped with a 337-nm N$_2$ laser.

Precursors Me-EDT-TTF **1** [3], (*S,S*)- and (*R,R*)-DM-EDT-TTF **2** [5] were synthesized according to the literature procedures, while the preparation of (*meso*)-**2** is described in this report.

(*meso*)-DM-DDDT **4**: Compound **3** (4.5 g, 14.8 mmol) was added into a 500 mL Schlenk round bottomed flask under argon, and then dry THF (400 mL) was poured into the flask. After 10 min, a solution of caesium hydroxide (7 g, 41.6 mmol) in dry methanol (50 mL) was added dropwise, and the color started to turn violet. Then, the resulting solution was left under stirring for one hour at room temperature, followed by the addition of (*meso*)-2,3-dibromobutane (5.5 g, 25 mmol) and reflux for one night. After cooling to room temperature, the solvent was removed under vacuum, and the solid residue was extracted twice with DCM and water (250/500 mL). The combined organic phases were concentrated, and the residue was purified by column chromatography (petroleum ether/DCM 8:2) to give compound **4** (0.36 g, 10%); ^{1}H NMR (300 MHz, CDCl$_3$):δ 3.57 (m, 2H, -SCH), 1.36 (d, 6H, -CH$_3$) ppm.

(*meso*)-DM-EDT-TTF-(COOMe)$_2$ **6**: Compound **4** (0.3 g, 1.19 mmol) and dimethyl 2-oxo-1,3-dithiole-4,5-dicarboxylate **5** (0.6 g, 2.5 mmol) were mixed under argon in freshly distilled trimethyl phosphite (10 mL), and the mixture was heated at 110 °C for 5 h. After this period, the solvent was evaporated in a rotary evaporator, and then toluene (20 mL) was added and evaporated. The last procedure was repeated twice. The product was solubilized in dichloromethane and passed through a silica column to remove the remaining phosphite and then purified by chromatography using petroleum spirit/dichloromethane 1/1 to afford a red-brown solid (0.255 g, 49%); ^{1}H NMR (300 MHz, CDCl$_3$) δ ppm: 3.83 (s, 6H, −OCH_3), 3.54 (m, 2H, -SCH), 1.36 (d, 6H, −CH_3); MS (MALDI-TOF) m/z: 437.4 (M_{th} = 437.92).

(*meso*)-DM-EDT-TTF (*meso*)-**1**: Compound **6** (0.25 g, 0.57 mmol) and LiBr (0.75 g, 8.6 mmol) were mixed in dimethylformamide (25 mL). The solution was stirred at 150 °C for 30 min, the formation of the product being monitored by TLC. The product was extracted with dichloromethane, and the organic phase was washed with brine and water and then dried over MgSO$_4$. The solvent was removed under vacuum, and the product was purified by chromatography on a silica gel column with petroleum spirit/dichloromethane 6/4 to afford a red solid (62 mg, 34 %); ^{1}H NMR (300 MHz, CDCl$_3$) δ ppm: 6.32 (s, 2H, −SCH=), 3.54 (m, 2H, -SCH), 1.36 (d, 6H, −CH_3); MS (MALDI-TOF) m/z: 321.89 (M_{th} = 321.91).

(**1**)$_2$AsF$_6$: Single crystals of (**1**)$_2$AsF$_6$ (*rac*), (*S*) and (*R*) were obtained by electrocrystallization. The electrolyte solution was prepared from 34.9 mg (5 eq.) of [(*n*-Bu)$_4$N]AsF$_6$ dissolved in 12 mL of tetrahydrofuran. The anodic chamber was filled with 5 mg of the corresponding donor dissolved in 6 mL of the previously prepared electrolyte solution, whereas the rest of the electrolyte solution (6 mL) was added in the cathodic compartment of the electrocrystallization cell. Single crystals, as black crystalline plates, of the salts were grown at 2–3 °C over a period of 5 days on a platinum wire electrode by applying a constant current of 1 µA.

(**1**)AsF$_6$·C$_4$H$_8$O: Single crystals of (**1**)AsF$_6$·C$_4$H$_8$O (*S*) and (*R*) were obtained by electrocrystallization. The electrolyte solution was prepared from 34.9 mg (5 eq.) of [(*n*-Bu)$_4$N]AsF$_6$ dissolved in 6 mL of tetrahydrofuran. The anodic chamber was filled with 5 mg of the corresponding donor dissolved in 6 mL of tetrahydrofuran, and the previously prepared electrolyte solution was added in the cathodic compartment of the electrocrystallization cell. Single crystals, as black crystalline plates, of the salt were grown at 2–3 °C over a period of 4 to 6 days on a platinum wire electrode by applying a constant current of 1 µA.

(**1**)I$_3$: 20 mg of [(*n*-Bu)$_4$N]I$_3$ were dissolved in 6 mL of acetonitrile, and the solution was poured in the cathodic compartment of an electrocrystallization cell. The anodic chamber was filled with 5 mg of the donor dissolved in 6 mL of acetonitrile. Single crystals of the salts [(*rac*)-**1**]I$_3$, [(*S*)-**1**]I$_3$ and [(*R*)-**1**]I$_3$ were grown at 20 °C over a period of two weeks on a platinum wire electrode by applying a constant current of 0.5 µA. Black crystalline plates were grown on the electrode.

(**2**)I$_3$: Single crystals of (**2**)I$_3$ (*rac*), (*S,S*) and (*R,R*) were obtained by electrocrystallization. The electrolyte solution was prepared from 48.3 mg (5 eq.) of [(*n*-Bu)$_4$N]I$_3$ dissolved in 12 mL of acetonitrile/chloroform 1/1. The anodic chamber was filled with 5 mg of the corresponding donor dissolved in 6 mL of the previously prepared electrolyte solution, whereas the rest of the electrolyte solution (6 mL) was added in the cathodic compartment of the electrocrystallization cell. Single crystals of the salts, as black crystalline blocks, were grown at 20 °C over a period of 5 days on a platinum wire electrode by applying a constant current of 1 µA.

[(*meso*)-**2**]PF$_6$: 20 mg of [(*n*-Bu)$_4$N]PF$_6$ were dissolved in 6 mL CHCl$_3$, and then the solution was poured into the cathodic compartment of an electrocrystallization cell. The anodic chamber was filled with 5 mg of [(*meso*)-**2**] dissolved in 6 mL CHCl$_3$. Single crystals of the salt were grown at 20 °C over a period of one week on a platinum wire electrode, by applying a constant current of 0.5 µA. Black crystalline needles were collected on the electrode.

[(*meso*)-**2**]ClO$_4$: 20 mg of [(*n*-Bu)$_4$N]ClO$_4$ were dissolved in 6 mL of a mixture of 1,1,2-trichloroethane: tetrahydrofuran 1:1, and then the solution was poured in the cathodic compartment of an electrocrystallization cell. The anodic chamber was filled with 5 mg of [(*meso*)-**2**] dissolved in 6 mL of a mixture of 1,1,2-trichloroethane/tetrahydrofuran 1/1.

Single crystals of the salt were grown at 3 °C over a period of one week on a platinum wire electrode by applying a constant current of 1 µA. Black crystalline needles were grown on the electrode.

[(*meso*)-2]ReO$_4$: The same procedure as previously was applied by using 20 mg of [(*n*-Bu)$_4$N]ReO$_4$ instead of [(*n*-Bu)$_4$N]ClO$_4$. Black crystalline needles were grown on the electrode.

Details about data collection and solution refinement are given in Tables S1, S4, S7, S9 and S11. Single crystals of the compounds were mounted on glass fibre loops using a viscous hydrocarbon oil to coat the crystal and then transferred directly to cold nitrogen stream for data collection. X-ray data collection was performed at 150 K on an Agilent Supernova with CuKα (λ = 1.54184 Å). The structures were solved by direct methods with the SHELXS-97 and SIR92 programs and refined against all F^2 values with the SHELXL-97 program using the WinGX graphical user interface. All non-H atoms were refined anisotropically. Hydrogen atoms were introduced at calculated positions (riding model), included in structure factor calculations but not refined. Crystallographic data for the structures have been deposited with the Cambridge Crystallographic Data Centre, deposition numbers CCDC 2085699 ([(*rac*)-1]$_2$AsF$_6$), 2085700 ([(*S*)-1]$_2$AsF$_6$), 2085701 ([(*R*)-1]$_2$AsF$_6$), 2085702 ([(*S*)-1]AsF$_6$·C$_4$H$_8$O), 2085703 ([(*R*)-1]AsF$_6$·C$_4$H$_8$O), 2085704 ([(*rac*)-1]I$_3$), 2085705 ([(*S*)-1]I$_3$), 2085706 ([(*R*)-1]I$_3$), 2085707 ([(*rac*)-2]I$_3$), 2085708 ([(*S,S*)-2]I$_3$), 2085709 ([(*R,R*)-2]I$_3$), 2085710 ((*meso*)-2), 2085711 ([(*meso*)-2]ClO$_4$), 2085712 ([(*meso*)-2]PF$_6$) and 2085713 ([(*meso*)-2]ReO$_4$). These data can be obtained free of charge from CCDC, 12 Union road, Cambridge CB2 1EZ, UK (e-mail: deposit@ccdc.cam.ac.uk or http://www.ccdc.cam.ac.uk).

Thermoelectric power and electrical conductivity measurements for [(*rac*)-1]$_2$AsF$_6$ and [(*R*)-1]$_2$AsF$_6$ were made along the longer axis of the crystals in the temperature range of 100–310 K. The measurement cell used was attached to the cold stage of a closed cycle helium refrigerator. The thermopower was measured using a slow AC (ca. 10^{-2} Hz) technique [33], by attaching two ⌀ = 25 µm diameter 99.99% pure Au wires (Goodfellow), thermally anchored to two quartz blocks, with Pt paint (Demetron 308A) to the extremities of an elongated sample using a previously described apparatus [34], controlled by a computer [35]. The oscillating thermal gradient was kept below 1 K and was measured with a differential Au-0.05 at. % Fe vs. chromel thermocouple. The absolute thermoelectric power of the samples was obtained after correction for the absolute thermopower of the Au leads by using the data of Huebener [36]. Electrical resistivity measurements were conducted in a four-in-line contact configuration where a low-frequency AC method (77 Hz) was used; the measurements were conducted with a SRS Model SR83 lock-in amplifier, and a 5 µA current was applied. Electrical resistivity of [(*meso*)-2]PF$_6$ and [(*meso*)-2]ReO$_4$ was measured in two points on needle-shaped single crystals 0.5 mm long. Gold wires were glued with silver paste on gold-evaporated contacts. Different techniques were used to measure resistivity, either applying a constant voltage (1–5 V) and measuring the current with a Keithley 486 or applying a DC current (0.1–0.01 µA) and measuring the voltage with a Keithley 2400. We have checked for each crystal that both techniques give the same resistance value at room temperature. A low temperature was provided by a homemade cryostat equipped with a 4 K pulse-tube.

Supplementary Materials: The following are available online at https://www.mdpi.com/article/10.3390/magnetochemistry7060087/s1: Figure S1: Layer of donors in the packing of [(*rac*)-1]$_2$AsF$_6$; Figure S2: C–H···F short contacts in the packing of [(*rac*)-1]$_2$AsF$_6$; Figure S3: View of the ordered donor molecule and anion in the crystal structure of [(*R*)-1]AsF$_6$·C$_4$H$_8$O; Figure S4: Packing of the donors in the structure of [(*S*)-1]AsF$_6$·C$_4$H$_8$O; Figure S5: Packing of the donors in the structure of [(*R*)-1]AsF$_6$·C$_4$H$_8$O; Figure S6: View of the structure of [(*R*)-1]AsF$_6$·C$_4$H$_8$O; Figure S7: View in the *bc* plane of the packing within [(*rac*)-2]I$_3$; Figure S8: View in the *bc* plane of the packing within [(*R,R*)-2]I$_3$; Figure S9: Packing of the donors and highlight of the S···S short contacts for [(*rac*)-2]I$_3$; Figure S10: Packing diagram for (*meso*)-2; Figure S11: Solid state structure of [(*meso*)-2]PF$_6$; Figure S12: Molecular structure of [(*meso*)-2]ClO$_4$; Figure S13: Molecular structure of [(*meso*)-2]ReO$_4$; Figure S14: Packing diagram for [(*meso*)-2]ReO$_4$; Table S1: Crystal Data and Structure Refinement for [(*rac*)-

1]$_2$AsF$_6$, [(S)-1]$_2$AsF$_6$ and [(R)-1]$_2$AsF$_6$; Table S2: C–H···F hydrogen bonding distances (Å) and angles in [(rac)-1]$_2$AsF$_6$; Table S3: C–H···F hydrogen bonding distances (Å) and angles in [(S)-1]$_2$AsF$_6$; Table S4: Crystal Data and Structure Refinement for [(S)-1]AsF$_6$·C$_4$H$_8$O and [(R)-1]AsF$_6$·C$_4$H$_8$O; Table S5: C–H···F hydrogen bonding distances (Å) and angles in [(S)-1]AsF$_6$·C$_4$H$_8$O; Table S6: C–H···F hydrogen bonding distances (Å) and angles in [(R)-1]AsF$_6$·C$_4$H$_8$O; Table S7: Crystal Data and Structure Refinement for [(rac)-1]I$_3$, [(S)-1]I$_3$ and [(R)-1]I$_3$; Table S8: Selected C = C and C–S internal bond lengths for [(S)-1]I$_3$; Table S9: Crystal Data and Structure Refinement for [(rac)-2]I$_3$, [(S,S)-2]I$_3$ and [(R,R)-2]I$_3$; Table S10: Selected C = C and C–S internal bond lengths for (2)I$_3$; Table S11: Crystal Data and Structure Refinement for (meso)-2, [(meso)-2]ClO$_4$, [(meso)-2]PF$_6$ and [(meso)-2]ReO$_4$; Table S12: Selected C = C and C–S internal bond lengths for (meso)-2, [(meso)-2]ClO$_4$, [(meso)-2]PF$_6$ and [(meso)-2]ReO$_4$.

Author Contributions: N.A. conceived and designed the experiments; N.M., A.B. and F.P. synthesized and characterized the materials; N.V. performed the chiral HPLC separation of precursor 1; P.A.-S., E.B.L. and M.A. investigated the electron transport properties; N.A. and F.P. wrote and/or reviewed the manuscript with contributions from all authors. All authors have read and agreed to the published version of the manuscript.

Funding: This research was partially funded in France by the National Agency for Research (ANR), Project 15-CE29-0006-01 ChiraMolCo, French Ministry of Europe and Foreign Affairs through the Eiffel Program (grant to A.B.), and in Portugal by FCT under contracts UIDB/04349/2020 and LISBOA-01-0145-FEDER-029666.

Institutional Review Board Statement: Not applicable.

Acknowledgments: This work was supported in France by the CNRS and the University of Angers. Magali Allain (MOLTECH-Anjou, University of Angers) is warmly thanked for help with the single crystal X-ray structure refinement. The collaboration between the Portuguese and French team members was also supported by a FCT–French Ministry of Foreign Affairs bilateral action FCT/PHC-PESSOA 2020-21 (Project 44647UB).

Conflicts of Interest: The authors declare no conflict of interest.

References

1. Avarvari, N.; Wallis, J.D. Strategies Towards Chiral Molecular Conductors. *J. Mater. Chem.* **2009**, *19*, 4061–4076. [CrossRef]
2. Pop, F.; Zigon, N.; Avarvari, N. Main-Group-Based Electro- and Photoactive Chiral Materials. *Chem. Rev.* **2019**, *119*, 8435–8478. [CrossRef]
3. Mroweh, N.; Auban-Senzier, P.; Vanthuyne, N.; Canadell, E.; Avarvari, N. Chiral EDT-TTF precursors with one stereogenic centre: Substituent size modulation of the conducting properties in the (R-EDT-TTF)$_2$PF$_6$ (R = Me or Et) series. *J. Mater. Chem. C* **2019**, *7*, 12664–12673. [CrossRef]
4. Mroweh, N.; Auban-Senzier, P.; Vanthuyne, N.; Lopes, E.B.; Almeida, M.; Canadell, E.; Avarvari, N. Chiral Conducting Me-EDT-TTF and Et-EDT-TTF Based Radical Cation Salts with the Perchlorate Anion. *Crystals* **2020**, *10*, 1069. [CrossRef]
5. Pop, F.; Auban-Senzier, P.; Frąckowiak, A.; Ptaszyński, K.; Olejniczak, I.; Wallis, J.D.; Canadell, E.; Avarvari, N. Chirality Driven Metallic versus Semiconducting Behavior in a Complete Series of Radical Cation Salts Based on Dimethyl-Ethylenedithio-Tetrathiafulvalene (DM-EDT-TTF). *J. Am. Chem. Soc.* **2013**, *135*, 17176–17186. [CrossRef] [PubMed]
6. Rikken, G.L.J.A.; Fölling, J.; Wyder, P. Electrical Magnetochiral Anisotropy. *Phys. Rev. Lett.* **2001**, *87*, 236602. [CrossRef] [PubMed]
7. Krstić, V.; Roth, S.; Burghard, M.; Kern, K.; Rikken, G.L.J.A. Magneto-Chiral Anisotropy in Charge Transport Through Single-Walled Carbon Nanotubes. *J. Chem. Phys.* **2002**, *117*, 11315–11319. [CrossRef]
8. Pop, F.; Auban-Senzier, P.; Canadell, E.; Rikken, G.L.J.A.; Avarvari, N. Electrical magneto-chiral anisotropy in a bulk chiral molecular conductor. *Nat. Commun.* **2014**, *5*, 3757. [CrossRef] [PubMed]
9. Pop, F.; Auban-Senzier, P.; Canadell, E.; Avarvari, N. Anion size control of the packing in the metallic versus semiconducting chiral radical cation salts (DM-EDT-TTF)$_2$XF$_6$ (X = P, As, Sb). *Chem. Commun.* **2016**, *52*, 12438–12441. [CrossRef]
10. Mroweh, N.; Mézière, C.; Allain, M.; Auban-Senzier, P.; Canadell, E.; Avarvari, N. Conservation of structural arrangements and 3:1 stoichiometry in a series of crystalline conductors of TMTTF, TMTSF, BEDT-TTF, and chiral DM-EDT-TTF with the oxo-bis[pentafluorotantalate(V)] dianion. *Chem. Sci.* **2020**, *11*, 10078–10091. [CrossRef]
11. Yagubskii, E.B.; Shchegolev, I.F.; Laukhin, V.N.; Kononovich, P.A.; Kartsovnik, M.V.; Zvarykina, A.V.; Buravov, L.I. Normal-pressure superconductivity in an organic metal (BEDT-TTF)$_2$I$_3$ [bis (ethylene dithiolo) tetrathiofulvalene triiodide]. *JETP Lett.* **1984**, *39*, 12–16.
12. Crabtree, G.W.; Carlson, K.D.; Hall, L.N.; Copps, P.T.; Wang, H.H.; Emge, T.J.; Beno, M.A.; Williams, J.M. Superconductivity at ambient pressure in di[bis(ethylenedithio)tetrathiafulvalene] triiodide, (BEDT-TTF)$_2$I$_3$. *Phys. Rev. B* **1984**, *30*, 2958–2960. [CrossRef]

13. Tokumoto, M.; Murata, K.; Bando, H.; Anzai, H.; Saito, G.; Kajimura, K.; Ishiguro, T. Ambient-pressure superconductivity at 8 K in the organic conductor β-(BEDT-TTF)$_2$I$_3$. *Solid State Commun.* **1985**, *54*, 1031–1034. [CrossRef]
14. Wallis, J.D.; Karrer, A.; Dunitz, J.D. Chiral metals? A chiral substrate for organic conductors and superconductors. *Helv. Chim. Acta* **1986**, *69*, 69–70. [CrossRef]
15. Pop, F.; Laroussi, S.; Cauchy, T.; Gómez-García, C.J.; Wallis, J.D.; Avarvari, N. Tetramethyl-Bis(ethylenedithio)-Tetrathiafulvalene (TM-BEDT-TTF) Revisited: Crystal Structures, Chiroptical Properties, Theoretical Calculations, and a Complete Series of Conducting Radical Cation Salts. *Chirality* **2013**, *25*, 466–474. [CrossRef] [PubMed]
16. Karrer, A.; Wallis, J.D.; Dunitz, J.D.; Hilti, B.; Mayer, C.W.; Bürkle, M.; Pfeiffer, J. Structures and Electrical Properties of Some New Organic Conductors Derived from the Donor Molecule TMET (*S*,*S*,*S*,*S*-Bis(dimethylethylenedithio) tetrathiafulvalene). *Helv. Chim. Acta* **1987**, *70*, 942–953. [CrossRef]
17. Galán-Mascarós, J.R.; Coronado, E.; Goddard, P.A.; Singleton, J.; Coldea, A.I.; Wallis, J.D.; Coles, S.J.; Alberola, A. A Chiral Ferromagnetic Molecular Metal. *J. Am. Chem. Soc.* **2010**, *132*, 9271–9273. [CrossRef] [PubMed]
18. Pop, F.; Mézière, C.; Allain, M.; Auban-Senzier, P.; Tajima, N.; Hirobe, D.; Yamamoto, H.M.; Canadell, E.; Avarvari, N. Unusual stoichiometry, band structure and band filling in conducting enantiopure radical cation salts of TM-BEDT-TTF showing helical packing of the donors. *J. Mater. Chem. C* **2021**, *9*. [CrossRef]
19. Matsumiya, S.; Izuoka, A.; Sugawara, T.; Taruishi, T.; Kawada, Y. Effect of Methyl Substitution on Conformation and Molecular Arrangement of BEDT-TTF Derivatives in the Crystalline Environment. *Bull. Chem. Soc. Jpn.* **1993**, *66*, 513–522. [CrossRef]
20. Mroweh, N.; Mézière, C.; Pop, F.; Auban-Senzier, P.; Alemany, P.; Canadell, E.; Avarvari, N. In Search of Chiral Molecular Superconductors: κ-[(*S*,*S*)-DM-BEDT-TTF]$_2$ClO$_4$ Revisited. *Adv. Mater.* **2020**, *32*, 2002811. [CrossRef]
21. Matsumiya, S.; Izuoka, A.; Sugawara, T.; Taruishi, T.; Kawada, Y.; Tokumoto, M. Crystal Structure and Conductivity of Chiral Radical Ion Salts (Me$_2$ET)$_2$X. *Bull. Chem. Soc. Jpn.* **1993**, *66*, 1949–1954. [CrossRef]
22. Kimura, S.; Maejima, T.; Suzuki, H.; Chiba, R.; Mori, H.; Kawamoto, T.; Mori, T.; Moriyama, H.; Nishio, Y.; Kajita, K. A new organic superconductor β-(*meso*-DMBEDT-TTF)$_2$PF$_6$. *Chem. Commun.* **2004**, *21*, 2454–2455. [CrossRef] [PubMed]
23. Kimura, S.; Suzuki, H.; Maejima, T.; Mori, H.; Yamaura, J.-I.; Kakiuchi, T.; Sawa, H.; Moriyama, H. Checkerboard-Type Charge-Ordered State of a Pressure-Induced Superconductor, β-(*meso*-DMBEDT-TTF)$_2$PF$_6$. *J. Am. Chem. Soc.* **2006**, *128*, 1456–1457. [CrossRef] [PubMed]
24. Hountas, A.; Terzis, A.; Papavassiliou, G.C.; Hilti, B.; Pfeiffer, J. Structures of the Conducting Salts of Ethylenedithiotetrathiafulvalene (EDTTTF) and Methylenedithiotetrathiafulvalene (MDTTTF): (EDTTTF)I$_3$ and (MDTTTF)I$_3$. *Acta Cryst. C* **1990**, *C46*, 220–223. [CrossRef]
25. Cauchy, T.; Pop, F.; Cuny, J.; Avarvari, N. Conformational Study and Chiroptical Properties of Chiral Dimethyl-Ethylenedithio-Tetrathiafulvalene (DM-EDT-TTF). *Chimia* **2018**, *72*, 389–393. [CrossRef]
26. Abhervé, A.; Mroweh, N.; Cauchy, T.; Pop, F.; Cui, H.; Kato, R.; Vanthuyne, N.; Alemany, P.; Canadell, E.; Avarvari, N. Conducting chiral nickel(II) bis(dithiolene) complexes: Structural and electron transport modulation with the charge and the number of stereogenic centres. *J. Mater. Chem. C* **2021**, *9*, 4119–4140. [CrossRef]
27. Atzori, M.; Pop, F.; Auban-Senzier, P.; Clérac, R.; Canadell, E.; Mercuri, M.L.; Avarvari, N. Complete Series of Chiral Paramagnetic Molecular Conductors Based on Tetramethyl bis(ethylenedithio)-tetrathiafulvalene (TM-BEDT-TTF) and Chloranilate-Bridged Heterobimetallic Honeycomb Layers. *Inorg. Chem.* **2015**, *54*, 3643–3653. [CrossRef]
28. Coronado, E.; Galán-Mascarós, J.R.; Gómez-García, C.; Laukhin, V. Coexistence of ferromagnetism and metallic conductivity in a molecule-based layered compound. *Nature* **2000**, *408*, 447–449. [CrossRef] [PubMed]
29. Rashid, S.; Turner, S.S.; Day, P.; Howard, J.A.K.; Guionneau, P.; McInnes, E.J.L.; Mabbs, F.E.; Clark, R.J.H.; Firth, S.; Biggs, T. New Superconducting Charge-Transfer Salts (BEDT-TTF)$_4$[A·M(C$_2$O$_4$)$_3$]·C$_6$H$_5$NO$_2$ (A = H$_3$O or NH$_4$, M = Cr or Fe, BEDT-TTF = Bis(Ethylenedithio)Tetrathiafulvalene). *J. Mater. Chem.* **2001**, *11*, 2095–2101. [CrossRef]
30. Uji, S.; Shinagawa, H.; Terashima, T.; Yakabe, T.; Terai, Y.; Tokumoto, M.; Kobayashi, A.; Tanaka, H.; Kobayashi, H. Magnetic-Field-Induced Superconductivity in a Two-Dimensional Organic Conductor. *Nature* **2001**, *410*, 908–910. [CrossRef] [PubMed]
31. Kurmoo, M.; Graham, A.W.; Day, P.; Coles, S.J.; Hursthouse, M.B.; Caulfield, J.L.; Singleton, J.; Pratt, F.L.; Hayes, W.; Ducasse, L.; et al. Superconducting and Semiconducting Magnetic Charge Transfer Salts: (BEDT-TTF)$_4$AFe(C$_2$O$_4$)$_3$·C$_6$H$_5$CN (A = H$_2$O, K, NH$_4$). *J. Am. Chem. Soc.* **1995**, *117*, 12209–12217. [CrossRef]
32. Martin, L.; Lopez, J.R.; Akutsu, H.; Nakazawa, Y.; Imajo, S. Bulk Kosterlitz–Thouless Type Molecular Superconductor β″-(BEDTTTF)$_2$[(H$_2$O)(NH$_4$)$_2$Cr(C$_2$O$_4$)$_3$]·18-crown-6. *Inorg. Chem.* **2017**, *56*, 14045–14052. [CrossRef]
33. Chaikin, P.M.; Kwak, J.F. Apparatus for thermopower measurements on organic conductors. *Rev. Sci. Instrum.* **1975**, *46*, 218–220. [CrossRef]
34. Almeida, M.; Alcácer, L.; Oostra, S. Anisotropy of thermopower in *N*-methyl-*N*-ethylmorpholinium bistetracyanoquinodimethane, MEM(TCNQ)$_2$, in the region of the high-temperature phase transitions. *Phys. Rev. B* **1984**, *30*, 2839–2844. [CrossRef]
35. Lopes, E.B. *INETI-Sacavém*; Internal Report; INETI Press: Sacavém, Portugal, 1991.
36. Huebener, R.P. Thermoelectric Power of Lattice Vacancies in Gold. *Phys. Rev.* **1964**, *135*, A1281–A1291. [CrossRef]

Article

Influence of the Size and Shape of Halopyridines Guest Molecules G on the Crystal Structure and Conducting Properties of Molecular (Super)Conductors of $(BEDT-TTF)_4A^+[M^{3+}(C_2O_4)_3]\cdot G$ Family

Tatiana G. Prokhorova [1,*], Eduard B. Yagubskii [1,*], Andrey A. Bardin [2,*], Vladimir N. Zverev [3,4], Gennadiy V. Shilov [1] and Lev I. Buravov [1]

[1] Institute of Problems of Chemical Physics, Russian Academy of Sciences, 142432 Chernogolovka, Russia; shilg@icp.ac.ru (G.V.S.); buravov@icp.ac.ru (L.I.B.)
[2] Institute of Microelectronics Technology and High Purity Materials, Russian Academy of Sciences, 142432 Chernogolovka, Russia
[3] Institute of Solid State Physics, Russian Academy of Sciences, 142432 Chernogolovka, Russia; zverev@issp.ac.ru
[4] Moscow Institute of Physics and Technology, 141700 Dolgoprudny, Russia
* Correspondence: prokh@icp.ac.ru (T.G.P.); yagubski@icp.ac.ru (E.B.Y.); dr.abardin@gmail.com (A.A.B.)

Citation: Prokhorova, T.G.; Yagubskii, E.B.; Bardin, A.A.; Zverev, V.N.; Shilov, G.V.; Buravov, L.I. Influence of the Size and Shape of Halopyridines Guest Molecules G on the Crystal Structure and Conducting Properties of Molecular (Super)Conductors of $(BEDT-TTF)_4A^+[M^{3+}(C_2O_4)_3]\cdot G$ Family. *Magnetochemistry* **2021**, *7*, 83. https://doi.org/10.3390/magnetochemistry7060083

Academic Editors: Carlos J. Gómez García, John Wallis, Lee Martin, Scott Turner and Hiroki Akutsu

Received: 3 May 2021
Accepted: 31 May 2021
Published: 4 June 2021

Publisher's Note: MDPI stays neutral with regard to jurisdictional claims in published maps and institutional affiliations.

Copyright: © 2021 by the authors. Licensee MDPI, Basel, Switzerland. This article is an open access article distributed under the terms and conditions of the Creative Commons Attribution (CC BY) license (https://creativecommons.org/licenses/by/4.0/).

Abstract: New organic (super)conductors of the β''-$(BEDT-TTF)_4A^+[M^{3+}(C_2O_4)_3]G$ family, where BEDT-TTF is bis(ethylenedithio)tetrathiafulvalene; M is Fe; A is the monovalent cation NH_4^+; G is 2-fluoropyridine (2-FPy) (**1**); 2,3-difluoropyridine (2,3-DFPy) (**2**); 2-chloro-3-fluoropyridine (2-Cl-3-FPy) (**3**); 2,6-dichloropyridine (2,6-DClPy) (**4**); 2,6-difluoropyridine (2,6-DFPy) (**5**), have been prepared and their crystal structure and transport properties were studied. All crystals have a layered structure in which the conducting layers of BEDT-TTF radical cations alternate with paramagnetic supramolecular anionic layers $\{A^+[Fe^{3+}(C_2O_4)_3]^{3-}\cdot G^0\}^{2-}$. Crystals **1** undergo a structural phase transition from the monoclinic ($C2/c$) to the triclinic ($P\bar{1}$) symmetry in the range 100–150 K, whereas crystals **2–5** have a monoclinic symmetry in the entire range of the X-ray experiment (100–300 K). The alternating current (ac) conductivity of salts **1–4** exhibits metallic behavior down to 1.4 K, whereas the salt **5** demonstrates the onset of a superconducting transition at 3.1 K. The structures and conducting properties of **1–5** are compared with those of the known monoclinic phases of the family containing different monohalopyridines as "guest" solvent molecules G.

Keywords: BEDT-TTF; molecular paramagnetic (super)conductors; radical cation salts; tris(oxalato)metallate anions

1. Introduction

The large family of layered molecular (super)conductors $(BEDT-TTF)_4A^+[M^{3+}(C_2O_4)_3]G$, where BEDT-TTF is bis(ethylenedithio)tetrathiafulvalene; M^{3+} is a magnetic or non-magnetic metal cation (M = Fe, Cr, Mn, Ru, Rh, Ga, Co, Al); G is a neutral "guest" molecule; A^+ is a small monovalent cation ($A^+ = NH_4^+$, K^+, H_3O^+, Rb^+), continues to be actively investigated [1–29]. This is due to the fact that structural and conducting properties of this four-component system can be widely modified by changing the components A^+, M^{3+} and G. All crystals have a layered structure in which the conducting radical cation layers of BEDT-TTF alternate with insulating supramolecular anionic layers $\{A^+[M^{3+}(C_2O_4)_3]^{3-}\cdot G^0\}^{2-}$. Cationic and anionic layers interact with each other through the formation of a large number of hydrogen bonds between the components of the anionic layer and terminal ethylene groups of BEDT-TTF. The anionic layers have a familiar honeycomb-like structure in which M^{3+} and A^+ cations linked by oxalate bridges form the hexagonal cavities in which neutral "guest" molecules G are incorporated. Cationic layers have different BEDT-TTF packing types depending on the chemical composition of the anionic layers.

The main feature of (BEDT-TTF)$_4$A$^+$[M^{3+}(C$_2$O$_4$)$_3$]G crystals is that the crystal symmetry, packing type of conducting BEDT-TTF layers, and therefore, the conducting properties of crystals depend mainly on the size and shape of the included "guest" molecule G [1,3,17–20]. The variation of M (keeping the G the same) leads to a noticeable change of conducting properties, but does not affect the crystal symmetry and the type of packing of cation layers. The nature of the singly charged cation A$^+$ affects the properties of the crystals weakly. Currently, four groups of (BEDT-TTF)$_4$A$^+$[M^{3+}(C$_2$O$_4$)$_3$]G crystals are known, which have different symmetry and BEDT-TTF packing type.

The largest and more interesting group of this family is the group of monoclinic (C2/c) crystals [1–17,19,21,23–28] with β″-packing type of BEDT-TTF, according to the structural classification of salts of BEDT-TTF and its analogues [30,31]. The β″-layers are composed of continuous stacks of radical cations, the planes of which are almost parallel and shifted with respect to the short axis of the BEDT-TTF molecule. The interplanar distances in the stacks are considerably shortened in comparison with the normal van der Waals distances. There are a large number of shortened S S contacts between adjacent stacks in the layer.

Monoclinic β″-(BEDT-TTF)$_4$A$^+$[M^{3+}(C$_2$O$_4$)$_3$]G crystals were obtained with a large number of guest molecules G (G = PhCN, PhNO$_2$, PhBr, PhCl, PhF, PhI, DMF, py, CH$_2$Cl$_2$, 1,2-dichloro- and 1,2-dibromobenzene, different isomers of halopyridines). The conducting properties of these crystals vary from semiconducting to metallic and superconducting depending on the size, shape, and orientation of G in the hexagonal cavity, while the dimensions of the hexagonal cavity weakly depend on the size of G and are determined by the size of the M^{3+} cation and the distance from the outer oxalate oxygen atoms to the cation A$^+$. As a result, the large molecules G occupy almost the entire volume of the cavity and their position is strictly fixed. The smaller molecules G can occupy different positions. This leads to an increase in the structural disorder in the anionic layers and terminal ethylene groups of BEDT-TTF and to the suppression of the conductivity of molecular crystals. It was shown later that the conductivity of these crystals depends not so much on the molecular volume of G, as on its length along the direction c of the unit cell, [17]. Not so long ago, a correlation between the value of the c parameter of the unit cell and conducting properties of BEDT-TTF crystals with tris(oxalato)metallate anions was found [29].

The majority of these crystals keep the monoclinic symmetry in the entire range of the X-ray experiment (90–300 K). An exception is several β″-(BEDT-TTF)$_4$(NH$_4^+$)[Fe(C$_2$O$_4$)$_3$]G crystals containing PhHal (Hal = F, Cl, Br) or a mixture of PhHal with PhCN as G. In these crystals, the superconducting transition is preceded by a structural phase transition from the monoclinic to triclinic state with decreasing temperature [21,22,25]. This structural transition arises from noticeable positional shifts of all components of the complex anion, giving rise to two nonequivalent organic β″-layers and the partial ordering of the ethylene groups of BEDT-TTF molecules. The consequence of these changes is the phase transitions of these crystals from metallic to mixed metallic/insulating states [32]. According to the theoretical concepts of Merino and McKenzie, the appearance of this state precedes the appearance of superconductivity in layered molecular conductors with a quarter-filled conduction band [33]. All of the aforementioned crystals containing PhHal as G support this conclusion.

In addition to β″-monoclinic crystals, three more groups of crystals were discovered in the (BEDT-TTF)$_4$A$^+$[M^{3+}(C$_2$O$_4$)$_3$]G family: orthorhombic crystals (space group Pbcn) with "pseudo-k"-type of BEDT-TTF-packing [1,4,19,23,27] and bi-layered triclinic crystals (space group $P\bar{1}$) in the structure in which conducting layers with α- and β″- or α- and "pseudo-κ"-types of BEDT-TTF packing alternate [11,20,24].

The group of orthorhombic "pseudo-k"-crystals includes several semiconducting crystals (M = Fe, Cr, Co, Al; A$^+$ = K$^+$, NH$_4^+$, H$_3$O$^+$; G = PhCN and its mixtures with PhNO$_2$ or C$_6$H$_4$Cl$_2$) [1,4,19]. These crystals grow in "dry" solvents G and also together with monoclinic crystals in the presence of traces of water in the reaction medium. The organic layer of these crystals is formed by charged [(BEDT-TTF)$_2$]$^{2+}$ dimers surrounded by neutral [BEDT-TTF] molecules that are perpendicular to the dimers. Due to the strong localization of a charge on dimers, all orthorhombic crystals are semiconductors.

Triclinic crystals are formed with large unsymmetrical (including chiral) solvent molecules (M/G = Fe/PhAc; Fe/(X)-PhCH(OH)Me, X = R/S (racemate) or S (enantiomer); Ga/PhN(Me)CHO; Ga/BnCN; Fe/$C_6H_4Br_2$, Ga/$C_6H_4Br_2$) [11,18–20,24]. Bulky unsymmetrical solvent molecules are arranged asymmetrically relative to neighboring conducting layers and interact with them in different ways. For this reason, in the structure of triclinic crystals, conducting layers alternate with two different packing types, α-β''- or α-"pseudo-κ". In the α-layer, the stacks of BEDT-TTF are inclined to one another. Triclinic α-"pseudo-κ"-crystals are metals down to low helium temperatures, while triclinic α-β''-crystals undergo a metal-semiconductor transition with decreasing temperature.

Despite the large number of investigations of the influence of guest molecules G on the crystals and conducting properties of family crystals, these studies are ongoing and bring interesting results. In our recent articles, we have reported on the synthesis of crystals containing Fe, Cr, Ga as M and various isomers of monohalopyridines (HalPy) as G in which the halogen atom occupied different positions with respect to the nitrogen atom of the pyridine ring [25,28]. It was shown that the conducting and structural properties of the resulting β''-(BEDT-TTF)$_4$A$^+$[Fe^{3+}(C$_2$O$_4$)$_3$](HalPy) crystals strongly depend on the position of the halogen atom in the pyridine ring and also on the nature of M (see Table 1).

Table 1. Structural and conducting properties of the β''-(BEDT-TTF)$_4$A$^+$[M^{3+}(C$_2$O$_4$)$_3$]·G crystals, where G is HalPy; M is Fe, Cr, Ga; A = H$_3$O$^+$, [K$_{0.8}$(H$_3$O)$_{0.2}$]$^+$.

G/M	Structural Properties	Conducting Properties	Ref.
2-ClPy/Fe	$C2/c$ to $P\bar{1}$ transition at 215 K	SC, T_c = 2.4–4.0 K	[25]
2-BrPy/Fe	$C2/c$ to $P\bar{1}$ transition at 190 K	SC, T_c = 4.3 K	[25]
3-ClPy/Fe	No structural transition	M > 0.5 K	[25]
3-BrPy/Fe	No structural transition	M > 0.5 K	[25]
2-ClPy/Cr	Incommensurate structure appears upon cooling	M > 0.5 K	[28]
2-BrPy/Cr	Incommensurate structure appears upon cooling	M > 0.5 K	[28]
2-ClPy/Ga	Incommensurate structure appears upon cooling	M > 0.5 K	[28]
2-BrPy/Ga	Incommensurate structure appears upon cooling	M > 0.5 K	[28]

Here, we report the synthesis, crystal structure and transport properties of new monoclinic crystals β''-(BEDT-TTF)$_4$(NH$_4^+$)[Fe(C$_2$O$_4$)$_3$]G in which various mono- and dihalopyridines are included as G: 2-fluoropyridine (2-FPy) (**1**), 2,3-difluoropyridine (2,3-DFPy) (**2**), 2-chloro-3-fluoropyridine (2-Cl-3-FPy) (**3**), 2,6-dichloropyridine (2,6-DClPy) (**4**), 2,6-difluoropyridine (2,6-DFPy) (**5**).

2. Results and Discussion

2.1. Synthesis

The reaction medium for the growth of (BEDT-TTF)$_4$A$^+$[M^{3+}(C$_2$O$_4$)$_3$]G crystal consists of the guest solvent either neat or mixed with other solvents (H$_2$O, CH$_3$OH, C$_2$H$_5$OH, CH$_3$CN, etc.) which are usually not incorporated in the structure but have a considerable effect on the electrocrystallization process. In particular, they increase the solubility of the inorganic electrolyte in the guest organic solvent and, hence, enable variation of the current and the donor/electrolyte molar ratio, i.e., parameters that largely determine the crystal structure, growth rate and quality of crystals formed.

For example, in the medium of anhydrous C$_6$H$_5$CN, only semiconducting orthorhombic crystals grow, while both orthorhombic and monoclinic superconducting crystals (T_c = 7–8.5 K) in the form of needles were obtained by P. Day et al. using C$_6$H$_5$CN with a small addition of water [1]. These were the first superconducting crystals in this large family. Until now, the T_c of these crystals remains the highest among the crystals of this family. However, usually, only orthorhombic crystals form in this medium. In addition, needle crystals are of poor quality.

Interesting results were obtained by us when we used the mixture of C_6H_5CN with 1,2,4-trichlorobenzene (or 1,3-dibromobenzene) and C_2H_5OH as a reaction medium [19]. The molecules of 1,2,4-trichlorobenzene and 1,3-dibromobenzene do not enter the structure of crystals due to their size (1,2,4-trichlorobenzene) or geometry (1,3-dibromobenzene), but their presence facilitates the formation of high-quality monoclinic crystals. In this case, monoclinic crystals grow in the form of thick plates or parallelepipeds. Unlike most crystals of BEDT-TTF salts, in which the conducting layers are arranged in the plate plane, in these crystals, the conducting layers are perpendicular to this plane.

The addition of 1,2,4-trichlorobenzene to the reaction medium made it possible to obtain high-quality (super)conducting monoclinic crystals where G is 1,2-dichlorobenzene (M = Fe), various isomers of monohalopyridines (Hal = Cl, Br; M = Fe, Cr, Ga) as well as crystals containing a 4d metal cation ($M^{3+} = Ru^{3+}$, G = PhBr) and crystals containing two guest solvents, β''-(BEDT-TTF)$_4$A[Fe(C$_2$O$_4$)$_3$][(G1)$_x$(G2)$_{1-x}$], where G1 = benzonitrile, G2 = 1,2-dichlorobenzene, nitrobenzene, fluorobenzene, chlorobenzene, bromobenzene [17,19,23,25,28].

Of note is that in the medium of 1,2,4-trichlorobenzene with acetophenone or chlorobenzene (G1) and fluorobenzene (G2), the new series of isomorphic crystals with tris(oxalato)metallate anions β''-(BEDT-TTF)$_2$[(H$_2$O)(NH$_4$)$_2$M(C$_2$O$_4$)$_3$]18-crown-6 (M = Rh, Cr, Ru, Ir) were synthesized by Lee Martin et al. [34–36]. Two of these crystals (M = Cr, Rh) are superconductors.

In this work, the new monoclinic β''-(BEDT-TTF)$_4$(NH$_4^+$)[Fe(C$_2$O$_4$)$_3$](HalPy) crystals (**1–5**) were also obtained by using the mixtures of 1,2,4-trichlorobenzene with different mono- and dihalopyridines as reaction medium (Table 2).

Table 2. The β''-(BEDT-TTF)$_4$(NH$_4^+$)[Fe(C$_2$O$_4$)$_3$]G salts obtained and their properties.

Salts	G	Structural Properties	Conducting Properties
1	2-FPy	$C2/c$ to $P\bar{1}$ transition at 100–150 K	M > 1.4 K
2	2,3-DFPy	$C2/c$ no structural transition	M > 1.4 K
3	2-Cl-3-FPy	$C2/c$ no structural transition	M > 1.4 K
4	2,6-DClPy	$C2/c$ no structural transition	M > 1.4 K
5	2,6-DFPy	$C2/c$ no structural transition	SC, T_c = 3.1 K

Note that unlike all previously used guest components G, which are liquids, 2,6-dichloropyridine (crystals **4**) is a solid substance. It is possible that some other solids with interesting physical properties can also be included in the structure of the anionic layers for obtaining new multifunctional compounds combining two or more physical properties in the same crystal lattice.

2.2. Structure

Salts **1–5** of common formula β''-(BEDT-TTF)$_4$(NH$_4^+$)[Fe(C$_2$O$_4$)$_3$](HalPy) crystallize in the monoclinic space group $C2/c$, with two crystallographically independent BEDT-TTF (A and B) molecules, half a [Fe(C$_2$O$_4$)$_3$]$^{3-}$ anion, half a NH$_4^+$ cation, and half a halopyridine guest molecule. The full cell contains two formula units. Salts **1–5** are isostructural with other monoclinic $C2/c$ BEDT-TTF tris(oxalato)metallates packed by β'' type [27,29]. The structure consists of β''-type packed BEDT-TTF conducting radical cation layers separated by complex insulating anion layers {(NH$_4^+$)[Fe(C$_2$O$_4$)$_3$]$^{3-}$(HalPy)}$^{2-}$ interleaved along the c axis (Figure 1). Within the layers, BEDT-TTF radical cations form dimerized stacks with nearly coplanar molecular mean planes, incorporating radical cations expanding in a AABBAABBAABB manner with a shift of around a half of molecular short axis. Radical cations from adjacent stacks are arranged into chains expanding in the same dimerized manner. In the chains, BEDT-TTF radical cations are placed side-by-side and are nearly coplanar (Figure 2).

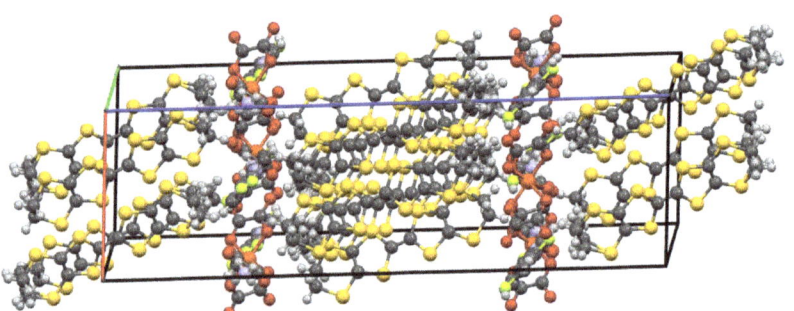

Figure 1. Crystal structure of salt **2** β''-(BEDT-TTF)$_4$(NH$_4$)[Fe(C$_2$O$_4$)$_3$](2,3-DFPy; the other salts are isostructural. Carbon is dark gray, hydrogen is light gray, sulfur is yellow, oxygen is red, nitrogen is light blue, fluorine is light green, and iron is orange.

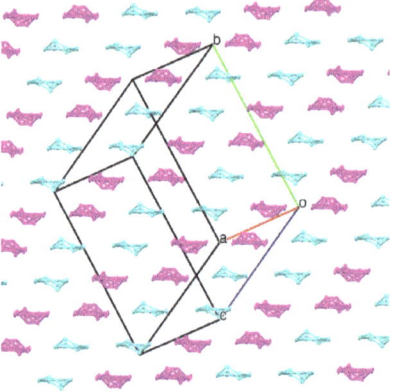

Figure 2. Conducting BEDT-TTF radical cation layers of salts **1–5** packed by β''-type. Crystallographically unique BEDT-TTF radical cations are denoted by color. Molecule A is colored magenta, B is colored cyan. AABBAABB packing arrangement for BEDT-TTF stacks and chains is shown.

An average charge on the BEDT-TTF is +1/2 obtained from the charge balance requirements. In the salt **5**, the radical cation A is completely ordered, both terminal ethylene groups of the radical cation B are disordered over two positions with the occupancies of main positions of 0.61 (atoms C17, C18) and 0.74 (atoms C19, C20) for each terminal correspondingly, Figure S1. Minor positions are complementary to the main positions with the occupancies of 0.39 (atoms C17′, C18′) and 0.26 (atoms C19′, C20′). Mutual arrangements of ethylene groups in a BEDT-TTF respect eclipsed conformation in an ordered molecule A and staggered conformation in a molecule B for both main and minor disordered configurations (Figure 3). The data of the other salts are summarized in Table 3.

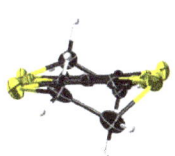 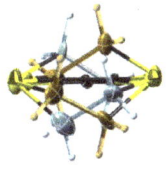

Figure 3. Relative positions in a chain and conformations of terminal ethylene groups for independent BEDT-TTF radical cation A (**left**) and B (**right**) in salt **5**. Main positions for disordered ethylene groups of B are colored cyan and complementary are colored navy. The conformation is eclipsed for *A* and staggered for *B*. Thermal ellipsoids are of 50% probability.

Table 3. BEDT-TTF radical cation terminal ethylene group occupancies and mutual conformation for the salts **1–5**. In the case of disordered groups, the occupancies of the main positions are presented. In the case of completely ordered groups, occupancies are equal to unity. Terminal *A1* is defined as the ethylene group closest to the Fe^{3+} ion, *A2*—an opposite side of the molecule. Terminals *B1/B2* are defined by the same manner in the relation to the NH^{4+} ion. Letter *e* denotes the eclipsed conformation while *s* denotes the staggered one. An example of *e/s* conformation is presented in Figure 3. Experiment temperature is 100 K unless otherwise noted.

Terminal	1a *	1b *	1c *	2	3	4	5
A1	1/1 $	1	1	1	1	1	1
A2	1/1 $	1	1	1	1	1	1
B1	1/0.52 &	0.74	0.63	1	d #	1	0.74
B2	0.86/0.63 &	0.60	0.69	0.69	d #	0.62	0.61
A1/A2	e/e	e	e	e	e	e	e
B1/B2	s/s	s	s	e	s	e	s

* **1a** = 100 K (triclinic), **1b** = 150 K (monoclinic), **1c** = 240 K (monoclinic); # severe disorder that is unable to be evaluated due to the low quality of the crystal; $ second value is for molecule C; & second value is for molecule D.

It is known that the order and configuration of BEDT-TTF radical cation terminal ethylene groups play an important role in the charge transport of BEDT-TTF-based molecular conductors. From Table 3, it becomes immediately clear that, in all salts **1–5**, the molecule A is fully ordered and adopts eclipsed conformation.

Another clear observation is that the configuration of salt **1** at 240 K, and, especially at 150 K, resembles that of **5** at 100 K. Upon cooling in the range 100–150 K, salt **1** experiences a crystal lattice transformation from higher (monoclinic) to lower (triclinic) symmetry. A similar phase transition was observed earlier in the family of halobenzene solvents (PhF, PhCl, PhBr) [21] or their mixtures with PhCN [22] as well as with the guest molecule types closest to the current research—monosubstituted halopyridines, namely, 2-Cl-Py and 2-Br-Py (Table 1). In these cases, the structural transition was associated with a superconductive one. However, no superconductive transition is observed in salt **1**. Let us speculate about possible reasons from the perspective of the crystal structure.

The structural transition results in halving of the unit cell volume and appearance of four independent BEDT-TTF radical cations (A, B, C, D). Molecule C is generated from A by the loss of symmetry, and molecule D is generated from B. It is seen from Table 3 that both molecules A and C are still ordered and adopt the same configuration while in the pair B/D the former is almost fully ordered but the latter, in contrast, demonstrates almost randomly distributed occupancies of the terminal group carbon atoms. Such a redistribution of occupancies is most likely accompanied by a charge redistribution that, in turn, results in weak charge localization and loss of superconductive transition. Indeed, analysis of the charge distribution over independent BEDT-TTF radical cations based on bond lengths enables us to roughly estimate the charges on A/B/C/D as +0.5/+0.33/+0.5/+0.67, respectively, indicating a charge redistribution in the pair B/D [37]. This formula, when applied to the structures with high R-values, often provides wrong absolute values and should not be taken too seriously. However, that gives us a good qualitative insight into the intermolecular charge distribution.

Within each anion layer, the Fe^{3+} and NH_4^+ cations are linked by oxalate bridges forming a hexagonal packing arrangement where Fe^{3+} is octahedrally coordinated by oxalate ligands. Outer oxygens of oxalates form hydrogen bonds with the NH_4^+ cations (Figure 4). This anionic network forms large hexagonal cavities where halopyridine guest molecules reside. Fe^{3+} cations and nitrogen of NH_4^+ cations reside in special positions. There are twofold symmetry axes connecting them.

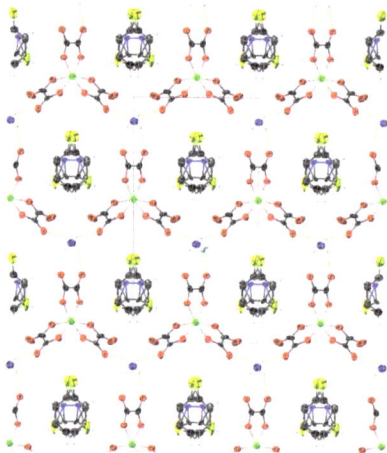

Figure 4. Projection of the anion layer along c axis showing the honeycomb packing arrangement of the tris(oxalato)ferrate anions for the salts **1–5**, using salt **5** as an example. Thermal ellipsoids are of 50% probability.

There are two chemically and structurally reasonable alternatives for a cation incorporating into an anion layer—ammonium (NH_4^+) or oxonium (H_3O^+). It should be noted that the unambiguous identification of the cation is a challenging task that has a long history. That is especially a source of doubt for compounds demonstrating superconductive properties, as there is an empirical rule that H_3O^+ favors the superconducting transition [1,27,29].

In the particular case of superconducting salt **5**, we consider the cation to be NH_4^+, guided by the following arguments:

1. Both symmetrically unequivalent hydrogen atoms were found on the electronic differential map;
2. Adding each of the hydrogens results in a subtle but sensible decrease of the *R*-value;
3. Replacing N by O does not improve the *R*-value;
4. NH_4^+ cation is residing on a twofold axis and adopts distorted tetrahedral geometry that has been restored to a symmetrical tetrahedron by geometrical constraints without loss of the *R*-value.

It should be emphasized that, while molecules of 2,6-DFPy and 2,6-DClPy have their own molecular twofold symmetry axes, they still reside in the cavities with the substantial displacement of approximately 0.8 Å with no atoms occupying special positions, resulting in a positional disorder over two positions around the lattice twofold axis (Table 4, Figures 5 and 6). It is most likely a consequence of the requirement to completely fill up the cavity void space by a smaller molecule. However, the larger chlorine-containing

2,6-DClPy is almost 0.1 Å less displaced than the fluorine-containing 2,6-DFPy but requires a larger tilt to be accommodated.

Table 4. The dimensions of the anion layer hexagonal cavity and the orientation of the HalPy guest molecule as annotated in Figure 5. δ is the angle between the pyridine ring plane and the plane of the hexagonal cavity. e is a mean guest molecule displacement defined as $(e_1 + e_2)/2$, where e_1 is a displacement of the outer halopyridine halogen atom nearest to the cusp Fe^{3+} ion around the twofold axis, e_2 is a displacement of the inner halopyridine ring carbon/nitrogen atom closest to the nadir nitrogen atom of NH_4^+ ion. e' is an alternative mean displacement value $(e_1 + e_2 + \ldots + e_n)/n$ where averaging is taken for all displaced atoms. S is a square unit of the hexagonal cavity taken as $w(b+h)/2$.

Salts	1a	1b	1c	2	3	4	5
S, Å^2	101.71	102.17	102.850	102.21	101.90	102.03	101.83
h, Å	13.417	13.453	13.549	13.474	13.463	13.449	13.406
w, Å	10.298	10.309	10.330	10.331	10.309	10.327	10.322
a, Å	6.203	6.254	6.291	6.285	6.584	6.277	6.258
b, Å	6.338	6.370	6.364	6.313	6.306	6.312	6.326
c, Å	4.625	4.656	4.748	4.484	4.414	4.473	4.813
d, Å	4.881	5.108	4.969	4.520	4.750	4.535	4.519
e, Å	0	0	0.651	0.822	0	0.895	0.808
e', Å	0	0	0.547	0.877	1.108	0.902	0.840
δ, °	31.73	34.58	34.97	38.14	32.37	39.46	36.36

1a = 100 K (triclinic), **1b** = 150 K (monoclinic), **1c** = 240 K (monoclinic).

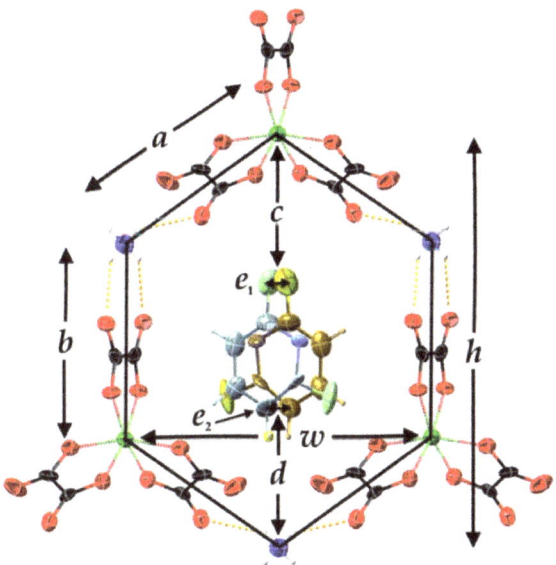

Figure 5. Scheme for the hexagonal unit dimensions and the guest molecule placement in the anion layers of salts **1–5**, using salt **5** as an example.

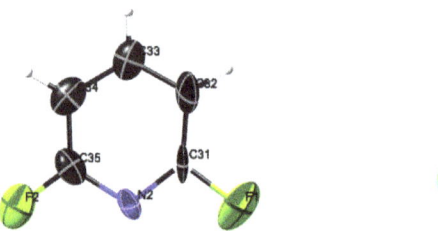
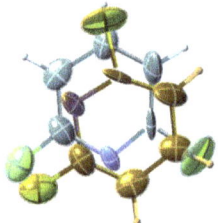

Figure 6. View of the 2,6-DFPy molecule in the crystal structure **5** along the molecular twofold axis (**left**) and along the lattice twofold axis $(-x, y, 1/2 - z)$ (**right**). The 2,6-DClPy molecule in the crystal structure **4** is arranged in the same manner.

It appears that the smallest 2-FPy guest molecule that lacks any molecular symmetry presents the most interesting case. At higher temperatures, the arrangement of 2-FPy in the cavity resembles 2,6-DFPy/2,6-DClPy cases with a much smaller displacement. However, upon cooling, the fluorine atom and the two pyridine carbon atoms (C41 and C44) are situated on the lattice twofold axis (Figure 7). This effect is already clear at 150 K—far away from the structural transition. Thus, 2-FPy localization precedes or even induces a complete structural transformation. Nitrogen and carbon positions of 2-FPy appear to be inseparable at temperatures 150 and 100 K; that is probably due to the formation of a domain superstructure. The elevated electric field strength along the domain grains may serve as a factor causing failure of the superconductive transition. However, such speculations require further deeper investigations.

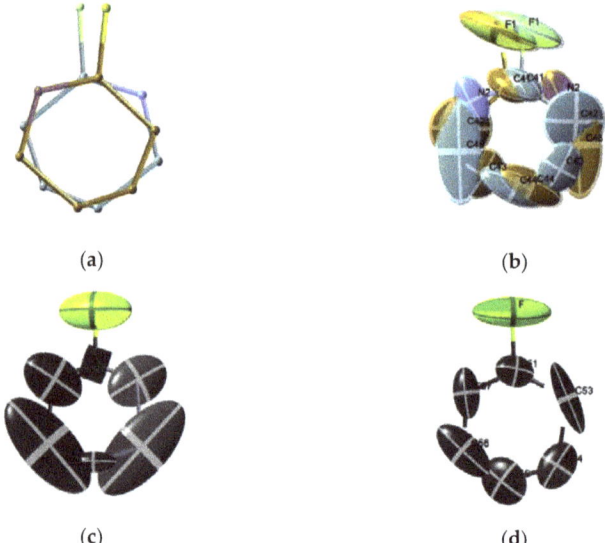

Figure 7. View of 2-FPy molecule in the crystal structure **1**. Wireframe (**a**) and ellipsoid (**b**) views at 240 K (**1c**), 150 K (**1b**) (**c**) and 100 K (**1a**) (**d**). Fluorine and central carbon atoms are placed on lattice twofold axis $(-x, y, 1/2 - z)$ at 150 K.

Unlike all other salts presented in the current work, salt **3**, with the most asymmetrical halopyridine molecule—2-Cl-3-FPy, contains an NH_4^+ cation displaced from the lattice twofold axis and shows an oxalic ligand disorder in the tris(oxalato)ferrate anion (Figure 8). Guest solvent molecules could not be identified by means of crystal structure tools and were refined as a set of separated carbon atoms plus fluorine with released occupancies. It appears that a highly asymmetric guest solvent molecule with substituents of different sizes does not favor growth of high-quality crystals.

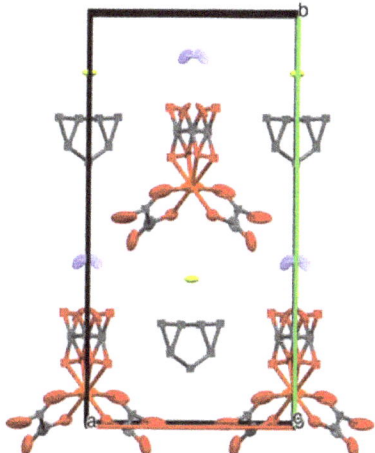

Figure 8. Anion layer of salt **3**.

Though the structure-property correlation is a very complex and ambiguous topic, some preliminary conclusions can be provided:

1. Use of highly nonsymmetrical halopyridine 2-Cl-3-FPy leads to low quality single crystals;
2. Guest molecules of higher symmetry and larger volume may produce superconductive single crystals;
3. Ordering of non-symmetrical guest molecules at lower temperatures may impede the SC transition by casting a constant dipole moment at the place of the localization.

2.3. Conducting Properties

The temperature dependences of the out-of-plane resistivity for the samples **1–5** are presented in Figure 9. All these samples are metals despite the existence of the negative slopes on the low-temperature parts on the $\rho_\perp(T)$ curves for the samples **5** and **1**. Moreover, the sample **5** demonstrates a sharp resistance drop at low temperature which one can surely attribute to the onset of a superconducting transition at $T_c = 3.1$ K. Unfortunately, we could not reach the real zero resistance state, even at our minimal temperature 1.4 K, because the superconducting transition is not very narrow. This is typical for organic metals, for which the transition width is usually more or about 2 K. The behavior of the sample **1** at $T < 20$ K is similar to that of the sample **5**, so we do not exclude the possibility that this sample could also be a superconductor at temperatures below 1 K, but we could not check this statement because our temperature region was restricted by 1.4 K.

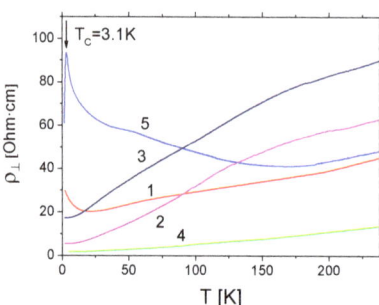

Figure 9. The temperature dependences of the out-of-plane resistivity for the samples **1–5**: 2-FPy (**1**); 2,3-DFPy (**2**); 2-Cl-3-FPy (**3**); 2,6-DClPy (**4**); 2,6-DFPy (**5**).

3. Materials and Methods

3.1. Synthesis

Crystals of five new radical cation salts of the $\beta''\text{-(BEDT-TTF)}_4\text{A}^+[\text{M}^{3+}(\text{C}_2\text{O}_4)_3]\text{G}$ family were obtained by electrochemical oxidation of BEDT-TTF at a platinum electrode in organic solvents, at constant current and temperature (25 °C), in the presence of supporting electrolytes, namely, ammonium tris(oxalato)metallates combined with a 18-crown ether. The electrocrystallization process was performed in conventional two-compartment U-shaped cells separated by a porous glass filter. BEDT-TTF, $(\text{NH}_4)_3[\text{Fe(ox)}_3]\cdot3\text{H}_2\text{O}$, 18-crown ether and solvents were placed in the cathode compartment and, then, the solution obtained was distributed between the two compartments of the cell.

BEDT-TTF, 1,2,4-trichlorobenzene, 2-fluoropyridine, 2,3-difluoropyridine, 2-chloro-3-fluoropyridine; 2,6-difluoropyridine; 2,6-dichloropyridine, $(\text{NH}_4)_3[\text{Fe(ox)}_3]\cdot3\text{H}_2\text{O}$ were used as received (Aldrich); 18-crown-6 (Aldrich) was purified by recrystallization from acetonitrile and dried in vacuum at 30 °C over P_2O_5.

The exact conditions for the synthesis of each salt are described below.

3.1.1. $\beta''\text{-(BEDT-TTF)}_4(\text{NH}_4^+)[\text{Fe(ox)}_3]$(2-FPy) (**1**)

15 mg of BEDT-TTF, 150 mg of $(\text{NH}_4)_3[\text{Fe(ox)}_3]\cdot3\text{H}_2\text{O}$, 450 mg of 18-crown-6 and the mixture of 2-FPy (10 mL) with 1,2,4-trichlorobenzene (10 mL) and 96% EtOH (2 mL); $J = 0.85$ µA. Many crystals in the form of thick plates were collected from the anode after 9 days.

3.1.2. $\beta''\text{-(BEDT-TTF)}_4(\text{NH}_4^+)[\text{Fe(ox)}_3]$(2,3-DFPy) (**2**)

8 mg BEDT-TTF, 75 mg of $(\text{NH}_4)_3[\text{Fe(ox)}_3]\cdot3\text{H}_2\text{O}$, 250 mg of 18-crown-6 and the mixture of 2,3-DFPy (4 mL) with 1,2,4-trichlorobenzene (4 mL) and 96% EtOH (1 mL); $J = 0.85$ µA. Several thick crystals were collected from the anode after 12 days.

3.1.3. $\beta''\text{-(BEDT-TTF)}_4(\text{NH}_4^+)\text{Fe(ox)}_3]$(2-Cl-3-FPy) (**3**)

8 mg BEDT-TTF, 75 mg of $(\text{NH}_4)_3[\text{Fe(ox)}_3]\cdot3\text{H}_2\text{O}$, 250 mg of 18-crown-6 and the mixture of 2-Cl-3-FPy (4 mL) with 1,2,4-trichlorobenzene (4 mL) and 96% EtOH (1 mL); $J = 0.9$ µA. Very many small thick crystals were collected from the anode after 2 weeks.

3.1.4. $\beta''\text{-(BEDT-TTF)}_4(\text{NH}_4^+)\text{Fe(ox)}_3]$(2,6-DClPy) (**4**)

A total of 14 mg of BEDT-TTF, 150 mg of $(\text{NH}_4)_3\text{Fe(ox)}_3\cdot3\text{H}_2\text{O}$, 450 mg of 18-crown-6, 5 g of 2,6-DClPy were dissolved in the mixture of 30 mL 1,2,4-trichlorobenzene with 3 mL of 96% ethanol. $J = 0.95$ µA. Several thick crystals in the form of plates were collected from the anode after 2 weeks.

3.1.5. β″-(BEDT-TTF)$_4$(NH$_4^+$)[Fe(ox)$_3$](2,6-DFPy) (5)

A total of 18 mg of BEDT-TTF, 150 mg of (NH$_4$)$_3$Fe(ox)$_3$, 450 mg of 18-crown-6 were dissolved in the mixture of 15 mL of 2,6-DFPy, 5 mL 1,2,4-trichlorobenzene and 2 mL of 96% ethanol. J = 0.9 µA. Several thick crystals in the form of plates were collected from the anode after 6 days.

3.2. Structure

X-ray diffraction analyses of the salts **1–5** were carried out on a CCD Agilent XCalibur diffractometer with an EOS detector (Agilent Technologies UK Ltd., Yarnton, Oxfordshire, England). Data collection, determination and refinement of unit cell parameters were carried out using the CrysAlis PRO program suite [38]. X-ray diffraction data at 100(1) K for the salts **1–5** were collected using MoKα (λ = 0.71073 Å) radiation. The same single crystal for the salt **1** was used for the data collection at different temperatures. Data at 100(1) K for the experiment **1a**, data at 150(1) K for the experiment **1b**, and data at 240(1) K for the experiment **1c** were seamlessly collected without removing the crystal from the diffractometer.

The structures **1–5** were solved by the direct methods. The positions and thermal parameters of non-hydrogen atoms were refined isotropically and then anisotropically by the full-matrix least-squares method. The positions of the hydrogen atoms were calculated geometrically. The guest molecule was found to be disordered over two positions. The geometry of guest molecule was recovered for salts **1–5** except salt **3**. Guest solvent molecules in salt **3** were refined as a set of separated carbon atoms plus fluorine with released occupancies.

The X-ray crystal structures data have been deposited with the Cambridge Crystallographic Data Center, with reference codes CCDC 2081357 (**1a**), 2081360 (**1b**), 2081356 (**1c**), 2081355 (**2**), 2081361 (**3**), 2081354 (**4**), 2079902 (**5**). All calculations were performed with the SHELX-97 program package [39].

3.3. Conducting Properties

The temperature dependences of the electrical resistance of single crystals were measured using a four-probe technique by a lock-in detector at 20 Hz alternating current J = 1 mkA in the temperature range (1.4–300 K). Two contacts were attached to each of the two opposite sample surfaces with conducting graphite paste. In the experiment, we have measured the out-of-plane resistance $R_\perp$ with the current running perpendicular to the conducting layers. Because of the high anisotropy of the samples, we did not succeed in the measurements of the in-plane sample resistance. The out-of-plane resistivity $\rho_\perp$ of the samples was calculated from $R_\perp$ taking into account that the out-of-plane current is uniform due to the high sample anisotropy.

4. Conclusions

Crystals of new layered molecular (super)conductors (**1–5**) based on radical cation salts of bis(ethylenedithio)tetrathiafulvalene (BEDT-TTF) with paramagnetic tris(oxalate)metallate anions, β″- BEDT-TTF)$_4$(A$^+$)[M^{3+}(C$_2$O$_4$)$_3$]G, where M is Fe; A is NH$_4^+$; G is an isomer of mono- or dihalopyridine: 2-fluoropyridine (2-FPy) (**1**), 2,3-difluoropyridine (2,3-DFPy) (**2**), 2-chloro-3-fluoropyridine (2-Cl-3-FPy) (**3**), 2,6-dichloropyridine (2,6-DClPy) (**4**) and 2,6-difluoropyridine (2,6-DFPy) (**5**), belong to the monoclinic group of the large family of (BEDT-TTF)$_4$(A$^+$)[M^{3+}(C$_2$O$_4$)$_3$]G crystals. In the structures of these crystals, the conducting layers of BEDT-TTF radical cations alternate with supramolecular insulating anionic layers {A$^+$[M^{3+}(C$_2$O$_4$)$_3$]G}$^{2-}$. Changing the number of halogen atoms in the "guest" molecule halopyridine (G) as well as their size and mutual arrangement has a significant effect on the structure and conducting properties of the crystals. So, salt **1** (G = 2-FPy) undergoes a structural phase transition from monoclinic to triclinic symmetry in the range 100–150 K. A similar transition in isostructural crystals with G = 2-ClPy and 2-BrPy precedes the superconducting transition [25]. However, crystals **1** show stable metallic properties down to 1.4 K and do not undergo a superconducting transition. In contrast to **1**, crystals **2–5**, where G are various isomers of dihalopyridine, retain monoclinic symmetry in the entire range of the X-Ray experiment (90–300 K). Crystals **2** and **3**, which contain the asymmetric "guest" molecules (2,3-DFPy and 2-Cl-3-FPy, respectively), exhibit the stable metallic properties down to 1.4 K without a superconducting transition. However, the quality of crystals **2** (R-factor = 4.95%) containing two identical substituents in the pyridine molecule is significantly higher than the quality of crystals **3**, where G contain two different substituents. Unlike all other salts, the salt **3** contains an NH_4^+ cation displaced from the lattice twofold axis and shows an oxalic ligand disorder in the tris(oxalato)ferrate anion (see Section Structure).

Crystals **4** (G = 2,6-DClPy) and **5** (2,6-DFPy) contain the highly symmetric G molecules. Each of these molecules contains two identical substituents. In this case, there were no problems with the quality of the crystals formed. Crystals **4** demonstrate metallic properties down to 1.4 K, while crystals **5** show the onset of a superconducting transition at 3.1 K. This is the first superconductor among crystals of the β''-$(BEDT-TTF)_4(A^+)[M^{3+}(C_2O_4)_3]G$ family containing dihalopyridines or dihalobenzenes as G.

Supplementary Materials: The following are available online at https://www.mdpi.com/article/10.3390/magnetochemistry7060083/s1, Figure S1: Molecular structure of crystals **5**, β''-(BEDT-TTF)$_4$(NH$_4$)[Fe(C$_2$O$_4$)$_3$](2,6-DFPy), with atom numbers.

Author Contributions: Idea, methodology, synthesis of crystals, writing sections "Abstract, Introduction, Synthesis, Conclusions"—T.G.P., E.B.Y.; the X-ray experiments—A.A.B., G.V.S.; analysis and description of the X-ray data—A.A.B.; the conductivity measurement on crystals, writing section "Conducting properties"—V.N.Z., L.I.B. All authors have read and agreed to the published version of the manuscript.

Funding: This research was done according to the state tasks of Ministry of Science and Higher Education of the Russian Federation (Grants No. AAAA-A19-119092390079-8 for IPCP RAS and No. 075-00920-20-00 for IMTHPM RAS) and supported by the Russian Foundation for Basic Research (Grant No. 21-52-12027).

Institutional Review Board Statement: Not applicable.

Informed Consent Statement: Not applicable.

Data Availability Statement: The crystallographic data have been deposited with the Cambridge Crystallographic Data Centre (https://www.ccdc.cam.ac.uk/structures/), deposition numbers 2079902, 2081354–2081357, 2081360, 2081361. See Section 3.2 for details.

Conflicts of Interest: The authors declare no conflict of interest. The funders had no role in the design of the study; in the collection, analyses, or interpretation of data; in the writing of the manuscript, or in the decision to publish the results.

References

1. Kurmoo, M.; Graham, A.W.; Day, P.; Coles, S.J.; Hursthouse, M.B.; Caufield, J.L.; Singleton, J.; Pratt, F.L.; Hayes, W.; Ducasse, L.; et al. Superconducting and Semiconducting Magnetic Charge Transfer Salts: (BEDT-TTF)$_4$AFe(C$_2$O$_4$)$_3$ C$_6$H$_5$CN; A = H$_2$O, K, NH$_4$. *J. Am. Chem. Soc.* **1995**, *117*, 12209–12217. [CrossRef]
2. Martin, L.; Turner, S.S.; Day, P.; Mabbs, F.E.; McInnes, J.L. New molecular superconductor containing paramagnetic chromium (III) ions. *Chem. Commun.* **1997**, 1367–1368. [CrossRef]
3. Turner, S.S.; Day, P.; Abdul Malik, K.M.; Hursthouse, M.B.; Teat, S.J.; MacLean, E.J.; Martin, L.; French, S.A. Effect of Included Solvent Molecules on the Physical Properties of the Paramagnetic Charge Transfer Salts β''-(BEDT-TTF)$_4$[(H$_3$O)Fe(C$_2$O$_4$)$_3$]Solvent (BEDT-TTF = Bis(ethylenedithio)tetrathiafilvalene). *Inorg. Chem.* **1999**, *38*, 3543–3549. [CrossRef]
4. Martin, L.; Turner, S.S.; Day, P.; Guionneau, P.; Howard, J.A.K.; Hibbs, D.E.; Light, M.E.; Hursthouse, M.B.; Uruichi, M.; Yakushi, K. Crystal Chemistry and Physical Properties of Superconducting and Semiconducting Charge Transfer Salts of the Type (BEDT-TTF)$_4$[A^{I}M^{III}(C$_2$O$_4$)$_3$] PhCN (A^I = H$_3$O, NH$_4$, K.; M^{III} = Cr, Fe, Co, Al; BEDT-TTF = Bis(ethylenedithio)tetrathiafulvalene). *Inorg. Chem.* **2001**, *40*, 1363–1371. [CrossRef]
5. Rashid, S.; Turner, S.S.; Le Pevelen, D.; Day, P.; Light, M.E.; Hursthouse, M.B.; Firth, S.; Clark, R.J.H. β''-(BEDT-TTF)$_4$[(H$_3$O)Cr(C$_2$O$_4$)$_3$]CH$_2$Cl$_2$: Effect of Included Solvent on the Structure and Properties of a Conducting Molecular Charge-Transfer Salt. *Inorg. Chem.* **2001**, *40*, 5304–5306. [CrossRef]

6. Rashid, S.; Turner, S.S.; Day, P.; Howard, J.A.K.; Guionneau, P.; McInnes, E.J.L.; Mabbs, F.E.; Clark, R.J.H.; Firth, S.; Biggse, T. New superconducting charge-transfer salts (BEDTTTF)$_4$[AM(C$_2$O$_4$)$_3$]C$_6$H$_5$NO$_2$(A = H$_3$O or NH$_4$, M = Cr or Fe, BEDT-TTF = bis(ethylenedithio)tetrathiafulvalene. Novel charge transfer salts of BEDT-TTF with metal oxalate counterions. *J. Mater. Chem.* **2001**, *11*, 2095–2101. [CrossRef]
7. Rashid, S.S.; Turner, P.; Day, M.E.; Light, M.B.; Hursthouse, P. Guionneau. *Synth. Met.* **2001**, *120*, 985–986. [CrossRef]
8. Akutsu, H.; Akutsu-Sato, A.; Turner, S.S.; Le Pevelen, D.; Day, P.; Laukhin, V.; Klehe, A.-K.; Singleton, J.; Tocher, D.A.; Probert, M.R.; et al. Effect of Included Guest Molecules on the Normal State Conductivity and Superconductivity of β''-(ET)$_4$[(H$_3$O)Ga(C$_2$O$_4$)$_3$]G (G = Pyridine, Nitrobenzene). *J. Am. Chem. Soc.* **2002**, *124*, 12430–12431. [CrossRef]
9. Prokhorova, T.G.; Khasanov, S.S.; Zorina, L.V.; Buravov, L.I.; Tkacheva, V.A.; Baskakov, A.A.; Morgunov, R.B.; Gener, M.; Canadell, E.; Shibaeva, R.P.; et al. Molecular Metals Based on BEDT-TTF Radical Cation Salts with Magnetic Metal Oxalates as Counterions: β''-(BEDT-TTF)$_4$A[M(C$_2$O$_4$)$_3$]DMF (A = K$^+$, NH$_4^+$; M = Fe^{III}, Cr^{III}). *Adv. Func. Mater.* **2003**, *13*, 403–411. [CrossRef]
10. Coldea, A.I.; Bangura, A.F.; Singleton, J.; Ardavan, A.; Akutsu-Sato, A.; Akutsu, H.; Turner, S.S.; Day, P. Fermi-surface topology and the effects of intrinsic disorder in a class of charge-transfer salts containing magnetic ions: β''-(BEDT-TTF)$_4$[(H$_3$O)M(C$_2$O$_4$)$_3$]Y (M = Ga, Cr, Fe; Y = C$_5$H$_5$N). *Phys. Rev. B* **2004**, *69*, 085112. [CrossRef]
11. Akutsu, H.; Akutsu-Sato, A.; Turner, S.S.; Day, P.; Canadell, E.; Firth, S.; Clark, R.J.N.; Yamada, J.; Nakatsuji, S. Superstructures of donor packing arrangements in a series of molecular charge transfer salts. *Chem. Commun.* **2004**, 18–19. [CrossRef]
12. Audouard, A.; Laukhin, V.N.; Brossard, L.; Prokhorova, T.G.; Yagubskii, E.B.; Canadell, E. Combination frequencies of magnetic oscillations in β''-(BEDT-TTF)$_4$(NH$_4$)[Fe(C$_2$O$_4$)$_3$]DMF. *Phys. Rev. B* **2004**, *69*, 144523. [CrossRef]
13. Akutsu-Saito, A.; Kobayashi, A.; Mori, T.; Akutsu, H.; Yamada, J.; Nakatsuji, S.; Turner, S.S.; Day, P.; Tocher, D.A.; Light, M.E.; et al. Structures and Physical Properties of New β'-BEDT-TTF Tris-Oxalatometallate (III) Salts Containing Chlorobenzene and Halomethane Guest Molecules. *Synth. Met.* **2005**, *152*, 373–376. [CrossRef]
14. Coronado, E.; Curelli, S.; Giménez-Saiz, C.; Gómez-García, C.J. New magnetic conductors and superconductors based on BEDT-TTF and BEDS-TTF. *Synth. Met.* **2005**, *154*, 245–248. [CrossRef]
15. Coronado, E.; Curelli, S.; Giménez-Saiz, C.; Gómez-García, C.J. A novel paramagnetic molecular superconductor formed by bis(ethylenedithio)tetrathiafulvalene, tris(oxalato)ferrate(III) anions and bromobenzene as guest molecule: (ET)$_4$[(H$_3$O)Fe(C$_2$O$_4$)$_3$]C$_6$H$_5$Br. *J. Mater. Chem.* **2005**, *15*, 1429–1436. [CrossRef]
16. Akutsu-Sato, A.; Akutsu, H.; Yamada, J.; Nakatsuji, S.; Turner, S.S.; Day, P. Suppression of superconductivity in a molecular charge transfer salt by changing quest molecule: β''-(BEDT-TTF)$_4$[(H$_3$O)Fe(C$_2$O$_4$)$_3$](C$_6$H$_5$CN)$_x$(C$_5$H$_5$N)$_{1-x}$. *J. Mater. Chem.* **2007**, *17*, 2497–2499. [CrossRef]
17. Zorina, L.V.; Prokhorova, T.G.; Simonov, S.V.; Khasanov, S.S.; Shibaeva, R.P.; Manakov, A.I.; Zverev, V.N.; Buravov, L.I.; Yagubskii, E.B. Structure and Magnetotransport Properties of the New Quasi-Two-Dimensional Molecular Metal β''-(BEDT–TFF)$_4$H$_3$O[Fe(C$_2$O$_4$)$_3$]C$_6$H$_4$Cl$_2$. *JETP* **2008**, *106*, 347–354. [CrossRef]
18. Martin, L.; Day, P.; Akutsu, H.; Yamada, J.; Nakatsuji, S.; Clegg, W.; Harrington, R.W.; Horton, P.N.; Hursthouse, M.B.; McMillan, P.; et al. Metallic molecular crystals containing chiral or racemic guest molecules. *CrystEngComm* **2007**, *9*, 865–867. [CrossRef]
19. Prokhorova, T.G.; Buravov, L.I.; Yagubskii, E.B.; Zorina, L.V.; Khasanov, S.S.; Simonov, S.V.; Shibaeva, R.P.; Korobenko, A.V.; Zverev, V.N. Effect of electrocrystallization medium on quality, structural features, and conducting properties of single crystals of the (BEDT-TTF)$_4$A^I[Fe^{III}(C$_2$O$_4$)$_3$]G family. *CrystEngComm* **2011**, *13*, 537–545. [CrossRef]
20. Zorina, L.V.; Khasanov, S.S.; Simonov, S.V.; Shibaeva, R.P.; Zverev, V.N.; Canadell, E.; Prokhorova, T.G.; Yagubskii, E.B. Coexistence of two donor packing motifs in the stable molecular metal α-'pseudo-κ'-(BEDT-TTF)$_4$(H$_3$O)[Fe(C$_2$O$_4$)$_3$] C$_6$H$_4$Br$_2$. *CrystEngComm* **2011**, *13*, 2430–2438. [CrossRef]
21. Coronado, E.; Curreli, S.; Giménez-Saiz, C.; Gómez-García, C.J. The Series of Molecular Conductors and Superconductors ET$_4$[AFe(C$_2$O$_4$)$_3$]·PhX (ET = bis(ethylenedithio)tetrathiafulvalene; (C$_2$O$_4$)$^{2-}$ = oxalate; A$^+$ = H$_3$O$^+$, K$^+$; X = F, Cl, Br, and I): Influence of the Halobenzene Guest Molecules on the Crystal Structure and Superconducting Properties. *Inorg. Chem.* **2012**, *51*, 1111–1126. [PubMed]
22. Zorina, L.V.; Khasanov, S.S.; Simonov, S.V.; Shibaeva, R.P.; Bulanchuk, P.O.; Zverev, V.N.; Canadell, E.; Prokhorova, T.G.; Yagubskii, E.B. Structural phase transition in the β''-(BEDT-TTF)$_4$H$_3$O[Fe(C$_2$O$_4$)$_3$]·G crystals (where G is a guest solvent molecule). *CrystEngComm* **2012**, *14*, 460–465. [CrossRef]
23. Prokhorova, T.G.; Zorina, L.V.; Simonov, S.V.; Zverev, V.N.; Canadell, E.; Shibaeva, R.P.; Yagubskii, E.B. The first molecular superconductor based on BEDT-TTF radical cation salt with paramagnetic tris(oxalato)ruthenate anion. *CrystEngComm* **2013**, *15*, 7048–7055. [CrossRef]
24. Prokhorova, T.G.; Buravov, L.I.; Yagubskii, E.B.; Zorina, L.V.; Simonov, S.V.; Shibaeva, R.P.; Zverev, V.N. New metallic bi-and monolayered radical cation salts based on BEDT-TTF with tris(oxalato)gallate anion. *Eur. J. Inorg. Chem.* **2014**, *2014*, 3933–3940. [CrossRef]
25. Prokhorova, T.G.; Buravov, L.I.; Yagubskii, E.B.; Zorina, L.V.; Simonov, S.V.; Zverev, V.N.; Shibaeva, R.P.; Canadell, E. Effect of the halopyridine guest molecules (G) on the structure and (super)conducting properties of the β''-(BEDT-TTF)$_4$(H$_3$O)[Fe(C$_2$O$_4$)$_3$]G crystals. *Eur. J. Inorg. Chem.* **2015**, *2015*, 5611–5620. [CrossRef]
26. Martin, L.; Morritt, A.L.; Lopez, J.R.; Nakazawa, Y.; Akutsu, H.; Imajo, S.; Ihara, Y.; Zhang, B.; Zhange, Y.; Guof, Y. Molecular conductors from bis(ethylenedithio)tetrathiafulvalene with tris(oxalato)rhodate. *Dalt. Trans.* **2017**, *46*, 9542–9548. [CrossRef] [PubMed]

27. Prokhorova, T.G.; Yagubskii, E.B. Organic conductors and superconductors based on bis(ethylenedithio)tetrathiafulvalene radical cation salts with supramolecular tris(oxalato)metallate anions. *Russ. Chem. Rev.* **2017**, *86*, 164–180. [CrossRef]
28. Prokhorova, T.G.; Yagubskii, E.B.; Zorina, L.V.; Simonov, S.V.; Zverev, V.N.; Shibaeva, R.P.; Buravov, L.I. Specific structural disorder in an anion layer and its influence on conducting properties of new crystals of the (BEDT-TTF)$_4$A$^+$[M^{3+}(ox)$_3$]G family, where G is 2-halopyridine; M is Cr, Ga; A$^+$ is [K$_{0.8}$(H$_3$O)$_{0.2}$]$^+$. *Crystals* **2018**, *8*, 92. [CrossRef]
29. Martin, L. Molecular conductors of BEDT-TTF with tris(oxalato)metallate anions. *Coord. Chem. Rev.* **2018**, *376*, 277–291. [CrossRef]
30. Mori, T. Structural Genealogy of BEDT-TTF-Based Organic Conductors I. Parallel Molecules: β and β″ Phases. *Bull. Chem. Soc. Jpn.* **1998**, *71*, 2509–2526. [CrossRef]
31. Mori, T.; Mori, H.; Tanaka, S. Structural Genealogy of BEDT-TTF-Based Organic Conductors II. Inclined Molecules: θ, α, and κ Phases. *Bull. Chem. Soc. Jpn.* **1999**, *72*, 179–197. [CrossRef]
32. Olejniczak, I.; Frackowiak, A.; Swietlik, R.; Prokhorova, T.G.; Yagubskii, E.B. Charge Fluctuations and Ethylene-Group-Ordering Transition in β″-(BEDT-TF)$_4$[(H3O)Fe(C$_2$O$_4$)$_3$]Y Molecular Charge-Transfer Salts. *ChemPhysChem* **2013**, *14*, 3925–3935. [CrossRef]
33. Merino, J.; McKenzie, R.H. Superconductivity Mediated by Charge Fluctuations in Layered Molecular Crystals. *Phys. Rev. Lett.* **2001**, *87*, 237002. [CrossRef]
34. Martin, L.; Morritt, A.; Lopez, J.R.; Akutsu, H.; Nakazawa, Y.; Imajo, S.; Ihara, Y. Ambient-pressure molecular superconductor with a superlattice containing layers of tris(oxalato)rhodate enantiomers and 18-crown-6. *Inorg. Chem.* **2017**, *56*, 717–720. [CrossRef]
35. Martin, L.; Lopez, J.R.; Nakazawa, Y.; Imajo, S. Bulk Kosterlitz–Thouless Type Molecular Superconductor β″-(BEDT-TTF)$_2$[(H$_2$O)(NH$_4$)$_2$Cr(C$_2$O$_4$)$_3$]18-crown-6. *Inorg. Chem.* **2017**, *56*, 14045–14052. [CrossRef] [PubMed]
36. Morritt, A.L.; Lopez, J.R.; Blundell, T.J.; Canadell, E.; Akutsu, H.; Nakazawa, Y.; Imajo, S.; Martin, L. 2D Molecular Superconductor to Insulator Transition in the β″-(BEDT-TTF)$_2$[(H$_2$O)(NH$_4$)$_2$M(C$_2$O$_4$)$_3$]·18-crown-6 Series (M = Rh, Cr, Ru, Ir). *Inorg. Chem.* **2019**, *58*, 10656–10664. [CrossRef] [PubMed]
37. Guionneau, P.; Kepert, C.J.; Chasseau, D.; Truter, M.R.; Day, P. Determining the charge distribution in BEDT-TTP salts. *Synth. Met.* **1997**, *86*, 1973–1974. [CrossRef]
38. Agilent. *CrysAlis PRO Version171.35.19*; Agilent Technologies UK Ltd.: Oxfordshire, UK, 2011.
39. Sheldrick, G.M. *SHELXL-97. Program for Crystal Structure Refinement*; University of Göttingen: Göttingen, Germany, 1997.

Article

Crystal-to-Crystal Transformation from $K_2[Co(C_2O_4)_2(H_2O)_2] \cdot 4H_2O$ to $K_2[Co(\mu-C_2O_4)(C_2O_4)]$

Bin Zhang [1,*], Yan Zhang [2], Guangcai Chang [3], Zheming Wang [4] and Daoben Zhu [1]

[1] Organic Solid Laboratory, BNLMS, CMS & Institute of Chemistry, Chinese Academy of Sciences, Beijing 100190, China; zhudb@iccas.ac.cn
[2] Department of Physics, Institute of Condensed Matter and Material Physics, Peking University, Beijing 100871, China; zhang_yan@pku.edu.cn
[3] BSRF, Institute of High Energy Physics, Chinese Academy of Sciences, Beijing 100047, China; changgc@ihep.ac.cn
[4] State Key Laboratory of Rare Earth Materials Chemistry and Applications, BNLMS, College of Chemistry and Molecular Engineering, Peking University, Beijing 100871, China; zmw@pku.edu.cn
* Correspondence: zhangbin@iccas.ac.cn

Citation: Zhang, B.; Zhang, Y.; Chang, G.; Wang, Z.; Zhu, D. Crystal-to-Crystal Transformation from $K_2[Co(C_2O_4)_2(H_2O)_2] \cdot 4H_2O$ to $K_2[Co(\mu-C_2O_4)(C_2O_4)]$. Magnetochemistry 2021, 7, 77. https://doi.org/10.3390/magnetochemistry 7060077

Academic Editors: Carlos J. Gómez García, Lee Martin, Scott Turner, John Wallis and Hiroki Akutsu

Received: 28 April 2021
Accepted: 24 May 2021
Published: 28 May 2021

Publisher's Note: MDPI stays neutral with regard to jurisdictional claims in published maps and institutional affiliations.

Copyright: © 2021 by the authors. Licensee MDPI, Basel, Switzerland. This article is an open access article distributed under the terms and conditions of the Creative Commons Attribution (CC BY) license (https:// creativecommons.org/licenses/by/ 4.0/).

Abstract: Crystal-to-crystal transformation is a path to obtain crystals with different crystal structures and physical properties. $K_2[Co(C_2O_4)_2(H_2O)_2] \cdot 4H_2O$ (**1**) is obtained from $K_2C_2O_4 \cdot 2H_2O$, $CoCl_2 \cdot 6H_2O$ in H_2O with a yield of 60%. It is crystallized in the triclinic with space group $P\bar{1}$ and cell parameters: a = 7.684(1) Å, b = 9.011(1) Å, c = 10.874(1) Å, α = 72.151(2)°, β = 70.278(2)°, γ = 80.430(2)°, V = 670.0(1) Å^3, Z = 2 at 100 K. **1** is composed of K$^+$, mononuclear anion $[Co(C_2O_4)_2(H_2O)_2]^{2-}$ and H_2O. Co^{2+} is coordinated by two bidentated oxalate anion and two H_2O in an octahedron environment. There is a hydrogen bond between mononuclear anion $[Co(C_2O_4)_2(H_2O)_2]^{2-}$ and H_2O. $K_2[Co(\mu-C_2O_4)(C_2O_4)]$ (**2**) is obtained from **1** by dehydration. The cell parameters of **2** are a = 8.460(5) Å, b = 6.906 (4) Å, c = 14.657(8) Å, β = 93.11(1)°, V = 855.0(8) Å^3 at 100 K, with space group in $P2/c$. It is composed of K$^+$ and zigzag $[Co(\mu-C_2O_4)(C_2O_4{}^{2-}]_n$ chain. Co^{2+} is coordinated by two bisbendentate oxalate and one bidentated oxalate anion in trigonal-prism. **1** is an antiferromagnetic molecular crystal. The antiferromagnetic ordering at 8.2 K is observed in **2**.

Keywords: oxalate; cobalt; crystal structure; magnetic property

1. Introduction

The change of the weak interaction of guest molecules, coordination geometry distortion, and coordination number in coordination compounds can effectively modulate the physical properties as magnetism, absorption, and chirality, so the dynamic molecular crystals have received great attention for their potential applications in molecular devices, as molecular sensors and switches become a powerful method for obtaining a specific compound with the yield of 100% by crystal-to-crystal transformation in crystal engineering [1–7]. The crystal-to-crystal transformations were observed between different dimensional coordination units as zero-dimensional (0D), one-dimensional chain (1D), two-dimensional (2D) layer, and three-dimensional (3D) coordination frameworks [8–10]. We are interested in dynamic crystals of MX_2–(1,4-dioxane)–H_2O system, and 0D to 2D, 1D to 2D, 1D to 3D crystal-to-crystal transformations were found [11–13]. Oxalate ($C_2O_4{}^{2-}$) is one of most popular used three-atoms ligands in the study of molecular-based magnet, its versatile abilities and intermediating efficient magnetic coupling among transition atoms have constructed 1D, 2D, and 3D magnetic materials [14–21]. However, the research on oxalate-based dynamic crystal is limited. Herein, we present a crystal-to-crystal transformation from 0D mononuclear compound $K_2[Co(C_2O_4)_2(H_2O)_2] \cdot 4H_2O$ (**1**) into a reported 1D coordination compound $K_2[Co(\mu-C_2O_4)(C_2O_4)]$ (**2**) accompanied by changes in crystal color, cell parameters, space group, coordination environment, crystal structure, and magnetic property.

2. Experiment and Discussion

1 was obtained from $K_2C_2O_4 \cdot 2H_2O$, $Co(NO_3)_2 \cdot 6H_2O$ in H_2O with yield of 60%.

When **1** was heated at an elevated temperature (at 120 °C for three minutes), it transferred to **2** after dehydration with a mass loss of 25.7%, crystal structure changed from mononuclear to one-dimensional chain (Scheme 1) and the crystal color changed from orange to pink. **2** remained stable until 300 °C (Figure 1). This is the second method to obtain **2** except the solvothermal method. The IR bands (Figure S1) between **1** and **2** is the strong broad band above 3000 cm^{-1} ν(O-H) as from H_2O in **1**. The existence weak broad band above 3000 cm^{-1} means **2** is unstable to air as reported [17].

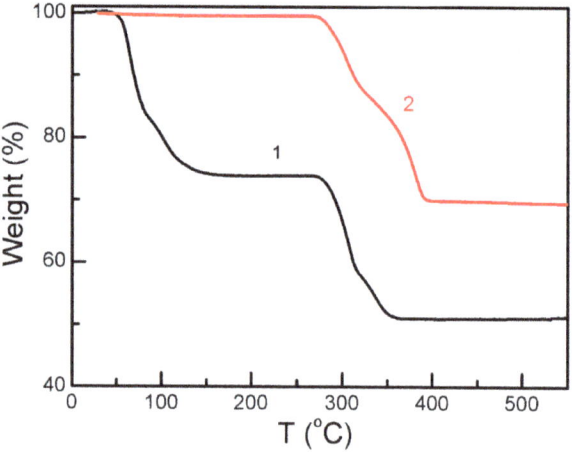

Figure 1. TGA plot of **1** (black) and **2** (red).

1 crystallizes in triclinic with space group of $P\bar{1}$: a = 7.684(1) Å, b = 9.011(1) Å, c = 10.874(1) Å, α = 72.151(2)°, β = 70.278(2)°, γ = 80.430(2)°, V = 670.0(1) Å^3, Z = 2. **1** is composed of K$^+$, mononuclear coordination anion Co(C$_2$O$_4$)$_2$(H$_2$O)$_2^{2-}$ and H$_2$O (Figure 2a). There are two K$^+$, one Co^{2+}, two oxalate (C$_2$O$_4^{2-}$), and six H$_2$O in an independent unit. K1 is surrounded by five O from three oxalato and four H$_2$O, K2 is surrounded by five O from three oxalato and four H$_2$O. K column formed by K1 and K2 host the vacancy of H-bond network formed by oxalate and H$_2$O along the b axis. K1$\cdots$K2 distances are 3.872(2) Å and 5.630(2) Å alternatively, and K1$\cdots$K1 and K2$\cdots$K2 distances are 9.011(1) Å. Each Co^{2+} is coordinated by two oxalate anions with Co-O 2.078(2)~2.099(2) Å on the equatorial plane, and two H$_2$O with Co-O 2.114(4) Å~2.120(3) Å to fulfill the octahedron environment. O-Co-O angles are among 88.4(2)~92.3(1)° from H$_2$O to equatorial plane and 176(1)° between two H$_2$O atoms. Viewed along the b axis, Co$\cdots$Co distances are of 9.011(1) Å and 7.684(2) Å alternatively along the a axis. There are hydrogen bonds between anions and coordinated H$_2$O: O9-H1$\cdots$O8(-x,1-y,1-z) 2.839(5) Å/170°, O10-H4$\cdots$O1(-x,2-y,-z) 3.178(5) Å/127°; between anions and solvent H$_2$O: O14-H12$\cdots$O1 2.795 Å/163°, O13-H9$\cdots$O6(1-x,y,z) 2.740 Å/169°, O9-H2$\cdots$O14(-x,1-y,-z) 2.760 Å/179°, O12-H8$\cdots$O2(1-x,1-y,z) 2.820 Å/167°, O11-H5$\cdots$O8(-x,1-y,1-z) 2.826 Å/156°, O10-H3$\cdots$O12(-x,1-y,z) 2.755 Å/176°, O11-H6$\cdots$O3(-x,1-y,-z) 2.709 Å/178°, O13-H10$\cdots$O7 2.803 Å/164°.

The crystal structure of **2** is the same as reported isostructural of K$_2$[Fe(μ-C$_2$O$_4$)(C$_2$O$_4$)] [17]. It consists of K$^+$ and zigzag chain [Co(μ-C$_2$O$_4$)(C$_2$O$_4$)$^{2-}$]$_n$ (Figure 2b). There are one and two half K$^+$, one Co^{2+}, one and two half oxalato in an independent unit. Co^{2+} is trigonal-prismatic coordinated by three oxalate anions with Co-O distance 2.058(3)~2.151(4) Å. K$^+$ is in the vacant formed by Co(C$_2$O$_4$)$_2^{2-}$ chain.

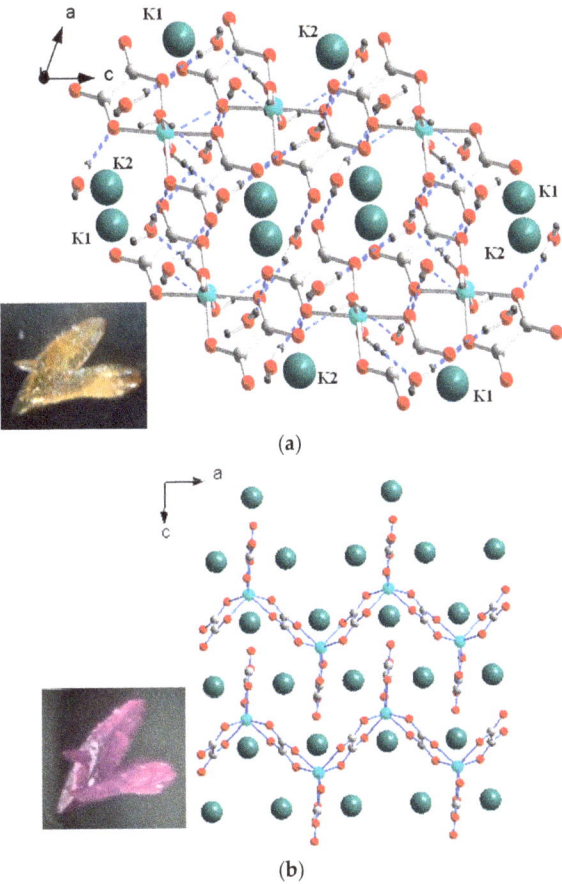

Figure 2. Crystal structure and appearance of **1** (**a**) and **2** (**b**). Color code: K, dark green; Co, cyan; C, light grey; O, red.; H, grey. Blue dashed lines are hydrogen bonds.

Depending on the extensive hydrogen bonds in **1** and zigzag chained structure of **2**, the magnetic properties of them were investigated. The transformation from **1** to **2** is irreversible. The sample was checked and remained the same before and after magnetic experiments.

1: χT is 3.41 cm^3 K mol^{-1} at 300 K. It is significantly larger than the value of 1.875 cm^3 K mol^{-1} expected for an isolated, spin-only ion with $S = 3/2$ and $g = 2.00$. This suggests a strong spin-orbit coupling. [20–22] The χT value decreased upon cooling and reached 1.30 cm^3 K mol^{-1} at 2 K. The susceptibility data above 50 K fit the Curie–Weiss law well, giving Curie and Weiss constants of $C = 3.613(6)$ cm^3 K mol^{-1} and $\theta = -21.2(2)$ K, respectively, with $R = 3.74 \times 10^{-5}$ (Figure 3). The negative Weiss constant means the antiferromagnetic interaction between Co^{2+} ions through hydrogen bonds. At 2 K, the isothermal magnetization is 2.24 Nβ at 65 kOe (Figure 4). No long-range magnetic ordering was observed in **1**.

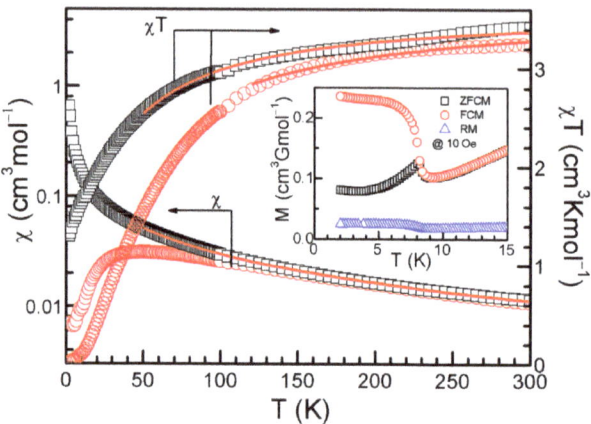

Figure 3. Magnetic susceptibility of **1** (black square) and **2** (red circle) under 1000 Oe. Inset: ZFCM/FCM/RM at 10 Oe of **2**. Red line: Curie–Weiss fitting.

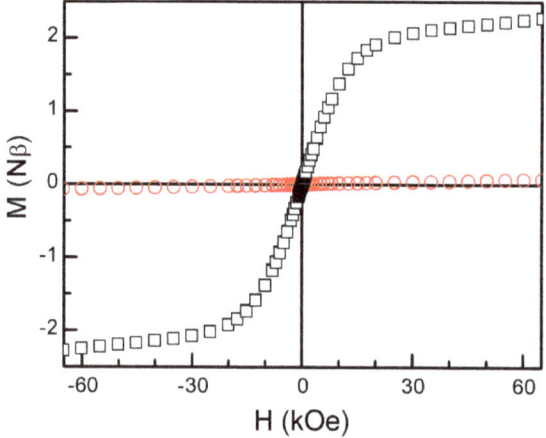

Figure 4. Isothermal magnetization of **1** (empty black square) and **2** (empty red circle) at 2 K.

2: On χ versus T plot, a broad maximum of 0.031 cm^3 mol^{-1} was observed around 50 K, which is similar to reported oxalate-bridged one-dimensional compounds [15,17,20]. Then χ value decreased upon cooling smoothly, it is 0.0070 cm^3 mol^{-1} at 2 K. At 300 K, χT is 3.24 cm^3 K mol^{-1}, this means a strong spin-orbit coupling of Co^{2+} as **1**. The χT value decrease upon cooling and reach 0.014 cm^3 K mol^{-1} at 2 K. The data above 120 K were fitted with Curie–Weiss law, giving Curie and Weiss constant C = 3.66(2) cm^3 K mol^{-1}, θ = $-$35(1) K, R = 4.96 $\times$ 10^{-5}. Field-cooled magnetization (FCM) and zero-field-cooled magnetization (ZFCM) measurements under a field of 10 Oe show a magnetic ordering at 8.2 K (Figure 3, inset). At 2 K, the isothermal magnetization increases smoothly and reaches 0.072 Nβ at 65 kOe. The Hysteresis loop (Hc) is 500 Oe.

3. Conclusions

Orange **1** transfer to pink **2** by dehydration. **1** is composed of K$^+$, mononuclear coordination anion Co(C$_2$O$_4$)$_2$(H$_2$O)$_2$$^-$ and H$_2$O with extensive hydrogen bond between anion and H$_2$O, H$_2$O, and H$_2$O. **2** is consisted of K$^+$ and zigzag chain anion [Co(μ-C$_2$O$_4$)(C$_2$O$_4$)$^{2-}$]$_n$.

The antiferromagnetic interaction in **1** from hydrogen bonds is weaker than oxalate-bridge in **2** [23]. **2** shows antiferromagnetic ordering at 8.2 K.

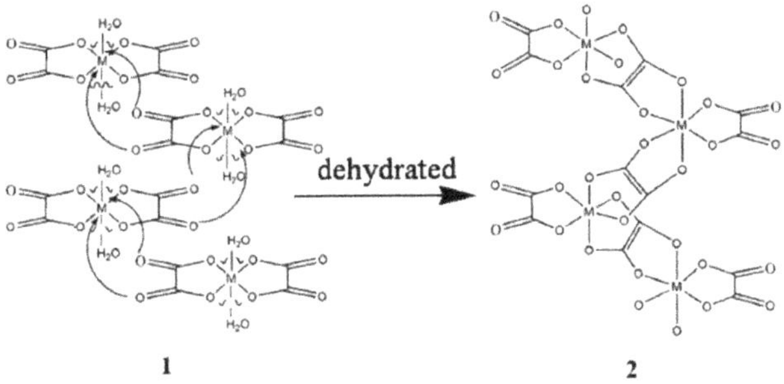

Scheme 1. Schematic drawing of the possible route of crystal-to-crystal transformation from **1** to **2**.

Supplementary Materials: The following are available online at https://www.mdpi.com/article/10.3390/magnetochemistry7060077/s1, sample preparation and characterization, Figure S1: IR spectra of 1 and 2, Figure S2: Powder X-ray diffraction patterns of **1** and **2**.

Author Contributions: Conceptualization, B.Z.; data curation, Y.Z.; G.C.; investigation, Z.W.; D.Z. All authors have read and agreed to the published version of the manuscript.

Funding: This research was funded by the National Natural Science Foundation of China (Grant Nos. 21172230, 21573242, and 22073106), Chinese Ministry of Science and Technology (Grant Nos. 2011CB932302 and 2013CB933402) and the Strategic Priority Research Program (B) of the Chinese Academy of Sciences (Grant No. XDB12030100).

Data Availability Statement: Sample preparation, characterization, IR spectra and powder X-ray diffraction pattern of 1 and 2 are available on Supplementary Materials via. www.mdpi.com/xxx/s1.

Acknowledgments: This work was financially supported by the National Natural Science Foundation of China (Grant Nos. 21172230, 21573242, and 22073106).

Conflicts of Interest: The authors declare no conflict of interest.

References

1. Lim, S.H.; Olmstead, M.M.; Balch, A.L. Molecular accordion: Vapoluminescence and molecular flexibility in the orange and green luminescent crystals of the dimer, Au2(μ-bis-(diphenylphosphino)ethane)2Br2. *J. Am. Chem. Soc.* **2011**, *133*, 10229–10238. [CrossRef] [PubMed]
2. Huang, Y.; Schoenecker, B.; Mu, P.; Carson, C.G.; Karra, J.R.; Cai, Y.; Walton, K.S. A porous flexible homochiral SrSi$_2$ array of single-stranded helical nanotubes exhibiting single-crystal-to-single-crystal oxidation transformation. *Angew. Chem. Int. Ed.* **2011**, *50*, 436–440. [CrossRef] [PubMed]
3. Chatterjee, P.B.; Audhya, A.; Bhattacharya, S.; Abtab, S.M.T.; Bhattacherya, K.; Chaudhury, M. Single crystal-to-single crystal irreversible transformation from a discrete Vanadium(V)-Alcoholate to an Aldehydic-Vanadium(IV) oligomer. *J. Am. Chem. Soc.* **2010**, *132*, 15842–15845. [CrossRef]
4. Huang, Z.; White, P.S.; Brookhart, M. Ligand exchanges and selective catalytic hydrogenation in molecular single crystals. *Nature* **2010**, *465*, 598–601. [CrossRef]
5. Kawasaki, T.; Hakoda, Y.; Mineki, H.; Suzuki, K.; Soai, K. Generation of absolute controlled crystal chirality by the removal of crystal water from achiral crystal of nucleobase cytosine. *J. Am. Chem. Soc.* **2010**, *132*, 2874–2875. [CrossRef]
6. Allan, P.K.; Xiao, B.; Teat, S.J.; Knight, J.W.; Morris, R.E. In situ single-crystal diffraction studies of the structural transition of metal-organic framework copper 5-sulfoisophthalate, Cu-SIP-3. *J. Am. Chem. Soc.* **2010**, *132*, 3605–3611. [CrossRef]
7. Thorarinsdottir, A.E.; Harris, T.D. Metal-Organic framework magnets. *Chem. Rev.* **2020**, *120*, 8716–8789. [CrossRef]
8. Kaneko, W.; Ohba, M.; Kitagawa, S. A flexible coordination polymer crystal providing reversible structural and magnetic conversions. *J. Am. Chem. Soc.* **2007**, *129*, 13706–13712. [CrossRef]
9. Zhang, Y.; Liu, T.; Kanegawa, S.; Sato, O. Reversible single-crystal-to-single-crystal transformation from achiral antiferromagnetic hexanuclears to a chiral ferrimagnetic double zigzag chain. *J. Am. Chem. Soc.* **2009**, *131*, 7942–7943. [CrossRef] [PubMed]
10. Zhang, Y.; Liu, T.; Kanegawa, S.; Sato, O. Interconversion between a nonporous nanocluster and a microporous coordination polymer showing selective gas adsorption. *J. Am. Chem. Soc.* **2010**, *132*, 912–913. [CrossRef]

11. Duan, Z.; Zhang, Y.; Zhang, B.; Zhu, D. Crystal-to-Crystal transformation from antiferromagnetic chains into a ferromagnetic diamondoid framework. *J. Am. Chem. Soc.* **2009**, *131*, 6934–6935. [CrossRef]
12. Zhang, B.; Zhu, D.; Zhang, Y. Crystal-to-Crystal transformation from a mononuclear compound in a hydrogen-bonded three-dimensional framework to a layered coordination polymer. *Chem. Eur. J.* **2010**, *16*, 9994–9997. [CrossRef]
13. Shi, J.; Zhang, Y.; Zhang, B.; Zhu, D. Crystal-to-crystal transformation from a chain compound to a layered coordination polymer. *Dalton Trans.* **2016**, *45*, 89–92. [CrossRef]
14. Mathoniere, C.; Carling, S.G.; Yusheng, D.; Day, P. Molecular-based Mixed Valency Ferrimagnets $(XR_4)Fe^{II}Fe^{III}(C_2O_4)_3$ (X = N, P; R = n-propyle, n-butyl, phenyl): Anomalous negative magnetisation in the tetra-n-butylammonium derivative. *J. Chem. Soc. Chem. Commun.* **1994**, 1551–1552. [CrossRef]
15. Clemente-Leon, M.; Coronado, E.; Marti-Gastaldo, C.; Romero, F.M. Multifunctionality in hybrid magnetic materials based on bimetallic oxalate complexes. *Chem. Soc. Rev.* **2011**, *40*, 473–497. [CrossRef]
16. Hernandez-Molina, M.; Lloret, F.; Ruiz-Perez, C.; Julve, M. Weak ferromagnetism in chiral 3-dimensional oxalate-bridged Cobalt(II) compounds. crystal structure of $[Co(bpy)_3][Co_2Ox_3]ClO_4$. *Inorg. Chem.* **1998**, *37*, 4131–4135. [CrossRef]
17. Hursthouse, M.B.; Light, M.E.; Price, D.J. One-dimensional magnetism in new anhydrous iron and cobalt ternary oxalates with rare trigonal prismatic metal coordination. *Angew. Chem. Int. Ed.* **2004**, *43*, 472–475. [CrossRef]
18. Coronado, E.; Galan-Mascaros, J.R.; Marti-Gastalo, C. Single chain magnets based on the oxalate ligand. *J. Am. Chem. Soc.* **2008**, *130*, 14987–14989. [CrossRef]
19. Glerup, J.; Goodson, P.A.; Hodgson, D.J.; Michelsen, K. Magnetic Exchange through Oxalate Bridges: Synthesis and Characterization of (μ-Oxalato)dimetal (II) Complexes of Magnanese, Iron, Cobalt, Nickel, Copper and Zinc. *Inorg. Chem.* **1995**, *34*, 6255–6264. [CrossRef]
20. Garcia-Couceiro, U.; Castillo, O.; Luque, A. A new hydrated phase of cobalt(II) oxalate: Crystal structure, thermal behavior and magnetic properties of $\{[Co(-Ox)(H_2O)_2]\cdot 2H_2O\}_n$. *Inorg. Chim. Acta* **2004**, *357*, 339–344. [CrossRef]
21. Duan, Z.; Zhang, Y.; Zhang, B.; Zhu, D. $Co(C_2O_4)(HO(CH_2)_3OH)$: An antiferromagnetic neutral zigzag chain compound showing long-range-ordering of spin canting. *Inorg. Chem.* **2008**, *47*, 9152–9154. [CrossRef]
22. Duan, D.; Zhang, Y.; Zhang, B.; Pratt, F. Two homometallic antiferromagnets based on oxalato-bridged honeycomb assemblies: $(A)_2[M^{II}{}_2C_2O_4)_3]$ (A = ammonium salt derived from diethylenetriamine; M^{II} = Fe^{2+}, Co^{2+}). *Inorg. Chem.* **2009**, *48*, 2140–2146. [CrossRef]
23. Desplanches, C.; Ruiz, E.; Rodrigue-Fortea, A.; Alvarez, S. Exchange coupling of transition-metal ions through hydrogen bonding: A theoretical investigation. *J. Am. Chem. Soc.* **2002**, *124*, 5197–5205. [CrossRef]

Article

Solvent-Induced Hysteresis Loop in Anionic Spin Crossover (SCO) Isomorph Complexes

Emmelyne Cuza [1], Samia Benmansour [2], Nathalie Cosquer [1], Françoise Conan [1], Carlos J. Gómez-García [2,*] and Smail Triki [1,*]

1. Univ Brest, CNRS, CEMCA, 6 Avenue Le Gorgeu, C.S. 93837, CEDEX 3, 29238 Brest, France; Emmelyne.Cuza@univ-brest.fr (E.C.); nathalie.cosquer@univ-brest.fr (N.C.); Francoise.Conan@univ-brest.fr (F.C.)
2. Instituto de Ciencia Molecular (ICMol), Departamento de Química Inorgánica, Universidad de Valencia, C/Catedrático José Beltrán 2, 46980 Paterna, Spain; sam.ben@uv.es
* Correspondence: carlos.gomez@uv.es (C.J.G.-G.); smail.triki@univ-brest.fr (S.T.); Tel.: +34-963-544-423 (C.J.G.-G.); +33-298-016-146 (S.T.)

Citation: Cuza, E.; Benmansour, S.; Cosquer, N.; Conan, F.; Gómez-García, C.J.; Triki, S. Solvent-Induced Hysteresis Loop in Anionic Spin Crossover (SCO) Isomorph Complexes. *Magnetochemistry* 2021, 7, 75. https://doi.org/10.3390/magnetochemistry7060075

Academic Editor: Marius Andruh

Received: 17 April 2021
Accepted: 20 May 2021
Published: 23 May 2021

Publisher's Note: MDPI stays neutral with regard to jurisdictional claims in published maps and institutional affiliations.

Copyright: © 2021 by the authors. Licensee MDPI, Basel, Switzerland. This article is an open access article distributed under the terms and conditions of the Creative Commons Attribution (CC BY) license (https://creativecommons.org/licenses/by/4.0/).

Abstract: Reaction of Fe(II) with the tris-(pyridin-2-yl)ethoxymethane (py$_3$C-OEt) tripodal ligand and in the presence of the pseudohalide ancillary NCSe$^-$ (E = S, Se, BH$_3$) ligand leads to the mononuclear complex [Fe(py$_3$C-OEt)$_2$][Fe(py$_3$C-OEt)(NCSe)$_3$]$_2$·2CH$_3$CN (**3**), which has been characterised as an isomorph of the two previously reported complexes, Fe(py$_3$C-OEt)$_2$][Fe(py$_3$C-OEt)(NCE)$_3$]$_2$·2CH$_3$CN, with E = S (**1**), BH$_3$ (**2**). X-ray powder diffraction of the three complexes (**1**–**3**), associated with the previously reported single crystal structures of **1**–**2**, revealed a monomeric isomorph structure for **3**, formed by the spin crossover (*SCO*) anionic [Fe(py$_3$C-OEt)(NCSe)$_3$]$^-$ complex, associated with the low spin (*LS*) [Fe(py$_3$C-OEt)$_2$]$^{2+}$ cationic complex and two solvent acetonitrile molecules. In the [Fe(py$_3$C-OEt)$_2$]$^{2+}$ complex, the metal ion environment involves two py$_3$C-OEt tridentate ligands, while the [Fe(py$_3$C-OEt)(NCSe)$_3$]$^-$ anion displays a hexacoordinated environment involving three N-donor atoms of one py$_3$C-OEt ligand and three nitrogen atoms arising from the three (NCSe)$^-$ coligands. The magnetic studies for **3** performed in the temperature range 300-5-400 K, indicated the presence of a two-step SCO transition centred around 170 and 298 K, while when the sample was heated at 400 K until its complete desolvation, the magnetic behaviour of the high temperature transition ($T_{1/2}$ = 298 K) shifted to a lower temperature until the two-step behaviour merged with a gradual one-step transition at ca. 216 K.

Keywords: tripodal ligands; pseudohalide coligands; iron complex; spin crossover; magnetic properties

1. Introduction

The spin crossover (*SCO*) materials are by far the most investigated molecular systems among switchable systems during the last decade due to their many possible applications for the development of new generations of electronic devices, such as displays [1–4], memory devices [4–8], and sensors [9–14]. Although the *SCO* behaviour can be essentially observed in octahedral complexes based on metal ions allowing spin state changes between the low spin (*LS*) and high spin (*HS*) states under external stimulus, such as temperature, pressure, light irradiation, or magnetic field, those based on Fe(II) ion exhibiting d^6 electronic configuration remain the most studied systems [15–29]. Nevertheless, such complexes are mostly either cationic or neutral, and the Fe(II) anionic complexes exhibiting *SCO* behaviour have been relatively scarcely reported [21–29]. Furthermore, the few anionic *SCO* examples are restricted to only three different systems. The first one is the series [Fe^{II}H$_3$L][Fe^{II}L]X, (X$^-$ = AsF$_6^-$, BF$_4^-$, ClO$_4^-$, PF$_6^-$ and SbF$_6^-$), based on the ligand tris-(2-(((2-methylimidazol-4-yl)methylidene)amino)ethyl) amine (H$_3$L) and on its deprotonated anionic form (L^{3-}) [21]. The second one consists of the trinuclear [Fe^{II}$_3$(μ-L)$_6$(H$_2$O)$_6$]$^{6-}$ complex involving the 4-(1,2,4-triazol-4-yl)ethanedisulfonate anion

(L^{2-}) [22], which displays a *HS-HS-HS* to *HS-LS-HS* transition around room temperature and a large hysteresis loop (>85 K). The last system concerns the series of mononuclear complexes involving the tris(2-pyridyl)methane (py$_3$C-R, R = C_nH_{2n+1}, aryl group, O-C_nH_{2n+1}, O-aryl, O-CO-C_nH_{2n+1}), tridentate functionalized ligands (Scheme 1a) [23–29]. Such complexes, of general formula {A[Fe((py$_3$C-R)(NCE)$_3$)]$_m$} (A = [(C_nH_{2n+1})$_4$N]$^+$, [Fe(py$_3$C-R)$_2$]$^{2+}$, E = S, Se, BH$_3$), are based on the mononuclear [Fe((py$_3$C-R)(NCE)$_3$)]$^-$ anion composed by an Fe(II) metal centre, one py$_3$C-R tridentate ligand, and three terminal κN-SCE linear coligands (Scheme 1b). The different studies, reported essentially by Ishida et al. and some of us [23–29], have concerned the study of different chemical effects, such as those of the cationic counter ion or of the functional group (R) covalently linked to the tripodal py$_3$C motif, on the transition temperatures and the cooperativity. More recently, some of us extended such effects to that of crystal packing by designing a series of polymorph complexes [29].

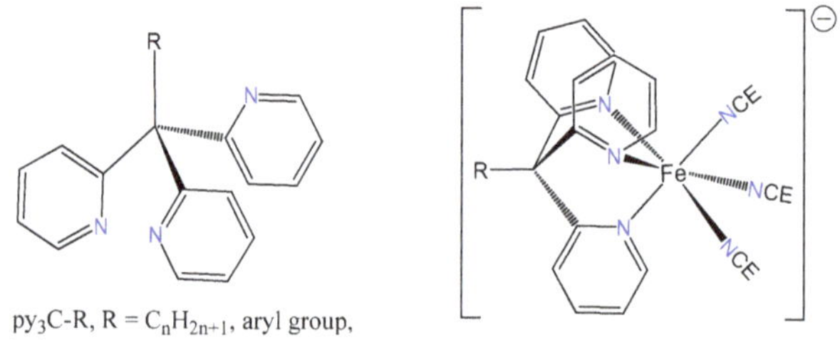

py$_3$C-R, R = C_nH_{2n+1}, aryl group, O-C_nH_{2n+1}, O-aryl, O-CO-C_nH_{2n+1}

(**a**)

[Fe(py$_3$C-R)(NCE), E = S, Se, BH$_3$

(**b**)

Scheme 1. (**a**) The tris-(2-pyridyl)methane (py$_3$C-R) ligands and their tridentate coordination mode (**b**).

In order to determine the effect of the ancillary anionic coligands (NCE$^-$ with E = S, BH$_3$, Se) on the transition temperatures and the cooperativity, we have reported recently two isomorphic complexes of general formula, [Fe(py$_3$C-OEt)$_2$][Fe(py$_3$C-OEt)(NCE)$_3$]$_2$·2CH$_3$CN (NCE$^-$ = NCS$^-$, NCBH$_3$$^-$), based on the *SCO* [Fe(py$_3$C-OEt)(NCE)$_3$]$^-$ anion and on the cationic *LS* complex, [Fe(py$_3$C-OEt)$_2$]$^{2+}$, as counter ion [28]. At the same time, the NCSe$^-$ analogue complex (E = Se), which completes such isomorphic series, has been also prepared. However, this complex, of a presumably chemical formula of [Fe(py$_3$C-OEt)$_2$][Fe(py$_3$C-OEt)(NCSe)$_3$]$_2$·2CH$_3$CN (**3**), could not be obtained as single crystals due to their instability. In addition, while magnetic behaviours of complexes **1** and **2** remained unchanged in the heating and cooling scan modes, complex **3** showed significant changes during the cooling/warming scan modes. These unexpected observations pushed us to explore in detail the peculiar switching behaviour of this compound. Here, we report the syntheses, structural characterization, infrared spectroscopy, and magnetic properties of the new isomorph [Fe(py$_3$C-OEt)$_2$][Fe(py$_3$C-OEt)(NCSe)$_3$]$_2$·2CH$_3$CN (**3**) exhibiting solvent-induced hysteresis loop of 50 K.

2. Results and Discussion

2.1. Syntheses

The py$_3$C-OEt (tris-(pyridin-2-yl)ethoxymethane) tripodal ligand was prepared as previously described [28–31]. Compound **3** was obtained as a red polycrystalline powder and as single crystals by mixing a solution of [N(C$_2$H$_5$)$_4$](NCSe) with a solution of FeCl$_2$ and tris(pyridin-2-yl)ethoxymethane at −32 °C (see details in Section 3).

2.2. Structural Characterization and Magnetic Properties

In contrast to complexes **1** and **2**, for which the crystal structures were determined using single crystal X-ray diffraction, complex **3**, which was expected to be isomorph to the structure observed for **1** and **2**, showed poor quality single crystal diffraction patterns that clearly precluded any correct single crystal structural characterization. However, after several attempts at 100 K, we succeeded in collecting some intensities, which led to unit cell parameters depicted in Table S1. Comparison of these parameters to those of the two isomorph complexes **1** and **2**, indicated that the structure of complex **3** was isomorphic to complexes **1** and **2** (Table S1). This conclusion was supported by the experimental X-ray powder diffraction pattern observed for the polycrystalline powder of complex **3**, which was very similar to the one observed for complex **1**, as well as to the simulated pattern derived from the single crystal structure of complex **1** (Figure 1 and Figure S1).

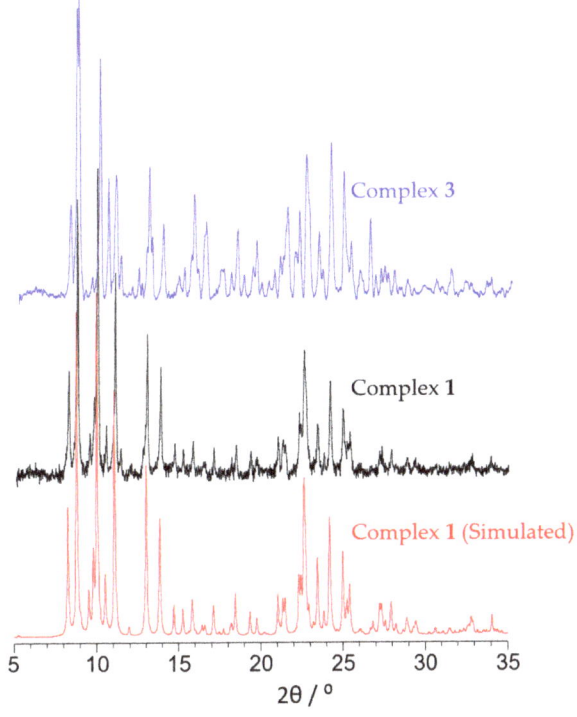

Figure 1. X-ray powder diffraction patterns for [Fe(py$_3$C-OEt)$_2$][Fe(py$_3$C-OEt)(NCE)$_3$]$_2$·2CH$_3$CN (E = S (**1**), Se (**3**)) and the simulated pattern derived from the crystal structure of complex **1**.

It is also worth mentioning that the elemental analyses of complex **3** agreed with the chemical formula, [Fe(py$_3$C-OEt)$_2$][Fe(py$_3$C-OEt)(NCSe)$_3$]$_2$·2CH$_3$CN, expected for a complex isomorph to **1** and **2**. Therefore, all the data strongly support that the crystal structure of complex **3** is isomorphic to those of complexes **1** and **2** [28]; and therefore its crystal structure consists of a low spin (LS) [Fe(py$_3$C-OEt)$_2$]$^{2+}$ cationic complex (Figure 2b), an anionic [Fe(py$_3$C-OEt)(NCE)$_3$]$^-$ (E = S (**1**), BH$_3$ (**2**) Se (**3**)) complex (Figure 2a) and two CH$_3$CN solvent molecules.

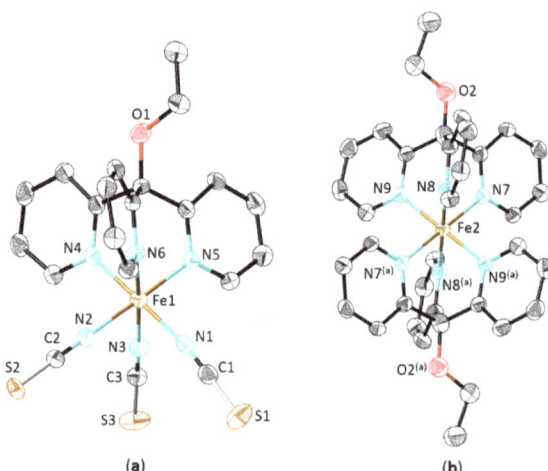

Figure 2. View of the anionic [Fe(py$_3$C-OEt)(NCS)$_3$]$^-$ (a) and of the cationic [Fe(py$_3$C-OEt)$_2$]$^{2+}$ (b) complexes in [Fe(py$_3$C-Oet)$_2$][Fe(py$_3$C-OEt)(NCS)$_3$]$_2$·2CH$_3$CN (1) [28]. Codes of equivalent positions: (a) = -x, -y, -z.

As previously described, complex **1** exhibited an incomplete SCO transition ($T_{1/2}$ = 205 K), while **2** displayed a complete two-step transition at 245 K and 380 K (Figure 3). Their magnetic properties were studied in warming and cooling modes (300-2-400-2 K for **1**; 300-2-500-2 K for **2**), but no significant hysteretic effects or any change due to possible desolvation were detected. Also, irradiation at 10 K with a green light for several hours revealed no noticeable increase of the thermal variation of the product of the molar magnetic susceptibility times the temperature ($\chi_m T$) for both compounds. As for complex **1**, susceptibility measurements for complex **3** were performed in the 300-5 K and 5-400 K temperature ranges. The thermal variation of the $\chi_m T$ product for the three complexes (**1–3**) are shown in Figure 3. For compound **3**, the $\chi_m T$ value per formula at 400 K ($\approx$6.71 cm^3 K mol^{-1}) was in agreement with the value expected for two isolated Fe(II) ions (S = 2 and g $\approx$ 2.1), revealing the presence of two magnetically isolated HS Fe(II) ions [15–29]. On cooling, the $\chi_m T$ value of **3** showed an initial abrupt drop, at around 298 K, reaching a value of ca. 3.40 cm^3 K mol^{-1} at 265 K. On further cooling, we observed a second drop, at around 170 K, to reach a plateau at ca. 0.56 cm^3 K mol^{-1} below 116 K. This low temperature $\chi_m T$ value implies the presence of a residual HS fraction of ca. 8 %. This behaviour indicates the presence of an incomplete HS to LS two-step transition centred at around 170 and 298 K.

In contrast to isomorphs **1** and **2**, where the magnetic properties did not show any change after successive cooling and heating scans in the ranges 2–400 K for **1** and 2–500 K for **2**, the two-step behaviour described above for complex **3** (Figure 3) was irreversible due to a gradual desolvation of the sample at 400 K, as previously observed in several solvated systems [32–40]. As a matter of fact, when the sample was maintained at 400 K until its complete desolvation, the second cycle (400-50-400 K) in **3** produced a shift to lower temperatures for the high temperature step, while the transition temperature of the low temperature step remained unchanged (See Figure 4). Similar trends were observed for the third and fourth cycles, until the fifth cycle where the initial high temperature step merged with the low temperature step (Figure 4), to lead to the gradual one-step transition depicted in Figure 5a. There were no hysteretic effects and the HS fraction of ca. 8% remained unchanged after the different heating and cooling cycles.

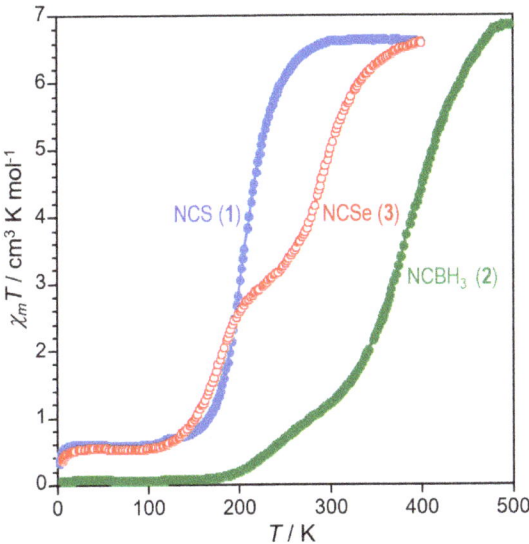

Figure 3. (a) Temperature dependence of the $\chi_m T$ product of **1** (●), **2** (●), and **3** (○).

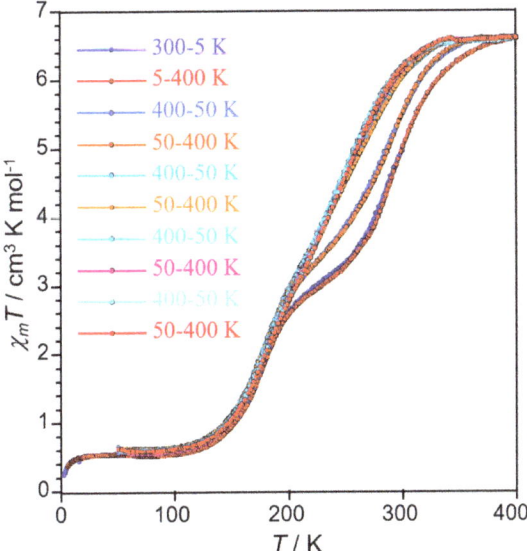

Figure 4. Thermal variation of $\chi_m T$ product of a freshly prepared sample of compound **3** for different consecutive heating and cooling scans.

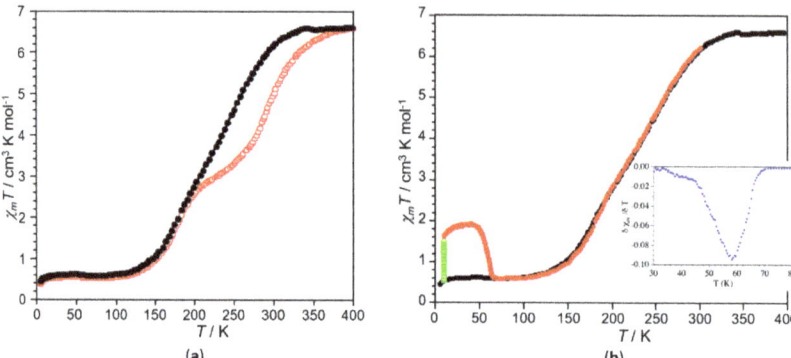

Figure 5. Magnetic properties of compound **3**: (a) thermal evolution of the $\chi_m T$ product for the solvated (○) and desolvated (●) samples; (b) thermal evolution of the $\chi_m T$ product for the desolvated sample in the dark (●), under 532 nm (△) light irradiation at 10 K, and the subsequent thermal relaxation (●) in the dark. Inset shows the thermal variation of the derivative of $\chi_m T$.

Irradiation of the sample with a green laser (λ = 532 nm) at 10 K produced an increase of the $\chi_m T$ product, indicative of the presence of a light-induced excited spin state trapping at low temperatures (LIESST effect). After ca. 4 h, the $\chi_m T$ product reached saturation at a value of ca. 1.6 cm^3 K mol^{-1} (Figure S3). This value indicates that around 1/3 of one of the two LS Fe(II) centres in the [Fe(py$_3$C-OEt)(NCS)$_3$]$^-$ anions were excited to the *HS* state. After switching off the light irradiation, heating the sample further increased the $\chi_m T$ value up to ca. 2.0 cm^3 K mol^{-1} (representing ca. 44 % of one of the two Fe(II) centres). On further heating, the sample relaxed to the *LS* state at a T_{LIESST} of ca. 58 K (Figure 5b).

One of the major points, deserving special attention in regards to the three isomorphic complexes, concerns the origin of the unexpected and singular magnetic behaviour that only occurred for isomorph **3** (see Figures 3 and 5a). To try to understand the process that occurred at high temperatures, we performed thermogravimetric analysis (TGA) and X-ray powder diffraction on the three isomorphs to know more about the desolvation and solvation processes of this system. Thus, TGA measurements were performed for the three isomorphs, which were heated at 5 °C min^{-1}, under nitrogen atmosphere, from room temperature to 390 K. In Figure 6, the mass evolution with temperature for the three complexes were gathered, showing clearly that the two isomorphs **1** and **2** remained stable and retained their solvent molecules up to 390 K, while complex **3** started to lose weight from room temperature and lost 4.37 % of its mass when heated up to 370 K, corresponding to two CH$_3$CN solvent molecules per formula unit. These measurements revealed that despite their isomorphic structures, complexes **1** and **2** retained their crystallization solvent molecules while complex **3** lost them even at moderate temperatures, suggesting that the crystal packing in **3** has larger cavities that allow an easy desolvation (see below).

To check for the reversibility of this desolvation process, we performed successive desolvation and resolvation cycles, by heating the solvated sample and by adding two drops of CH$_3$CN on the desolvated sample, respectively. After resolvation, magnetic measurements (Figure S4) and X-ray powder diffraction (Figure 7) showed that the sample recovered its original behaviour, supporting the reversibility of the desolvation/resolvation process.

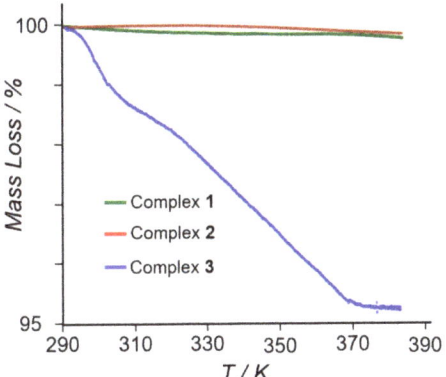

Figure 6. TGA analyses of compounds **1–3**.

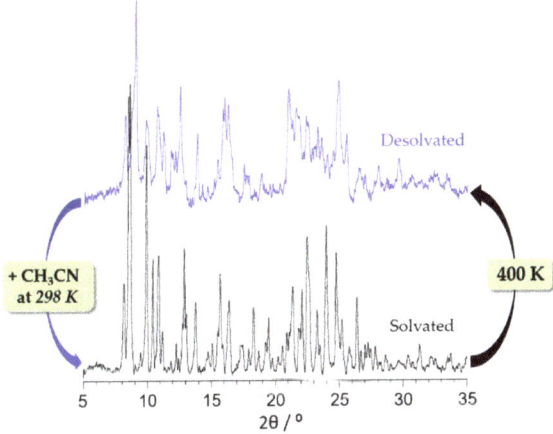

Figure 7. Experimental X-ray powder diffraction patterns of solvated and desolvated samples of **3**, confirming the reversibility of the desolvation/resolvation process.

2.3. Variable Temperature Magnetic Properties and Infrared Spectroscopy

In order to confirm the Fe(II) spin state at high and low temperatures and the presence of the incomplete *HS* to *LS* transition for **3** (see Figure 3), we measured the infrared spectrum at 100 and 300 K in the range of the fundamental stretching vibration of the NCSe⁻ units (1975–2130 cm^{-1}), since it has been clearly established that the intensities of these stretching vibrations are very sensitive to the spin state of the Fe(II) metal ion [28,29,41–47]. We thus recorded the infrared spectra for **3** at 350 and 100 K, according to the thermal evolution of the $\chi_m T$ product depicted in Figure 5a. The infrared spectra for **3** in the C≡N frequency region (1975–2130 cm^{-1}) at 350 and 100 K are displayed in Figure 8. At 350 K, two $\nu_{C\equiv N}$ stretching broad bands, characteristic of the *HS* state, appeared at 2050 and 2075 cm^{-1}, while at 100 K, four strong bands, characteristic of the *LS* state, appeared at 2044, 2075, 2082, and 2109 cm^{-1}. In agreement with the presence of an 8 % *HS* fraction at low temperatures (see magnetic section), the three bands observed at 2044, 2075, and 2082 cm^{-1} can be viewed as resulting from the decrease in intensity of the two broad and strong bands observed for the *HS* state, while the band observed at a higher frequency (2109 cm^{-1}) appeared as the specific band of the *LS* state.

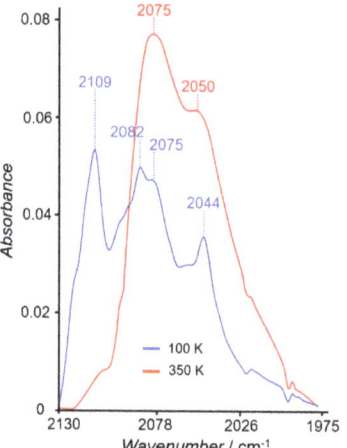

Figure 8. The principal infrared bands in the 1975–2130 cm^{-1} region for complex **3** at 350 and 100 K.

2.4. Magneto-Spectroscopic Relationships

Based on the magnetic behaviours (Figure 5a) and on the main bands that were temperature sensitive (Figure 8) of both solvated and desolvated phases of complex **3**, we investigated the thermal evolution of the infrared spectra of the stretching vibration of the NCSe$^-$ in the range 1975–2130 cm^{-1}. For both phases (solvated and desolvated), we recorded the infrared spectra in the vicinity of the *SCO* transitions from 100 to 350 K. First, we recorded the infrared spectra for the freshly prepared complex **3**, heating the sample from 100 to 350 K to avoid the partial desolvation of the sample (Figure 9a). Then, the same sample was heated during one hour at 400 K to ensure its complete desolvation, and the corresponding infrared spectra were then recorded cooling the sample from 350 to 100 K (Figure 9b).

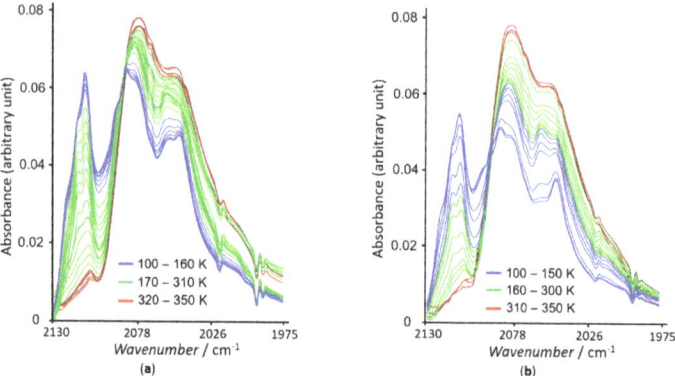

Figure 9. Temperature dependence of the infrared spectra, in the temperature range 350–100 K, for compound **3**: solvated (**a**) and desolvated (**b**) samples.

The intensities of the two ν(NCSe) broad bands (2050 and 2075 cm^{-1}) attributed to the *HS* state decreased gradually with decreasing temperature from 350 to 100 K, but persisted even at 100 K, supporting the presence of a fraction of *HS* Fe(II) centres, as revealed by the magnetic data. In parallel, a new band, characteristic of the *LS* state, appeared at higher frequencies (2109 cm^{-1}), whose intensity gradually increased with decreasing the temperature. However, as can be easily observed in both Figure 9a,b, the infrared spectra did not show any clear difference between the infrared band evolutions of the solvated (Figure 9a) and desolvated (Figure 9b) phases of complex **3**, as revealed by the magnetic data. Thus, in order to show more clearly this difference and to appreciate, at least qualitatively, the consistency of the experimental infrared data, we correlated the thermal variation of the $\chi_m T$ product derived from magnetic study and the thermal evolution of the intensity of the infrared bands. The results, depicted in Figure 10, showed that the intensities of the characteristic infrared bands (2109, 2075, 2050 cm^{-1}) fit perfectly with the thermal evolution of the $\chi_m T$ product, in agreement with the presence of a two-step *SCO* transition in the solvated sample (Figure 10a) and a gradual one-step *SCO* behaviour in the desolvated sample (Figure 10b).

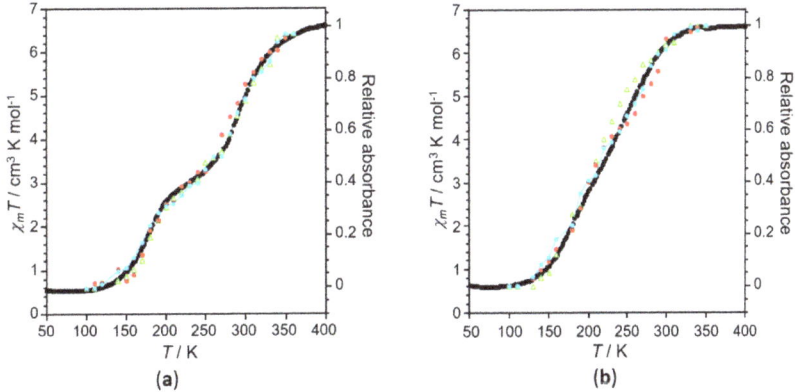

Figure 10. Temperature dependences of the $\chi_m T$ product and of the relative absorbance of the ν(CN) bands observed at 2109 cm^{-1} (△), 2075 cm^{-1} (•), and at 2050 cm^{-1} (○), for the solvated (**a**) and desolvated (**b**) samples of **3**.

The final question to be answered by this study was why the isomorph based on the NCSe$^-$ ligand (complex **3**) displayed a reversible solvation/desolvation process with CH$_3$CN while the two other isomorphs (complexes **1** and **2**) retained the solvent molecules at temperatures exceeding 390 K. The answer could only come from the crystal packing of this triad of isomorphs. Indeed, examination of the crystal packing revealed that the CH$_3$CN solvent molecules are located in tetragonal-like channels running along the [010] direction, which are generated by eclipsed stacks of the -Fe-NCE ... Fe ... ECN-Fe-NCE ... Fe- metallacycles (see Figure 11). Even though the three isomorphs display, as expected, similar crystal packing, the fact that they differ by the nature of the NCE$^-$ ancillary ligands (E = S (**1**), BH$_3$ (**2**), Se (**3**)), induces strong differences in the sizes of the metallacycles, due to the different lengths of the three linear anions (the N-E distance increases from 2.72 Å for E = BH$_3$, to 2.80 Å for E = S and 2.94 Å for E = Se). The largest tetragonal-like channels are expected to be those of the isomorph with the longest ancillary linear ligand (i.e., compound **3**). Unfortunately, this difference could not be quantified due to the lack of single crystal structural data of complex **3**. However, in order to provide a reasonable estimation of the effect of the NCE$^-$ ligands on the size of the metallacycles, we have calculated the relative increase of their dimensions when passing from complex **2** to complex **1** (see Figure 11 and Table 1 with the different Fe···Fe distances (d_1–d_4) determining the size of the channels).

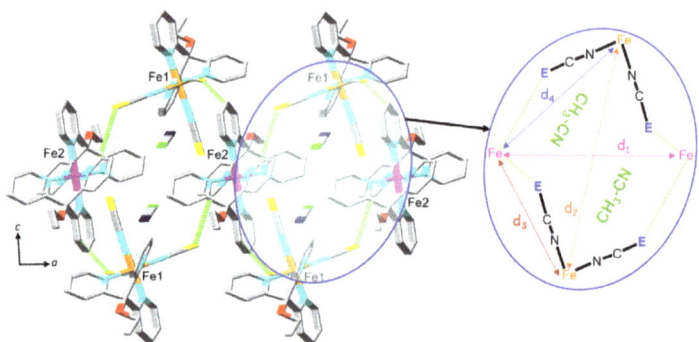

Figure 11. Structure and dimensions of the tetragonal-like channels where the CH$_3$CN molecules are located in compounds **1–3**.

Table 1. Relative increase Fe$\cdots$Fe distances (d_1 and d_2) as function of the nature of the NCE$^-$ ancillary ligands ($d_{N\cdots E}$) in the two isomorphs **1** and **2**.

	2 (E = BH$_3$)	1 (E = S)	3 (E = Se)
$d_{N\cdots E}$ (relative increase)	2.72 Å	2.80 Å (+2.9 %)	2.94 Å (+8.1 %)
T/Spin State	200 K/LS	100 K/LS	—
d_1 (relative increase)	12.757	14.774 Å (+15.8%)	—
d_2 (relative increase)	11.683 Å	12.653 Å (+8.3%)	—
d_3 (relative increase)	8.146 Å	10.481 Å (+28.6%)	—
d_4 (relative decrease)	9.125 Å	8.907 Å (−2.4%)	—

As can be seen in Table 1, when passing from complex **2** (with NCBH$_3^-$ and N$\cdots$B distance of 2.72 Å) to complex **1** (with NCS$^-$ and N$\cdots$S distance of 2.80 Å), there is an increase of +2.9 % in the size of the anion, which led to increases in the Fe$\cdots$Fe distances in the metallacycle of up to 28.6 % (see d_1 to d_4 in Table 1). Therefore, if we consider complex **3** based on the NCSe$^-$ linear ligand, which corresponds to the biggest linear anion (with N$\cdots$Se distance of 2.94 Å), the expected metallacycle should be significantly larger than those observed for the isomorphs **1** and **2**. These observations explain clearly why complex **3** involves larger tetragonal-like channels, allowing the easy and reversible solvation and desolvation processes.

3. Experimental Section

3.1. Starting Materials

All starting reagents and solvents were purchased and used as received. The tris-(pyridin-2-yl)ethoxymethane (py$_3$C-OEt) ligand was prepared under nitrogen atmosphere as described previously [28–31].

3.2. Synthesis of [Fe(py$_3$C-OEt)$_2$][Fe(py$_3$C-OEt)(NCSe)$_3$]$_2$·2CH$_3$CN (3)

Tris-(pyridin-2-yl)ethoxymethane (50.0 mg, 0.17 mmol) and FeCl$_2$ (20.0 mg, 0.16 mmol) were dissolved in methanol (5 mL) in the presence of a few mg of ascorbic acid. The mixture was stirred at room temperature for 15 min. To the resulting solution was added a solution of acetonitrile (5 mL) containing the [(C$_2$H$_5$)$_4$N]NCSe salt (162 mg, 0.69 mmol). After 30 min stirring, the resulting solution was filtered and quickly cooled at −32 °C. After three days, the bright red polycristalline powder of (**3**) as well as a few red single crystals were obtained. Anal. Calcd. (%) for [Fe(py$_3$C-OEt)$_2$][Fe(py$_3$C-OEt)(NCSe)$_3$]$_2$·2CH$_3$CN (C$_{82}$H$_{74}$Fe$_3$N$_{20}$O$_4$Se$_6$) **3**: C, 48.2; H, 3.7; N, 13.7; found (%): C, 47.9; H, 3.8; N, 14.0. IR data (ν/cm^{-1}) for the freshly filtered sample (powder and single crystals, Figure S5): 410 (w), 423 (w), 477 (w), 500 (w), 513 (w), 530 (w), 659 (m), 726 (w), 758 (w), 886 (w), 1011 (m), 1086 (w), 1108 (m), 1143 (m), 1205 (w), 1252 (w), 1291 (w), 1389 (w), 1434 (m), 1462 (s), 1593 (m), 2060 (s), 2244 (w), 2871 (w), 2901 (w), 2972.11 (w), 3076 (w), 3442 (br).

3.3. Characterization of the Materials

Infrared spectra of complex **3** were performed using a platinum ATR Vertex 70 BRUKER spectrometer with variable temperature cell holder (VT Cell Holder typer P/N GS21525). ^{1}H and ^{13}C NMR spectra were performed using BRUKER DRX 300 MHz, Advance 400 MHz and Advance III HD 500 MHz equipment. TGA measurements were performed on ATG-LabsysTM, Setaram (see details in Supplementary Information).

3.4. Magnetic Measurements

Magnetic susceptibility measurements were performed using a Quantum Design MPMS-XL-5 SQUID susceptometer (San Diego, CA, USA). The susceptibility data were corrected for the diamagnetic contributions using Pascal's constant tables [48]. The photomagnetic studies were performed by irradiating the sample at 10 K with a green Diode Pumped Solid State Laser DPSS-532-20 from Chylas (see details in Supplementary Information).

4. Conclusions

We have shown that the compound [Fe(py$_3$C-OEt)$_2$][Fe(py$_3$C-OEt)(NCSe)$_3$]$_2$·2CH$_3$CN (**3**), based on the [Fe(py$_3$C-OEt)$_2$]$^{2+}$ *LS* cation and the *SCO* [Fe(py$_3$C-OEt)(NCSe)$_3$]$^-$ anion, displays a reversible desolvation process that affects the *SCO* behaviour. This compound has been prepared as a polycrystalline powder and as single crystals using a similar protocol to that used previously for the syntheses of the two isomorphic complexes [Fe(py$_3$C-OEt)$_2$][Fe(py$_3$C-OEt)(NCE)$_3$]$_2$·2CH$_3$CN (E = S (**1**), BH$_3$ (**2**)) based on the two ancillary linear ligands NCS$^-$ and NCBH$_3$$^-$ [28]. Despite being obtained in the form of prismatic-shaped single crystals, complex **3** could not be characterized by X-ray single crystal diffraction, because of the low stability of its single crystals, in contrast to complexes **1** and **2**, which have been structurally characterized. However, combined X-ray powder diffraction, infrared spectra, and CHN elemental analyses clearly revealed that complex **3** exhibits an isomorphic structure to those of complexes **1** and **2**. TGA analyses performed on the single crystal samples of the three isomorphs revealed clearly that the two isomorphs **1** and **2**, for which the corresponding single crystals are stable, retained their solvent molecules up to 390 K, while complex **3** began to lose its CH$_3$CN solvent molecules from room temperature. The magnetic studies for **3** performed in cooling and heating scans in the temperature ranges 300–5 K and 5–400 K, respectively, indicated the presence of an incomplete *HS* to *LS* two-step like transition centred around 170 and 298 K, while when the sample was heated at 400 K until its complete desolvation, the magnetic behaviour of the high temperature transition ($T_{1/2}$ = 298 K) shifted to a lower temperature until the two-step behaviour merged with a gradual one-step transition at ca. 216 K. Such behaviour, which can be viewed as a solvent-induced hysteresis loop of 50 K [41–47], was confirmed by infrared spectra recorded in the vicinity of the *SCO* transition for both solvated and desolvated samples. Furthermore, successive desolvation and solvation cycles, tracked by SQUID measurements and X-ray powder diffraction, showed that the desolvation process is fully reversible. As shown by the TGA analysis, compound **3** exhibited a different magnetic behaviour due to its easy desolvation and solvation process, in contrast to the two other isomorphs, which retained their solvent molecule up to 390 K, despite their isomorphic structures. This unexpected behaviour was elucidated by careful examination of the crystal packing of these isomorph complexes, which clearly revealed that the solvent molecules are located in tetragonal-like channels generated by the eclipsed stacks of the "-Fe-NCE . . . Fe . . . ECN-Fe-NCE . . . Fe-" metallacycles. Therefore, the isomorph based on the bigger NCSe$^-$ ancillary ligand, should display the larger tetragonal-like channels, allowing easier solvation and desolvation as observed in isomorph **3**.

Supplementary Materials: The following are available online at https://www.mdpi.com/article/10.3390/magnetochemistry7060075/s1, Table S1: Crystal data of [Fe(py$_3$C-OEt)$_2$][Fe(py$_3$C-OEt)(NCE)$_3$]$_2 \cdot$ 2CH$_3$CN (E = S (**1**), BH$_3$ (**2**), Se (**3**)), Figure S1: Experimental and simulated XRPD patterns for compounds **1** and **2**, and experimental one for complex **3**, Figure S2: Thermal variation of χ_m of a freshly prepared sample of compound **3** for different consecutive heating and cooling scans, Figure S3: Time dependence of $\chi_m T$ at 10 K for compound **3** under laser irradiation with green light (λ = 532 nm, switched on at time ca. 10 min), Figure S4: Different heating and cooling scans of the $\chi_m T$ product performed on the de-solvated compound **3** after re-solvation with two drops of CH$_3$CN, Figure S5: IR spectra of single crystals and polycrystalline powder of complex **3**.

Author Contributions: E.C. performed the chemical syntheses (ligand and the complex) under supervision of S.T. and F.C., and recorded and analysed the infrared spectra under supervision of S.T. and N.C.; F.C. interpreted the NMR spectra; S.B. performed the X-ray powder diffraction; C.J.G.-G. performed and interpreted the magnetic measurements; S.T. supervised the experimental work and wrote the manuscript on which all the authors have contributed. All authors have read and agreed to the published version of the manuscript.

Funding: This research was funded by "Mission pour les Initiatives Transverses et Interdisciplinaires (MITI, CNRS)", Mol-CoSM ANR Project N° ANR-20-CE07-0028-01 the "Université de Brest" (IBSAM institute), the Région Bretagne (EC), the Generalidad Valenciana (Prometeo2019/076 project) and the Spanish MINECO (Project CTQ2017-87201-P AEI/FEDER, EU).

Institutional Review Board Statement: Not applicable.

Informed Consent Statement: Not applicable.

Data Availability Statement: Data is contained within the article or Supplementary Material.

Conflicts of Interest: The authors declare no conflict of interest.

References

1. Cobo, S.; Molnár, G.; Real, J.-A.; Bousseksou, A. Multilayer Sequential Assembly of Thin Films That Display Room-Temperature Spin Crossover with Hysteresis. *Angew. Chem. Int. Ed.* **2006**, *45*, 5786–5789. [CrossRef]
2. Kahn, O.; Martinez, C.J. Spin-transition polymers: From molecular materials toward memory devices. *Science* **1998**, *279*, 44–48. [CrossRef]
3. Kumar, K.S.; Ruben, M. Sublimable Spin-Crossover Complexes: From Spin-State Switching to Molecular Devices. *Angew. Chem. Int. Ed.* **2021**, *60*, 7502–7521. [CrossRef]
4. Murray, K.S.; Kepert, C.J. Cooperativity in Spin Crossover Systems: Memory, Magnetism and Microporosity. In *Spin Crossover in Transition Metal Compounds I*; Gütlich, P., Goodwin, H.A., Eds.; Topics in Current Chemistry; Springer: Berlin/Heidelberg, Germany, 2004; Volume 233, pp. 195–228.
5. Létard, J.-F.; Guionneau, P.; Goux-Capes, L. Towards Spin Crossover Applications. In *Spin Crossover in Transition Metal Compounds III*; Gütlich, P., Goodwin, H.A., Eds.; Topics in Current Chemistry; Springer: Berlin/Heidelberg, Germany, 2004; Volume 235, pp. 221–249.
6. Bousseksou, A.; Molnar, G.; Salmon, L.; Nicolazzi, W. Molecular spin crossover phenomenon: Recent achievements and prospects. *Chem. Soc. Rev.* **2011**, *40*, 3313–3335. [CrossRef] [PubMed]
7. Rueckes, T.; Kim, K.; Joselevich, E.; Tseng, G.Y.; Cheung, C.-L.; Lieber, C.M. Carbon Nanotube-Based Nonvolatile Random Access Memory for Molecular Computing. *Science* **2000**, *289*, 94–97. [CrossRef]
8. Galet, A.; Gaspar, A.B.; Carmen Muñoz, M.; Bukin, G.V.; Levchenko, G.; Real, J.-A. Tunable Bistability in a Three-Dimensional Spin-Crossover Sensory- and Memory-Functional Material. *Adv. Mater.* **2005**, *17*, 2949–2953. [CrossRef]
9. Bartual-Murgui, C.; Akou, A.; Thibault, C.; Molnar, G.; Vieu, C.; Salmon, L.; Bousseksou, A. Spin-crossover metal–organic frameworks: Promising materials for designing gas sensors. *J. Mater. Chem. C* **2015**, *3*, 1277–1285. [CrossRef]
10. Lapresta-Fernandez, A.; Titos-Padilla, S.; Herrera, J.M.; Salinas-Castillo, A.; Colacio, E.; Capitan-Vallvey, L.F. Photographing the synergy between magnetic and colour properties in spin crossover material [Fe(NH$_2$trz)$_3$](BF$_4$)$_2$: A temperature sensor perspective. *Chem. Commun.* **2013**, *49*, 288–290. [CrossRef] [PubMed]
11. Linares, J.; Codjovi, E.; Garcia, Y. Pressure and temperature spin crossover sensors with optical detection. *Sensors* **2012**, *12*, 4479–4492. [CrossRef] [PubMed]
12. Cuéllar, M.P.; Lapresta-Fernández, A.; Herrera, J.M.; Salinas-Castillo, A.; del Carmen Pegalajar, M.; Titos-Padilla, S.; Colacio, E.; Capitán-Vallvey, L.F. Thermochromic sensor design based on Fe(II) spin crossover/polymers hybrid materials and artificial neural networks as a tool in modelling. *Sens. Actuators B* **2015**, *208*, 180–187. [CrossRef]
13. Rodriguez-Jimenez, S.; Feltham, H.L.C.; Brooker, S. Non-Porous Iron(II)-Based Sensor: Crystallographic Insights into a Cycle of Colorful Guest-Induced Topotactic Transformations. *Angew. Chem. Int. Ed.* **2016**, *55*, 15067–15071. [CrossRef]

14. Benaicha, B.; Van Do, K.; Yangui, A.; Pittala, N.; Lusson, A.; Sy, M.; Bouchez, G.; Fourati, H.; Gómez-García, C.J.; Triki, S.; et al. Interplay between Spin-Crossover and Luminescence in a Multifunctional Single Crystal Iron(II) complex: Towards a New Generation of Molecular Sensors. *Chem. Sci.* **2019**, *10*, 6791–6798. [CrossRef]
15. Halcrow, M.A. (Ed.) *Spin-Crossover Materials, Properties and Applications*; John Wiley & Sons Ltd: Oxford, UK, 2013.
16. Phan, H.; Hrudka, J.J.; Igimbayeva, D.; Lawson Daku, L.M.; Shatruk, M. A Simple Approach for Predicting the Spin State of Homoleptic Fe(II) Tris-diimine Complexes. *J. Am. Chem. Soc.* **2017**, *139*, 6437–6447. [CrossRef] [PubMed]
17. Pittala, N.; Thétiot, F.; Charles, C.; Triki, S.; Boukheddaden, K.; Chastanet, G.; Marchivie, M. An unprecedented trinuclear Fe^{II} triazole-based complex exhibiting a concerted and complete sharp spin transition above room temperature. *Chem. Commun.* **2017**, *53*, 8356–8359. [CrossRef] [PubMed]
18. Shatruk, M.; Phan, H.; Chrisostomo, B.A.; Suleimenova, A. Symmetry-breaking structural phase transitions in spin crossover complexes. *Coord. Chem. Rev.* **2015**, *289–290*, 62–73. [CrossRef]
19. El Hajj, F.; Sebki, G.; Patinec, V.; Marchivie, M.; Triki, S.; Handel, H.; Yefsah, S. Macrocycle-based spin-crossover materials. *Inorg. Chem.* **2009**, *48*, 10416–10423. [CrossRef]
20. Setifi, F.; Milin, E.; Charles, C.; Thétiot, F.; Triki, S.; Gómez-García, C.G. Spin Crossover Iron(II) Coordination Polymer Chains: Syntheses, Structures, and Magnetic Characterizations of [Fe(aqin)$_2$(μ-M(CN)$_4$)] (M = Ni(II), Pt(II), aqin = Quinolin-8-amine). *Inorg. Chem.* **2014**, *53*, 97–104. [CrossRef] [PubMed]
21. Yamada, M.; Ooidemizu, M.; Ikuta, Y.; Osa, S.; Matsumoto, N.; Iijima, S.; Kojima, M.; Dahan, F.; Tuchagues, J.-P. Interlayer Interaction of Two-Dimensional Layered Spin Crossover Complexes [Fe^{II}H$_3$LMe][Fe^{II}LMe]X (X$^-$ = ClO$_4^-$, BF$_4^-$, PF$_6^-$, AsF$_6^-$, and SbF$_6^-$; H$_3$LMe = Tris[2-(((2-methylimidazol-4-yl)methylidene)amino)ethyl]amine). *Inorg. Chem.* **2003**, *42*, 8406–8416. [CrossRef]
22. Gómez, V.; Sáenz de Pipaón, C.; Maldonado-Illescas, P.; Waerenborgh, J.C.; Martin, E.; Benet-Buchholz, J.; Galán-Mascarós, J.-R. Easy Excited-State Trapping and Record High TTIESST in a Spin-Crossover Polyanionic Fe(II) Trimer. *J. Am. Chem. Soc.* **2015**, *137*, 11924–11927. [CrossRef] [PubMed]
23. Hirosawa, N.; Oso, Y.; Ishida, T. Spin crossover and light-induced excited spin-state trapping observed for an iron (II) complex chelated with tripodal tetrakis(2-pyridyl) methane. *Chem. Lett.* **2012**, *41*, 716–718. [CrossRef]
24. Yamasaki, M.; Ishida, T. Spin-crossover thermal hysteresis and light-induced effect on iron (II) complexes with tripodal tris (2-pyridyl) methanol. *Polyhedron* **2015**, *85*, 795–799. [CrossRef]
25. Yamasaki, M.; Ishida, T. Heating-rate dependence of spin-crossover hysteresis observed in an iron (II) complex having tris (2-pyridyl) methanol. *J. Mater. Chem. C* **2015**, *3*, 7784–7787. [CrossRef]
26. Ishida, T.; Kaneto, T.; Yamasaki, M. An iron(II) complex tripodally chelated with 1,1,1-tris(pyridine-2-yl)ethane showing room-temperature spin-crosssover behaviour. *Acta Cryst. Sect. C* **2016**, *72*, 797–801. [CrossRef] [PubMed]
27. Kashiro, A.; Some, K.; Kobayashi, Y.; Ishida, T. Iron(II) and 1,1,1-Tris(2-pyridyl)nonadecane Complex Showing an Order–Disorder-Type Structural Transition and Spin-Crossover Synchronized over Both Conformers. *Inorg. Chem.* **2019**, *58*, 7672–7676. [CrossRef]
28. Cuza, E.; Benmansour, S.; Cosquer, N.; Conan, F.; Pillet, S.; Gómez-García, C.J.; Triki, S. Spin Cross-Over (SCO) Anionic Fe(II) Complexes Based on the Tripodal Ligand Tris(2-pyridyl)ethoxymethane. *Magnetochemistry* **2020**, *6*, 26. [CrossRef]
29. Cuza, E.; Mekuimemba, C.D.; Cosquer, N.; Conan, F.; Pillet, S.; Chastanet, G.; Triki, S. Spin Crossover and High-Spin State in Fe(II) Anionic Polymorphs Based on Tripodal Ligands. *Inorg. Chem.* **2021**, *60*, 6536–6549. [CrossRef]
30. White, D.L.; Faller, J.W. Preparation and Reactions of the C_{3v} Ligand Tris(2-pyridyl)methane and Its Derivatives. *Inorg. Chem.* **1982**, *21*, 3119–3122. [CrossRef]
31. Jonas, R.T.; Stack, T.D.P. Synthesis and Characterization of a Family of Systematically Varied Tris(2-pyridyl)methoxymethane Ligands: Copper(I) and Copper(II) Complexes. *Inorg. Chem.* **1998**, *37*, 6615–6629. [CrossRef]
32. Benmansour, S.; Gómez-Claramunt, P.; Gómez-García, C.G. Effects of water removal on the structure and spin-crossover in an anilato-based compound. *J. Appl. Phys.* **2021**, *129*, 123904. [CrossRef]
33. Kulmaczewski, R.; Bamiduro, F.; Shahid, N.; Cespedes, O.; Halcrow, M.A. Structural Transformations and Spin-Crossover in [FeL$_2$]$^{2+}$ Salts (L=4-{tert-Butylsulfanyl}-2,6-di{pyrazol-1-yl}pyridine): The Influence of Bulky Ligand Substituents. *Chem. Eur. J.* **2021**, *27*, 2082–2092. [CrossRef]
34. Mondal, D.J.; Roy, S.; Yadav, J.; Zeller, M.; Konar, S. Solvent-Induced Reversible Spin-Crossover in a 3D Hofmann-Type Coordination Polymer and Unusual Enhancement of the Lattice Cooperativity at the Desolvated State. *Inorg. Chem.* **2020**, *59*, 13024–13028. [CrossRef] [PubMed]
35. Phonsri, W.; Davies, C.G.; Jameson, G.N.L.; Moubaraki, B.; Murray, K.S. Spin Crossover, Polymorphism and Porosity to Liquid Solvent in Heteroleptic Iron(III) {Quinolylsalicylaldimine/Thiosemicarbazone-Salicylaldimide} Complexes. *Chem. Eur. J.* **2016**, *22*, 1322–1333. [CrossRef]
36. Lennartson, A.; Southon, P.; Sciortino, N.F.; Kepert, C.J.; Frandsen, C.; Mørup, S.; Piligkos, S.; McKenzie, C. Reversible Guest Binding in a Non-Porous Fe^{II} Coordination Polymer Host Toggles Spin Crossover. *J. Chem. Eur. J.* **2015**, *21*, 16066–16072. [CrossRef] [PubMed]
37. Roberts, T.D.; Tuna, F.; Malkin, T.L.; Kilner, C.A.; Halcrow, M.A. An iron(II) complex exhibiting five anhydrous phases, two of which interconvert by spin-crossover with wide hysteresis. *Chem. Sci.* **2012**, *3*, 349–354. [CrossRef]

38. Clemente-León, M.; Coronado, E.; Giménez-López, M.C.; Romero, F.M. Structural, Thermal, and 52Magnetic Study of Solvation Processes in Spin-Crossover [Fe(bpp)$_2$][Cr(L)(ox)$_2$]$_2$·nH$_2$O Complexes. *Inorg. Chem.* **2007**, *46*, 11266–11276. [CrossRef]
39. Hayami, S.; Gu, Z.-Z.; Yoshiki, H.; Fujishima, A.; Sato, O. Iron(III) Spin-Crossover Compounds with a Wide Apparent Thermal Hysteresis around Room Temperature. *J. Am. Chem. Soc.* **2001**, *123*, 11644–11650. [CrossRef]
40. Garcia, Y.; Van Koningsbruggen, P.J.; Codjovi, E.; Lapouyade, R.; Kahn, O.; Rabardel, L. Non-classical FeII spin-crossover behaviour leading to an unprecedented extremely large apparent thermal hysteresis of 270 K: Application for displays. *J. Mater. Chem.* **1997**, *7*, 857–858. [CrossRef]
41. Sorai, M.; Seki, S. Phonon coupled cooperative low-spin $^1A_1$ ↔ high-spin $^5T_2$ transition in [Fe(phen)$_2$(NCS)$_2$] and [Fe(phen)$_2$(NCSe)$_2$] crystals. *J. Phys. Chem. Solids* **1974**, *35*, 555–570. [CrossRef]
42. Brehm, G.; Reiher, M.; Le Guennic, B.; Leibold, M.; Schindler, S.; Heinemann, F.W.; Schneider, S. Investigation of the low-spin to high-spin transition in a novel [Fe(pmea)(NCS)$_2$] complex by IR and Raman spectroscopy and DFT calculations. *J. Raman Spectrosc.* **2006**, *37*, 108–122. [CrossRef]
43. Park, Y.; Jung, Y.M.; Sarker, S.; Lee, J.-J.; Lee, Y.; Lee, K.; Oh, J.J.; Joo, S.-W. Temperature-dependent infrared spectrum of (Bu$_4$N)$_2$[Ru(dcbpyH)$_2$(NCS)$_2$] on nanocrystalline TiO$_2$ surfaces. *Solar Energy Mater. Solar Cells* **2010**, *94*, 857–864. [CrossRef]
44. Varma, V.; Fernandes, J.-R. An Infrared Spectroscopic Study of the Low-Spin-High-spin transition in in Fe$_x$Mn$_{1-x}$(Phen)$_2$(NCS)$_2$: A Composition-Induced Change in the Order of the Spin-State Transition. *Chem. Phys. Let.* **1990**, *167*, 367–370. [CrossRef]
45. Sankar, G.; Thomas, J.M.; Varma, V.; Kulkani, G.U.; Rao, C.N.R. An investigation of the first-order spin-state transition in the Fe(phen)$_2$(NCS)$_2$ EXAFS and infrared spectroscopy. *Chem. Phys. Lett.* **1996**, *251*, 79–83. [CrossRef]
46. Smit, E.; de Waal, D.; Heyns, A.M. The spin-transition complexes [Fe(Htrz)$_3$](ClO$_4$)$_2$ and [Fe(NH$_2$trz)$_3$](ClO$_4$)$_2$ I. FT-IR spectra of a low pressure and a low temperature phase transition. *Mater. Res. Bull.* **2000**, *35*, 1697–1707. [CrossRef]
47. Durand, P.; Pillet, S.; Bendeif, E.-E.; Carteret, C.; Bouazaoui, M.; El Hamzaoui, H.; Capoen, B.; Salmon, L.; Hébert, S.; Ghanbaja, J.; et al. Room temperature bistability with wide thermal hysteresis in a spin crossover silica nanocomposite. *J. Mater. Chem. C* **2013**, *1*, 1933–1942. [CrossRef]
48. Bain, G.A.; Berry, J.F. Diamagnetic corrections and Pascal's constants. *J. Chem. Educ.* **2008**, *85*, 532–536. [CrossRef]

magnetochemistry

Article

Neutron Studies of a High Spin Fe$_{19}$ Molecular Nanodisc

Francis L. Pratt [1,*], Tatiana Guidi [1], Pascal Manuel [1], Christopher E. Anson [2], Jinkui Tang [3], Stephen J. Blundell [4] and Annie K. Powell [2]

1. ISIS Neutron and Muon Source, STFC Rutherford Appleton Laboratory, Didcot OX11 0QX, UK; tatiana.guidi@stfc.ac.uk (T.G.); pascal.manuel@stfc.ac.uk (P.M.)
2. Institute of Inorganic Chemistry, Karlsruhe Institute of Technology, D76131 Karlsruhe, Germany; christopher.anson@kit.edu (C.E.A.); annie.powell@kit.edu (A.K.P.)
3. Changchun Institute of Applied Chemistry, Changchun 130022, China; tang@ciac.ac.cn
4. Department of Physics, University of Oxford, Clarendon Laboratory, Oxford OX1 3PU, UK; stephen.blundell@physics.ox.ac.uk
* Correspondence: francis.pratt@stfc.ac.uk

Abstract: The molecular cluster system [Fe$_{19}$(metheidi)$_{10}$(OH)$_{14}$O$_6$(H$_2$O)$_{12}$]NO$_3$·24H$_2$O, abbreviated as Fe$_{19}$, contains nineteen Fe(III) ions arranged in a disc-like structure with the total spin $S = 35/2$. For the first order, it behaves magnetically as a single molecule magnet with a 16 K anisotropy barrier. The high spin value enhances weak intermolecular interactions for both dipolar and superexchange mechanisms and an eventual transition to antiferromagnetic order occurs at 1.2 K. We used neutron diffraction to determine both the mode of ordering and the easy spin axis. The observed ordering was not consistent with a purely dipolar driven order, indicating a significant contribution from intermolecular superexchange. The easy axis is close to the molecular Fe1–Fe10 axis. Inelastic neutron scattering was used to follow the magnetic order parameter and to measure the magnetic excitations. Direct transitions to at least three excited states were found in the 2 to 3 meV region. Measurements below 0.2 meV revealed two low energy excited states, which were assigned to $S = 39/2$ and $S = 31/2$ spin states with respective excitation gaps of 1.5 and 3 K. Exchange interactions operating over distances of order 10 Å were determined to be on the order of 5 mK and were eight-times stronger than the dipolar coupling.

Keywords: magnetic molecular cluster; high spin molecule; single molecule magnet; magnetic nanodisc; neutron diffraction; inelastic neutron scattering; dipolar driven magnetism; Monte Carlo simulation; weak superexchange interactions

1. Introduction

Coupled clusters of transition metal ions can provide examples of high-spin single molecule magnets (SMMs) [1,2]. Since the field of SMMs opened with the discovery of the slow relaxation of the magnetisation in the Mn12 molecule Mn$_{12}$O$_{12}$(CH$_3$COO)$_{16}$(H$_2$O)$_4$ [3], a large number of further Mn(III)-based SMMs have been discovered, among them the high anisotropy system Mn6 [4] and the high-spin system Mn17 with $S = 37$ [5]. In addition to further examples based on other 3d metals, such as Fe(III), the incorporation of 4f ions has also proven a useful strategy for introducing large anisotropy to high spin systems in 3d–4f molecules, mostly utilising Mn(III) or Fe(III) [6,7].

These SMM systems attract interest both from the viewpoint of fundamental science and in terms of their potential applications in areas, such as data storage and quantum computing. One of the key factors for such applications is the effect of finite intermolecular interactions that can lead to decoherence of the spin state [8] and, hence, a loss of information. While careful chemical control of the intermolecular environment can, in principle, reduce the contribution of exchange interactions to the decoherence, dipolar interactions are always present for such molecules and depend only on the spin orientation and the geometry of the lattice. Weak residual dipolar interactions between the spins on different molecules can, therefore, ultimately lead to dipolar ordering in SMM systems at low temperatures [9,10].

The Fe$_{19}$ molecular cluster SMM system [11], which is studied here, has the full formula [Fe$_{19}$(metheidi)$_{10}$(OH)$_{14}$O$_6$(H$_2$O)$_{12}$]NO$_3$·24H$_2$O. This provides an interesting example combining high spin with moderate anisotropy. The high spin makes it a good candidate to observe dipolar-driven long range magnetic ordering. The magnetic core of the molecule (Figure 1) has a disc-like geometry; therefore, it can be described as a magnetic nanodisc. The spin state has been identified as 35/2 [12,13], resulting from a ferrimagnetic intramolecular arrangement of the $S = 5/2$ spins from the Fe sites, with thirteen of the spins pointing in one direction and the remaining six spins antiparallel to these thirteen.

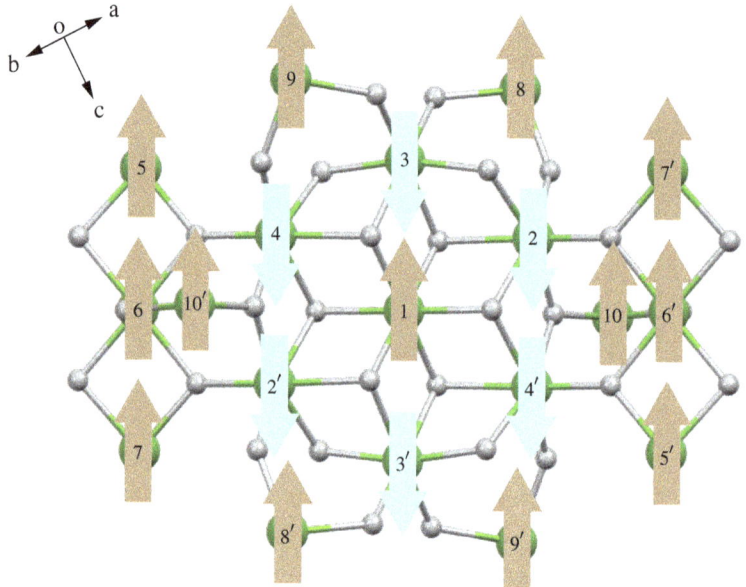

Figure 1. The core of the Fe$_{19}$ molecule viewed along the (110) crystal axis, showing only the Fe atoms and the bridging O atoms. The site labels for the Fe atoms are shown, and site 1 is a centre of inversion. The $S = 35/2$ ground state of the molecule is the result of strong intramolecular exchange interactions that place the six Fe sites surrounding the central Fe on one spin sublattice with the remaining thirteen Fe sites, including the central site on a second sublattice with opposite spin. The easy axis is shown as vertical in this figure for clarity, the direction previously suggested by EPR [13] is a vector from site 1 to site 2, whereas, in this study, we found a direction closer to the site 1 to site 10 vector.

Two variants of Fe$_{19}$ have been reported [11], both having the same core structure, but with slightly different ligands. We study here the metheidi form. Antiferromagnetic (AF) ordering at $T_N = 1.19$ K for this material was previously observed with the specific heat [14,15], and this was confirmed by μSR [16]. Evidence was also found for a static internal field and AF spin waves below 1 K using proton NMR [17].

Previous single crystal EPR studies [13] obtained anisotropy parameters for the spin Hamiltonian and identified an easy spin axis pointing in the Fe1–Fe2 direction; however, that study also found evidence for the presence of excited states that could not be fully characterised. On the basis of these reported ground state properties, it was shown that the observed ordering temperature was consistent with a predominantly dipolar driven scenario, and this produced a clear prediction that the ordering mode would have the wave vector $k = (0\ 0\ 0.5)$ [16]. The present studies were carried out firstly to test this prediction using neutron diffraction and secondly to determine the spectrum of excited spin states using inelastic neutron scattering.

2. Materials and Methods

Samples were prepared using the method given in the report of Goodwin et al. [11] using either H_2O or D_2O in the synthesis. The sample identity and purity was checked using powder X-ray diffraction at 300 K and confirmed to be consistent with the structure reported at 150 K [11]. The neutron diffraction measurements were carried out on the WISH instrument at the ISIS Neutron and Muon Source [18].

This instrument is optimised for the powder diffraction of systems with large unit cells, such as the molecular magnet studied here. Some of the particular features of the neutron powder diffraction for molecule-based systems were previously reviewed by Peter Day [19]. The sample was 750 mg of polycrystalline powder, and it was cooled in a helium-3 sorption cryostat, which allowed measurements to be made down to a temperature of 0.4 K. Data collection was made in a 5 Hz double frame mode, allowing access to a d spacing range up to 100 Å.

For the inelastic neutron scattering measurements, a partially deuterated sample was used in which the lattice water and hydroxide groups were deuterated but not the metheidi ligands. In this case, 86 out of 186 hydrogen atoms were replaced by deuterium, thus, producing the formula $[Fe_{19}(metheidi)_{10}(OD)_{14}O_6(D_2O)_{12}]NO_3 \cdot 24D_2O$, where metheidi is $C_7H_{10}NO_5$. The partial deuteration of the molecule allowed us to considerably reduce the incoherent scattering and to clearly observe the magnetic excitations up to a 5 meV energy transfer with only a small contribution from the vibrational modes at high Q.

The spectra were measured on the LET instrument at ISIS using a helium dilution refrigerator, which enabled measurements to be made in temperatures down to 35 mK. The LET instrument is a multi-chopper cold neutron spectrometer [20]. Data were taken with the focusing of the instrument centred on an incident energy of 2 meV with choppers operating at speeds of 280 Hz (chopper 5) and 70 Hz (chopper 3) and using the high resolution disk slot.

This configuration allows the simultaneous collection of data with incident energies E_i and nominal resolutions at the elastic line of 5.15 meV (90 μeV), 2 meV (24 μeV), and 1.06 meV (10 μeV). Energy transfer spectra were obtained by integrating over the momentum range 0 to 1.4 Å^{-1} for $E_i = 1.06$ meV, 0 to 1.8 Å^{-1} for $E_i = 2.00$ meV and 0 to 2.5 Å^{-1} for $E_i = 5.15$ meV. Data reduction was done using Mantid [21], and Wimda [22] was used for the global fitting of the data via its user modelling feature.

3. Results

The results are presented in two parts. First, we report the neutron diffraction results determining the low temperature cell parameters, the mode of magnetic ordering, and the orientation of the easy spin axis. The second part of the data reports the inelastic neutron results, which allow us to determine the internal magnetic field in the ordered state and the anisotropy parameters for the spin states. In addition to the $S = 35/2$ ground state multiplet, six excited state multiplets were identified from the data with spin values ranging from 31/2 to 39/2.

3.1. Diffraction

Cell parameters were obtained from the neutron data at 1.5 K and were found to be broadly consistent with those reported at 150 K with X-rays [11], after allowing for a modest thermal contraction that reduces the cell volume by around 2% (Table 1).

New magnetic diffraction peaks appeared below the AF ordering transition, as can be seen in Figure 2a, where four extra peaks are seen, and these are made clearer in the difference plots of Figure 2b,c. Information about the spin structure and the easy axis orientation may be obtained from the relative intensities of these magnetic peaks. The intensity of the magnetic peaks can be written here as a product of three terms

$$I(\mathbf{Q}) \propto (1 - (\hat{\mathbf{Q}} \cdot \hat{\mathbf{m}})^2) |F_M^{cell}(\mathbf{Q})|^2 |F_M^{mol}(\mathbf{Q})|^2. \quad (1)$$

The first term in Equation (1) is a factor from the component of the scattering vector $\mathbf{Q}$ that is perpendicular to the magnetic moment axis $\hat{\mathbf{m}}$. The second term in Equation (1) is the structure factor for magnetic ordering, which is summed over the lattice vectors $\mathbf{L}_i$ defining the magnetic cell, i.e.,

$$F_M^{cell}(\mathbf{Q}) = \sum_{cell} e^{i\mathbf{Q} \cdot \mathbf{L}_i}. \quad (2)$$

The third term in Equation (1) is the structure factor for the spatial arrangement of the up and down magnetic moments of the Fe sites within the molecule, as shown in Figure 1, i.e.,

$$F_M^{mol}(\mathbf{Q}) = \sum_{\uparrow} e^{i\mathbf{Q} \cdot \mathbf{r}_i} - \sum_{\downarrow} e^{i\mathbf{Q} \cdot \mathbf{r}_i}. \quad (3)$$

Note that no atomic form factors are needed in Equation (3), as we only have one type of magnetic site, and the Q values are too low to produce any difference in form factor between the magnetic peaks. The low symmetry triclinic cell also ensures that the magnetic peaks correspond to unique scattering planes, rather than to a superposition of magnetically inequivalent symmetry-related planes as is often found in higher symmetry structures.

Table 1. The cell parameters obtained with neutrons at 1.5 K compared with those previously obtained at 150 K with X-rays [11].

	150 K (X-rays)	1.5 K (Neutrons)
a (Å)	13.309 (3)	13.116 (1)
b (Å)	17.273 (5)	17.189 (1)
c (Å)	17.600 (4)	17.503 (2)
α (°)	65.201 (10)	65.397 (8)
β (°)	74.514 (16)	74.300 (8)
γ (°)	81.30 (2)	81.753 (7)
space group	$P\bar{1}$	$P\bar{1}$

The four magnetic peaks were found to be consistent only with an AF ordering whose propagation wave vector was (0.5 0 0.5). Thus, the (0 0 0.5) mode and all other modes that might be associated with purely dipolar-driven ordering are excluded (see Table 2). This indicates that there is a significant contribution from superexchange to the magnetic coupling between the molecules. The relative scale of the interactions will be discussed in Section 4.

The calculated intensities for the magnetic peaks are shown in Figure 2b for the $S = 35/2$ arrangement of spins in the molecule as defined in Figure 1, with the ordering mode set to (0.5 0 0.5) and $\hat{\mathbf{m}}$ aligned in the Fe1–Fe2 easy axis direction, which was reported in a previous EPR study [13]. These intensities are seen to be a relatively poor overall match to the measured data with the intensity of the (−0.5 1 0.5) peak being strongly underestimated. A significant improvement was obtained by allowing the easy axis to rotate toward the long axis of the molecule, and this provided a very good match with the intensities, as shown in Figure 2c.

Table 2. Dipolar stabilisation energies for the $S = 35/2$ ground state of Fe$_{19}$ calculated for different modes of magnetic ordering using, firstly, the previously proposed Fe1-Fe2 easy spin axis, for which (0 0 0.5) is the most stable mode, and secondly the axis determined from this study. A spherical sample is assumed for the case of the (0 0 0) ferromagnetic ordering mode. Negative values are unstable.

Ordering Mode			Dipolar Energy (K) $(-1.12, 1, 0.06)$ Easy Axis	Dipolar Energy (K) $(-0.45, 1, -0.34)$ Easy Axis
0	0	0.5	0.809	0.436
0.5	0.5	0	0.575	0.219
0	0	0	0.561	0.662
0.5	0.5	0.5	0.515	0.533
0.5	0	0.5	−0.271	−0.337
0	0.5	0	−0.425	−1.028
0	0.5	0.5	−0.556	−0.024
0.5	0	0	−0.792	−0.144

The easy axis vector obtained from fitting the data is shown in Table 3. We found this to be quite close to the Fe1–Fe10 axis and making an angle of 12° to that axis, whereas it is at 26° to the originally proposed Fe1–Fe2 axis. Although the molecule only has crystallographic inversion symmetry, the molecular symmetry conforms closely to 2/m, with the twofold axis defined by Fe3–Fe1–Fe3′. It is, thus, very likely that the orientation of the easy axis also conforms to this symmetry (either parallel to the twofold axis or perpendicular to it). The axis orientation found here, lying close to the molecular mirror plane (defined by Fe10–Fe2–Fe1–Fe2′–Fe10′) is, thus, more consistent with this expectation than that suggested by the earlier EPR study.

Table 3. For an easy axis vector specified in cell units (S_a S_b S_c), the direction is fully determined by two parameters. The orientation obtained here is compared with the Fe1–Fe10 direction and also the Fe1–Fe2 direction suggested by the previous EPR study [13]. The fitted orientation is close to the plane formed by the Fe1, Fe2, and Fe10 sites and is close to the Fe1–Fe10 axis as shown by the angle given in the final column.

	S_a	S_b	S_c	Angle (°)
This study	−0.45(2)	1	−0.34(2)	
Fe1–Fe10	−0.18	1	−0.42	12
Fe1–Fe2 (EPR)	−1.12	1	0.06	26

3.2. Inelastic Neutron Scattering

In addition to the appearance of new magnetic diffraction peaks, the magnetic phase transition can also be observed via changes in the low energy inelastic neutron scattering spectra with temperature. This provides a measurement of the effective internal field that appears below the transition. Using measurements at low energies, taken just above the transition at 1.2 K, the low lying excitations are determined. Further excitations are also found at higher energies, and the combined measurements allow the seven lowest energy spin states to be identified. The data obtained are analysed with a spin Hamiltonian for each spin state S taking the form

$$H = DS_z^2 + E(S_x^2 - S_y^2) + g\mu_B(S_x B_x + S_y B_y + S_z B_z), \tag{4}$$

where D and E are axial and rhombic anisotropy parameters defining the zero field splitting (z is the direction of the anisotropy axis) and B is a spontaneous static internal magnetic field that is only present in the magnetically ordered state. The energy levels E_{S,S_z} are determined by the eigenvalues of each spin state, offset by Δ, the energy gap of the lowest level of the spin state multiplet with respect to the $S = 35/2$, $S_z = 35/2$ ground state. When

B is present, a shift in the energy levels occurs, which is mainly determined by the z component of the field, which we will label B_0.

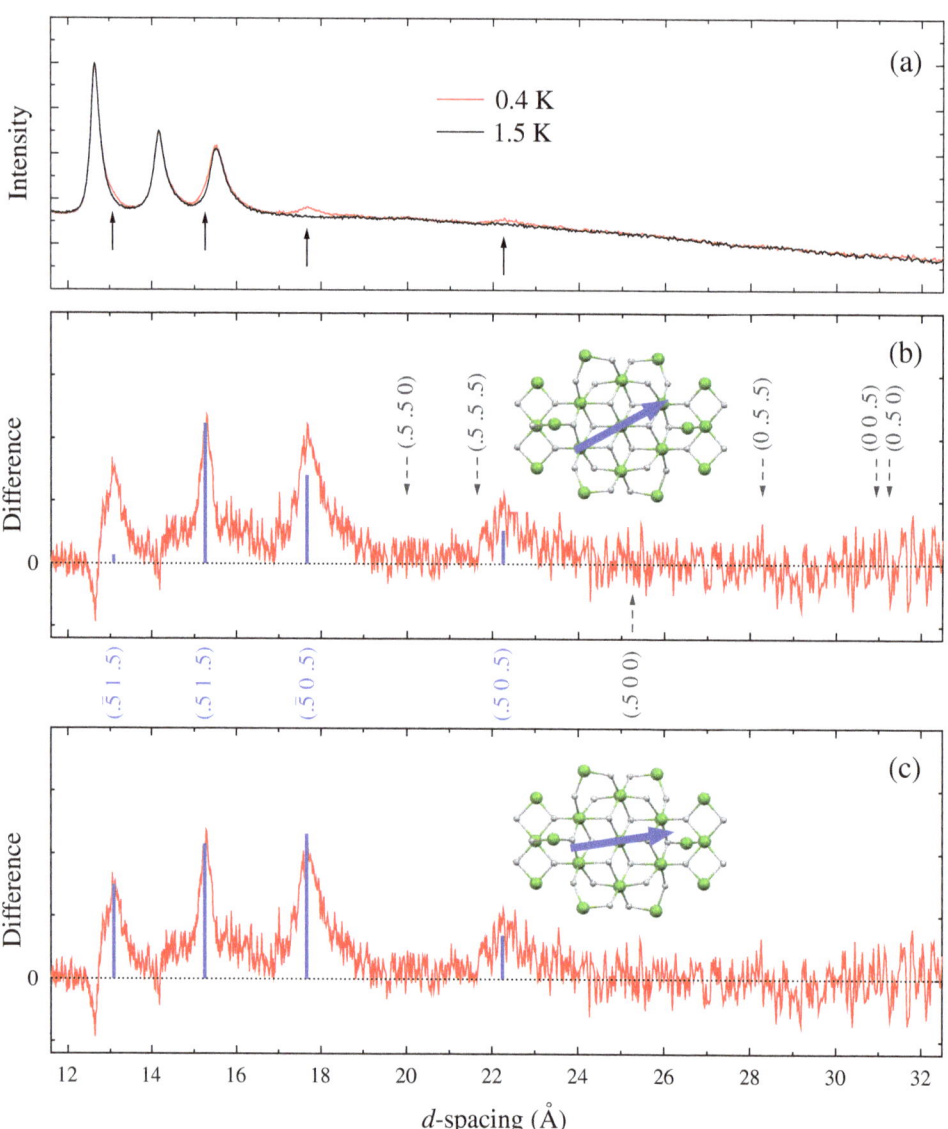

Figure 2. Neutron powder diffraction spectra measured on the WISH instrument. (**a**) Comparison of the data above and below the 1.2 K magnetic ordering temperature. The arrows highlight four new peaks that appear with d spacings in the region between 13 and 23 Å. (**b**) The difference spectrum shows the magnetic peaks more clearly. These new peaks are only consistent with an ordering that has the propagation vector (0.5 0 0.5). The positions of absent peaks that would be present for other modes of ordering are marked in grey. The predicted relative intensities of the observed peaks for (0.5 0 0.5) ordering are marked with blue lines, taking the Fe1–Fe2 orientation of the easy axis that was suggested by EPR [13] (shown as the blue arrow). (**c**) An improved match to the measured intensities is obtained with a revised spin orientation that is close to the Fe1–Fe10 axis of the molecule (see Table 3).

For the current simplified model, we have treated each S-state independently within the Giant-Spin-Approximation using Equation (4). We note that, besides the anisotropy contribution from crystal field and dipole–dipole interactions, S-state mixing [23] can also contribute to the effective D and E terms in the Spin Hamiltonian.

3.2.1. Internal Field

The temperature dependence of the position of the lowest energy transition is shown in Figure 3. The magnetic transition just below 1.2 K is clearly seen in this data as an upward shift in the excitation energy. This shift is due to the different splittings of the $S_z = 35/2$ and $S_z = 33/2$ levels in the effective internal field of the ordered state B_0, which is obtained from the energy shift as

$$B_0 = \frac{\Delta E}{g\mu_B}. \tag{5}$$

The energy shift reflects intermolecular interactions that can have contributions from both dipolar and exchange mechanisms. Since the dipolar contribution can be determined exactly from knowledge of the spin structure, the measured value of B_0 allows the exchange contribution to be precisely quantified. This will be explored further in Section 4.

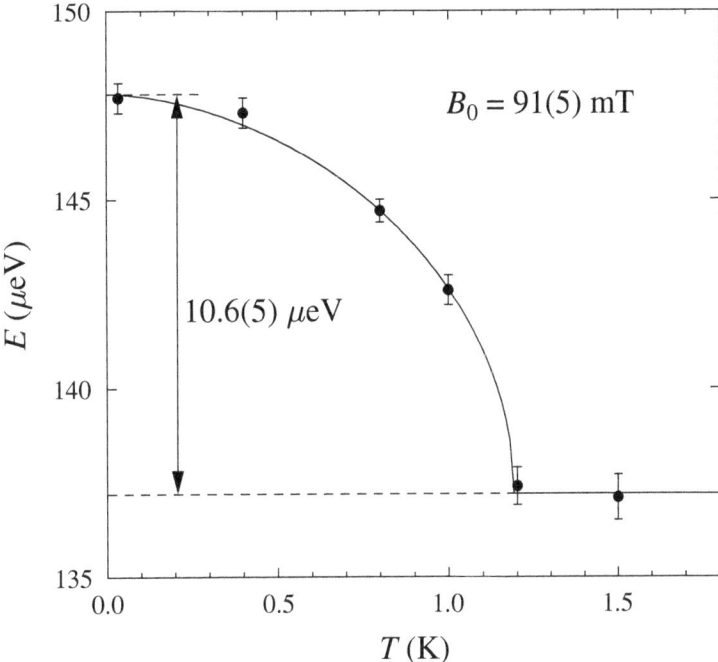

Figure 3. Temperature dependence of the lowest energy excitation of the $S = 35/2$ ground state (from $S_z = 35/2$ to $S_z = 33/2$). The measurements were made on the LET instrument with an incident energy of 1.06 meV, and the line is a guide to the eye. The internal field in the ordered state splits the levels and shifts the transition energy. The maximum value of the internal field B_0 was determined to be 91(5) mT.

3.2.2. Low Temperature Excitation Spectrum

The excitation spectrum measured at a temperature of 35 mK is shown in Figure 4a. This plot combines data for $E_i = 2.00$ meV and $E_i = 5.15$ meV. At this low temperature, all of the excitations originate from the ground state. There are four excitation bands labelled

I to IV in Figure 4a. Labelling the states as (S, S_z), band I is assigned to the transition from the $(35/2, 35/2)$ ground state to the next level in the $35/2$ spin state, i.e., $(35/2, 33/2)$ (Figure 4b). Gaussian fits of bands II to IV are shown in Figure 4a, with the parameters listed in Table 4.

Consideration of the selection rules for allowed magnetic dipole transitions can assist in the assignment of these bands. These are $\Delta S = 0 \pm 1$ and $\Delta S_z = 0 \pm 1$. Hence, a single transition is expected for $\Delta S = -1$, whereas two closely spaced transitions would be expected for $\Delta S = 0$ and three closely spaced transitions for $\Delta S = +1$. The relative intensities within the sets of closely spaced transitions depend on matrix elements whose ratios within a given multiplet can be determined via the Wigner–Eckart theorem, leading to the following expression for the intensity

$$I(\Delta S, \Delta S_z) \propto \begin{pmatrix} 35/2 & 1 & 35/2 + \Delta S \\ -35/2 & \Delta S_z & 35/2 - \Delta S_z \end{pmatrix}^2, \tag{6}$$

where the six element array is the Wigner 3-j symbol. The relative intensity values are shown in Table 5, where it can be seen that the lowest energy transition within the set is always the strongest. The second transition, when allowed, is reduced from the first transition by a factor of S_f and the third transition, when allowed, is smaller than the first transition by a factor of S_f^2. The transitions with high intensity according to Equation (6) and Table 5 are signified by the thicker lines in Figure 4b.

Since the fitted widths of the Gaussians are comparable for bands II and III and 50% larger for band IV, we take this as an indication of unresolved multiple transitions in band IV, leading to the assignments shown in Table 4 and Figure 4b with bands II and III corresponding to transitions to two $S = 35/2$ excited states and band IV corresponding to a superposition of two transitions having $\Delta S = \pm 1$. In Figure 4b, we illustrate one possibility for band IV with transitions to $S = 33/2$ (IVa) and to $S = 37/2$ (IVb). The four states associated with bands II to IV span a region of 25 to 33 K above the ground state.

The Q dependence (Figure 5) provides further support for this assignment [24–26], with the intensities of bands II and III tracking each other and remaining large at low Q, consistent with $\Delta S = 0$, whereas band IV is weaker at low Q, consistent with $\Delta S = \pm 1$. We also considered whether bands II and III could be two transitions into a single $S = 35/2$ excited state; however, their separation would require the magnitude of the D parameter to be more than double that of the ground state, which we believe to be less likely than having two separate $S = 35/2$ excited states.

Table 4. Parameters obtained for fitting the inelastic neutron transitions observed in the 2 to 3 meV region with Gaussian peaks. The final column provides the assignment for the transition band.

Band	Position (meV)	Width (meV)	Assignment
II	2.163(2)	0.120(3)	$\Delta S = 0$
III	2.492(3)	0.129(4)	$\Delta S = 0$
IV	2.841(7)	0.185(10)	$\Delta S = \pm 1$

Table 5. Intensity ratios for the magnetic dipole transitions from a ground state level $(S, S_z) = (35/2, -35/2)$ to the three possible excited spin states with spin S_f. Note that the intensity ratios only apply to the different ΔS_z transitions within a given multiplet. Comparing the intensities between different multiplets requires knowledge of their respective reduced matrix elements.

			ΔS_z	
ΔS	S_f	−1	0	1
−1	33/2			1
0	35/2		0.9459	0.0541
1	37/2	0.9474	0.0512	0.0014

3.2.3. Low Lying Thermally Accessible Spin Excitations

An EPR study [13] suggested the presence of an excited spin state at 8 K rather than at 25 K; however, we observed here no direct transitions in the associated 0.7 meV spectral region (Figure 4a). A thermal population of low lying excited states will produce additional low energy transitions, which allows such states to be identified, even when direct transitions from the ground state are not allowed.

Further information is contained in the relative intensities of the transitions, which are determined by the matrix elements for magnetic dipole transitions under Equation (4). The temperature dependence of the spectra in the band I region are shown in Figure 4c. These spectra were analysed using the energy levels and matrix elements derived from Equation (4), taking into account the thermal occupancies of the levels given by

$$n_{S,S_z} = \frac{e^{-\beta E_{S,S_z}}}{\sum_{S,S_z} e^{-\beta E_{S,S_z}}}, \tag{7}$$

where $\beta = 1/k_B T$. The relatively low anisotropy of the Fe_{19} system prevents us from being able to resolve the individual transitions within a spin state multiplet in the way that was possible for previous inelastic neutron studies on systems with larger anisotropy, such as Fe_8 [27] and Mn_{12} [28]. Nevertheless, the characteristic asymmetric shape of the broadened scattering profile still enables us to extract important parameters from the data.

A global fit of the spectra obtained between 1.2 and 10 K below 0.2 meV (Figure 4c) assigns the spectral features to the superposition of internal transitions within three spin states corresponding to the $S = 35/2$ ground state and two low lying states for which direct transitions from the $S = 35/2$ ground state are forbidden—namely $S = 31/2$ and $S = 39/2$. The spectral fits are shown as the blue curves in Figure 4c, and the parameters obtained from this analysis are shown in Table 6. The gaps obtained for these two states are 1.5 and 3 K, significantly lower than the 8 K gap to the $S = 33/2$ state suggested from the EPR data. The D parameter of the ground state at -51 mK is a little larger in magnitude than the value of -43 mK obtained from the EPR.

The $E/|D|$ ratios here were found to be around 0.2, four times larger than the ratio of 0.05 that was suggested in the EPR study. The primary effect of a large $E/|D|$ ratio is to increase the minimum level separation in the upper levels of the spin multiplet (Figure 4b). This shows up most clearly in the data as a relatively pronounced low energy cutoff for the 5 and 10 K spectra (Figure 4c). The value of χ^2 per degree of freedom for the global fit shown in Figure 4c is 1.65. If the excited states are not included at all, then the χ^2 value becomes increased by a factor of 8 compared to the two excited state fit. If only one excited state is included, then the χ^2 value increases by a factor of 1.5 compared to the two excited state fit.

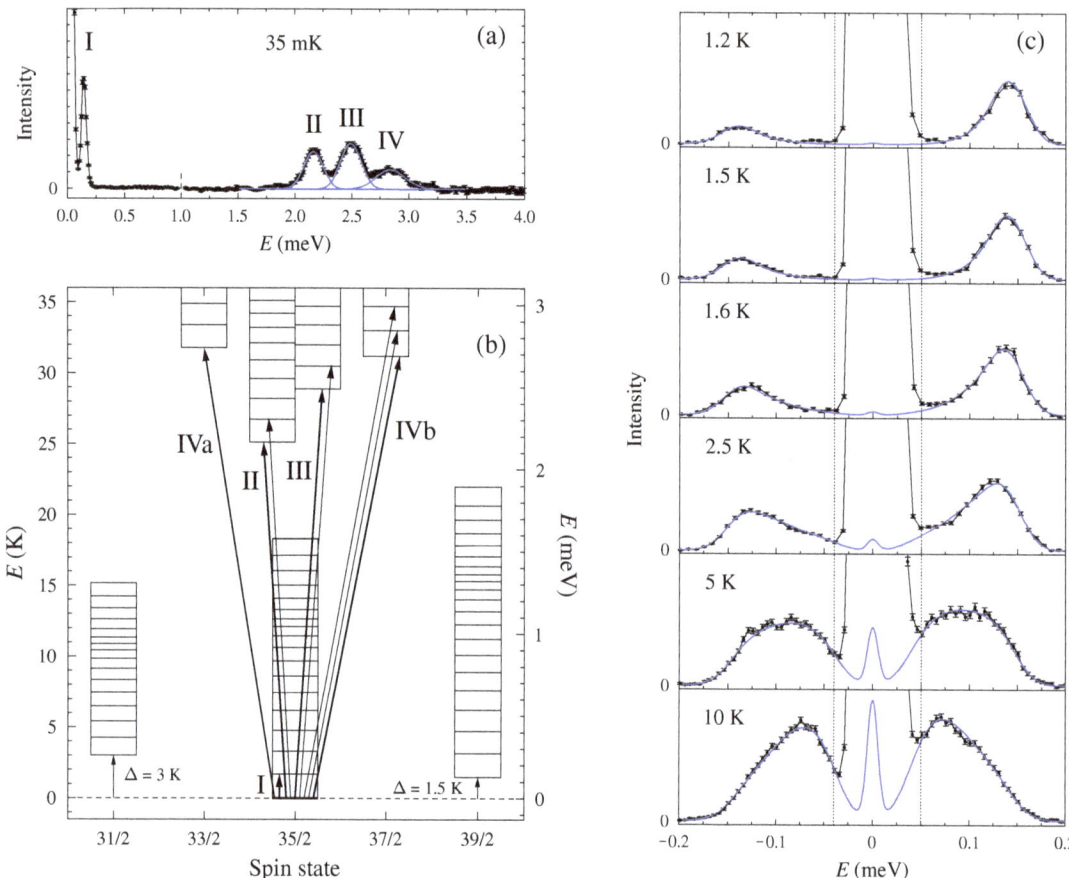

Figure 4. (a) Inelastic spectra measured on LET at 35 mK. Data below 1 meV are for incident energy E_i = 2.00 meV and data above 1 meV are for E_i = 5.15 meV (with a linear background subtracted). Gaussian fits to bands II to IV are indicated, with fit parameters given in Table 4. (b) The assignment of band I is the lowest $\Delta S = 0$ transition of the $S = 35/2$ multiplet. Bands II to IV are due to magnetic-dipole-allowed transitions with $\Delta S = 0, +1$, or -1. On the basis of width, amplitude, and Q dependence, bands II and III are assigned to two distinct $S = 35/2$ excited states and band IV to unresolved transitions into excited states with $\Delta S = \pm 1$. One possibility is shown here for band IV with transitions to the $S = 33/2$ and $S = 37/2$ states. (c) The temperature dependence of the low energy spectral region allows two low lying excited spin states to be identified. The blue lines show the fitted thermal dependence that is well described by three spin states, with the corresponding fit parameters given in Table 6.

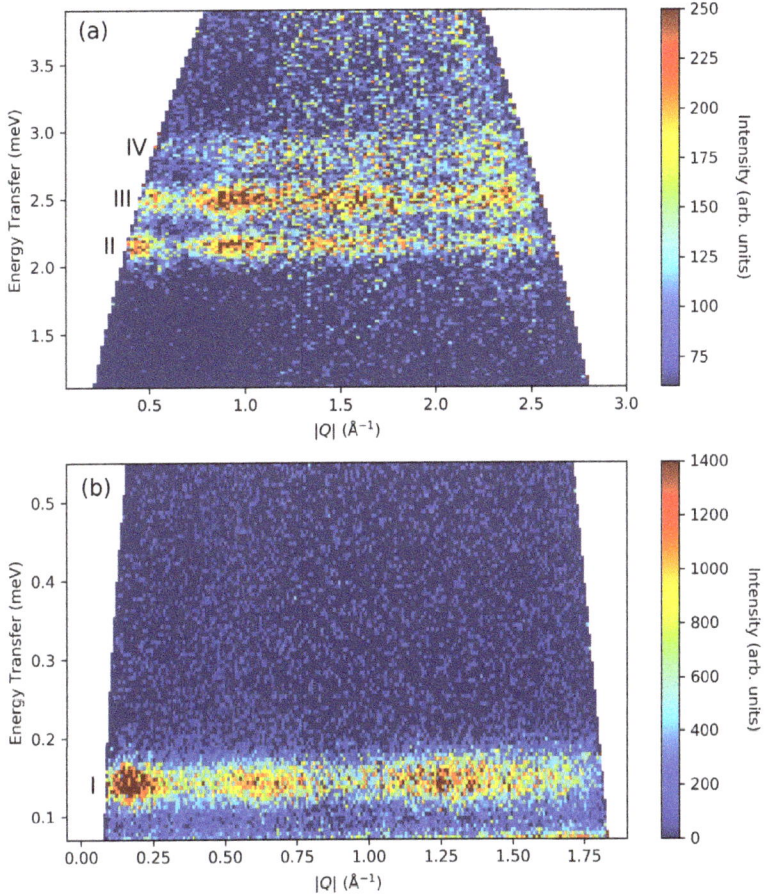

Figure 5. The Q dependence of the scattering intensity measured at 35 mK with an incident neutron energy of (**a**) 5.15 meV, showing bands II, III, and IV, and (**b**) 2.00 meV, showing band I.

Table 6. Parameters for the ground state and two lowest lying excited states obtained from a global fit of the temperature dependent inelastic neutron spectra (Figure 4c). The data were analysed with the zero field version of the model Hamiltonian, Equation (4). The offset of the minimum energy of the excited state with respect to the $S = 35/2$ ground state is shown in the final column as the excitation gap Δ.

Spin State S	D (mK)	$E/|D|$	Δ (K)
35/2	−51.0 (1)	0.20 (1)	0
39/2	−44.3 (2)	0.16 (1)	1.5 (2)
31/2	−42.6 (3)	0.22 (2)	3.0 (1)

4. Discussion

Having established the key properties of the Fe_{19} system, namely the mode of ordering, the easy spin axis, the internal field in the ordered state, the ground state, and the spectrum of excited spin states, we are now in a good position to assess the intermolecular magnetic interactions leading to the magnetic ordering. In particular, we can determine the relative importance of dipolar and superexchange interactions for this high spin system.

4.1. Dipolar Interactions versus Superexchange

The relative stabilities of different modes of magnetic ordering originating from only dipolar interactions are compared in Table 2. For an easy spin axis in the Fe1–Fe2 direction, the ordering mode with wave vector (0 0 0.5) would be expected for a transition driven purely by dipolar interactions [16]. The observation of the (0.5 0 0.5) ordering mode in this study however rules out that scenario, and the easy axis orientation was also found to be different. For the newly found easy axis, ferromagnetic ordering is the most stable dipolar driven mode (assuming a spherical sample) and the (0.5 0.5 0.5) mode is the most stable AF state (Table 2). The observed (0.5 0 0.5) mode is unstable for the revised easy axis with dipolar interactions alone. This points clearly towards the presence of significant AF superexchange interactions between the molecules.

To assess which interaction paths are the most likely to contribute, we identified the shortest Fe–Fe distances between adjacent molecules, and the six closest interactions are listed in Table 7. These are in a range from just below 9 Å to around 10 Å. The magnitude of the coupling will ultimately depend on the detailed electronic structure of the exchange path; however, the weakest link is most likely to be the hydrogen bond within the exchange path.

The shortest length for these direct intermolecular hydrogen bonds is given in the final column of Table 7, and these links are illustrated in Figure 6 for the b and c axes. From this, it can be seen that the c axis interaction is expected to be the weakest. In order to estimate the strength of the exchange interactions in relation to the dipolar coupling, we used a simple empirical method. This involves tuning the interactions to match the three experimental characteristics that we obtained, namely the internal field, the mode of ordering, and the transition temperature.

As a starting point, we consider the five interactions with an overall Fe–Fe distance around 9 Å or less (Table 7) and set them all to the same value of AF exchange coupling. The scaling factor of the coupling J_0 is then tuned so that the B_0 value of the internal field from the sum of the dipolar field in the easy axis direction and the exchange field matches the experimental value.

Table 7. The short overall Fe–Fe distances between neighbouring molecules in the lattice and the shortest intermolecular H-bonds not involving lattice water.

Intermolecular Vector	d_{Fe-Fe}(Å)	Linked Sites	Shortest H-Bond (Å)
b	8.655	Fe8–Fe10′, Fe10–Fe8′	2.388
b-c	8.741	Fe5–Fe9, Fe9–Fe5	2.387
a-b	8.798	Fe7–Fe7′	2.533
c	8.846	Fe5–Fe8′, Fe8–Fe5′	3.147
a	9.047	Fe7–Fe7′	2.344
a-c	10.081	Fe8–Fe9	2.481

When this initial uniform exchange coupling model is used, we confirmed that the magnetic ordering mode (0.5 0 0.5) was the most stable. The ordering temperature with this set of interactions was then estimated from a Monte Carlo simulation (see the fifth column in Table 8). The T_N found in this case at 2.7 K was more than two times larger than that observed experimentally. A high ratio of T_N to the characteristic coupling energy was found when the network of interactions had a three dimensional nature. The lower experimental value for this ratio for Fe_{19}, therefore, suggests that the interactions are actually weaker in at least one of the crystal axes.

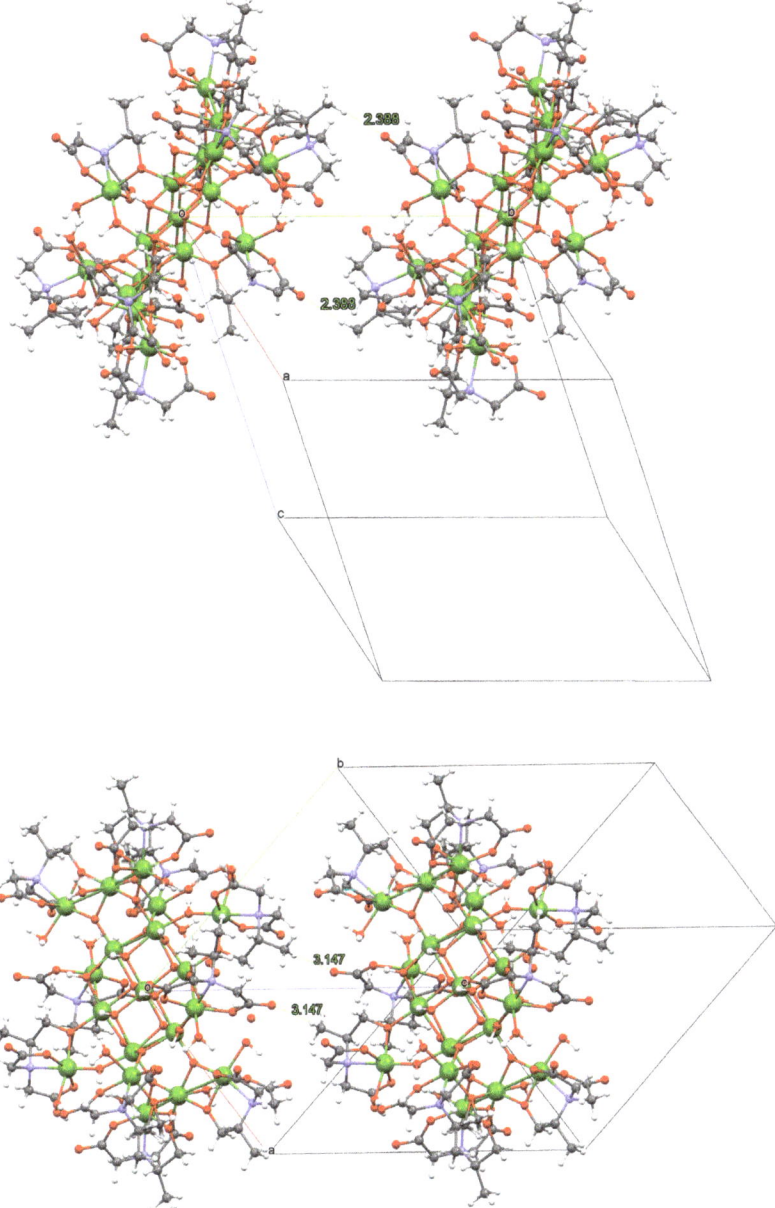

Figure 6. The shortest direct hydrogen bonds between molecules shown as green dotted lines for molecules displaced in the b direction (**top**) and the c direction (**bottom**). Oxygen atoms are shown as red and hydrogen atoms as light grey. The lengths are marked in Å (see Table 7).

As noted earlier, the direct H-bonded coupling along the c axis is particularly weak, and Appendix A shows that the indirect coupling via lattice water was also absent for the c axis. Consequently, we explore the effect of reducing the exchange interactions for directions involving the c axis in the second and third rows of Table 8. Reducing the interactions in both the c and $b-c$ directions by a factor of four is seen to bring T_N much closer to the experimental value; however, this also has the effect of making the (0.5 0.5 0.5) mode more stable than the (0.5 0 0.5) mode. By making J_c slightly smaller than J_{b-c}, the experimentally observed ordering mode of (0.5 0 0.5) is however restored (fourth row in Table 8). An additional relative strengthening of J_b and J_{a-b} versus J_a brings T_N down to the experimental value of 1.2 K (last row in Table 8).

Bearing in mind the large parameter space for the set of exchange couplings, the final parameters in Table 8 are not expected to be unique, but they provide a representative set that is consistent with the experimental data and they serve to give a good estimate of the order of magnitude of the exchange coupling in this system. The final result is that the stabilising contribution of the exchange terms to the magnetic energy was found to be eight-times larger in magnitude compared with the destabilising term from the dipolar coupling.

Table 8. The ordering temperature obtained from Monte Carlo simulation versus superexchange couplings between the nearest neighbour molecules. J_0 is obtained from the requirement to match the observed B_0 in the ordered state.

J_a (J_0)	J_{a-b}, J_b (J_0)	J_c, J_{b-c} (J_0)	J_0 (mK)	T_N (K)	Ordering Mode
1	1	1	2.6	2.7	(0.5 0 0.5)
1	1	0.5	4.0	2.0	(0.5 0 0.5)
1	1	0.25	5.2	1.4	(0.5 0.5 0.5)
1	1	0.22, 0.28	5.4	1.4	(0.5 0 0.5)
1	1.18	0.22, 0.28	5.4	1.2	(0.5 0 0.5)

5. Conclusions

Finding two spin states that are 1.5 and 3 K above the ground state energy is an important result of this study. The presence of such low lying states was suspected in previous studies, and their existence has complicated the interpretation of previous data on this system, leading to inconsistent conclusions from different studies that were measured with different techniques and at different temperatures. The present results should enable some clarification and reinterpretation of the properties of Fe_{19}.

Whilst the high spin value of Fe_{19} originally suggested that it was a strong candidate for having its AF ordering below 1.2 K driven by dipolar interactions, the present results clearly show that the magnetic ordering is primarily due to intermolecular exchange interactions on the order of 5 mK. These exchange interactions between neighbouring molecules operate via superexchange paths operating through ligands and hydrogen bonds. The hydrogen bonds reflect both direct intermolecular contacts and also lattice water mediated paths. The dipolar interactions in this case act to destabilise the ordering, and the magnitude of the dipolar energy is smaller than the magnitude of the exchange energy by around a factor of eight.

When considering the use of SMMs for quantum information processing, weak intermolecular interactions contribute to decoherence. The results of this study show that the exchange contribution to the intermolecular interaction can be surprisingly robust in this type of system even when the magnetic atoms of different molecules are well separated in the crystal structure by bulky spacer ligands and water molecules.

The present study also demonstrates that low temperature neutron scattering studies of high spin molecular systems provide a unique opportunity to study very weak residual superexchange interactions that can act over distances of order 10 Å. A better understanding of the subtleties of interactions in such molecule based system can be obtained via long range magnetic ordering. This understanding may also allow us to steer the organisation of molecular magnets into arrays that have sufficiently weak coupling to allow information processing.

Author Contributions: Sample preparation and characterisation J.T., C.E.A., and A.K.P.; neutron measurements and analysis F.L.P., T.G., and P.M.; writing—original draft preparation, F.L.P.; writing—review and editing, F.L.P., T.G., P.M., J.T., C.E.A., S.J.B. and A.K.P. All authors have read and agreed to the published version of the manuscript.

Funding: This research was supported by STFC via beam time allocations at the ISIS Neutron and Muon Source.

Institutional Review Board Statement: Not applicable.

Informed Consent Statement: Not applicable.

Data Availability Statement: Neutron diffraction data from the WISH instrument is available from https://doi.org/10.5286/ISIS.E.RB1510283 (accessed on 20 May 2021). Inelastic neutron data from the LET instrument is available from https://doi.org/10.5286/ISIS.E.RB1620177 (accessed on 20 May 2021).

Acknowledgments: We dedicate this paper to the memory of Peter Day, who was an early advocate of using neutrons to learn about molecular magnets and a great supporter of collaboration between physicists and chemists.

Conflicts of Interest: The authors declare no conflict of interest.

Abbreviations

The following abbreviations are used in this manuscript:

SMM	Single molecule magnet
AF	Antiferromagnetic
EPR	Electron paramagnetic resonance
μSR	Muon spin rotation
NMR	Nuclear magnetic resonance

Appendix A. Hydrogen-Bonded Exchange Network through the Lattice Water Sites

A likely scenario to consider is that AF intercluster interactions could be mediated by hydrogen bonding involving lattice water. Although each Fe_{19} cluster has several nearest neighbours in the crystal lattice, we are concerned here with those neighbours to which there is a clear pathway that can mediate such interactions. Inspection of the crystal structure shows that the shortest hydrogen-bonded pathways are of the types $Fe_3(\mu_3-OH)\cdots OH_2 \cdots O_{carb}$ or $Fe-OH_2 \cdots OH_2 \cdots O_{carb}$, in which either a triply-bridging hydroxo or a singly-bound aquo ligand makes a hydrogen bond to a lattice water, which, in turn, makes a further hydrogen bond to the outer oxygen (O_{carb}) of a carboxylate group coordinated to an iron centre in an adjacent cluster.

Three such pathways were identified. The first of these links a cluster with its symmetry equivalent at {1 + x, y, z} corresponding to an interaction in the a-direction (Figure A1). The three (μ_3–OH) ligands on one face of the disc each make hydrogen bonds to one lattice water, and each of these forms a hydrogen bond to the same outer carboxylate oxygen on the next molecule. A further such system of hydrogen bonds, related to the first by inversion, links the two clusters to two further clusters in the opposite direction along a with the overall effect being to link cluster molecules into chains running parallel to the a direction through the crystal.

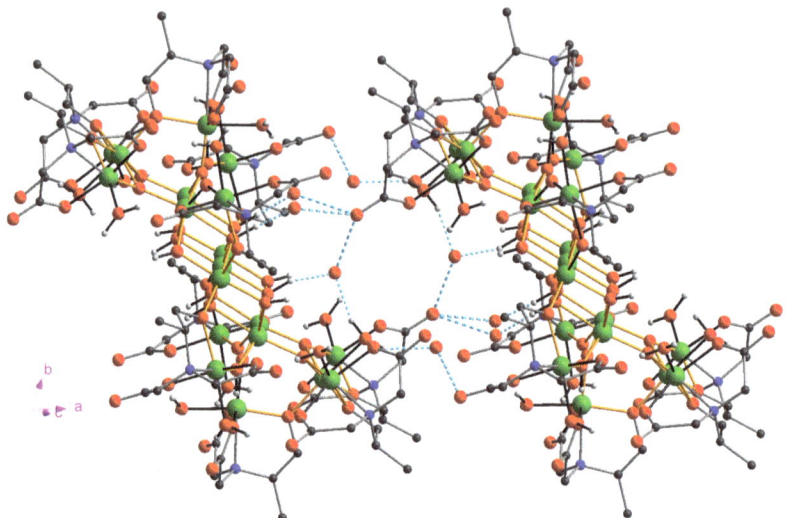

Figure A1. Intercluster hydrogen-bonding (shown as light blue dashed lines) between the Fe_{19} cluster and its symmetry-equivalent at {x + 1, y, z}. Note that H-atoms on the lattice waters were not located in the structural determination.

Two further pathways involve aquo ligands. The pathway between a cluster and its neighbour at {x, y + 1, z}, mediating interactions in the *b* direction, is much simpler than the *a*-axis case; two inversion-related lattice waters each accept a hydrogen bond from an aquo ligand while making a further hydrogen bond to a carboxylate oxygen of the adjacent cluster (Figure A2).

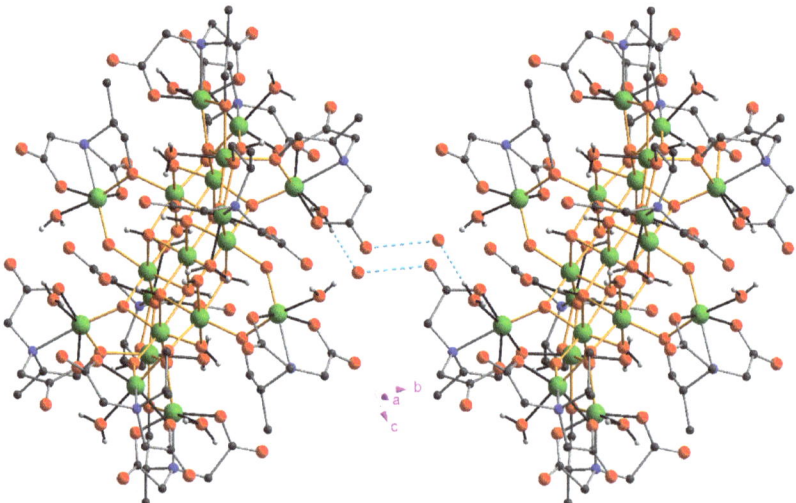

Figure A2. Intercluster hydrogen-bonding (shown as light blue dashed lines) between the Fe_{19} cluster and its symmetry-equivalent at {x, y + 1, z}.

The interaction between a cluster and its equivalent at {x + 1, y, z − 1}, corresponding to the vector **a-c**, is also simple (Figure A3), with two inversion-related lattice waters each forming hydrogen bonds to carboxylate oxygens from adjacent clusters. The relative magnitudes of the AF superexchange interactions mediated by these three pathways are, thus, expected to be $J_a \gg J_b > J_{a\text{-}c}$.

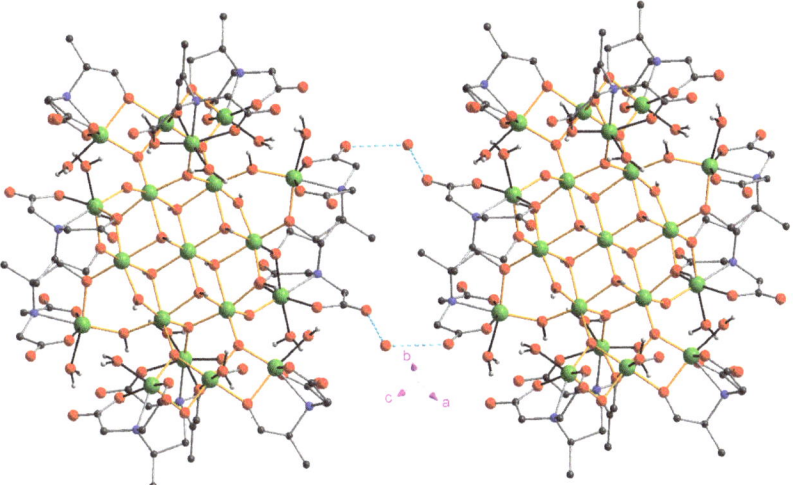

Figure A3. Intercluster hydrogen-bonding (shown as light blue dashed lines) between the Fe$_{19}$ cluster and its symmetry-equivalent at {x + 1, y, z − 1}.

The simple picture that is explored here of lattice water paths dominating the intermolecular exchange is not, however, consistent with the observed ordering. The strong AF interactions along the *a* direction, when taken in combination with the dipolar coupling, stabilise the (0.5 0.5 0.5) mode of ordering rather than the observed (0.5 0 0.5) mode. Introducing the weaker lattice-water-mediated interactions along the *b* and *a-c* directions further stabilises the (0.5 0.5 0.5) mode against the actual (0.5 0 0.5) mode. We, therefore, conclude that additional direct intermolecular interactions that do not involve the lattice water also provide a significant contribution to the exchange couplings.

References

1. Gatteschi, D.; Sessoli, R.; Villain, J. *Molecular Nanomagnets*; Oxford University Press: New York, NY, USA, 2006.
2. Blundell, S.J.; Pratt, F.L. Organic and molecular magnets. *J. Phys. Condens. Matter* **2004**, *16*, R771–R828. [CrossRef]
3. Sessoli, R.; Gatteschi, D.; Caneschi, A.; Novak, M.A. Magnetic bistability in a metal-ion cluster. *Nature* **1993** *365*, 141–143. [CrossRef]
4. Milios, C.J.; Vinslava, A.; Wernsdorfer, W.; Moggach, S.; Parsons, S.; Perlepes, S.P.; Christou, G.; Brechin, E.K. A Record Anisotropy Barrier for a Single-Molecule Magnet. *J. Am. Chem. Soc.* **2007** *129*, 2754–2755. [CrossRef]
5. Moushi, E.E.; Stamatatos, T.C.; Wernsdorfer, W.; Nastopoulos, V.; Christou, G.; Tasiopoulos, A.J. A Mn17 Octahedron with a Giant Ground-State Spin: Occurrence in Discrete Form and as Multidimensional Coordination Polymers. *Inorg. Chem.* **2009** *48*, 5049–5051. [CrossRef]
6. Mishra, A.; Wernsdorfer, W.; Abboud, K.A.; Christou, G. Initial Observation of Magnetization Hysteresis and Quantum Tunneling in Mixed Manganese-Lanthanide Single-Molecule Magnets. *J. Am. Chem. Soc.* **2004** *126*, 15648–15649. [CrossRef]
7. Ghulam Abbas, G.; Lan, Y.; Mereacre, V.; Wernsdorfer, W.; Clerac, R.; Buth, G.; Sougrati, M.T.; Grandjean, F.; Long, G.J.; Anson, C.E.; et al. Magnetic and ^{57}Fe Mössbauer Study of the Single Molecule Magnet Behavior of a Dy$_3$Fe$_7$ Coordination Cluster. *Inorg. Chem.* **2009** *48*, 9345–9355. [CrossRef]
8. Wedge, C.J.; Timco, G.A.; Spielberg, E.T.; George, R.E.; Tuna, F.; Rigby, S.; McInnes, E.J.L.; Winpenny, R.E.P.; Blundell, S.J.; Ardavan, A. Chemical engineering of molecular qubits. *Phys. Rev. Lett.* **2012**, *108*, 107204. [CrossRef] [PubMed]
9. Morello, A.; Mettes, F.L.; Luis, F.; Fernández, F.J.; Krzystek, J.; Aromí, G.; Christou, G.; de Jongh, L.J. Long-range dipolar ferromagnetic ordering of high-spin molecular clusters. *Phys. Rev. Lett.* **2003**, *90*, 017206. [CrossRef] [PubMed]
10. Burzuri, E.; Luis, F.; Barbara, B.; Ballou, R.; Ressouche, E.; Montero, O.; Campo, J.; Maegawa, S. Magnetic dipolar ordering and quantum phase transition in an Fe$_8$ molecular magnet. *Phys. Rev. Lett.* **2011**, *107*, 097203. [CrossRef] [PubMed]

11. Goodwin, J.C.; Sessoli, R.; Gatteschi, D.; Wernsdorfer, W.; Powell, A.K; Heath, S.L. Towards nanostructured arrays of single molecule magnets: New Fe_{19} oxyhydroxide clusters displaying high ground state spins and hysteresis. *J. Chem. Soc. Dalton Trans.* **2000**, 1835–1840. [CrossRef]
12. Ruiz, E.; Rodríguez-Forte, A.; Cano, J.; Alvarez, S. Theoretical study of exchange coupling constants in an Fe_{19} complex. *J. Phys. Chem. Sol.* **2004**, *65*, 799–803. [CrossRef]
13. Castelli, L.; Fittipaldi, M.; Powell, A.K.; Gatteschi, D.; Sorace, L. Single crystal EPR study at 95 GHz of a large Fe based molecular nanomagnet: toward the structuring of magnetic nanoparticle properties. *Dalton Trans.* **2011**, *40*, 8145–8155. [CrossRef] [PubMed]
14. Affronte, M.; Lasjaunias, J.C.; Wernsdorfer, W.; Sessoli, R.; Gatteschi, D.; Heath, S.L.; Fort, A.; Rettori, A. Magnetic ordering in a high-spin Fe_{19} molecular nanomagnet. *Phys. Rev. B* **2002**, *66*, 064408. [CrossRef]
15. Affronte, M.; Sessoli, R.; Gatteschi, D.; Wernsdorfer, W.; Lasjaunias, J.C.; Heath, S.L.; Powell, A.K.; Fort, A.; Rettori, A. Effects of intercluster coupling in high spin molecular magnets. *J. Phys. Chem. Sol.* **2004**, *65*, 745–748. [CrossRef]
16. Pratt, F.L.; Micotti, E.; Carretta, P.; Lascialfari, A.; Arosio, P.; Lancaster, T.; Blundell, S.J.; Powell, A.K. Dipolar ordering in a molecular nanomagnet detected using muon spin relaxation. *Phys. Rev. B* **2014**, *89*, 144420. [CrossRef]
17. Belesi, M.; Borsa, F.; Powell, A.K. Evidence for spin-wave excitations in the long-range magnetically ordered state of a Fe_{19} molecular crystal from proton NMR. *Phys. Rev. B* **2006**, *74*, 184408. [CrossRef]
18. Chapon, L.C.; Manuel, P.; Radaelli, P.G.; Benson, C.; Perrott, L.; Ansell, S.; Rhodes, N.J.; Raspino, D.; Duxbury, D.; Spill, E.; et al. Wish: The New Powder and Single Crystal Magnetic Diffractometer on the Second Target Station. *Neutron News* **2011**, *22*, 22–25. [CrossRef]
19. Day, P. Neutron powder diffraction in molecule-based magnetic materials: Long and short-range magnetic order. *Inorg. Chim. Acta* **2008**, *361*, 3365–3370. [CrossRef]
20. Bewley, R.I.; Taylor, J.W.; Bennington, S.M. LET, a cold neutron multi-disk chopper spectrometer at ISIS. *Nucl. Instrum. Methods Phys. Res. A* **2011**, *637*, 128–134. [CrossRef]
21. Arnold, O.; Bilheux, J.C.; Borreguero, J.M.; Buts, A.; Campbell, S.I.; Chapon, L.; Doucetc, M.; Draperab, N.; Leald, R.F.; Gigg, M.A.; et al. Mantid—Data analysis and visualization package for neutron scattering and μSR experiments. *Nucl. Instrum. Methods Phys. Res. Sect. A Accel. Spectrometers Detect. Assoc. Equip.* **2014**, *764*, 156–166. [CrossRef]
22. Pratt, F.L. WIMDA: A muon data analysis program for the Windows PC. *Physica B* **2000**, *289–290*, 710–714. [CrossRef]
23. Liviotti, E.; Carretta, S.; Amoretti, G. S-mixing contributions to the high-order anisotropy terms in the effective spin Hamiltonian for magnetic clusters. *J. Chem. Phys.* **2002**, *117*, 3361–3368. [CrossRef]
24. Chaboussant, G.; Sieber, A.; Ochsenbein, S.; Güdel, H.-U.; Murrie, M.; Honecker, A.; Fukushima, N.; Normand, B. Exchange interactions and high-energy spin states in Mn_{12}-acetate. *Phys. Rev. B* **2004**, *70*, 104420. [CrossRef]
25. Furrer, A.; Güdel, H.U. Neutron inelastic scattering from isolated clusters of magnetic ions. *J. Magn. Magn. Mater.* **1979**, *14*, 256–264. [CrossRef]
26. Chiesa, A.; Guidi, T.; Carretta, S.; Ansbro, S.; Timco, G.A.; Vitorica-Yrezabal, I.; Garlatti, E.; Amoretti, G.; Winpenny, R.E.P.; Santini, P. Magnetic Exchange Interactions in the Molecular Nanomagnet Mn_{12}. *Phys. Rev. Lett.* **2017** *119*, 217202 [CrossRef]
27. Caciuffo, R.; Amoretti, G; Murani, A. Sessoli, R.; Caneschi, A.; Gatteschi, D. Neutron spectroscopy for the Magnetic Anisotropy of Molecular Clusters. *Phys. Rev. Lett.* **1998**, *81*, 4744–4747. [CrossRef]
28. Mirebeau, I.; Hennion, M.; Casalta, H.; Andres, H.; Güdel, H.U.; Iridova, A.V.; Caneschi, A. Low-Energy Magnetic Excitations of the Mn_{12}-Acetate Spin Cluster Observed by Neutron Scattering. *Phys. Rev. Lett.* **1999**, *83*, 628–631. [CrossRef]

Article

New Spin-Crossover Compounds Containing the [Ni(mnt)] Anion (mnt = Maleonitriledithiolate)

Scott S. Turner [1,*], Joanna Daniell [1], Hiroki Akutsu [2], Peter N. Horton [3], Simon J. Coles [3] and Volker Schünemann [4]

1. Department of Chemistry, University of Surrey, Guildford GU2 7XH, UK; j.daniell@surrey.ac.uk
2. Department of Chemistry, Graduate School of Science, Osaka University, 1-1 Machikaneyama, Toyonaka, Osaka 560-0043, Japan; akutsu@chem.sci.osaka-u.ac.jp
3. EPSRC National Crystallographic Service, School of Chemistry, Faculty of Engineering and Physical Sciences, University of Southampton, Southampton SO17 1BJ, UK; P.N.Horton@soton.ac.uk (P.N.H.); S.J.Coles@soton.ac.uk (S.J.C.)
4. Technische Universität Kaiserslautern, Erwin Schrödinger-Str 46, D-67663 Fachbereich Physik, Germany; schuene@physik.uni-kl.de
* Correspondence: s.s.turner@surrey.ac.uk

Citation: Turner, S.S.; Daniell, J.; Akutsu, H.; Horton, P.N.; Coles, S.J.; Schünemann, V. New Spin-Crossover Compounds Containing the [Ni(mnt)] Anion (mnt = Maleonitriledithiolate). *Magnetochemistry* 2021, 7, 72. https://doi.org/10.3390/magnetochemistry7050072

Academic Editor: Marius Andruh

Received: 26 April 2021
Accepted: 14 May 2021
Published: 19 May 2021

Publisher's Note: MDPI stays neutral with regard to jurisdictional claims in published maps and institutional affiliations.

Copyright: © 2021 by the authors. Licensee MDPI, Basel, Switzerland. This article is an open access article distributed under the terms and conditions of the Creative Commons Attribution (CC BY) license (https://creativecommons.org/licenses/by/4.0/).

Abstract: Two novel salts containing the anion $[Ni(mnt)_2]^-$ (mnt = maleonitriledithiolate) have been synthesized. The counter-ions, $[Fe(II)(L^1 \text{ or } L^2)_2]$, are cationic complexes where L^1 and L^2 are methylated derivatives of 2,6-bis(pyazolyl)pyridine or pyrazine, which are similar to ligands found in a series of spin-crossover (SCO) complexes. Both salts are characterized by variable temperature single crystal X-ray diffraction and bulk magnetization measurements. Compound **1**, $[Fe(II)(L^1)_2][Ni(mnt)_2]_2$ displays an incomplete and gradual SCO up to 300 K, followed by a more rapid increase in the high-spin fraction between 300 and 350 K. Compound **2**, $[Fe(II)(L^2)_2][Ni(mnt)_2]_2 \cdot MeNO_2$, shows a gradual, but more complete SCO response centered at 250 K. For compound **2**, the SCO is confirmed by variable temperature Mössbauer spectroscopy. In both cases, the anionic moieties are isolated from each other and so no electrical conductivity is observed.

Keywords: spin-crossover; molecular magnets; magnetic materials; molecular materials

1. Introduction

The phenomenon of spin-crossover (SCO) has been known since the 1930s, following the serendipitous work by Cambi et al. [1,2]. Contemporary interest in SCO is due to potential applications in data storage devices [3], sensors and displays. Their utility relies on electronic instability between high-spin (HS) and low-spin (LS) configurations in transition metal complexes with octahedral geometries and 4 to 7 valence d-electrons. Exploitable switching properties can be associated with SCO, for example color, magnetization [4,5], dielectric constant [6,7], photo-physical properties [8–10], electrical conductivity [11] and structural parameters [12]. The most studied complexes contain Fe(II) and N-donor heterocyclic ligands, since spin state conversion in these compounds tend to give relatively large changes in properties [13,14]. In these cases, HS-LS switching also involves the maximal change in electron multiplicity, $S = 0$ (LS) to $S = 2$ (HS). A popular series of ligands in Fe(II) complexes, relevant to this work, has been 2,6-bis(pyrazol-1-yl)pyridine (dpp) and its derivatives, pioneered by Halcrow et al. [15–17].

In addition, SCO cationic complexes are attractive components of multifunctional hybrid salts when combined with anions that bring other properties, such as electrical conductivity or magnetic ordering [18]. This goes together with the development of molecular conductors where the conducting network is made from anionic moieties [19–21]. The goal is to obtain a molecular conductor that also exhibits SCO, such that the induced structural changes affect the materials ability to conduct electricity. Indeed, it is well-known that the electrical properties of molecular conductors are sensitive to small modifications of their crystal structure. There have been some successes in this area. For example, Fe(III) complexes of salycyl-type ligands [Fe(III)(sal$_2$-trien)][Ni(dmit)$_2$]$_n$ with n = 1 [22–24] were the first materials to contain [Ni(dmit)] anions and SCO cations, although the compound did not conduct electricity, and when n = 3 SCO, it disappears with the appearance of semi-conductivity [25]. The same group also made [Fe(III)(salten)-Mepepy][Ni(dmit)$_2$]$_3$, which had a gradual SCO associated with photoisomerization of the Mepepy ligand and moderated conductivity [26]. Takahashi et al. first observed a clear synergy between SCO and electrical conduction [27] in the related compound [Fe(III)(qsal)$_2$][Ni(dmit)$_2$]$_3$·CH$_3$CN·H$_2$O, qsalH = N-(8-quinolyl)-salicyladimine. Generally, these early approaches focused on dmit-based anionic components with Fe(III) SCO complex cations, whereas, more recently, mnt-based anions (Scheme 1) have been used. Most notably, a compound closely related to those in this work, [Fe(II)(dpp)$_2$][Ni(III)(mnt)$_2$]$_2$·MeNO$_2$ [28] was characterized by multi-stage spin-state conversions with mixed HS-LS states. A related Fe(II)/Ni(III) compound, [Fe(dppTTF)$_2$][Ni(mnt)$_2$]$_2$(BF$_4$)$_2$·PhCN, included a derivatized dpp with tetrathifulvalene (TTF) resulting in synergy between SCO and electrical conductivity [29]. In the current work, we focus on Ni(III) mnt-based anions with Fe(II) complex cations with methylated derivatives of dpp (Scheme 1, X = C) and the related pyrazine (X = N). This expansion is useful to investigate if the chemical flexibility, when derivatizing the dpp ligand, allows or disrupts SCO when metal complex anions are used.

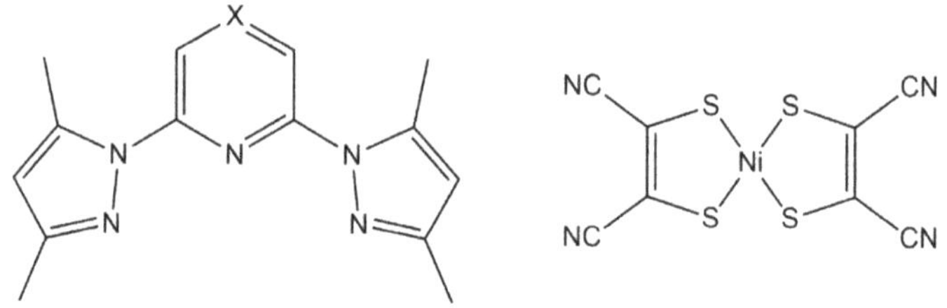

Scheme 1. Ligands (left) used in this work, where X = C (ligand **L^1**) and X = N (**L^2**). Mono-anionic complex, [Ni(III)(mnt)$_2$], (right) used in this work.

2. Results and Discussion

2.1. Description of Structures of Compounds 1 and 2

Crystals amenable to single X-ray diffraction were obtained of compound **1**, [Fe(II)(**L^1**)$_2$][Ni(mnt)$_2$]$_2$ and compound **2**, [Fe(II)(**L^2**)$_2$][Ni(mnt)$_2$]$_2$·MeNO$_2$, as detailed in Sections 3.1 and 3.2. For both compounds, the structures were solved at 100 and ca. 290 K. Both compounds crystalize in the P-1 (No. 2) space group. Table 1 contains the relevant crystallographic data and collection parameters. Figures 1 and 2 show the molecular structures of both compounds at 100 K, with thermal ellipsoids and the atom numbering scheme referred to in the text. The asymmetric unit of compound **1** contains one crystallographically independent Fe(II) cation, two [Ni(mnt)$_2$]$^-$ anions and no included solvent. By contrast, compound **2** contains the same 1:2 Fe(II): Ni(III) metal complex ratio with the additional inclusion of one CH$_3$NO$_2$ molecule.

For compound **1**, the coordination geometry about the Fe(II) cation has a distorted octahedral FeN$_6$ environment. From the point of view of SCO, it is important to look at the Fe-N bond lengths, which reflect the spin-state [12]. In compound **1**, at 100 K, these bond lengths range from 1.889 to 1.987 Å (average 1.957 Å). At 293 K, there is little change, with the bond lengths varying between 1.891 and 1.987 Å (average 1.954 Å). At both temperatures, this is indicative of a LS-state Fe(II) center. In addition, a change in the distortion from ideal local octahedral geometry can indicate a spin-state change. The OctoDist program [30] was used to determine the distortion parameters. The first parameter, Σ, is the sum of the deviations from 90° of the twelve *cis* N-Fe-N angles. A second parameter, Θ, is the sum of the deviations from 60° of the twenty-four N-Fe-N angles, six per pseudo three-fold axis, measured on a projection of opposite triangular faces of the FeN$_6$ octahedron, orientated by superimposing the face centroids. Typically, Σ and Θ are lower for LS and higher for more distorted HS complexes [12]. For compound **1**, Σ is 87.09° at 100 K and 87.16° at 293 K, essentially unchanged. Similarly, Θ is 287.00° at 100 K and 285.14° at 293 K is also unchanged. These structural parameters indicate LS complexes at both temperatures and do not reflect the SCO found from magnetization data (Section 2.2), most likely due to the substantially incomplete nature of the transition.

Table 1. Crystallographic data and collection parameters for compounds **1** and **2**.

Parameter	Compound 1		Compound 2	
Empirical formula	$C_{46}H_{34}N_{18}S_8FeNi_2$		$C_{45}H_{35}N_{21}S_8O_2FeNi_2$	
Molecular Mass	1268.66		1331.64	
T/K	100(2)	293(2)	100(2)	290(2)
CCDC number	2080108	2080109	2080110	2080111
Crystal color & shape	Brown Plate		Black Prism	
Crystal system	Triclinic		Triclinic	
Space Group	P-1 (2)		P-1 (2)	
a/Å	11.1318(3)	11.2229(5)	13.3643(6)	13.4983(8)
b/Å	12.1322(4)	12.3046(6)	14.7900(6)	15.1017(10)
c/Å	19.4594(6)	19.6208(7)	14.9875(8)	15.1633(9)
α/°	94.543(2)	93.625(3)	79.731(6)	79.756(6)
β/°	90.216(2)	89.349(3)	81.844(6)	81.882(6)
γ/°	92.986(2)	90.932(4)	69.962(5)	68.796(3)
Volume/Å^3	2616.16(14)	2703.6(2)	2728.01(15)	2825.8(3)
Wavelength/Å	0.71075		0.71075	
Radiation Type	Mo Kα		Mo Kα	
Z	2	2	2	2
μ/mm^{-1}	1.360	1.316	1.312	1.267
Measured reflections	33346	35619	26321	27746
Independent reflections	11866	12326	12418	12853
Reflections I $\geq$ 2 σ(I)	7764	6982	7709	6893
wR_1 (all data)	0.0956	0.1109	0.0828	0.1435
R_1, I $\geq$ 2 σ(I)	0.0496	0.0528	0.0456	0.0772

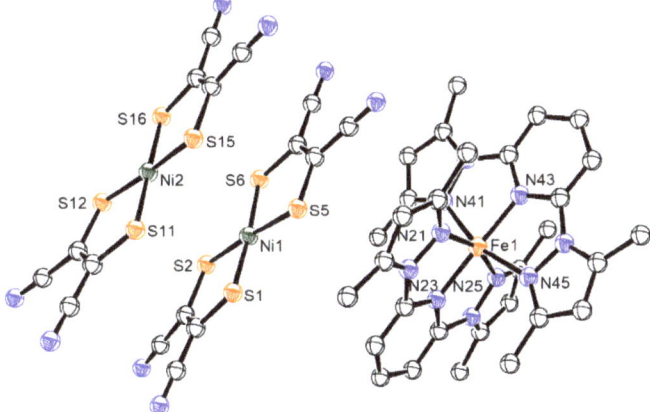

Figure 1. The molecular structure of compound **1** at 100 K, in the ORTEP [29] style with 50% thermal ellipsoids and atom numbering scheme. H atoms have been omitted for clarity.

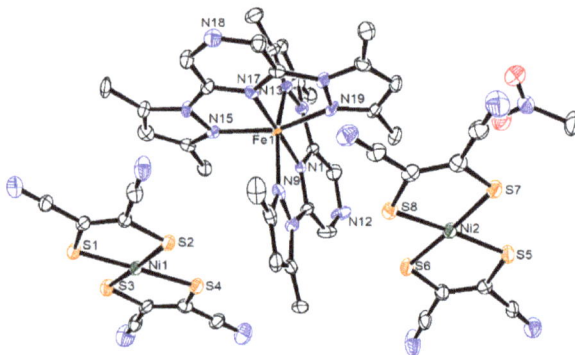

Figure 2. The molecular structure of compound **2** at 100 K, in the ORTEP [29] style with 50% thermal ellipsoids and atom numbering scheme. H atoms have been omitted for clarity.

At both temperatures, the overall structure of compound **1** consists of alternating layers of cations and anions when viewed down the crystallographic *ab* plane, as seen in Figure 3. The anionic Ni layer consists of discrete dimers that occupy cavities created by the cations, with close S ... S contacts between dimers. At 100 K, these are 3.393 Å between S1 and S11, and 3.526 Å between S6 and S16 with significantly longer distances to the next nearest dimer. The anionic charge, and by extension the Ni oxidation state, can be correlated with the geometry, specifically the Ni-S bond lengths since these are affected by the electron population of the molecular orbitals. For compound **1**, the two crystallographically independent anions have similar bond lengths that are compatible with mono-anions and Ni(III). The bond lengths range from 2.135 to 2.143 Å at 293 K and 2.139 to 2.151 Å at 100 K, whereas large bond lengths of ca. 2.17 Å are found for $[Ni(II)(mnt)_2]^{2-}$ [31].

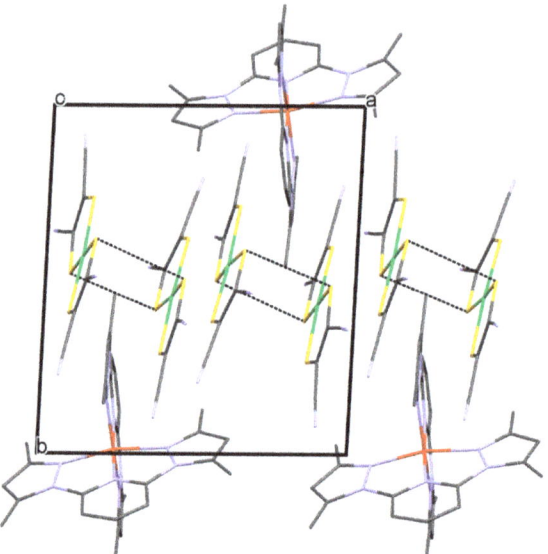

Figure 3. The crystal packing of compound **1** at 100 K. Short contacts between S atoms are indicated by black dotted lines. H atoms have been omitted for clarity.

For compound **2**, the complex geometry about Fe(II) is also a distorted octahedron, but in this case, there is a variation in Fe-N bond lengths as a function of temperature. At 100 K, the bond lengths vary from 1.967 to 2.070 Å with an average of 2.030 Å. However, there is an increase at 290 K to an average of 2.102 Å with a minimum bond length of 2.047 and a maximum of 2.136 Å. This change reflects the spin-state change that is seen in the bulk magnetism and Mössbauer data vide infra. In addition, there is a clear distortion of the octahedral geometry between the two temperatures. The parameter Σ is 109.14° at 100 K, meaning that the LS state of compound **2** has an intrinsically more distorted geometry than the LS state of compound **1**. However, at 290 K, this significantly increases to 132.94°, which is indicative of a spin-state change. Additionally, an increase was seen in the distortion parameter Θ of 354.66° at 100 K, but is 428.34° at 290 K. Furthermore, Halcrow suggested that whole molecule deviation from ideal D_{2d} symmetry in HS [Fe(dpp)$_2$]$^{2+}$-like cations indicates a propensity to undergo SCO [12]. The contention is that HS structures that deviate too much from the typical distortion seen at LS are unable to undergo SCO, although this analysis is not definitive. The distortion from D_{2d} is parametized by the *trans*-N(pyridyl)-Fe-N(pyridyl) angle (θ) and the dihedral angle between ligand mean planes (ϕ). For compound **2**, parameter θ is 178.43 (100 K) and 178.11° (293 K), whereas ϕ is 84.84° (100 K) and 85.20° (24 = 93 K). The HS and LS values do not deviate far from each other, and therefore compound **2** is not an outlier in undergoing SCO [11].

In comparison to compound **1**, the anions and cations of compound **2** are not clearly separated into layers, since the two ions tend to interdigitate with each other. Correspondingly, there are no short S . . . S contacts between anions, the shortest S . . . S distance being 3.926 Å, so that each Ni anion is isolated from the nearest neighbor (Figure 4). The Ni-S bond lengths are similar for both crystallographically independent anions, indicating that they have the same charge and metal oxidation state. The values range from 2.138 to 2.155 Å at 290 K and 2.141 to 2.161 Å at 100 K, typical for a Ni(III) monoanion [31].

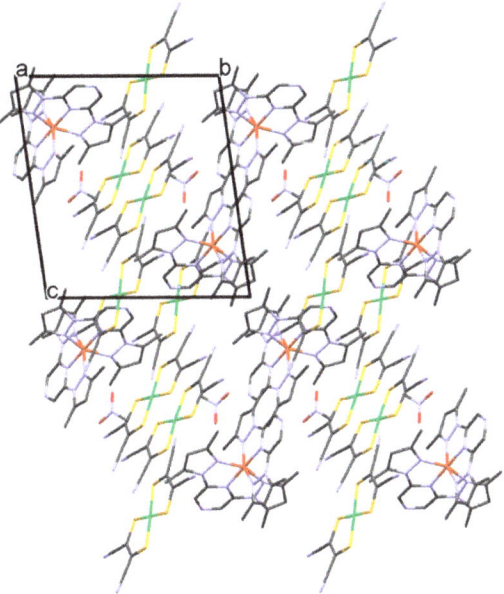

Figure 4. The crystal packing of compound **2** at 100 K. H atoms have been omitted for clarity.

2.2. Magnetic Properties

Figure 5 shows the magnetic data for compound **1**, [Fe(L^1)$_2$][Ni(mnt)$_2$]$_2$ and **2**, [Fe(L^2)$_2$][Ni(mnt)$_2$]$_2$·MeNO$_2$, expressed as the temperature (T) dependence of the product $\chi_m T$, where χ_m is the molar magnetic susceptibility. An initial analysis of the data can assume no magnetic exchange at higher temperatures. The magnetic data would then be a summation of contributions from HS Fe(II) (S = 2), LS Fe(II) (S = 0) and two [Ni(III)(mnt)$_2$] anions (S = 1/2 each). At lower temperatures there may also be a magnetic exchange that reduces or increases the contribution.

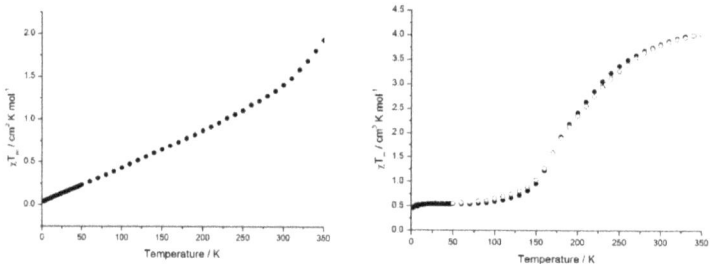

Figure 5. Temperature dependence of the product $\chi_m T$ for compound **1** (left) and compound **2** (right). Full circles are measurements of warmings from 2 to 350 K, while open circles are measurements and there is cooling across the same temperature range.

For compound **1**, at 2 K the value of $\chi_m T$ is 0.04 cm^3 K mol^{-1}, which implies that Fe(II) is in the LS state and there is also little contribution from the Ni anionic moieties. The reason for the small contribution from the anions is most likely linked to their dimerization and subsequent antiferromagnetic coupling. On warming, there is a steady, almost linear increase in $\chi_m T$ up to about 300 K (1.41 cm^3 K mol^{-1}), followed by a more rapid rise to 350 K where the maximum $\chi_m T$ is 1.93 cm^3 K mol^{-1}. These higher values cannot be achieved by a simple loss of antiferromagnetic coupling between anions, which would lead to a maximum $\chi_m T$ value of 0.75 cm^3 K mol^{-1}. Therefore, there must be an increased contribution from HS Fe(II) as the temperature rises. Nevertheless, this is an incomplete SCO and is not reflected in the variable temperature structural data of compound **1**. Such behavior has been previously seen in [FeL$_2$](ClO$_4$)$_2$, where L = 2,6-bis(3-methylpyrazol-1-yl)pyrazine [32].

Compound **2** distinctly shows different behavior. At 2 K, the value of $\chi_m T$ is 0.46 cm^3 K mol^{-1}, which rises slightly and reaches 0.5 cm^3 K mol^{-1} after 100 K, after which there is a more rapid rise to a maximum value of 4.02 cm^3 K mol^{-1} at 350 K. We can estimate the g-value for the Fe complex using the Mossbauer data and bulk magnetic data at 300 K. The Mossbauer data (Section 2.3) indicates that 83% of the Fe(II) complex is in the HS state at 300 K, whereas the bulk magnetic data gives a total $\chi_m T$ of 3.82 cm^3 K mol^{-1} at the same temperature. Assuming this high temperature $\chi_m T$ value also includes contributions from two independent [Ni(mnt)$_2$] anions (total 0.75 cm^3 K mol^{-1}), then 3.07 cm^3 K mol^{-1} is the contribution from 83% of a S = 2 HS Fe(II) and 17% S = 0 LS Fe(II). This gives an estimate of g of 2.2 using $\chi_m T = 0.83 \times 0.12505$ g^{2}S(S + 1). Using this g value and the same calculation, with the % fraction HS as the unknown, we can estimate the HS fraction to be 90% at 350 K and 46% at 200 K. Below about 100 K, the Fe(II) centers are mostly in the LS state and this does not contribute to the magnetism. The small value of $\chi_m T$ further implies an antiferromagnetic coupling between [Ni(mnt)$_2$] anions, albeit weaker than that shown by compound **1**. Indeed, the data below 80 K follows a Curie-Weiss law, with a negative Weiss constant of -0.6 K. However, bulk magnetization data is not an ideal method of decoupling the contributions from the Ni and Fe ions, but the above calculation of the HS fraction at 200 K agrees well with Mössbauer data *vida infra*. Finally, on measuring while cooling and warming the curves for compound **2** are essentially identical with no hysteresis, it reflects the lack of inter-cation short contacts in the structure. We can make some speculations as to the origin of the difference in abruptness of the SCO between the two salts. The compounds have cations with a very small molecular change; a CH in compound **1** is replaced by a N atom in compound **2**. This is unlikely to be directly responsible for such a difference in SCO, where more abrupt transitions are correlated to increased intermolecular interactions. However, the inclusion of the solvent in compound **2**, along with the overall change in packing, is likely to be more relevant. Compound **1**, with a much more gradual SCO, has a clear 2D layered structure, but the more abrupt SCO for compound **2** is associated with a more 3D structure, although both do not have any substantive inter-cation interactions. It is suggested that the more abrupt SCO for compound **2** is a result of the 3D organization, whereas the very gradual SCO for compound **1** is a response to more isolation of cations into layers.

2.3. Mössbauer Spectroscopy

The variable temperature ^{57}Fe Mössbauer data for compound **2** is presented in Figure 6. The associated hyperfine parameters were found by a least-squares fit, assuming Lorentzian shaped peaks, and are given in Table 2. Measurements were taken at several temperatures between 300 and 100 K but Figure 6 shows the fitted data at 300 K, 200 K and 100 K. All spectra were found to have signals that are attributable to both HS and LS Fe(II) complexes.

The experimental data is given as black dots. The signal for the HS Fe(II) having the largest quadrupole splitting (Table 2) is modelled by the green curve. The signal for the LS Fe(II) with the smallest quadrupole splitting is modelled by the blue curve. In each case, the hyperfine parameters compare well with the expected values. The red curve is the sum of the contributions from both the HS and LS Fe(II) curves and maps well onto the experimental data. The portion of the HS and LS complexes vary, as expected from the SQUID magnetic data. The measurements at 300 and 200 K clearly show two quadrupole doublets, indicating the presence of significant quantities of both HS and LS Fe(II) centers. An analysis of the integrals of the modelled curves indicates, at 300 K, that there is 83% of the HS fraction and 17% of the LS fraction. At 200 K, there is approximately 50% of each, and only 1.8% of the HS complex remains at 100 K. These percentage fractions are very close to those proposed from bulk magnetic data. Figure 7 shows all the Mössbauer data replotted to show the temperature variation of the HS fraction. This data confirms the conversion from almost complete LS Fe(II) at 100 K to a HS/LS mixture at 300 K, albeit with an excess of HS Fe(II). The transition temperature seen in Figure 7 also compares well with the bulk magnetic data in Figure 5.

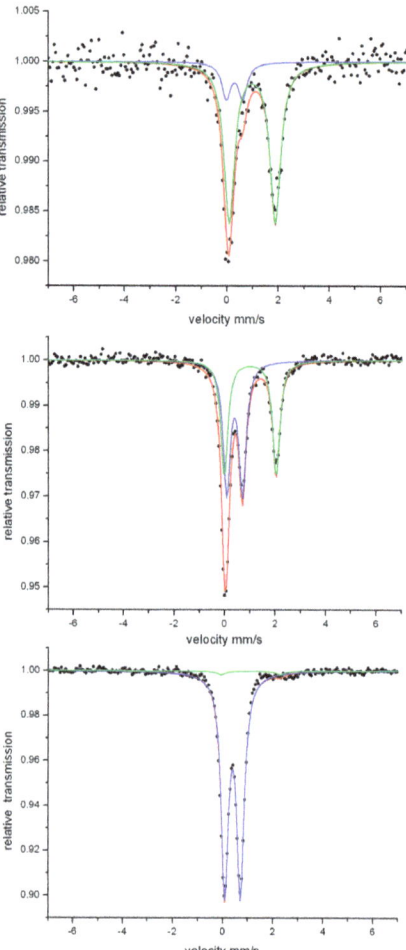

Figure 6. Mössbauer spectra of compound **2**, at 300 K (**top**), 200 K (**middle**) and 100 K (**bottom**). Experimental data are in black dots and the lines are data fitted, giving the parameters listed in Table 2. Fitting curves are low-spin (blue), high-spin (green) and total (red).

Table 2. Fitting parameters for variable temperature Mössbauer spectra [1] for compound **2**.

Temperature/K	Spin State	δ	ΔE_Q	Γ
300	HS	0.98	1.78	0.49
	LS	0.29	0.64	0.42
200	HS	1.01	2.05	0.34
	LS	0.39	0.63	0.34
100	HS	1.10	2.38	0.40
	LS	0.40	0.64	0.35

[1] δ is the isomer shift (± 0.02 mm s^{-1}), ΔE_Q is the quadrupole splitting (± 0.02 mm s^{-1}) and Γ is the full width and has the maximum of the peaks (± 0.03 mm s^{-1}). HS = high spin. LS = low spin.

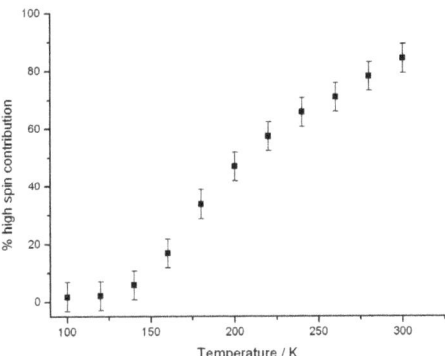

Figure 7. Graph derived from integral analysis of the Mössbauer data, showing temperature variation of the high-spin Fe(II) fraction in compound **2**.

3. Materials and Methods

All reactions were completed in air, unless otherwise stated, using commercial grade chemicals. $(C_4H_9)_4N[Ni(mnt)_2]$ was purchased from Tokyo Chemicals Industry (TCI) and was used as received. All other chemicals were purchased from Sigma-Aldrich and used as received. Magnetization measurements were completed using a MPMS-5 SQUID magnetometer (Quantum Design, CA, USA) with powdered samples held in gelatin capsules in an external field of 0.1 T. Each sample was cooled to 2 K and measured while being warmed to 300 K. Since compound **2** had a more abrupt SCO, this sample was also measured while cooling in order to determine hysteresis. Diamagnetic corrections were made for the sample holder and for the compound, the latter estimated using Pascals constants [33]. Resistance measurements were made using a two-probe method, at ambient temperature, with gold wire (diam 1 μm) attached to the opposite sides of a crystal with carbon paste. This was measured using a Keithley Instruments 6517A multimeter (Tektronics, UK). For compound **1**, single crystal X-ray diffraction was performed using a Rigaku AFC12 goniometer and enhanced sensitivity Saturn724+ detector with Superbright Mo rotating anode generator. Structure refinement was completed using SHELXL [34]. For compound **2**, single crystal X-ray diffraction was performed using a Rapid II imaging plate system with MicroMax-007 HF/VariMax rotating anode X-ray generator and confocal monochromated Mo-Kα radiation. Structure refinement was completed with SHELXL [34] or CRYSTALS [35] software. All non-hydrogen atoms were refined anisotropically. Hydrogen atom positions were calculated geometrically and refined using the riding model. Mössbauer spectra were recorded using a conventional spectrometer in the constant-acceleration mode. Isomer shifts are given relative to α-Fe. Temperature dependent Mössbauer spectra were recorded with a closed-cycle cryostat (CRYO Industries of America Inc., USA) and were analyzed by the least-square fits using Lorentzian line shapes. CHN elemental analysis was performed using an CE-440 Elemental Analyzer (Exeter Analytical Inc., Coventry, UK).

3.1. Synthesis of Ligands and Fe(II) Complexes

The ligand **L¹** was prepared by the literature procedure [36], although using the work-up methods later described by Halcrow et al [15]. The ligand **L²** was prepared following the procedure by Halcrow et al. [32]. The reaction of $Fe(BF_4)_2 \cdot 6H_2O$ was done by stirring for two hours in acetone with two equivalents of **L¹** or **L²**, followed by precipitation by adding diethyl ether yielded brown solids. The color indicates that both $[Fe(L^1)_2](BF_4)_2$ and $[Fe(L^2)_2](BF_4)_2$ are in the LS state at ambient temperature.

3.2. Synthesis of Compounds 1 and 2

Compounds **1** and **2** were produced using a moderated procedure from Oshio et al. [11]. For compound **1**, $[Fe(L^1)_2](BF_4)_2$ (38.2 mg, 0.05 mmol) in $MeNO_2$ (5 mL) was added dropwise to a solution of $(C_4H_9)_4N[Ni(mnt)_2]$ (69.7 mg, 0.12 mmol) in $MeNO_2$ (10 mL). The resulting dark green/brown solution was kept at $-20\ °C$ overnight to yield a dark brown microcrystalline powder, which was collected and washed with cold $MeNO_2$. Recrystallisation was achieved by redissolving the solid in hot $MeNO_2$, followed by cooling at $-20\ °C$ to give thin brown plate-like crystals. Yield 64%. Found C, 43.18; H, 2.02; N, 19.42. Calcd for $[Fe(L^1)_2][Ni(mnt)_2]_2$, $C_{46}H_{34}N_{18}S_8FeNi_2$: C, 43.55; H, 2.00; N, 19.87.

Compound 2 was prepared by a similar procedure except using $[Fe(L^2)_2](BF_4)_2$ in place of $[Fe(L^1)_2](BF_4)_2$. In this case, small black prism-shaped crystals precipitated after standing for 2 hours, without the need for cooling. Yield 58%. Found C, 40.85; H, 2.60; N, 22.19. Calcd for $[Fe(L^2)_2][Ni(mnt)_2]_2 \cdot MeNO_2$, $C_{45}H_{35}N_{21}S_8O_2FeNi_2$: C, 40.59; H, 2.65; N, 22.09.

4. Conclusions

Two novel salts have been synthesized: compound 1 with formula $[Fe(L^1)_2][Ni(mnt)_2]_2$ and compound 2 with formula $[Fe(L^2)_2][Ni(mnt)_2]_2 \cdot MeNO_2$. L^1 and L^2, which are methylated derivatives of a series of ligands that are present in several Fe(II) complexes that exhibit SCO, albeit such salts typically have simple counterions such as BF_4 and ClO_4. Bulk magnetization studies indicate that both compounds 1 and 2 show SCO. However, for compound 1, the SCO is incomplete up to 350 K and only represents a small conversion from the LS to HS state. Consequently, there is little evidence from variable temperature single crystal structural data that the typical structural transitions that accompany SCO are fully developed. By contrast, bulk magnetization measurements on compound 2 show that the compound is in the LS state below 100 K and almost completely HS at 350 K. The associated Fe-ligand bond length changes and geometry distortions are clearly observed even at 290 K, when most of the Fe complexes are in the HS state. Furthermore, the SCO in compound 2 is confirmed using Mossbauer spectroscopy. The introduction of the $[Ni(mnt)_2]$ anion illustrates the chemical flexibility and robust nature of SCO in these types of Fe(II) complexes. It also points to the possibility of incorporating secondary functionality into SCO materials.

Author Contributions: Conceptualization, methodology, analysis and article preparation S.S.T.; synthesis and magnetization measurements J.D.; X-ray crystallography (compound 1) P.N.H. & S.J.C. X-ray crystallography (compound 2) and discussions on magnetic data H.A.; Mössbauer spectroscopy V.S. All authors have read and agreed to the published version of the manuscript.

Funding: V.S. acknowledges the support of the research initiative NANOKAT. The PhD studies of J.D. were supported by the University of Surrey and EPSRC.

Data Availability Statement: Data available on request from the corresponding author. Data available includes raw SQUID magnetization and Mössbauer data. The crystallographic data has been submitted to the Cambridge Crystallographic Data Centre. The deposition references are CCDC 2080108 (cpd 1 100 K), 2080109 (cpd 2 293 K). 2080110 (cpd 2 100 K) and 2080111 (cpd 2 290 K) Available online: http://www.ccdc.cam.ac.uk/conts/retrieving.html.

Acknowledgments: S.S.T. would like to express his gratitude to Peter Day FRS with whom he worked for 10 years at the Royal Institution of Great Britain. Peter's passion, drive and scientific insight continues to provide inspiration to study the fascinating topic of molecular materials.

Conflicts of Interest: The authors declare no conflict of interest. The funders had no role in the design of the study; in the collection, analyses, or interpretation of data; in the writing of the manuscript, or in the decision to publish the results.

References

1. Cambi, L.; Szegö, L. The magnetic susceptibility of complex compounds. *Ber. Deutsch. Chem. Ges.* **1931**, *64*, 2591–2598. [CrossRef]
2. Cambi, L.; Malatesta, L. Magnetism and polymorphy of internal complex salts—Iron salts of dithio-carbonicamide acids. *Ber. Deutsch. Chem. Ges.* **1937**, *70*, 2067–2078. [CrossRef]
3. Kahn, O.; Martinez, C.J. Spin transition polymers: From materials towards memory devices. *Science* **1998**, *179*, 44–48. [CrossRef]
4. Hayami, S.; Holmes, S.M.; Halcrow, M.A. Spin-state switching in molecular materials. *J. Mater. Chem. C* **2015**, *3*, 7775–7778. [CrossRef]

5. Gütlich, P.; Gaspar, A.B.; Garcia, Y. Spin-state switching in iron coordination compounds. *Beilstein J. Org. Chem.* **2013**, *9*, 342–381. [CrossRef]
6. Bonhommeau, S.; Guillon, T.; Daku, L.M.L.; Demont, P.; Costa, J.S.; Létard, J.F.; Molnár, G.; Bousseksou, A. Photoswitching of the dielectric constant of the spin-crossover complex [Fe(L)(CN)$_2$].H$_2$O. *Angew. Chem. Int. Ed.* **2006**, *45*, 1625–1629. [CrossRef]
7. Guilon, T.; Bonhommeau, S.; Costa, J.S.; Zwick, A.; Létard, J.F.; Demont, P.; Molnár, G.; Bousseksou, A. On the dielectric properties of the spin-crossover complex [Fe(bpp)$_2$][BF$_4$]$_2$. *Phys. Status Solidi A Appl. Mat. Sci.* **2006**, *203*, 2974–2980. [CrossRef]
8. Schäfer, B.; Bauer, T.; Faus, I.; Wolny, J.A.; Dahms, F.; Fuhr, O.; Lebedkin, S.; Wille, H.C.; Schlage, K.; Chevalier, K.; et al. A luminescent P$_{12}$Fe spin-crossover complex. *Dalton Trans.* **2017**, *46*, 2289–2302. [CrossRef]
9. Wang, C.F.; Li, R.F.; Chen, X.Y.; Wei, R.J.; Zheng, L.S.; Tao, J. Synergistic spin-crossover and fluorescence in one-dimensional hybrid complexes. *Angew. Chem. Int. Ed.* **2015**, *54*, 1574–1577. [CrossRef]
10. Jiao, Y.; Zhu, J.P.; Guo, Y.; He, W.J.; Guo, Z.J. Synergetic effect between spin-crossover and luminescence in the [Fe(bpp)$_2$][BF$_4$]$_2$ (bpp=2,6-bis(pyrazol-1-yl)pyridine complex. *J. Mater. Chem. C* **2017**, *5*, 5214–5222. [CrossRef]
11. Nihei, M.; Takahashi, H.; Nishikawa, H.; Oshio, H. Spin-crossover behavior and electrical conduction property in iroj(II) complexes with tetrathiafulvalene derivatives. *Dalton Trans.* **2011**, *40*, 2154–2156. [CrossRef]
12. Halcrow, M.A. Structure-function relationships in molecular spin-crossover complexes. *Chem. Soc. Rev.* **2011**, *40*, 4119–4142. [CrossRef] [PubMed]
13. Boillot, M.L.; Weber, B. Mononuclear ferrous and ferric complexes. *Compte. Rendus. Chim.* **2018**, 1196–1208. [CrossRef]
14. Kumar, K.S.; Ruben, M. Emerging trends in spin-crossover (SCO) based functional materials. *Coord. Chem. Rev.* **2017**, *346*, 176–205. [CrossRef]
15. Halcrow, M.A. The synthesis and coordination chemistry of 2, 6-bis(pyrazolyl)pyridines and related ligands—Versatile terpyridine analogues. *Coord. Chem. Rev.* **2005**, *249*, 2880–2908. [CrossRef]
16. Halcrow, M.A. Iron(II) complexes of 2,6-di(pyrazol-1-yl)pyridines—A versatile system for spin-crossover research. *Coord. Chem. Rev.* **2009**, *253*, 2493–2514. [CrossRef]
17. Cook, L.J.K.; Mohammed, R.; Sherborne, G.; Roberts, T.D.; Alvarez, S.; Halcrow, M.A. Spin state behavior of iron(II)/dipyrazolylpyridine complexes. New insights from crystallographic and solution measurements. *Coord. Chem. Rev.* **2015**, *289*, 2–12. [CrossRef]
18. Coronado, E.; Galán-Mascarós, J.R.; Giménez-López, M.C.; Almeida, M.; Waerenborgh, J.C. Spin-crossover Fe(II) complexes as templates for bimetallic oxalate-based 3D magnets. *Polyhedron* **2007**, *26*, 1838–1844. [CrossRef]
19. Cassoux, P.; Valade, L.; Kobayashi, H.; Kobayashi, A.; Clark, R.A.; Underhill, A.E. Molecular metals and superconductors derived from metal complexes of 1,3-dithiol-2-thione-4,5-dithiolate (dmit). *Coord. Chem. Rev.* **1991**, *110*, 115–160. [CrossRef]
20. Robertson, N.; Cronin, L. Metal bis-1,2-dithiolene complexes in conducting or magnetic crystalline assemblies. *Coord. Chem. Rev.* **2002**, *227*, 93–127. [CrossRef]
21. Cassoux, P. Molecular (super)conductors derived from bis-dithiolate metal complexes. *Coord. Chem. Rev.* **1999**, *185*, 213–232. [CrossRef]
22. Dorbes, S.; Valade, L.; Real, J.A.; Faulmann, C. [Fe(sal$_2$-trien)][Ni(dmit)$_2$]; towards switchable spin crossover molecular conductors. *Chem. Comm.* **2005**, 69–71. [CrossRef] [PubMed]
23. Szilágyi, P.A.; Dorbes, S.; Molnár, G.; Real, J.A.; Homonnay, Z.; Faulmann, C.; Bousseksou, A. Temperature and pressure effects on the spin state of ferric ions in the [Fe(sal$_2$-trien)][Ni(mnt)$_2$] spin-crossover complex. *J. Phys. Chem. Solids* **2008**, *69*, 2681–2686. [CrossRef]
24. Faulmann, C.; Szilágyi, P.A.; Jacob, K.; Chahine, J.; Valade, L. Polymorphism and its effects on the magnetic behaviour of the [Fe(sal$_2$-trien)][Ni(dmit)$_2$] spin-crossover complex. *New J. Chem.* **2009**, *33*, 1268–1276. [CrossRef]
25. Faulmann, C.; Dorbes, S.; Real, J.A.; Valade, L. Electrical conductivity and spin crossover: Towards the first achievement with a metal bis-dithiolene complex. *J. Low Temp. Phys.* **2006**, *142*, 261–266. [CrossRef]
26. Faulmann, C.; Dorbes, S.; de Bonneval, W.G.; Molnár, G.; Bousseksou, A.; Gomez-Garcia, C.J.; Coronado, E.; Valade, L. Towards molecular conductors with a spin-crossover phenomenon: Crystal structures, magnetic properties and Mössbauer spectra of [Fe(salten)Mepepy][M(dmit)$_2$]. *Eur. J. Inorg. Chem.* **2005**, *16*, 3261–3270. [CrossRef]
27. Takahashi, K.; Cui, H.B.; Okano, Y.; Kaboyashi, H.; Einaga, Y.; Sato, O. Electrical conductivity modulation coupled to a high-spin-low-spin conversion in the molecular system [Fe-III(qsal)$_2$][Ni(dmit)$_2$]$_3$.CH$_3$CN.H$_2$O. *Inorg. Chem.* **2006**, *45*, 5739–5741. [CrossRef] [PubMed]
28. Nihei, M.; Tahira, H.; Takahashi, N.; Otake, Y.; Yamamura, Y.; Saito, K.; Oshio, H. Multiple bistability and tristability with dual spin-state conversions in [Fe(dpp)$_2$][Ni(mnt)$_2$].MeNO$_2$. *J. Am. Chem. Soc.* **2010**, *132*, 3553–3560. [CrossRef]
29. Farrugia, L.J. WinGX and ORTEP for Windows: An update. *J. Appl. Cryst.* **2012**, *45*, 849–854. [CrossRef]
30. Ketkaew, R.; Tabtirungrotechai, Y.; Harding, P.; Chastanet, G.; Guionneau, P.; Machivie, M.; Harding, D.J. OctaDist: A tool for calculating distortion parameters in spin crossover and coordination complexes. *Dalton Trans.* **2021**, *50*, 1086–1096. [CrossRef]
31. Ribera, E.; Rovira, C.; Veciana, J.; Tarrés, J.; Canadell, E.; Rousseau, R.; Molins, E.; Mas, M.; Schoeffel, J.P.; Pouget, J.P.; et al. The [(DT-TTF)$_2$M(mnt)$_2$] family of radical ion salts: From a spin ladder to delocalized conduction electrons that interact with localized magnetic moments. *Chem. Eur. J.* **1999**, *5*, 2025. [CrossRef]
32. Elhaïk, J.; Money, V.A.; Barrett, S.A.; Kilner, C.A.; Evans, I.R.; Halcrow, M.A. The spin states and spin-crossover behaviour of iron(II) complexes of 2,6-dipyrazol-1-ylpyrazine derivatives. *Dalton Trans.* **2003**, *10*, 2053–2060. [CrossRef]
33. Bain, G.A.; Berry, J.F. Diamagnetic corrections and Pascal's constants. *J. Chem. Educ.* **2008**, *85*, 532–536. [CrossRef]

34. Sheldrick, G.M. Crystal structure refinement with SHELXL. *Acta. Cryst. C* **2015**, *71*, 3–8. [CrossRef] [PubMed]
35. Betteridge, P.W.; Carruthers, J.R.; Cooper, R.I.; Prout, K.; Watkin, D.J. CRYSTALS version 12: Software for guided crystal structure analysis. *J. Appl. Cryst.* **2003**, *36*, 1487. [CrossRef]
36. Jameson, D.L.; Goldsby, K.A. 2,6-bis(N-pyrazolyl)pyridines—The convenient synthesis of a family of planar tridentate N_3 ligands that are terpyridine analogs. *J. Org. Chem.* **1990**, *55*, 4992–4994. [CrossRef]

Article

The Internal Field in a Ferromagnetic Crystal with Chiral Molecular Packing of Achiral Organic Radicals

Stephen J. Blundell [1,*], Tom Lancaster [2], Peter J. Baker [3], Francis L. Pratt [3], Daisuke Shiomi [4], Kazunobu Sato [4] and Takeji Takui [4]

1. Clarendon Laboratory, Department of Physics, University of Oxford, Parks Road, Oxford OX1 3PU, UK
2. Center for Materials Physics, Department of Physics, Durham University, Durham DH1 3LE, UK; tom.lancaster@durham.ac.uk
3. ISIS Facility, STFC Rutherford Appleton Laboratory, Chilton, Oxfordshire OX11 0QX, UK; peter.baker@stfc.ac.uk (P.J.B.); francis.pratt@stfc.ac.uk (F.L.P.)
4. Department of Chemistry and Molecular Materials Science, Graduate School of Science, Osaka City University, Sugimoto, Sumiyoshi-ku, Osaka 558-8585, Japan; daisukeshiomi@osaka-cu.ac.jp (D.S.); sato@sci.osaka-cu.ac.jp (K.S.); takui@sci.osaka-cu.ac.jp (T.T.)
* Correspondence: stephen.blundell@physics.ox.ac.uk

Abstract: The achiral organic radical dinitrophenyl nitronyl nitroxide crystallizes in two enantiomorphs, both being chiral tetragonal space groups that are mirror images of each other. Muon-spin rotation experiments have been performed to study the magnetic properties of these crystals and demonstrate that long-range magnetic order is established below a temperature of 1.10(1) K. Two oscillatory components are detected in the muon data, which show two different temperature dependences.

Keywords: organic magnet; nitronyl nitroxide; chirality; muon-spin rotation

1. Introduction

The impetus to synthesize purely organic ferromagnets [1,2] arises from the aim to achieve a goal that was once thought to be impossible: that of realising ferromagnetism in materials containing atoms that only have s and p electrons. Heisenberg's celebrated theory of ferromagnetism [3], which was formulated back in the 1920s and that first introduced the concept of the exchange interaction, provided a rationalisation for the apparent mandatory requirement for atoms containing d and f electrons. Purely organic materials can contain unpaired spins, and organic radical molecules are relatively common, however few are stable enough to be assembled into crystalline structures. Moreover, even when that is possible, it is another matter to attempt to ferromagnetically align these spins . Ferromagnets are, in fact, rather rare, even among the elements (and, of course, the elemental ferromagnets that do exist are only in the d- or f-blocks).

Organic ferromagnetism was first achieved using a nitronyl nitroxide organic radical [4]. The unpaired electron density in nitronyl nitroxides is predominantly distributed over the two NO moieties with only some smaller spin density being distributed over the rest of the molecule. The central carbon atom of the ONCNO moiety is a node of the singly-occupied molecular orbital. Nitronyl nitroxides are chemically stable, but the vast majority of them do not show a long range ferromagnetic order. Therefore, the discovery of long-range ferromagnetism in the β phase of *para*-nitrophenyl nitronyl nitroxide ($C_{13}H_{16}N_3O_4$, abbreviated to p-NPNN), was a major milestone, even though the transition temperature proved to be a disappointingly low 0.65 K [4]. A λ-type peak in the heat capacity at the critical temperature and a divergence in the ac susceptibility [4–6] indicated the transition to ferromagnetic order. μSR experiments on p-NPNN show the development of coherent spin precession oscillations below T_C [7,8]. A number of other nitronyl nitroxide systems were studied while using this technique [9–13], but the transition temperatures are all below 1 K.

Chirality [14] has recently been used in designing new chiral catalysts [15], developing organic spin filters [16], and investigating spin-dependent tunnelling through chiral molecules [17–19]. It has an important role in photochemistry that may have been important in the origin of life [20]. Therefore, the phenomenon of structural or magnetic chirality is an attractive feature to engineer in nitronyl nitroxide magnets, however preparing a chiral molecular crystal from achiral radicals in a controlled manner is a formidable task. It was first achieved ten years ago through the preparation of a crystal containing nitronyl nitroxide radicals, in which there is a chiral packing of the molecular units [21]. The compound is 3,5-dinitrophenyl nitronyl nitroxide (DNPNN), a molecule that is itself achiral (see the molecular structure depicted in Figure 1a) and optically inactive when dissolved in a solvent, but that is found to crystallize in two enantiomorphs, one of which is a chiral tettragonal space group of $P4_3$ (see Figure 1b). This has a left-handed, counterclockwise stacking of radical molecules along the fourfold screw axes. The other enantiomorph has the $P4_1$ space group and is its mirror image. The magnetic susceptibility fits to a one-dimensional Heisenberg ferromagnetic model with intrachain $2J/k_B = 12\,\text{K}$ (and weak interchain $2zJ'/k_B \approx 1\,\text{K}$). On the basis of magnetic susceptibility and heat capacity measurements [21], the system is believed to undergo a ferromagnetic phase transition at $T_C = 1.1\,\text{K}$, and the entropy that is found by heat capacity is equal to $R\ln 2$, as expected for a system of $S = \frac{1}{2}$ spins. The magnetic behaviour of the left-handed form and right-handed form are identical, but it is possible to prepare crystals of one type or the other [21]. In this paper, we report the results of a muon-spin rotation (μSR) study of this compound that shed further light on the magnetic properties of this material.

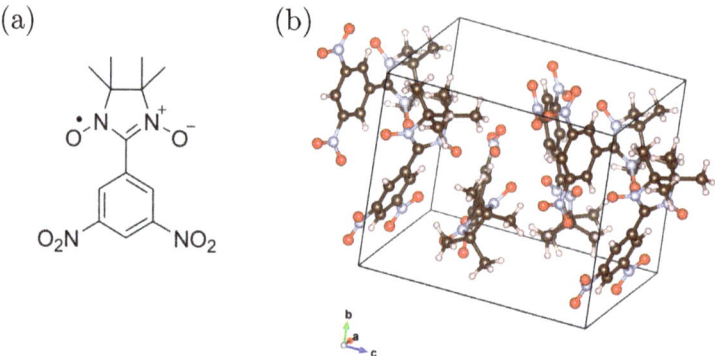

Figure 1. (a) The molecular structure of DNPNN. (b) The crystal structure of one enantiomorph of DNPNN (the other enantiomorph is the mirror image of this). The atoms are colour-coded, as follows: carbon (brown), oxygen (red), nitrogen (grey), and hydrogen (pale pink and small).

2. μSR Experiment

The technique of muon-spin rotation [22–25] is very effective in establishing three-dimensional ordering in low-dimensional magnets [26,27]. This is because, below the critical temperature, a spontaneous precession of the muon spin-polarization in zero-applied field can be observed (see e.g., Refs. [28–31] for examples in organic and molecular magnets), with its frequency being proportional to the order parameter. Of course, ordering can be detected using thermodynamic measurements, but these are often dominated by the effect of intrachain interactions in low-dimensional magnets. For example, three-dimensional ordering in a very anisotropic spin chain is only associated with a tiny fraction of the total entropy. As the sample is cooled, very long correlated segments begin to develop on the individual chains well in advance of the appearance of long range order [32]). Therefore, μSR experiments provide a clear and unambiguous signal of long range order. Furthermore, the magnetic susceptibility measurements can be dominated by magnetic impurities and, so, it is desirable to have a test of intrinsic magnetic order. µSR provides this, as it is a *volume probe*; muons stop throughout the bulk of a sample and, therefore, can provide volume fraction information. Moreover, muons do not require hydrogen-containing samples to be deuterated, which makes µSR a more convenient technique for studying nitronyl nitroxide magnets in comparision to neutron scattering.

The MuSR spectrometer at the ISIS Pulsed Neutron and Muon Facility based at the Rutherford Appleton Laboratory was used to perform the µSR experiments. This spectrometer is equipped with a dilution refrigerator. In the experiment, spin-polarized positive muons (μ^+, momentum 28 MeV/c) were implanted into an array of small crystals of DNPNN that we prepared according to the method described in [21]. Our sample contained a mixture of crystals with different handedness, however one would not expect a µSR experiment to be able to distinguish between the samples that are mirror images of each other. The muons stop quickly (in $<10^{-9}$ s), without a significant loss of spin-polarization. The observed quantity is the time evolution of the average muon spin polarization $P_z(t)$, which can be inferred [22–25] via the asymmetry in the angular distribution of emitted decay positrons, being parameterized by an asymmetry function $A(t)$ that is proportional to $P_z(t)$. Figure 2 shows two representative spectra for our sample of DNPNN. In the higher temperature data, a relaxing signal is observed, which is consistent with spin fluctuations, but not magnetic order. However, below T_C, an oscillatory signal develops that contains two distinct frequency components (that are apparent from the beating pattern observed in the 0.65 K data in Figure 2). This identifies the presence of bulk long range magnetic order in the sample.

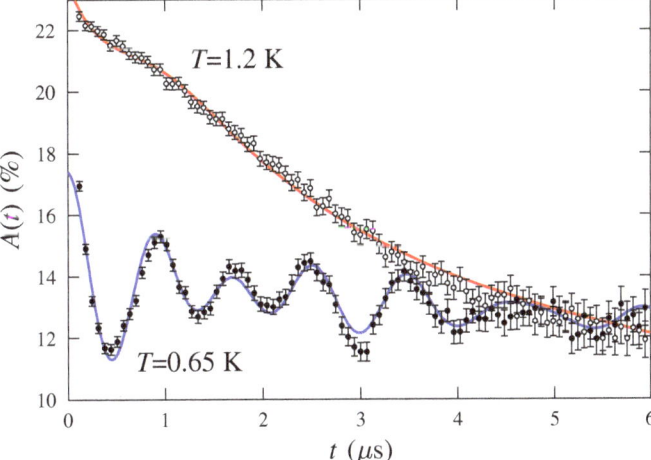

Figure 2. Example µSR spectra for DNPNN measured above and below the magnetic transition.

The µSR data are well described by the fitting function

$$A(t) = \left[\sum_{i=1}^{2} A_i e^{-\Lambda_i t} \cos(2\pi \nu_i t)\right] + A_3 e^{-\sigma^2 t^2} + A_{\text{bg}}, \quad (1)$$

where Λ_i is a relaxation rate and ν_i is a muon precession frequency (that is equal to $\gamma_\mu B_i/2\pi$, with B_i being the magnetic field at the ith muon site, and $\gamma_\mu = 2\pi \times 135.5$ MHz T^{-1} being the muon gyromagnetic ratio), and A_{bg} as the background contribution from those muons that stop outside the sample. The ratio of A_1 to A_2 was fixed across the temperature range (with $A_2/A_1 \approx 3$) although there is some temperature dependence in the total oscillatory amplitude $A_1 + A_2$ in our fits. In addition to the two oscillatory components, there is also a small Gaussian contribution (amplitude A_3 and relaxation rate σ) of unknown origin, although such components are commonly found in organic and molecular magnets [26]. These fits allow the extraction of the precession frequencies as a function of temperature, and these are plotted in Figure 3a. Their magnitudes, both reaching ≈ 1 MHz as $T \to 0$, are typical for precession signals that are measured in nitronyl nitroxide magnets [8–13], which result from a relatively dilute array of spin-$\frac{1}{2}$ moments, one per radical molecule. Both of the precession frequencies follow a typical temperature dependence for a magnetic order parameter and that can be fitted using the phenomenological function $\nu(T) = \nu(0)(1 - (T/T_C)^\alpha)^\beta$ (the best fit parameters given in Table 1) and they provide a well constrained estimate of the critical temperature as $T_C = 1.105(1)$ K, which is consistent with the earlier measurements using magnetic susceptibility and heat capacity [21]. The value of α, which is averaged over the two frequencies, is ≈ 1.5, as expected for magnons in a three-dimensional ferromagnet, so one can speculate that there are some structure factors that affect the sensitivity of two different sites to an additional antiferromagnetic component. Although the statistical error on our determination of T_C is 1 mK, the possible thermal offsets and calibration errors could lead to an uncertainty, which is an order of magnitude greater, and so our final determination of T_C from these measurements is $1.10(1)$ K.

Table 1. Parameters for fitting the temperature dependence of the precession frequencies to $\nu(T) = \nu(0)(1 - (T/T_C)^\alpha)^\beta$.

Parameter	Frequency 1	Frequency 2
T_C (K)	1.105(1)	1.104(1)
$\nu(0)$ (MHz)	1.21(2)	1.37(2)
α	1.03(6)	1.96(14)
β	0.41(1)	0.37(1)

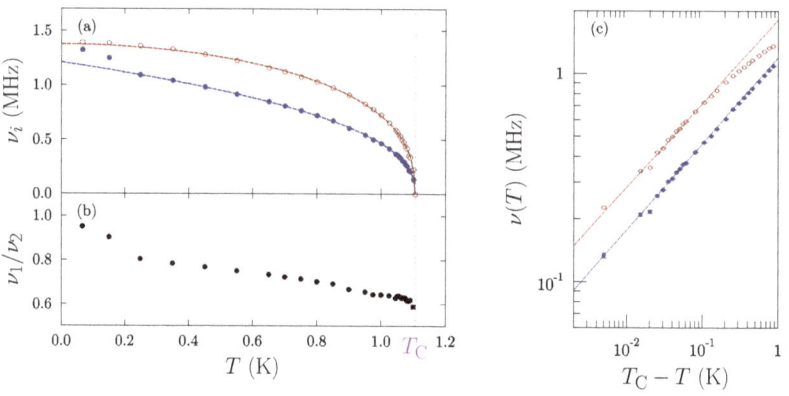

Figure 3. (a) Temperature evolution of the precession frequencies of the μSR spectra in DNPNN for $T < T_C$ from fits to Equation (1). (b) The ratio of the frequencies ν_1 and ν_2. (c) A scaling plot of the frequencies ν_1 and ν_2 in the critical regime.

Figure 3a also reveals a surprising feature, namely that the two precession frequencies do not follow exactly the same temperature dependence. Although they both collapse to zero at T_C, they approach the same limit as $T \to 0$, but become increasingly different as $T \to T_C$. This can be very clearly seen by plotting the ratio of the two frequencies as a function of temperature, as shown in Figure 3b. It is common in μSR data to find more than one frequency in the ordered state, but the individual frequencies usually follow the same temperature dependence, since they all are expected to be proportional to the order parameter, and so should only differ by a temperature-independent scaling factor. We return to this point in the following section.

Close to the transition, the two frequencies are expected to follow a scaling law $\nu(T) \propto M(T) \propto (T_C - T)^\beta$, where $M(T)$ is the magnetization (or staggered magnetization for an antiferromagnet) and β is a critical exponent. The value of β can be crudely extracted using the phenomenological function that is discussed above, but a better estimate is obtained by focussing on the data measured close to the transition using the scaling plot that is shown in Figure 3c. This shows that scaling behaviour is followed well in the region $T_C - T \ll 1$ and it yields $\beta = 0.40(2)$ for the higher frequency and $\beta = 0.41(2)$ for the lower frequency; hence, the two values are consistent with one another within error (and slightly larger than 0.37, the value obtained for frequency 2 by the less reliable phenomenological fit listed in Table 1). This is close to the value that is expected for the three-dimensional Heisenberg model ($\beta = 0.369$ [33]). It is also much larger than the value that is found for a chiral antiferromagnetic layered molecular magnet [34]. We note that our measured value is even closer to the theoretical value for the O(4) model ($\beta = 0.39$ [35]), although the link with such a model is not clear.

We carried out density functional theory (DFT) calculations to understand the nature of the muon stopping sites [36]. The calculations were carried out using the MuFinder software [37] and the plane-wave-based code CASTEP [38] while using the local density approximation. Muons, which were modelled by an ultrasoft hydrogen pseudopotential, were initialised in range of low-symmetry positions and the structure was allowed to relax (keeping the unit cell fixed) until the change in energy per ion was less than 2×10^{-5} eV. We used a cutoff energy of 545 eV and a $1 \times 1 \times 1$ Monkhorst-Pack grid for k-point sampling. The calculations were made on both the unit cell and a supercell in order to check for any self-interaction effects between the muons.

The lowest-energy diamagnetic muon sites are found around 1 Å from those oxygen atoms that are bonded to the five-membered ring in the DNPNN molecule. Sites that are close to the nitrogen atoms on these rings are also stabilised, but they are found around 0.55 eV higher in energy. In addition, sites close to the aromatic ring are also stable, although these are found around 0.45 eV high in energy than the oxygen sites. The calculations were repeated with muonium implanted instead of the bare muon and leading to a very similar range of sites with slightly different energies. The oxygen site is still found to be the lowest-energy configuration.

3. Discussion

The demonstration of long range order below 1.10(1) K in DNPNN is the main result of this study. This value can be compared with other quasi-one-dimensional organic ferromagnets (all of which are driven into long-range three-dimensional order due to the weak interchain interactions), and some selected values are included in Table 2. These show that DNPNN can be considered a to be fairly good approximation to a one-dimensional ferromagnet due to its relatively low value of $k_B T_C / J$ (the quoted value of J is obtained from heat capacity measurements [21]), but, when compared to other examples, the interchain interactions are still significant.

Table 2. Magnetic properties of selected one-dimensional ferromagnets. *p*-NPNN = para-nitrophenyl nitronyl nitroxide, Me = CH_3, DMSO = C_2H_6SO, *p*-CDTV = 3-(4-chlorophenyl)-1,5-dimethyl-6-thioxoverdazyl, TMSO = C_4H_8SO and CHAC = $C_6H_{11}NH_3CuCl_3$, *p*-CDpOV = 3-(4-chlorophenyl)-1,5-diphenyl-6-oxoverdazyl, TMCuC = tetramethylammonium copper trichloride, CHAB = $C_6H_{11}NH_3CuBr_3$, 2-BiMNN = 2-benzimidazolyl nitronyl nitroxide, F4BiMNN = 2-(4,5,6,7-tetrafluorobenzimidazol-2-yl)-nitronyl nitroxide (F4BImNN).

Compound	Reference	J/k_B (K)	T_c (K)	$k_B T_c/J$
γ-NPNN	[6]	2.15	0.65	0.30
$Me_3NHCuCl_3 \cdot 2H_2O$	[39]	0.85	0.165	0.19
$CuCl_2$(DMSO)	[40,41]	45	4.8	0.11
p-CDTV	[42]	6.0	0.67	0.11
$CuCl_2$(TMSO)	[40,41]	39	3	0.08
CHAC	[43,44]	45–53	2.18	0.04–0.05
p-CDpOV	[45]	5.5	0.21	0.038
TMCuC	[41,46,47]	30, 45	1.24	0.03–0.04
CHAB	[48]	55	1.50	0.027
2-BImNN	[49]	22	1.0	0.045
F4BImNN	[50]	22	0.72	0.033
DNPNN	[21], this work	5.6	1.015	0.18

The different temperature dependence of the two precession frequencies is an intriguing feature of the data. This is in stark contrast to many μSR studies of ordered magnets, where multiple precession frequencies can be observed but all follow the same temperature dependence, albeit with a different scaling factor. A good example of this is found in [FeCp$_2^*$][MnCr(ox)$_3$], which consists of layers of bimetallic oxalate that are separated by paramagnetic decamethylferrocenium and which that ferromagnetically below 5.17(3) K [51,52]. In this compound, μSR data show three distinct precession signals, presumably originating from three different muon sites in this rather complex structure, but, apart from different scaling factors, they all follow the same temperature dependence [53]. This scaling arises because the observed precession frequency is given by $\gamma_\mu B_{\text{dip}}(r_\mu)/(2\pi)$, where γ_μ is the muon gyromagnetic ratio and the dipolar field at the muon site is given by $B_{\text{dip}}^\alpha(r_\mu) = \sum_i D_i^{\alpha\beta}(r_\mu) m_i^\beta$, a sum over the magnetic ions in the crystal (see e.g., [25]); the magnetic moment of the ith ion is m_i and $D_i^{\alpha\beta}(r_\mu)$ is the dipolar tensor that is given by

$$D_i^{\alpha\beta}(r_\mu) = \frac{\mu_0}{4\pi R_i^3}\left(\frac{3R_i^\alpha R_i^\beta}{R_i^2} - \delta^{\alpha\beta}\right), \qquad (2)$$

where $R_i \equiv (R_i^x, R_i^y, R_i^z) = r_\mu - r_i$ and r_i is the position of the ith ion. Thus, if the magnetic moments in the sample at temperature T take the value $m_i(T) = m_i(0)f(T)$, where $f(T)$ is the temperature dependence of the order parameter, the precession frequency that is corresponding to the jth muon site $r_{\mu,j}$ will take the value

$$\nu_j(T) = \nu_j(0)f(T), \qquad (3)$$

the value of $\nu_j(0)$ being a function of $r_{\mu,j}$, but $f(T)$ being independent of j. Therefore, the only way in which one can explain two precession frequencies following different temperature dependences is if the magnetic structure (i.e. the arrangement of magnetic moments) is itself temperature dependent.

The magnetic moment in nitronyl nitroxide magnets results from the spin density delocalized over the O–N–C–N–O moiety [54,55]. The calculations of the dipolar field $B_{\text{dip}}^\alpha(r_\mu) = \sum_i D_i^{\alpha\beta}(r_\mu) m_i^\beta$ for the three candidate muon sites were performed, while assuming ferromagnetic order along the *c*-direction and with a $1\mu_B$ moment being delocalized over the O–N–C–N–O moiety (with equal spin density on the nitrogen and oxygen atoms, and weak negative spin density on the central carbon, as found in p-NPNN [54]). These calculations resulted in fields that were an order of magnitude higher than we observe, due to the fact that there is a very large coupling between the muon and molecule closest to it. If instead, as found in p-NPNN [8], the muon brings an electron that forms a singlet state with the molecule to which it is closest, then the magnetic moment on that molecule will be switched off. Repeating the calculation while assuming this local molecular singlet state gives values that are closer to the experimental values, although still too large. Including the contribution from the Lorentz field ($B_L = \mu_0 M/3$) and demagnetizing field ($B_{demag} = -NM$, where N is the demagnetizing factor, probably somewhere between $\approx 1/2$ and 1), gives a precession frequency $\approx$ 4.6–4.8 MHz for the lowest-energy state close to the oxygen, $\approx$3–4 MHz for the site close to the nitrogen atom, and $\approx$1.5–3 MHz for the site that is close to the aromatic ring (the uncertainty in these values reflecting the uncertainty in N, which is less important for the oxygen site for which the local field is dominated by a component of the dipolar field perpendicular to the magnetization and, hence, to the demagnetizing field B_{demag}). These frequencies are of the right order of magnitude, but they are all still too large, perhaps reflecting a magnetic state that is more complex than simple ferromagnetic order.

It is believed that the dominant exchange interaction in DNPNN is between the spin density on this O–N–C–N–O moiety on one molecule and the ortho-carbon atom of the phenyl group in the adjacent, $\pi/2$-rotated molecule [21]. This gives rise to one-dimensional spin chains along the helical molecular packing. In order to produce three-dimensional long range order, it is the role of the weak coupling between the chains that produces correlations whose length becomes infinite at T_C. It is possible that these weaker interactions, which, as noted earlier, are still significant in size, might be competing to produce the three-dimensional ordered state in zero-field, thereby resulting in a helical component to the ordered state, albeit with a strong ferromagnetic component. If these weaker interactions were temperature dependent, then the balance of competition could shift, resulting in changes to the pitch of the helix and, therefore, altering the field measured at the two different muon sites. Anisotropic (Dzyaloshinskii–Moriya) exchange could be another possible source of this effect, which can favour spin canting (and, hence, a ferromagnetic component) in an otherwise antiferromagnetic system. However, in a material that is composed of only carbon, hydrogen, oxygen, and nitrogen atoms, one would not normally expect anisotropic exchange (which is mainly due to the spin-orbit interaction) to be particularly strong. It is also possible that the chiral packing of the molecular units in crystals of DNPNN may result in noncollinear order. We speculate that it is the finely balanced competition between these different interchain interactions that results in a strong temperature sensitivity of the detailed magnetic structure, which results in the differing temperature dependences being measured at the two distinct muon sites.

4. Conclusions

The μSR results that are presented in this paper confirm the magnetic transition previously observed [21] in DNPNN, a nitronyl nitroxide magnet with a chiral crystal structure, to be a bulk effect, a transition to three-dimensional order below 1.10 K. Our results reveal a two-component precession signal that develops below T_C, and only spin relaxation is observed above the transition. The temperature dependence of the precession frequencies is unusual and the two frequencies follow markedly different temperature dependences. Although many nitronyl nitroxide magnets have been studied since the discovery of ferromagnetism in p-NPNN [4], these results highlight the special nature of DNPNN that appears to be a particularly interesting example of this family and deserving of further study.

Author Contributions: The muon experiments were performed by S.J.B., T.L., P.J.B. & F.L.P. The samples were prepared by D.S., K.S. and T.T. The site calculations and dipolar field calculations were performed by T.L. and S.J.B. respectively. The data were analysed by S.J.B. and T.L. All authors have read and agreed to the published version of the manuscript.

Funding: We acknowledge financial support from EPSRC (EP/N023803/1 and EP/N024028/1) and STFC (UK). This work was also supported by Grant-in-Aid for Scientific Research from MEXT, Japan. The muon experiments were performed at the STFC ISIS Pulsed Muon Souce. This work made use of the facilities of the Hamilton HPC Service of Durham University and the STFC Scientific Computing Department's SCARF cluster.

Institutional Review Board Statement: Not applicable.

Informed Consent Statement: Not applicable.

Data Availability Statement: The data are available at https://doi.org/10.5286/ISIS.E.RB1110363 (accessed on 13 May 2021).

Acknowledgments: Peter Day was an inspirational chemist with whom some of the present authors had the pleasure of working. He was an early adopter of experimental techniques, in particular neutron scattering which hitherto had been the preserve of physicists. He also made important contributions to realising interesting magnetic ground states in molecular architectures. We hope that he would have approved of the current article, particularly that it has resulted from a collaboration between chemists and physicists, a tradition he actively promoted.

Conflicts of Interest: The authors declare no conflict of interest. The funders had no role in the design of the study; in the collection, analyses, or interpretation of data; in the writing of the manuscript, or in the decision to publish the results.

References

1. Lahti, P.M. (Ed.) *Magnetic Properties of Organic Materials*; Dekker: New York, NY, USA, 1999.
2. Blundell, S.J.; Pratt, F.L. Organic and molecular magnets. *J. Phys. Condens. Matter* **2004**, *16*, R771–R828. [CrossRef]
3. Heisenberg, W. Zur Theorie des Ferromagnetismus. *Z. Phys.* **1928**, *49*, 619–636. [CrossRef]
4. Tamura, M.; Nakazawa, Y.; Shiomi, D.; Nozawa, K.; Hosokoshi, Y.; Ishikawa, M.; Takahashi, M.; Kinoshita, M. Bulk ferromagnetism in the β-phase crystal of the p-nitrophenyl nitronyl nitroxide radical. *Chem. Phys. Lett.* **1991**, *186*, 401–404. [CrossRef]
5. Kinoshita, M. p-nitrophenyl nitronyl nitroxideL the first organic ferromagnet. *Phil. Trans. R. Soc. Lond. A* **1999**, *357*, 2855–2872. [CrossRef]
6. Nakazawa, Y.; Tamura, M.; Shirakawa, N.; Shiomi, D.; Takahashi, M.; Kinoshita, M.; Ishikawa, M. Low-temperature magnetic properties of the ferromagnetic organic radical, p-nitrophenyl nitronyl nitroxide radical. *Phys. Rev. B* **1992**, *46*, 8906–8914. [CrossRef] [PubMed]
7. Le, L.P.; Keren, A.; Luke, G.M.; Wu, W.D.; Uemura, Y.J.; Tamura, M.; Ishikawa, M.; Kinoshita, M. Searching for spontaneous magnetic order in an organic ferromagnet. μSR studies of β-phase p-NPNN. *Chem. Phys. Lett.* **1993**, *206*, 405–408. [CrossRef]
8. Blundell, S.J.; Pattenden, P.A.; Pratt, F.L.; Valladares, R.M.; Sugano, T.; Hayes, W. μ^+SR of the organic ferromagnet p-NPNN: diamagnetic and paramagnetic states. *Europhys. Lett.* **1995**, *31*, 573–578. [CrossRef]
9. Blundell, S.J.; Sugano, T.; Pattenden, P.A.; Pratt, F.L.; Valladares, R.M.; Chow, K.H.; Uekusa, H.; Ohashi, Y.; Hayes, W. Magnetism in the nitronyl nitroxide isomers 1-NAPNN and 2-NAPNN studied by μ^+SR. *J. Phys. Condens. Matter* **1996**, *8*, L1–L6. [CrossRef]
10. Blundell, S.J.; Pattenden, P.A.; Pratt, F.L.; Chow, K.H.; Hayes, W.; Sugano, T. Organic magnetism in nitronyl nitroxides studied by μSR. *Hyp. Int.* **1997**, *104*, 251–256. [CrossRef]
11. Blundell, S.J. μSR studies of organic magnets. *Phil. Trans. R. Soc. Lond. A* **1999**, *357*, 2923–2937. [CrossRef]
12. Blundell, S.J.; Husmann, A.; Jestadt, T.; Pratt, F.L.; Marshall, I.M.; Lovett, B.W.; Kurmoo, M.; Sugano, T.; Hayes, W. Muon studies of molecular magnetism. *Physica B* **2000**, *289*, 115–118. [CrossRef]
13. Blundell, S.J.; Marshall, I.M.; Lovett, B.W.; Pratt, F.L.; Husmann, A.; Hayes, W.; Takagi, S.; Sugano, T. Organic magnetic materials studied by positive muons. *Hyp. Int.* **2001**, *133*, 169–177. [CrossRef]
14. Barron, L.D. Symmetry and molecular chirality. *Chem. Soc. Rev.* **1986**, *15*, 189–223. [CrossRef]
15. Yoon, T.P.; Jacobsen, E.N. Privileged Chiral Catalysts. *Science* **2003**, *299*, 1691–1693. [CrossRef]
16. Bullard, G.; Tassinari, F.; Ko, C.-H.; Mondal, A. K,; Wang, R.; Mishra, S.; Naaman, R.; Therien, M.J. Low-Resistance Molecular Wires Propagate Spin-Polarized Currents. *J. Am. Chem. Soc.* **2019**, *141*, 14707–14711. [CrossRef]
17. Naaman, R.; Waldeck, D.H. Chiral Supramolecular Structures as Spin Filters. *Annu. Rev. Phys. Chem.* **2015**, *66*, 263–281. [CrossRef]
18. Michaeli, K.; Naaman, R. Origin of spin-dependent tunneling through chiral molecules. *J. Phys. Chem. C* **2019**, *123*, 17043–17048. [CrossRef]

19. Mondal, A.K.; Brown, N.; Mishra, S.; Makam, P.; Wing, D.; Gilead, S.; Wiesenfeld, Y.; Leitus, G.; Shimon, L.J.W.; Carmieli, R.; Ehre, D.; Kamieniarz, G.; Fransson, J.; Hod, O.; Kronik, L.; Gazit, E.; Naaman, R. Long-Range Spin-Selective Transport in Chiral Metal–Organic Crystals with Temperature-Activated Magnetization. *ACS Nano* **2020**, *14*, 16624–16633. [CrossRef] [PubMed]
20. Rikken, G.L.J.A.; Raupach, E. Enantioselective magnetochiral photochemistry. *Nature* **2000**, *405*, 932–935. [CrossRef] [PubMed]
21. Shiomi, D.; Kanzaki, Y.; Okada, S.; Arima, R.; Miyazaki, Y.; Inaba, A.; Tanaka, R.; Sato, K.; Takui, T. An Enantiopair of Organic Ferromagnet Crystals Based on Helical Molecular Packing of Achiral Organic Radicals. *J. Phys. Chem. Lett.* **2011**, *2*, 3036–3039. [CrossRef]
22. Cox, S.F.J. Implanted muon studies in condensed matter science. *J. Phys. C* **1987**, *20*, 3187–3319. [CrossRef]
23. Blundell, S.J. Spin-polarized muons in condensed matter physics. *Contemp. Phys.* **1999**, *40*, 175–192. [CrossRef]
24. Dalmas de Réotier, P.; Yaouanc, A. Muon spin rotation and relaxation in magnetic materials. *J. Phys. C* **1997**, *9*, 9113–9166.
25. Blundell, S.J.; De Renzi, R.; Lancaster, T.; Pratt, F.L. (Eds.) *Introduction to Muon Spectroscopy*; Oxford University Press: Oxford, UK, 2021; to be published.
26. Lancaster, T.; Blundell, S.J.; Pratt, F.L. Another dimension: investigations of molecular magnetism using muon-spin relaxation. *Phys. Scr.* **2013**, *88*, 068506. [CrossRef]
27. Blundell, S.J. Molecular magnets. *Contemp. Phys.* **2007**, *48*, 275–290. [CrossRef]
28. Lancaster, T.; Blundell, S.J.; Brooks, M.L.; Baker, P.J.; Pratt, F.L.; Manson, J.L.; Landee, C.P.; Baines, C. Magnetic order in the quasi-one-dimensional spin-1/2 molecular chain compound copper pyrazine dinitrate. *Phys. Rev. B* **2006**, *73*, R020410. [CrossRef]
29. Manson, J.L.; Lancaster, T.; Chapon, L.C.; Blundell, S.J.; Schlueter, J.A.; Brooks, M.L.; Pratt, F.L.; Nygren, C.L.; Qualls, J.S. Cu(HCO$_2$)$_2$(pym) (pym=pyridine): Low-dimensional magnetic behavior and long-range ordering in a quantum-spin lattice. *Inorg. Chem.* **2005**, *44*, 989–995. [CrossRef] [PubMed]
30. Blundell, S.J.; Pattenden, P.A.; Valladares, R.M.; Pratt, F.L.; Sugano, T.; Hayes, W. Observation of a magnetic transition in *para*-pyridyl nitronyl nitroxide using zero-field μSR. *Solid State Commun.* **1994**, *92*, 569–572. [CrossRef]
31. Steele, A.J.; Lancaster, T.; Blundell, S.J.; Baker, P.J.; Pratt, F.L.; Baines, C.; Conner, M.M.; Southerland, H.I.; Manson, J.L.; Schlueter, J.A. Magnetic order in quasi-two-dimensional molecular magnets investigated with muon-spin relaxation. *Phys. Rev. B* **2011**, *84*, 064412. [CrossRef]
32. Blundell, S.J.; Lancaster, T.; Pratt, F.L.; Baker, P.J.; Brooks, M.L.; Baines, C.; Manson, J.L.; Landee, C.P. μ^+SR as a probe of anisotropy in low-dimensional molecular magnets. *J. Phys. Chem. Solids* **2007**, *68*, 2039–2043. [CrossRef]
33. Pelissetto, A.; Vicari, E. Critical phenomena and renormalization-group theory. *Phys. Rep.* **2002**, *368*, 549–727. [CrossRef]
34. Pratt, F.L.; Baker, P.J.; Blundell, S.J.; Lancaster, T.; Green, M.A.; Kurmoo, M. Chiral-Like Critical Behavior in the Antiferromagnet Cobalt Glycerolate. *Phys. Rev. Lett.* **2007**, *99*, 017202. [CrossRef]
35. Mukamel, D. Physical Realizations of $n \gtrsim 4$ Vector Models. *Phys. Rev. Lett.* **1975**, *34*, 481–485. [CrossRef]
36. Möller, J.S.; Bonfá, P.; Ceresoli, D.; Bernardini, F.; Blundell, S.J.; Lancaster, T.; De Renzi, R.; Marzari, N.; Watanabe, I.; Sulaiman, S.; Mohamed-Ibrahim, M.I. Playing quantum hide-and-seek with the muon: localizing muon stopping sites. *Phys. Scr.* **2013**, *88*, 068510. [CrossRef]
37. Huddart, B.M. A Program to Classify and Analyse Muon Stopping Sites. 2020. Available online: https://gitlab.com/BenHuddart/mufinder/ (accessed on 17 May 2021).
38. Clark, S.J.; Segall, M.D.; Pickard, C.J.; Hasnip, P.J.; Probert, M.I. J.; Refson, K.; Payne, M. First principles methods using CASTEP. *Z. Kristall.* **2005**, *220*, 567. [CrossRef]
39. Algra, H.A.; de Jongh, L.J.; Huiskamp, W.J.; Carlin, R.L. Magnetic behavior of [(CH$_3$)$_3$NH]CuCl$_3$·2H$_2$O. Evidence for lattice-dimensionality crossovers in a quasi one-dimensional ferromagnet. *Physica* **1977**, *92B*, 187–200. [CrossRef]
40. Swank, D.D.; Landee, C.P.; Willett, R.D. Crystal structure and magnetic susceptibility of copper (II) chloride tetramethylsulfoxide [CuCl$_2$ (TMSO)] and copper (II) chloride monodimethylsulfoxide [CuCl$_2$ (DMSO)]: Ferromagnetic spin-1/2 Heisenberg linear chains. *Phys. Rev. B* **1979**, *20*, 2154–2162. [CrossRef]
41. Willett, R.D., Landee, C.P. Ferromagnetism in one dimensional systems: synthesis and structural characterization. *J. Appl. Phys.* **1981**, *52*, 2004–2009. [CrossRef]
42. Takeda, K.; Konishi, K.; Nedachi, K.; Mukai, K. Experimental Study of Quantum Statistics for the $S = 1/2$ Quasi-One-Dimensional Organic Ferromagnet. *Phys. Rev. Lett.* **1995**, *74*, 1673–1676. [CrossRef]
43. Willett, R.D.; Landee, C.P.; Gaura, R.M.; Swank, D.D.; Groenendijk, H.A.; van Duyneveldt, A.J. Magnetic properties of one-dimensional spin 1/2 ferromagnets: Metamagnetic behavior of (C$_6$H$_{11}$NH$_3$)CuCl$_3$. *J. Magn. Magn. Mat.* **1980**, *15–18*, 1055–1056. [CrossRef]
44. Schouten, J.C.; van der Geest, G.J.; de Jonge, W.J. M.; Kopinga, K. Specific heat of (C$_6$H$_{11}$NH$_3$) CuCl$_3$ (CHAC), a system of ferromagnetic chains. *Phys. Lett. A* **1980**, *78*, 398–400. [CrossRef]
45. Takeda, K.; Hamano, T.; Kawae, M.; Hidaka, M., Takahashi, M.; Kawasaki, S.; Mukai, K. Experimental Check of Heisenberg Chain Quantum Statistics for a Ferromagnetic OrganicRadical Crystal. *J. Phys. Soc. Jpn.* **1995**, *64*, 2343–2346. [CrossRef]
46. Landee, C.P.; Willett, R.D. Tetramethylammonium Copper Chloride and tris (Trimethylammonium) Copper Chloride: $S = 1/2$ Heisenberg One-Dimensional Ferromagnets. *Phys. Rev. Lett.* **1979**, *43*, 463–466. [CrossRef]
47. Dupas, C.; Renard, J.P.; Seiden, J.; Cheikh-Rouhou, A. Static magnetic properties of (CH$_3$)$_4$NMn$_x$Cu$_{1-x}$Cl$_3$, a quantum ferromagnetic chain with classical impurities: Experiment and theory *Phys. Rev. B* **1982**, *25*, 3261–3272. [CrossRef]

48. Kopinga, K.; Tinus, A.M.C.; de Jonge, W.J.M. Magnetic behavior of the ferromagnetic quantum chain systems ($C_6H_{11}NH_3$)$CuCl_3$ (CHAC) and ($C_6H_{11}NH_3$)$CuBr_3$ (CHAB). *Phys. Rev. B* **1982**, *25*, 4685–4690. [CrossRef]
49. Sugano, T.; Blundell, S.J.; Lancaster, T.; Pratt, F.L.; Mori, H. Magnetic order in the purely organic quasi-one-dimensional ferromagnet 2-benzimidazolyl nitronyl nitroxide. *Phys. Rev. B* **2010**, *82*, 180401(R). [CrossRef]
50. Blundell, S.J.; Möller, J.S.; Lancaster, T.; Baker, P.J.; Pratt, F.L.; Seber, G.; Lahti, P.M. μSR study of magnetic order in the organic quasi-one-dimensional ferromagnet F4BImNN. *Phys. Rev. B* **2013**, *88*, 064423. [CrossRef]
51. Clemente-León, M.; Coronado, E.; Galán-Mascarós, J.R.; Gómez-García, C.J. Intercalation of decamethylferrocenium cations in bimetallic oxalate-bridged two-dimensional magnets. *Chem. Commun.* **1997**, 1727–1728. [CrossRef]
52. Coronado, E.; Galan-Mascaros, J.R.; Gomez-Garcia, C.J.; Burriel, R. A molecular chemical approach to the magnetic multilayers. *J. Magn. Magn. Mater.* **1999**, *197*, 558–560. [CrossRef]
53. Lancaster, T.; Blundell, S.J.; Pratt, F.L.; Coronado, E.; Galan-Mascaros, J.R. Magnetic order and local field distribution in the hybrid magnets [FeCp$_2^*$][MnCr(ox)$_3$] and [CoCp$_2^*$][FeFe(ox)$_3$]: a muon spin relaxation study. *J. Mater. Chem.* **2004**, *14*, 1518–1520. [CrossRef]
54. Zheludev, A.; Barone, V.; Bonnet, M.; Delley, B.; Grand, A.; Ressouche, E.; Rey, P.; Subra, R.; Schweizer, J. Spin density in a nitronyl nitroxide free radical. Polarized neutron diffraction investigation and ab initio calculations. *J. Am. Chem. Soc.* **1994**, *116*, 2019–2027. [CrossRef]
55. Heise, H.; Kohler, F.H.; Mota, F.; Novoa, J.J.; Veciana, J. Determination of the Spin Distribution in Nitronylnitroxides by Solid-State ^{1}H, ^{2}H, and ^{13}C NMR Spectroscopy. *J. Am. Chem. Soc.* **1999**, *121*, 9659–9667. [CrossRef]

Article

Magnetic Switching in Vapochromic Oxalato-Bridged 2D Copper(II)-Pyrazole Compounds for Biogenic Amine Sensing †

Nadia Marino [1,*], María Luisa Calatayud [2], Marta Orts-Arroyo [2], Alejandro Pascual-Álvarez [2], Nicolás Moliner [2], Miguel Julve [2,*], Francesc Lloret [2], Giovanni De Munno [1], Rafael Ruiz-García [2] and Isabel Castro [2,*]

1 Dipartimento di Chimica e Technologie Chemiche, Università della Calabria, 87036 Rende, Italy; demunno@unical.it
2 Instituto de Ciencia Molecular (ICMol), Universitat de València, C/Catedrático José Beltrán 2, 46980 Valencia, Spain; marialuisacalatayud@gmail.com (M.L.C.); martaorts.a@gmail.com (M.O.-A.); alejandro.pascual@uv.es (A.P.-Á.); Fernando.moliner@uv.es (N.M.); francisco.lloret@uv.es (F.L.); rafael.ruiz@uv.es (R.R.-G.)
* Correspondence: nadia.marino@unical.it (N.M.); miguel.julve@uv.es (M.J.); isabel.castro@uv.es (I.C.)
† In memoriam Professor Peter Day, one of the most respected inorganic chemists whose impact as academic and scientist can only be described in the highest praise, on his passing last May 2020. We are sure his legacy will have long-lasting effects.

Citation: Marino, N.; Calatayud, M.L.; Orts-Arroyo, M.; Pascual-Álvarez, A.; Moliner, N.; Julve, M.; Lloret, F.; De Munno, G.; Ruiz-García, R.; Castro, I. Magnetic Switching in Vapochromic Oxalato-Bridged 2D Copper(II)-Pyrazole Compounds for Biogenic Amine Sensing. *Magnetochemistry* 2021, 7, 65. https://doi.org/10.3390/magnetochemistry7050065

Academic Editors: Lee Martin, Scott Turner, John Wallis, Hiroki Akutsu and Carlos J. Gómez García

Received: 28 April 2021
Accepted: 8 May 2021
Published: 12 May 2021

Publisher's Note: MDPI stays neutral with regard to jurisdictional claims in published maps and institutional affiliations.

Copyright: © 2021 by the authors. Licensee MDPI, Basel, Switzerland. This article is an open access article distributed under the terms and conditions of the Creative Commons Attribution (CC BY) license (https://creativecommons.org/licenses/by/4.0/).

Abstract: A new two-dimensional (2D) coordination polymer of the formula $\{Cu(ox)(4\text{-Hmpz})\cdot 1/3H_2O\}_n$ (**1**) (ox = oxalate and 4-Hmpz = 4-methyl-1H-pyrazole) has been prepared, and its structure has been determined by single-crystal X-ray diffraction. It consists of corrugated oxalato-bridged copper(II) neutral layers featuring two alternating bridging modes of the oxalate group within each layer, the symmetric bis-bidentate (μ-$\kappa^2 O^1, O^2$:$\kappa^2 O^{2'}, O^{1'}$) and the asymmetric bis(bidentate/monodentate) (μ_4-κO^1:$\kappa^2 O^1, O^2$:$\kappa O^{2'}$:$\kappa^2 O^{2'}, O^{1'}$) coordination modes. The three crystallographically independent six-coordinate copper(II) ions that occur in **1** have tetragonally elongated surroundings with three oxygen atoms from two oxalate ligands, a methylpyrazole-nitrogen defining the equatorial plane, and two other oxalate-oxygen atoms occupying the axial positions. The monodentate 4-Hmpz ligands alternatively extrude above and below each oxalato-bridged copper(II) layer, and the water molecules of crystallization are located between the layers. Compound **1** exhibits a fast and selective adsorption of methylamine vapors to afford the adsorbate of formula $\{Cu(ox)(4\text{-Hmpz})\cdot 3MeNH_2\cdot 1/3H_2O\}_n$ (**2**), which is accompanied by a concomitant color change from cyan to deep blue. Compound **2** transforms into $\{Cu(ox)(4\text{-Hmpz})\cdot MeNH_2\cdot 1/3H_2O\}_n$ (**3**) under vacuum for three hours. The cryomagnetic study of **1**–**3** revealed a unique switching from strong (**1**) to weak (**2** and **3**) antiferromagnetic interactions. The external control of the optical and magnetic properties along this series of compounds might make them suitable candidates for switching optical and magnetic devices for chemical sensing.

Keywords: copper; oxalate; pyrazole; crystal structure; 2D coordination polymers; magnetic properties; sorption properties; amine sensing

1. Introduction

The design and synthesis of molecule-based multifunctional magnetic materials has opened new possibilities in the field of molecular magnetism [1,2]. The goal of this research is to explore new classes of compounds that combine two (or more) chemical and physical properties, besides the magnetic ones, which are of fundamental importance for industrial and technological applications [3–11]. New magnetic phenomena would arise from the synergy between these coexisting properties (magnetic second harmonic generation, magneto-chiral dichroism, or multiferroicity), which could eventually be further modified by the application of external stimuli (temperature, pressure, light, or chemical analytes) [10]. Such a fruitful avenue of molecular magnetism was initiated half a decade ago by Peter Day's pioneering work on polyhalide metal salts with inorganic and organic (TTF and BEDT-TTF derivatives) cations as examples of purely inorganic and inorganic–organic hybrid molecular magnetic materials that led to a unique class of magnetic molecular conductors [3,4,7]. Several research groups are currently exploring this method of acquiring new types of potentially switchable, molecule-based, multifunctional magnetic materials. Chiral and luminescent magnets; non-linear optics (NLO) magnets and multiferroics; protonic and electronic magnetic conductors; spin crossover or valence tautomeric magnets; and thermo-, piezo-, photo-, or chemoswitchable porous magnets constitute some illustrative examples. The addressing of these multifunctional molecular magnetic materials on thin films and surfaces, or their shaping as nanoparticles or nanocrystals, is mandatory for their use in nanoscience and nanotechnology in the near future. In fact, they should be integrated into devices to realize their potential applications, crossing the bridge between fundamental science and cutting-edge technological products [10].

Magnetic coordination polymers have become one of the most challenging issues among this diverse class of multifunctional molecule-based magnetic materials due to their structural tunability and potentially switchable chemical (sorption, sensing, redox, or catalytic) and physical (optical, thermal, magnetic, or conducting) properties [12–22]. This particular class of inorganic–organic hybrid porous materials, also referred to as metal–organic frameworks (MOFs), have emerged as suitable sensory materials for the detection and monitoring of gases and vapors from volatile organic compounds (VOCs) of industrial, medical, or environmental interest [23–34]. A major goal in the area of magnetic MOFs is to tune their optical, electronic, and/or magnetic properties by the inclusion of selected guests that are adsorbed through simple chemi- or physisorption processes [35–47]. Among the target molecules, ammonia, or its biogenic amine (BA) derivatives, from the simple methyl- and trimethylamine to the more complex tetramethylene- (putrescine) and pentamethylenediamine (cadaverine), are of particular interest for the future applications of magnetic coordination polymers in the chemical sensing of VOCs resulting from industrial procedures or food degradation [48–60]. Due to their unique multiresponsive and multifunctional character, magnetic MOFs are good alternatives to organic polymeric materials and their metal composites, envisaging the substitution of metal- and metal oxide-based commercial chemiresistive BA sensors, which are currently used in industrial process management and food quality control [54].

Our strategy in this field is based on the use of oxalate and pyrazole derivatives as bridging and terminal ligands, respectively, toward copper(II) ions for the preparation of heteroleptic copper(II) coordination polymers [61–63]. In fact, oxalate is a versatile polyatomic ligand because of the great number of coordination modes that it can adopt in its heteroleptic copper(II) complexes with pyrazole derivatives, depending on the steric and/or electronic effects of the pyrazole substituents [61]. Moreover, it is an efficient mediator of magnetic interactions between copper(II) ions when acting as a bridge, with the strength and nature of these interactions, either ferro- or antiferromagnetic, depending on the coordination mode of the oxalate bridge and on the geometry at the copper(II) ions [62,63]. Herein we focus on the synthesis and spectroscopic and magneto-structural characterization of a novel oxalato-bridged two-dimensional (2D) copper(II) coordination polymer of formula $\{Cu(ox)(4\text{-Hmpz})\cdot 1/3H_2O\}_n$ (**1**) (4-Hmpz = 4-methyl-1H-pyrazole), together with a preliminary study on its sorption properties toward polymethyl-substituted amines with different steric and/or electronic effects. Our goal is to investigate the influence of the number of methyl substituents on the adsorption behavior and eventually on the optical and magnetic properties of the resulting methylamine adsorbates of the formulae $\{Cu(ox)(4\text{-Hmpz})\cdot 3MeNH_2\cdot 1/3H_2O\}_n$ (**2**) and $\{Cu(ox)(4\text{-Hmpz})\cdot MeNH_2\cdot 1/3H_2O\}_n$ (**3**). The color change and modification of the magnetic properties that **1** exhibits upon the selective adsorption of methylamine makes it a new prototype for a bimodal optical (colorimetric) and magnetic sensor for the selective vapor detection of biogenic amines.

2. Materials and Methods

2.1. Materials

Oxalic acid (H_2ox), sodium oxalate (Na_2ox), 4-Hmpz, copper(II) perchlorate hexahydrate, 33% methylamine, dimethylamine, and trimethylamine solutions in absolute ethanol, and triethylamine were of laboratory grade and were used as received.

2.2. Preparations of 1 and Its Methylamine Adsorbates 2 and 3

2.2.1. {Cu(ox)(4-Hmpz)·1/3H$_2$O}$_n$ (1)

An aqueous solution (15 mL) of 4-Hmpz (0.123 g, 1.5 mmol) was added dropwise to an aqueous solution (20 mL) of copper(II) perchlorate hexahydrate (0.370 g, 1.0 mmol). Na_2ox (0.134 g, 1.0 mmol) dissolved in a hot aqueous solution (10 mL) was added dropwise to the above solution. The resulting sky blue mixture was stirred for 30 min under gentle warming. A pale blue polycrystalline solid of **1** that separated was filtered off and air-dried (0.151 g, 65% yield). Anal. calcd for $C_6H_6N_2CuO_4 \cdot 1/3H_2O$ (MW = 239.7 g mol^{-1}): C, 30.07; H, 2.80; N, 11.69%. Found: C, 29.27; H, 2.75; N, 11.65%. IR (KBr/cm^{-1}): 3496w [υ(O–H) from water], 3404m [υ(N–H) from 4-Hmpz], 3189w and 2926w [υ(C–H) from 4-Hmpz], 1706s, 1651s and 1599vs [υ_{as}(CO) from ox], 1358m, 1317w and 1297s [υ_s(CO) from ox] and 820m and 803m [δ(OCO) from ox]. When using an aqueous solution (10 mL) of H_2ox (0.045 g, 0.5 mmol) and Et_3N (0.14 mL, 1.0 mmol) instead of sodium oxalate, a small amount of tiny pale greenish blue platelets of **1**, suitable for X-ray analysis, were grown by slow evaporation of the filtered aqueous solutions upon standing at room temperature for several days.

2.2.2. {Cu(ox)(4-Hmpz)·3MeNH$_2$·1/3H$_2$O}$_n$ (2)

A polycrystalline sample of **1** (0.120 g, 0.5 mmol) was placed in a Schlenk flask connected to an argon current that had previously flowed through a bubbler filled with a 33% $MeNH_2$ solution in absolute ethanol (25 mL), and transformed into a deep blue powder of **2** after 3 h of exposure at room temperature. Anal. calcd for $C_9H_{21}N_5CuO_4 \cdot 1/3H_2O$ (MW = 332.7 g mol^{-1}): C, 32.48; H, 6.56; N, 21.04%. Found: C, 32.79; H, 6.59; N, 20.99%; IR (KBr/cm^{-1}): 3497w [(υ(O–H) from water] 3405m [υ(N–H) from 4-Hmpz and $MeNH_2$], 3101w and 2971w [υ(C–H) from 4-Hpmz and $MeNH_2$], 1674vs, 1652s and 1637s [υ_{as}(CO) from ox], 1418s [υ(C–H)$_{bending}$ from $MeNH_2$], 1370m and 1289s [υ_s(CO) from ox], 928m [υ(C–N)$_{stretching}$ from $MeNH_2$], and 829m and 807m [δ(OCO) from ox].

2.2.3. {Cu(ox)(4-Hmpz)·MeNH$_2$·1/3H$_2$O}$_n$ (3)

Treatment of **2** (0.166 g, 0.5 mmol) under vacuum for 3 h gave a grayish deep blue powder of **3**. Anal. calcd for $C_7H_{11}N_3CuO_4 \cdot 1/3H_2O$ (MW = 270.7 g mol^{-1}): C, 31.05; H, 4.34; N, 15.52%. Found: C, 31.03; H, 4.40; N, 15.37%; IR (KBr/cm^{-1}): 3497w [(υ(O–H) from water], 3407m [υ(N–H) from 4-Hmpz and $MeNH_2$], 3103w and 2950w [υ(C–H) from Hpmz and $MeNH_2$], 1675vs, 1652s and 1637s [υ_{as}(CO) from ox], 1418s cm^{-1} [υ(C–H)$_{bending}$ from $MeNH_2$], 1372m and 1289s [υ_s(CO) from ox], 927m [υ(C–N)$_{stretching}$ from $MeNH_2$], and 830m and 806m [δ(OCO) from ox].

2.3. Vapor Adsorption Studies

The vapor adsorption kinetic measurements were carried out on different aliquots containing powdered polycrystalline samples of **1** (0.024 g, 0.1 mmol) at room temperature in a Schlenk flask connected to an argon current that had previously flowed through a bubbler filled with either 33% $MeNH_2$, Me_2NH, or Me_3N solutions in absolute ethanol. The amine contents were determined by elemental analysis of the corresponding samples after 1, 6, 20, 60, and 180 min of exposure to the amine-saturated argon flow.

2.4. Physical Techniques

Elemental analyses (C, H, N) were performed by the Servei Central de Suport a la Investigació Experimental de la Universitat de València. FT-IR spectra were recorded on a Nicolet-5700 spectrophotometer as KBr pellets. X-ray powder diffraction (XRPD) patterns of powdered polycrystalline samples were collected at room temperature on a D8 Avance A25 Bruker diffractometer by using graphite-monochromated Cu-Kα radiation (λ = 1.54056 Å). Variable-temperature (2.0–300 K) magnetic susceptibility measurements were carried out with a SQUID magnetometer under applied fields of 5.0 kOe (T > 20 K) and 250 Oe (T < 20 K) to prevent any saturation effect at low temperature. The powdered polycrystalline samples were embedded on n-eicosane and placed in small sealed plastic bags to prevent any solvent loss during the magnetic measurements. The experimental magnetic susceptibility data were corrected for the diamagnetic contributions of the constituent atoms and the sample holder, as well as for the temperature-independent paramagnetism (tip) of the Cu^{II} ion (60 $\times$ 10^{-6} cm^3 mol^{-1}).

2.5. X-ray Crystallographic Data Collection and Structure Refinement

X-ray crystallographic data for **1** were collected with a Bruker-AXS SMART CCD diffractometer at 98 K using graphite monochromated Cu-Kα radiation (λ = 1.54178 Å). The crystal selected for data collection, a tiny single-laminar fragment with approximate dimensions of 0.005 $\times$ 0.050 $\times$ 0.110 mm, was coated with Paraton oil to prevent any potential solvent loss, attached to a glass fiber, and quickly transferred under the cold nitrogen stream of the diffractometer. The Bruker SMART and SAINT softwares were employed for data collection and integration, respectively. Empirical absorption corrections were calculated using SADABS [64–66]. The structures were solved by direct methods and subsequently completed by Fourier recycling using the SHELXTL software packages [67,68] and refined by the full-matrix least-squares refinements based on F^2 with all observed reflections. All non-hydrogen atoms were refined anisotropically. The hydrogen atoms of the 4-Hmpz ligands were set in calculated positions and refined using a riding model. The hydrogen atoms on the water molecule of crystallization were located on the ΔF map and refined with restraints on the O–H and H$\cdots$H distances, with the thermal factors fixed to 0.05 Å^2. The final geometrical calculations and graphical manipulations were performed using the XP utility within SHELX and the Diamond program [69]. Crystal data for **1** are summarized in Table 1. Selected bond distances and angles and hydrogen bonds for **1** are listed in Tables 2 and 3, respectively. The value of Z reported in Table 1 (Z = 12) refers to the formula {Cu(ox)(4-Hmpz)·1/3H$_2$O}$_n$; the asymmetric unit for **1** comprises three crystallographically independent copper(II) ions with analogous coordination environments, and one water molecule of crystallization [Z = 4 for {[Cu(ox)(4-Hmpz)]$_3$·H$_2$O}$_n$]. The CCDC reference number is 2079936.

Table 1. Summary of crystal data and structure refinement for **1**.

Formula	C$_6$H$_{6.67}$CuN$_2$O$_{4.33}$
Fw	239.66
Crystal system	Monoclinic
Space group	$P2_1/c$
a/Å	9.9554(5)
b/Å	9.3037(4)
c/Å	25.2695(12)
β/°	94.670(3)
V/Å^3	2332.74(19)
Z	12
D_c/g cm^{-3}	2.047
T/K	90(2)
μ/mm^{-1}	3.949
F(000)	1444
Refl. Collected	22,317
Refl. indep. [R_{int}]	3947 [0.0592]
Refl. obs. [I > 2σ(I)]	2956
Goodness-of-fit on F^2	1.079

Table 1. Cont.

R_1 [a] $[I > 2\sigma(I)]$ (all)	0.0497 (0.0690)
wR_2 [b] $[I > 2\sigma(I)]$ (all)	0.1309 (0.1422)
$\Delta\rho_{max, min}/e\text{ Å}^{-3}$	0.973 and −0.432

[a] $R_1 = \sum ||F_o| - |F_c||/\sum |F_o|$. [b] $wR_2 = \{\sum w(F_o^2 - F_c^2)^2/\sum [w(F_o^2)^2]\}^{1/2}$ and $w = 1/[\sigma^2(F_o) + (mP)^2 + nP]$ with $P = (F_o^2 + 2F_c^2)/3$, $m = 0.0633$ and $n = 7.8872$.

Table 2. Selected bond distances (Å) and angles (°), and intra- and inter-chain Cu···Cu distances (Å) for **1** [1].

Cu(1) environment		Cu(2) environment		Cu(3) environment	
Cu(1)-N(1)	1.985(4)	Cu(2)-N(3)	1.983(4)	Cu(3)-N(5)	1.982(4)
Cu(1)-O(1)	1.969(3)	Cu(2)-O(6)	1.961(3)	Cu(3)-O(8d)	1.968(3)
Cu(1)-O(3)	1.984(3)	Cu(2)-O(10)	1.998(3)	Cu(3)-O(11)	1.993(3)
Cu(1)-O(4b)	2.010(3)	Cu(2)-O(9)	1.978(3)	Cu(3)-O(12)	1.991(3)
Cu(1)-O(5)	2.388(3)	Cu(2)-O(7c)	2.561(3)	Cu(3)-O(2)	2.527(3)
Cu(1)-O(2a)	2.428(3)	Cu(2)-O(5)	2.317(3)	Cu(3)-O(7d)	2.287(3)
O(1)-Cu(1)-N(1)	97.70(15)	O(6)-Cu(2)-N(3)	93.96(15)	O(8d)-Cu(3)-N(5)	95.18(15)
O(1)-Cu(1)-O(2a)	75.08(12)	O(6)-Cu(2)-O(5)	78.54(12)	O(8d)-Cu(3)-O(7d)	79.09(12)
O(1)-Cu(1)-O(5)	94.19(12)	O(6)-Cu(2)-O(7c)	88.96(12)	O(8d)-Cu(3)-O(2)	91.07(12)
O(1)-Cu(1)-O(3)	90.90(13)	O(6)-Cu(2)-O(9)	91.76(13)	O(8d)-Cu(3)-O(12)	89.94(13)
O(1)-Cu(1)-O(4b)	172.64(13)	O(6)-Cu(2)-O(10)	176.00(13)	O(8d)-Cu(3)-O(11)	173.95(13)
O(3)-Cu(1)-N(1)	170.77(15)	O(9)-Cu(2)-N(3)	173.60(14)	O(12)-Cu(3)-N(5)	174.88(14)
O(3)-Cu(1)-O(5)	88.91(12)	O(9)-Cu(2)-O(7c)	83.13(11)	O(12)-Cu(3)-O(2)	87.89(11)
O(3)-Cu(1)-O(2a)	86.76(12)	O(9)-Cu(2)-O(5)	86.99(12)	O(12)-Cu(3)-O(7d)	91.32(12)
O(3)-Cu(1)-O(4b)	83.28(13)	O(9)-Cu(2)-O(10)	84.32(13)	O(11)-Cu(3)-O(12)	84.10(13)
N(1)-Cu(1)-O(5)	93.76(14)	N(3)-Cu(2)-O(7c)	94.10(13)	N(5)-Cu(3)-O(2)	91.81(14)
N(1)-Cu(1)-O(2a)	92.22(14)	N(3)-Cu(2)-O(5)	96.95(14)	N(5)-Cu(3)-O(7d)	89.84(14)
N(1)-Cu(1) O(4b)	87.87(15)	N(3)-Cu(2)-O(10)	90.00(14)	N(5)-Cu(3)-O(11)	90.78(15)
O(4b)-Cu(1)-O(2a)	100.00(12)	O(10)-Cu(2)-O(5)	100.36(12)	O(11)-Cu(3) O(7d)	102.05(12)
O(4b)-Cu(1)-O(5)	90.22(12)	O(10)-Cu(2)-O(7c)	91.39(12)	O(11)-Cu(3)-O(2)	87.66(11)
O(5)-Cu(1)-O(2a)	168.34(11)	O(7c)-Cu(2)-O(5)	163.82(11)	O(2)-Cu(3)-O(7d)	170.13(11)
Intrachain Cu-(μ-ox)-Cu		Intrachain Cu-(μ₄-ox)-Cu		Interchain shortest Cu···Cu	
Cu(1)···Cu(1b)	5.2216(13)	Cu(1)···Cu(1a)	5.6913(13)	Cu(1)···Cu(2)	4.1148(9)
Cu(2)···Cu(3)	5.1824(9)	Cu(2)···Cu(3e)	5.5166(9)	Cu(1a)···Cu(3)	4.3324(9)

[1] Symmetry code: (a) = −x + 1, −y + 1, −z + 1; (b) = −x + 1, −y, −z + 1; (c) = −x + 1, y + 1/2, −z + 1/2; (d) = x, y + 1, z; (e) = x, y − 1, z.

Table 3. Hydrogen bond distances (Å) and angles (°) for **1** [1,2].

D-H···A	d(D-H)	d(H···A)	d(D···A)	<(DHA)
N(2)-H(2)···O(1w)	0.88	1.8	2.662(5)	166.1
O(1w)-H(1w1)···O(12)	0.956(10)	1.923(19)	2.828(5)	157(4)
O(1w)-H(1w2)···O(3a)	0.958(10)	1.796(12)	2.752(5)	176(4)
N(4)-H(4)···O(11f)	0.88	1.99	2.800(5)	152.1
N(6)-H(6)···N(1a)	0.88	2.46	3.296(6)	158.2

[1] D = donor and A = acceptor. [2] Symmetry code: (a) = −x + 1, −y + 1, −z + 1; (f) = −x + 1, y − 1/2, −z + 1/2.

3. Results and Discussion

3.1. Synthesis and General Physicochemical Characterization of 1–3

1 was prepared by the straightforward reaction of sodium oxalate with copper(II) perchlorate and 4-Hmpz in a 1:1:1.5 ox:Cu(II):4-Hmpz molar ratio at room temperature. It was isolated as a pale blue-greenish polycrystalline powder in a reasonable yield. A few

X-ray quality crystals of **1** as thin light blue plates were grown by the slow evaporation of aqueous solutions where a mixture of oxalic acid and triethylamine in a 1:2 molar ratio was used instead of sodium oxalate. When a polycrystalline sample of **1** was reacted with MeNH$_2$ vapor, it transformed into **2**, which, in turn, became **3** after placing **2** under a vacuum (see Experimental Section). The chemical identity of **1** was determined by X-ray diffraction on single crystals, and the purity of the bulk was confirmed by XRPD (Figure S1 in the Supplementary Material). Unfortunately, all our attempts to solve the X-ray crystal structure of **2** were unsuccessful because the crystallinity retention upon methylamine adsorption was very poor (see discussion below). The methylamine adsorbates were then characterized by elemental analysis and IR spectroscopy. In fact, the IR spectra of **2** and **3** are quasi-identical, suggesting a very close structure for both methylamine adsorbates (Figure S2 in the Supplementary Material).

Absorptions at 3496m [υ(O–H)], 3404m [υ(N–H)], 3190w, and 2926w cm^{-1} [υ(C–H)] in the IR spectrum of **1** indicated the presence of water and 4-Hpmz in this compound (Figure S2a). The set of peaks assigned to the oxalate group at 1705s, 1651s, and 1599vs [υ_{as}(CO)]; 1358m, 1317w, and 1297s [υ_s(CO)]; and 820m and 803m cm^{-1} [δ(OCO)] pointed out the coexistence of different bridging modes of this ligand in **1**, as confirmed by the X-ray structure (see below). The occurrence of a strong absorption peak at 1418 cm^{-1} and a medium intensity peak at 927 cm^{-1} [υ(C–H)$_{bending}$ and υ(C–N)$_{stretching}$ modes] in the IR spectra of **2** and **3** were indicative of the presence of MeNH$_2$ in them (Figure S2b,c) [70]. Finally, the set of quasi-identical absorptions peaks attributed to the oxalate group in **2** [υ_{as}(CO) = 1674vs, 1652s and 1637s cm^{-1}, υ_s(CO) = 1370m and 1289s cm^{-1}, and δ(OCO) = 829m and 807m cm^{-1}] and **3** [υ_{as}(CO) = 1675vs, 1652s and 1637s cm^{-1}, υ_s(CO) = 1372m and 1289s cm^{-1}, and δ(OCO) = 839m and 806m cm^{-1}] in their IR spectra suggests that this ligand exhibits the same bridging modes in them.

3.2. Description of the Structure of **1**

1 crystallizes in the monoclinic space group $P2(1)/c$, with three crystallographically independent copper(II) ions bound to a 4-Hmpz ligand of each one, two complete and two half-oxalato bridges, and one water molecule of crystallization in the asymmetric unit (Figure 1a). Its structure consists of corrugated oxalato-bridged copper(II) layers growing in the crystallographic *bc* plane (Figure 1b). The monodentate 4-Hmpz ligands alternatively extrude above and below each oxalato-bridged copper(II) layer (Figure 1c); the water molecules of crystallization are anchored to the layers via quite strong H-bonds and confined into small hydrophobic cavities arising from the peculiar relative orientation of the three crystallographically independent 4-Hmpz moieties (Figures 1–3).

Each layer is made up by zig-zag chains of six-coordinate, axially elongated octahedral copper(II) ions, bridged by regularly alternated symmetric bis-bidentate (μ-ox-$\kappa^2O^1,O^2:\kappa^2O^{2'},O^{1'}$) and asymmetric bis-(bidentate/monodentate) (μ_4-ox-$\kappa O^1:\kappa^2O^1,O^2:\kappa O^{2'}:\kappa^2O^{2'},O^{1'}$) oxalate anions, featuring four short and two short/four long copper–oxygen bond distances, respectively (see Figure 1a,b and Table 2). The four short [with the μ-ox, values in the range 1.978(3)–2.010(3) Å] and two short/two long [with the μ_4-ox, values in the ranges 1.968(3)–1.969(3) and 2.287(3)–2.428(3) Å, respectively] copper-to-oxygen distances regularly alternate along the chain growing direction of the chain (the crystallographic *b* axis), while the two remaining long Cu–O bonds [with the μ_4-ox, values in the range 2.388(3)–2.561(3) Å] interconnect neighboring chains along the crystallographic *c* axis, following an *ABBABB* sequence (Figure 1b). *A* type chains only contain Cu(1) ions and the two centrosymmetric oxalate ions [μ_4-O(1,2)C(1/1a)O(1a,2a), symmetry code: (a) = $-x + 1$, $-y + 1$, $-z + 1$] and [μ-O(3,4)C(2/2b)O(3b,4b); (b) = $-x + 1$, $-y$, $-z + 1$], while those of *B* type feature regularly alternating Cu(2) and Cu(3) ions, bridged by the non-centrosymmetric oxalate groups, [μ_4-O(5,8)C(7/8)O(6,7)] and [μ-O(9,11)C(13/14)O(10,12)].

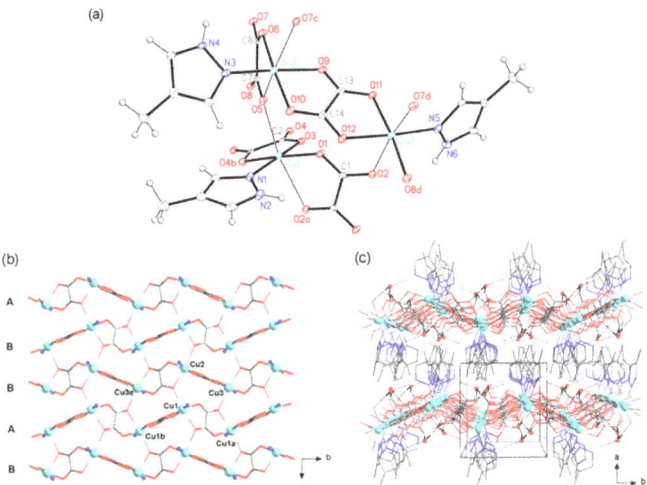

Figure 1. (**a**) Perspective view of a fragment of the neutral copper(II) layer of **1** with selected atom numbering. Thermal ellipsoids are drawn at the 30% probability level. (**b**) Projection of one corrugated oxalato-bridged copper(II) layer of **1** along the crystallographic a axis (terminal pyrazole ligands are omitted for clarity). The thinner solid lines represent the long Cu–O bond distances. (**c**) Perspective view of the crystal packing of **1** along the crystallographic c axis showing the zipper-type interpenetration of two parallel disposed layers. Hydrogen bonds are shown as dashed lines. Symmetry code: (a) = $-x + 1, -y + 1, -z + 1$; (b) = $-x + 1, -y, -z + 1$; (c) = $-x + 1, y + 1/2, -z + 1/2$; (d) = $x, y + 1, z$; (e) = $x, y - 1, z$.

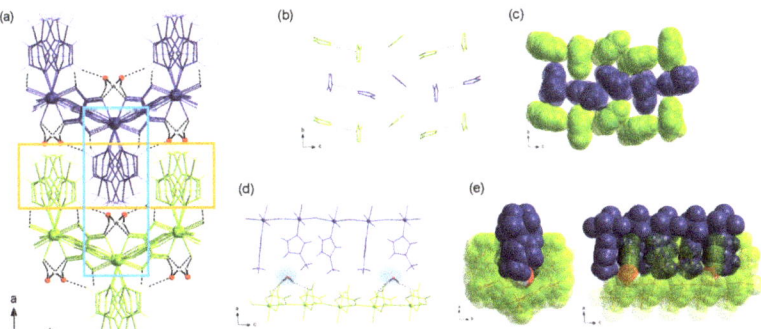

Figure 2. (**a**) Projection of a fragment of the crystal packing of **1** along the crystallographic c axis, showing two interdigitated layers in different colors. N-H···O$_w$, N-H···O$_{ox}$ and O$_w$···O$_{ox}$ type H-bonds are depicted as dashed lines. (**b**) Projection of the hydrophobic interlayer region [orange box in (**a**)] along the crystallographic a axis, with the weak inter-pyrazole intralayer N-H···N H-bond in evidence. (**c**) Same as (**b**), space-filling representation. (**d**) Detailed side view of a fragment of the interlayer region [sky blue box in (**a**)] showing the existence of very small hydrophobic cavities arising from the peculiar relative orientation of the three crystallographically independent 4-Hmpz moieties in **1**. (**e**) Another zoomed-in view of the interlayer region along the crystallographic c (left) and b (right) axes, evidencing the well-defined position of the water molecules of crystallization; space-filling representation.

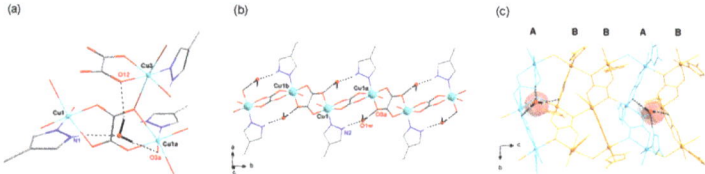

Figure 3. (**a**) A view of the three-fold H-bonding motif involving the water molecule of crystallization in **1**. (**b**) A view of the H-bonds established between the water molecules of crystallization and each *A* type chain in **1**. (**c**) Projection along the crystallographic *a* axis of a fragment of one oxalate-copper(II) layer, with *A* and *B* type chain fragments shown in different colors for clarity, aiming at illustrating both intra- and inter-chain (intralayer) H-bonding interactions that involve the water molecule of crystallization. H-bonds are depicted as dashed lines.

The tetragonally elongated surroundings of each six-coordinate copper(II) ion in **1** comprise three oxygen atoms from two oxalate ligands and a methylpyrazole-nitrogen atom, defining the equatorial plane and two other oxalate-oxygen atoms occupying the axial positions (Figure 1a). The two axial positions at each copper(II) center correspond, thus, to either an intra- or an inter-chain Cu–O bond (Figure 1b and Table 2).

Overall, the corrugated 2D array noted in **1** is analogous to that earlier reported for the parent anhydrous compounds with either the unsubstituted 1*H*-pyrazole (Hpz) or ammonia of formula {Cu(ox)(Hpz)}$_n$ (**4**) [61] or {Cu(ox)(NH$_3$)}$_n$ (**5**) [71], respectively. Within each layer of **1**, the average intrachain copper–copper distances (*r*) through the µ-ox and µ$_4$-ox type bridges are ~5.20 and 5.60 Å, respectively, while the average interchain copper–copper separation (*r'*) across µ$_4$-ox is ~4.22 Å (see Table 2). These values agree with those noted in the crystal structures of **4** (*r* ~ 5.52 and 5.55 Å; *r'* ~ 4.24 Å) [7] and **5** (*r* ~ 5.22 and 5.63 Å; *r'* ~ 4.19 Å) [71].

The close zipper-type packing of parallel disposed corrugated layers in the crystal lattice of **1** shows an interlayer separation (*d*) of 9.9575(4) Å, corresponding to the length of the crystallographic *a* axis (Figure 1c). This value is somewhat greater than that observed in **4** [*d* = 8.5694(6) Å] [61], a feature which is as expected because of the presence of the bulkier 4-Hmpz ligand in **1** vs. Hpz in **4**; the interlayer separation is as small as 4.19 Å in **5**, where ammonia is the terminal ligand [71]. Indeed, the value of the calculated density for **1** is slightly lower (~2.05 g cm^{-3}) than that reported for **4** (~2.16 g cm^{-3}), in spite of the fact that the X-ray data collection of **1** was done at 98 K (its density at RT would likely be less than 2.05 g cm^{-3}); moreover, the structure of **1** is capable of accommodating one water molecule of crystallization every three copper(II) ions. With respect to **4**, the bulkier 4-Hmpz moiety in **1** would induce some sort of improvement on the porosity of this family of structures, although to a quasi-negligible extent.

The closely packed 4-Hmpz ligands extruding above and below each oxalato-bridged copper(II) layer form a thick hydrophobic interlayer region (Figure 2). One of the three acidic N–H groups is tightly hydrogen bonded to the water molecule of crystallization; the remaining two are involved in moderate to very weak intralayer N–H···O$_{ox}$ and N–H···N type interactions, respectively (Figure 2 and Table 3). No interlayer H-bonds involving the acidic N–H moieties are noted, the close zipper-type packing of the parallel disposed neighboring layers being due only to hydrophobic interactions (Figure 2b,c).

The location and crystallographic occupation of the water molecules in the structure deserve particular attention. A projection of the crystal packing of **1** along the crystallographic *c* axis (Figures 1c and 2a) appears to indicate the possible existence of tiny channels running along the same axis, suitable for hosting the solvent of crystallization. A deeper look at those supposed channels, however, confirms the non-porous character of the structure (Figure 2d). Only one of the three independent 4-Hmpz orientations is compatible with the co-existence of the terminal pyrazole ligand and the water molecules of crystallization within the hydrophobic interlayer region. As evidenced in Figure 2d,e, there are no channels along the c axis, but only small cavities where one water molecule every three copper atoms would (and do) fit just perfectly.

The well-defined location of these tiny cavities coincides with the alternating up and down sites very close to each oxalato-copper(II) layer. Indeed, the confined water molecules of crystallization in **1** are bound to the hydrophilic layers via three strictly intralayer hydrogen bonds (Table 3 and Figure 3a,c). Two of such interactions, involving the water oxygen atom as either a donor [O1w–H1w2···O3a = 2.752(5) Å; (a) = $-x + 1$, $-y + 1$, $-z + 1$] or acceptor [N2–H2···O1w = 2.662(5) Å] toward a bis-bidentate oxalate or an acidic N-H moiety, respectively, are specifically associated to only A type chains (Figure 3b). However, the third interaction involving the water oxygen atom as a donor toward a bis-(bidentate/monodentate) oxalate group [O1w–H1w1···O12 = 2.828(5) Å] is established between the adjacent A and B chains.

3.3. Sorption Properties

The sorption properties of **1** toward methylamine (MA), dimethylamine (DMA), and trimethylamine (TMA) have been investigated (see the Experimental Section). The total loading after 3 h of exposure to an argon flow saturated with amine vapor at room temperature, expressed as the amine-to-copper molar ratio, decreased from three (MA) to one (DMA), and then to 0.1 (TMA) (Figure 4a). A remarkable color change under amine vapor was noted for both MA and DMA, being basically imperceptible for TMA (even after days of exposure), a feature which is consistent with the calculated amine loading. The kinetics of the methylamine adsorption is rather rapid for MA, and slightly slower for DMA (Figure 4a). In the case of MA, a fast loading occurs within the first minutes of exposure, with the saturation almost being reached after ~20 min.

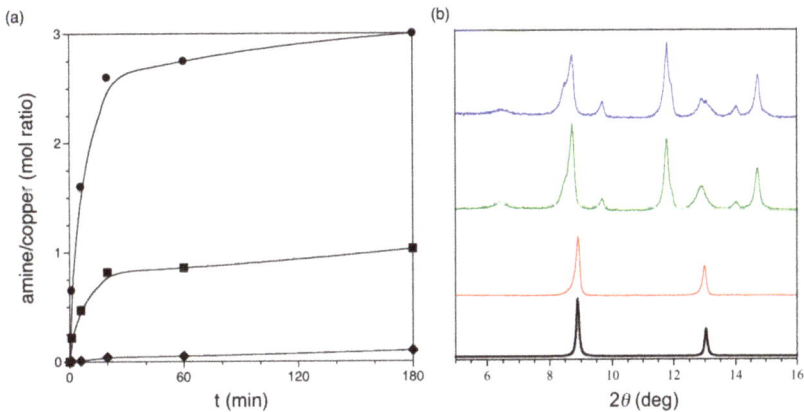

Figure 4. (a) Time profiles for the adsorption of MA (●), DMA (■), and TMA (◆) vapors by **1** at room temperature. The solid lines are only eye-guides. (b) XRPD of **1** (red line) and the MA adsorbates **2** (green line) and **3** (blue line). The bold black line represents the calculated XRPD of **1** from the single-crystal X-ray analysis.

The variation in the adsorption thermodynamics and kinetics along this series contrasts with those of the basicity and volatility of each amine, as expressed by the values of the basicity constant [pK_b = 3.35 (MA), 3.27 (DMA), and 4.20 (TMA) at 25 °C] and vapor pressure [vp = 2650 (MA), 1520 (DMA), and 1610 mm Hg (TMA) at 25 °C] [72]. The observed trend in the adsorption efficiencies (TMA < DMA < MA) does not coincide with the reported ones for either the basicity (TMA < MA < DMA) or the volatility (DMA < TMA < MA). This selective amine adsorption behavior, depending on the number of methyl substituents, would likely reflect both the steric constraints and the different coordination properties of the different polymethyl-substituted amines.

Focusing on the MA adsorbates, a dramatic color change of **1** from sky to deep blue rapidly occurs upon methylamine adsorption (see Figure 5) to finally give the methylamine adsorbate **2** after 3 h of exposure to an argon flow saturated with MA vapor at room temperature. This transformation is accompanied by an important change in their XRPD patterns, which suggests the occurrence of a major structural rearrangement upon the adsorption of methylamine (Figure 4b). Otherwise, **2** easily loses two of the three MA molecules under vacuum for 3 h at room temperature to give the grayish deep blue MA adsorbate **3**, the process being accompanied by no significant change in the XRPD pattern of **3** vs. **2** (Figure 4b).

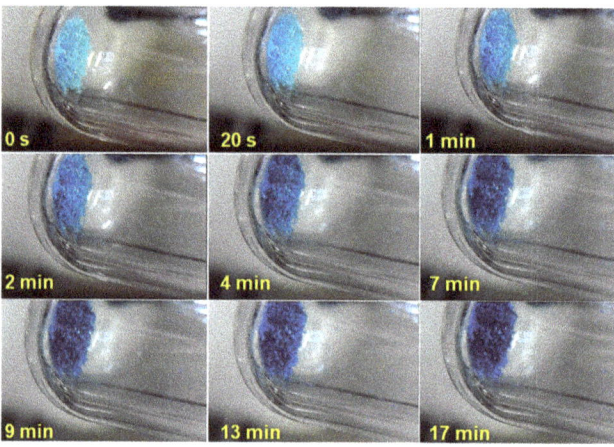

Figure 5. Sequential snapshots showing the color change of **1** under MA vapors.

Overall, these features suggest that one up from three methylamine molecules in **2** would coordinate each copper(II) ion, triggering a rearrangement of its original coordination environment, while the remaining two would possibly interact with the acidic N–H pyrazole moieties through hydrogen bonding and/or with the 4-Hmpz ligands as a whole via hydrophobic interactions.

According to this analysis, **2** would be alternatively formulated as {Cu(ox)(4-Hmpz)(MeNH$_2$)·2MeNH$_2$·1/3H$_2$O}$_n$. The uptake capability of **1** toward MA is quite astonishing, considering the non-porous nature of the compound. The adsorption process is irreversible and semi-disruptive, meaning that **2** and **3** are not as crystalline as the precursor **1**. This is not surprising, given that the coordination of one MA molecule would be incompatible with the original oxalato-copper(II) layered network, while the fairly stable incorporation of two extra MA molecules would contribute to the unsettling of the original zipper-type close packing. On the other hand, the fact that the water molecule of crystallization is kept would indicate the persistence of relatively strong hydrogen bonds with the oxalate oxygen atoms and/or the 4-Hmpz acidic N-H moieties, even after structural rearrangement.

3.4. Magnetic Properties of 1–3

The χ_M and $\chi_M T$ vs. T plots for **1–3** [χ_M being the molar magnetic susceptibility per copper(II) ion] show a concomitant change of the magnetic properties from strong (**1**) to weak (**2** and **3**) antiferromagnetic coupling (Figure 6). Therefore, the $\chi_M T$ values at room temperature increased from 0.21 cm^3 mol^{-1} K (**1**) to 0.40 (**2**) and 0.39 cm^3 mol^{-1} K (**3**). They were well or slightly below than that expected for a magnetically isolated spin doublet ($\chi_M T = 0.41$ cm^3 mol^{-1} K with $g = 2.1$). Upon cooling, $\chi_M T$ decreased continuously for **1**, and it vanished around 50 K. In the case of **2** and **3**, the values of $\chi_M T$ remained constant down to 50 (**2**) and 25 K (**3**), and then decreased smoothly to reach 0.15 (**2**) and 0.34 cm^3 mol^{-1} K (**3**) at 2.0 K (Figure 6a). Besides, χ_M showed a broad maximum around room temperature for **1**, whereas no maximum of the magnetic susceptibility occured for **2** and **3** down to 2.0 K (Figure 6b). Overall, these features support the occurrence of a fast and complete, irreversible solid-state transformation during the methylamine sorption process.

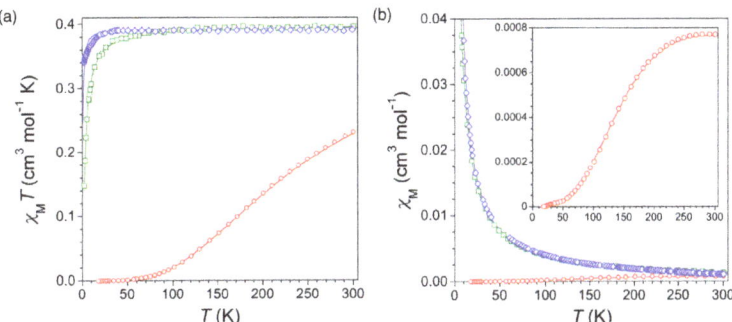

Figure 6. Temperature dependence of $\chi_M T$ (**a**) and χ_M (**b**) of **1** (○) and the MA adsorbates **2** (□) and **3** (◊). The inset in (**b**) aims to show how the maximum of χ_M of **1** (○) occured in the vicinity of 300 K. The solid lines are the best-fit curves (see text).

Bearing in mind the above results, the analysis of the magnetic susceptibility data for **1**–**3** was carried out by means of the Hatfield expression derived from the spin Hamiltonian for an alternating copper(II) chain, $H = -J\sum(S_{Cu2i} \cdot S_{Cu2i-1} + \alpha S_{Cu2i} \cdot S_{Cu2i+1}) + g\beta H\sum S_i$ ($S_{Cu2i} = S_{Cu} = \frac{1}{2}$) (see Scheme 1), where J and αJ are the two different antiferromagnetic intrachain coupling parameters (α being the alternation parameter) and g is the average Landé factor of the copper(II) ions ($g = g_{2i} = g_{Cu}$) [73]. Least-squares fits of the experimental data gave $-J$ = 322(2) (**1**), 3.20(2) (**2**), and 0.56(4) cm^{-1} (**3**), with α = 0.02(1) (**1**), 0.92(1) (**2**) and 0.90(1) (**3**), and g = 2.07(1) (**1**), 2.06(1) (**2**) and 2.05(1) (**3**). The theoretical curves match very well the experimental ones for all three compounds (solid lines in Figure 6).

The calculated $-J$ value for **1** is comparable to those found for the aforementioned pyrazole and ammonia analogues **4** and **5** ($-J$ = 312 and 265 cm^{-1}), which possess similar structural parameters [61,71]. Otherwise, the large decrease of the $-J$ value for **2** and **3** relative to **1** indicates the occurrence of a magnetic orbital reversal upon methylamine adsorption [74–76], as illustrated in Scheme 1. Hence, the overlap between the d($x^2 - y^2$)-type magnetic orbitals of the axially elongated octahedral Cu^{II} ions varies from very strong (**1**) to weak but non-negligible (**2** and **3**) for a coplanar and a perpendicular disposition of the metal equatorial planes with respect to the mean plane of the symmetric or asymmetric bridging oxalate, respectively (Scheme 1a,b). This magnetic switching behavior can likely be attributed to methylamine coordination to the metal ion in both adsorbates **2** and **3**, as earlier found in the related pair of regular µ-oxalatocopper(II) chains of formulae {Cu(ox)·1/3H$_2$O}$_n$ (**6**) ($J = -291$ cm^{-1}) and {Cu(ox)(NH$_3$)$_2$·2H$_2$O}$_n$ (**7**) ($J = -15.4$ cm^{-1}) [74].

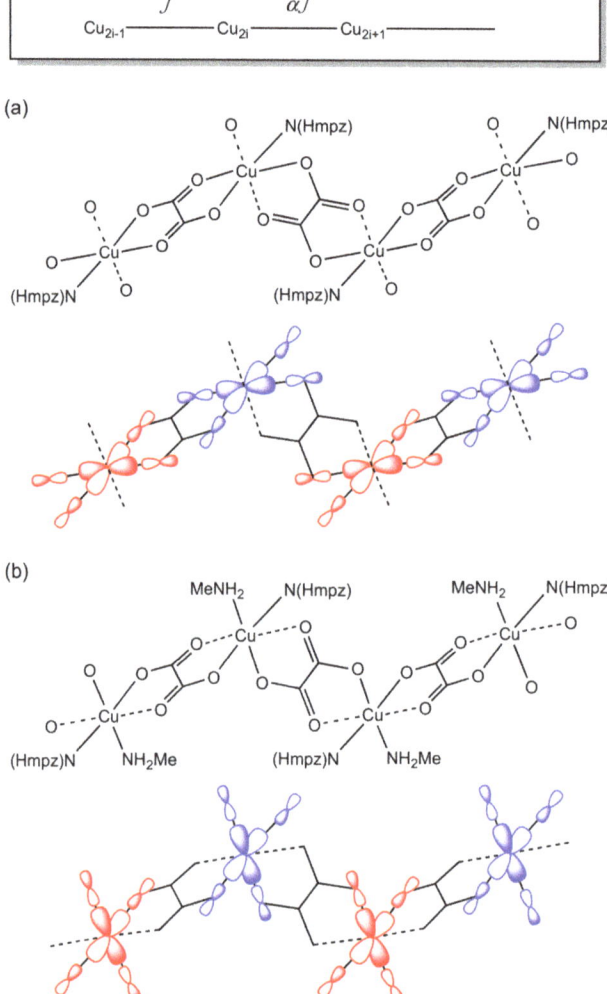

Scheme 1. Illustration of the magnetic coupling model for an alternating copper(II) chain showing the relative orientation of the magnetic orbitals centered on each copper(II) ion for **1** (**a**), relative to **2** and **3** (**b**). The solid and dashed lines represent short and long metal-ligand bonds, respectively.

4. Conclusions

In summary, a new oxalato-bridged copper(II)-pyrazole coordination polymer has been obtained from the copper(II)-mediated self-assembly of oxalate and 4-Hmpz in water under mild conditions. This new heteroleptic 2D copper(II) coordination polymer features a non-porous, interdigitated, zipper-type layered structure with acidic N–H sites from the terminal 4-methyl substituted pyrazole ligands within the interlayer space, in close proximity to the main hydrophilic oxalate-copper(II) region. In spite of its very dense crystal packing, **1** exhibits a fast and selective adsorption for polymethyl-substituted amines, which is accompanied by a dynamic switching behavior with dramatic changes in both the color and magnetic coupling following a structural rearrangement upon the adsorption of methylamine. Such features (structural modulation together with external control of the optical and magnetic properties), combined with the easy self-assembling and potential addressing as thin films over a variety of surfaces, make this novel class of 2D multifunctional magnetic materials suitable candidates for obtaining switching optical and magnetic devices for chemical sensing. This might allow for system optimization with a special focus on improving the porosity of the supramolecular 3D network and/or expanding the coordination capabilities of the metal ion.

Supplementary Materials: The following figures are available online at https://www.mdpi.com/article/10.3390/magnetochemistry7050065/s1, Figure S1: XRPD pattern from a bulk sample of **1** and the generated pattern from its X-ray single-crystal diffraction data; Figure S2: IR spectra of **1** (a), **2** (b), and **3** (c).

Author Contributions: I.C., N.M. (Nadia Marino), and M.J. conceived the project; I.C., N.M. (Nadia Marino), and R.R.-G. designed and discussed the experiments; M.L.C., M.O.-A., and A.P.-Á. synthesized and characterized the complexes; N.M. (Nadia Marino) and G.D.M. carried out the crystallographic study; N.M. (Nicolás Moliner), F.L., and R.R.-G. performed the magnetic study; R.R.-G., N.M. (Nadia Marino), and M.J. wrote and/or reviewed the manuscript with contributions. All authors have read and agreed to the published version of the manuscript.

Funding: This research was funded by the Spanish MICIU (Project PID2019-109735GB-I00) and the Generalitat Valenciana (AICO/2020/183). We also acknowledge the financial support from the European Commission (FSE, Fondo Sociale Europeo) and the Calabria Region for a fellowship grant N.M. (Nadia Marino), M.O.-A. and A.P.-Á. thank the MICIU for doctoral grants.

Institutional Review Board Statement: Not applicable.

Informed Consent Statement: Not applicable.

Data Availability Statement: The data are available by corresponding authors.

Acknowledgments: Thanks are also due to the Servei Central de Suport a la Investigació Experimental de la Universitat de València (SCSIE-UV) for its assistance with the analytical characterization of the compounds.

Conflicts of Interest: The authors declare no conflict of interest.

Sample Availability: Samples of **1–3** are available from the authors.

References

1. Ouahab, L. (Ed.) *Multifunctional Molecular Materials*, Pan Standford Publishing: Singapore, 2013.
2. Sieklucka, B.; Pinkowicz, D. (Eds.) *Molecular Magnetic Materials: Concepts and Applications*; Wiley-VCH: Weinheim, Germany, 2017.
3. Day, P. New transparent ferromagnets. *Acc. Chem. Res.* **1979**, *12*, 236–243. [CrossRef]
4. Day, P.; Kurmoo, M. Molecular magnetic semiconductors, metals and superconductors: BEDT-TTF salts with magnetic anions. *J. Mater. Chem.* **1997**, *7*, 1291–1295. [CrossRef]
5. Alivisatos, P.; Barbara, P.F.; Castleman, A.W.; Chang, J.; Dixon, D.A.; Klein, M.L.; McLendon, G.L.; Miller, J.S.; Ratner, M.A.; Rossky, P.J.; et al. From molecules to materials: Current trends and future directions. *Adv. Mater.* **1998**, *10*, 1297–1336. [CrossRef]
6. Dujardin, E.; Mann, S. Morphosynthesis of molecular magnetic materials. *Adv. Mater.* **2004**, *16*, 1125–1129. [CrossRef]
7. Coronado, E.; Day, P. Magnetic molecular conductors. *Chem. Rev.* **2004**, *104*, 5419–5448. [CrossRef]
8. Coronado, E.; Gatteschi, D. Trends and challenges in molecule-based magnetic materials. *J. Mater. Chem.* **2006**, *16*, 2513–2515.
9. Pinkowicz, D.; Czarnecki, B.; Reczynski, M.; Arczinski, M. Multifunctionality in molecular magnetism. *Sci. Prog.* **2015**, *98*, 346–378. [CrossRef]
10. Ferrando-Soria, J.; Vallejo, J.; Castellano, M.; Martínez-Lillo, J.; Pardo, E.; Cano, J.; Castro, I.; Lloret, F.; Ruiz-García, R.; Julve, M. Molecular magnetism, quo vadis? A historical perspective from a coordination chemist viewpoint. *Coord. Chem. Rev.* **2017**, *339*, 17–103. [CrossRef]
11. Coronado, E. Molecular magnetism: From chemical design to spin control in molecules, materials and devices. *Nat. Rev. Mater.* **2020**, *5*, 87–104. [CrossRef]
12. Kepert, C.J. Advanced functional properties in nanoporous coordination framework materials. *Chem. Commun.* **2006**, 695–700. [CrossRef]
13. Maspoch, D.; Ruiz-Molina, D.; Veciana, J. Old materials with new tricks: Multifunctional open-framework materials. *Chem. Soc. Rev.* **2007**, *36*, 770–818. [CrossRef]
14. Pardo, E.; Ruiz-García, R.; Cano, J.; Ottenwaelder, X.; Lescouëzec, R.; Journaux, Y.; Lloret, F.; Julve, M. Ligand design for multidimensional magnetic materials: A metallosupramolecular perspective. *Dalton Trans.* **2008**, 2780–2805. [CrossRef]
15. Kurmoo, M. Magnetic metal-organic frameworks. *Chem. Soc. Rev.* **2009**, *38*, 1353–1379. [CrossRef]

16. Dul, M.-C.; Pardo, E.; Lescouëzec, R.; Journaux, Y.; Ferrando-Soria, J.; Ruiz-García, R.; Cano, J.; Julve, M.; Lloret, F.; Cangussu, D.; et al. Supramolecular coordination chemistry of aromatic polyoxalamide ligands: A metallosupramolecular approach toward functional magnetic materials. *Coord. Chem. Rev.* **2010**, *254*, 2281–2296. [CrossRef]
17. Dechambenoit, P.; Long, J.R. Microporous magnets. *Chem. Soc. Rev.* **2011**, *40*, 3249–3265. [CrossRef]
18. Muñoz, M.C.; Real, J.A. Thermo-, piezo-, phot- and chemo-switchable spin crossover iron(II)-metallocyanate based coordination polymers. *Coord. Chem. Rev.* **2011**, *255*, 2068–2093. [CrossRef]
19. Coronado, E.; Mínguez-Espallargas, G. Dynamic magnetic MOFs. *Chem. Soc. Rev.* **2013**, *42*, 1525–1539. [CrossRef]
20. Grancha, T.; Ferrando-Soria, J.; Castellano, M.; Julve, M.; Pasán, J.; Armentano, D.; Pardo, E. Oxamato-based coordination polymers: Recent advances in multifunctional magnetic materials. *Chem. Commun.* **2014**, *50*, 7569–7585. [CrossRef]
21. Coronado, E.; Mínguez-Espallargas, G. magnetic functionalities in MOFs: From the framework to the pore. *Chem. Soc. Rev.* **2018**, *47*, 533–557.
22. Thorarinsdottir, A.E.; Harris, T.D. Metal-organic framework magnets. *Chem. Rev.* **2020**, *120*, 8716–8789. [CrossRef]
23. Kreno, L.E.; Leong, K.; Farha, O.K.; Allendorf, M.; Van Duyne, R.P.; Hupp, J.T. Metal-organic framework materials as chemical sensors. *Chem. Rev.* **2012**, *112*, 1105–1125. [CrossRef] [PubMed]
24. Kaushik, A.; Kumar, R.; Arya, S.K.; Nair, M.; Malhotra, B.D.; Bhansali, S. Organic-inorganic hybrid nanocomposite-based gas sensors for environmental monitoring. *Chem. Rev.* **2015**, *115*, 4571–4606. [CrossRef]
25. Kumar, P.; Deep, A.; Kim, K.-H.; Brown, R.J.C. Coordination polymers: Opportunities and challenges for monitoring volatile organic compounds. *Prog. Polym. Sci.* **2015**, *45*, 102–118. [CrossRef]
26. Yi, F.-Y.; Chen, D.; Wu, M.-K.; Han, L.; Jiang, H.-L. Chemical sensors based on metal-organic frameworks. *ChemPlusChem* **2016**, *81*, 675–690. [CrossRef] [PubMed]
27. Kumar, V.; Kim, K.-H.; Kumar, P.; Jeon, B.-H.; Kim, J.-C. Functional hybrid nanostructure materials: Advanced strategies for sensing applications toward volatile organic compounds. *Coord. Chem. Rev.* **2017**, *342*, 80–105. [CrossRef]
28. Zhang, Y.; Yuan, S.; Day, G.; Wang, X.; Yang, X.; Zhou, H.-C. Luminescent sensors based on metal-organic frameworks. *Coord. Chem. Rev.* **2018**, *354*, 28–45. [CrossRef]
29. Dolgopolova, E.A.; Rice, A.M.; Martin, C.R.; Shustova, N.B. Photochemistry and photophysics of MOFs: Steps towards MOF-based sensing enhancements. *Chem. Soc. Rev.* **2018**, *47*, 4710–4728. [CrossRef]
30. Wang, H.; Lusting, W.P.; Li, J. Sensing and capture of toxic and hazardous gases and vapors by metal-organic frameworks. *Chem. Soc. Rev.* **2018**, *47*, 4729–4756. [CrossRef] [PubMed]
31. Zhao, S.-N.; Wang, G.; Poelman, D.; Van Der Voort, P. Luminescent lanthanide MOFs: A unique platform for chemical sensing. *Materials* **2018**, *11*, 572. [CrossRef]
32. Rasheed, T.; Nabeel, F. Luminescent metal-organic frameworks as potential sensory materials for various environmental toxic agents. *Coord. Chem. Rev.* **2019**, *401*, 213065. [CrossRef]
33. Wang, P.-L.; Xie, L.-H.; Joseph, E.A.; Li, J.-R.; Su, X.-O.; Zhou, H.-C. Metal-organic frameworks for food safety. *Chem. Rev.* **2019**, *119*, 10638–10690. [CrossRef] [PubMed]
34. Lai, C.; Wang, Z.; Qin, L.; Fu, Y.; Li, B.; Zhang, M.; Liu, S.; Li, L.; Yi, H.; Liu, X.; et al. Metal-organic frameworks as burgeoning materials for the capture and sensing of indoor VOCs and radon gases. *Coord. Chem. Rev.* **2021**, *427*, 213565. [CrossRef]
35. Ohba, M.; Yoneda, K.; Agustí, G.; Muñoz, M.C.; Gaspar, A.B.; Real, J.A.; Yamasaki, M.; Ando, H.; Nakao, Y.; Sakaki, S.; et al. Bidirectional chemo-switching of spin state in a microporous framework. *Angew. Chem. Int. Ed.* **2009**, *48*, 4767–4771. [CrossRef] [PubMed]
36. Agustí, G.; Ohtani, R.; Yoneda, K.; Gaspar, A.B.; Ohba, M.; Sánchez-Royo, J.F.; Muñoz, M.C.; Kitagawa, S.; Real, J.A. Oxidative addition of halogens on open metal sites in a microporous spin-crossover coordination polymer. *Angew. Chem. Int. Ed.* **2009**, *48*, 8944–8947. [CrossRef]
37. Ohba, M.; Yoneda, K.; Kitagawa, S. Guest-responsive porous magnetic frameworks using polycyanometallates. *CrystEngComm* **2010**, *12*, 159–165. [CrossRef]
38. Ohtani, R.; Yoneda, K.; Furukawa, S.; Horike, N.; Kitagawa, S.; Gaspar, A.B.; Muñoz, M.C.; Real, J.A.; Ohba, M. Precise control and consecutive modulation of spin transition temperature using chemical migration in porous coordination polymers. *J. Am. Chem. Soc.* **2011**, *133*, 8600–8605. [CrossRef]
39. Coronado, E.; Giménez-Marqués, M.; Mínguez-Espallargas, G.; Brammer, L. Tuning the magneto-structural properties of non-porous coordination polymers by HCl chemisorption. *Nat. Commun.* **2012**, *3*, 1827. [CrossRef]
40. Ferrando-Soria, J.; Ruiz-García, R.; Cano, J.; Stiriba, S.-E.; Vallejo, J.; Castro, I.; Julve, M.; Lloret, F.; Amorós, P.; Pasán, J.; et al. Reversible solvatomagnetic switching in a spongelike manganese(II)-copper(II) open framework with a pillared square/octagonal layer architecture. *Chem. Eur. J.* **2012**, *18*, 1608–1617. [CrossRef]
41. Ferrando-Soria, J.; Serra-Crespo, P.; de Lange, M.; Gascon, J.; Kapteijn, F.; Julve, M.; Cano, J.; Lloret, F.; Pasán, J.; Ruiz-Pérez, C.; et al. Selective gas and vapor sorption and magnetic sensing by an isoreticular mixed-metal-organic framework. *J. Am. Chem. Soc.* **2012**, *134*, 15301–15304. [CrossRef] [PubMed]
42. Ferrando-Soria, J.; Khajavi, H.; Serra-Crespo, P.; Gascon, J.; Kapteijn, F.; Julve, M.; Lloret, F.; Pasán, J.; Ruiz-Pérez, C.; Journaux, Y.; et al. Highly selective chemical sensing in a luminiscent nanoporous magnet. *Adv. Mater.* **2012**, *24*, 5625–5629. [CrossRef]
43. Vallejo, J.; Fortea-Pérez, F.R.; Pardo, E.; Benmansour, S.; Castro, I.; Krzystek, J.; Armentano, D.; Cano, J. Guest-dependent single-ion magnet behaviour in a cobalt(II) metal-organic framework. *Chem. Sci.* **2016**, *7*, 2286–2293. [CrossRef] [PubMed]

44. Zhang, W.; Wang, D.; Zhu, L.; Zhai, F.; Weng, L.; Sun, J.; Ling, Y.; Chen, Z.; Zhou, Y. HCl chemisorption-induced drastic magneto-structural transformation in a layered cobalt-phosphonotriazolate coordination polymer. *Dalton Trans.* **2016**, *45*, 10510–10513. [CrossRef] [PubMed]
45. Perlepes, P.; Oyarzabal, I.; Mailman, A.; Yquel, M.; Platunov, M.; Dovgaliuk, I.; Rouzières, M.; Négrier, P.; Mondieig, D.; Suturina, E.A.; et al. Metal-organic magnets with large coercivity and ordering temperatures up to 242 °C. *Science* **2020**, *370*, 587–592. [CrossRef]
46. Turo-Cortes, R.; Bartual-Murgui, C.; Castells-Gil, J.; Muñoz, M.C.; Martí-Gastaldo, C.; Real, J.A. Reversible guest-induced gate-opening with multiplex spin crossover responses in two-dimensional Hofmann clathrates. *Chem. Sci.* **2020**, *11*, 11224–11234. [CrossRef]
47. Piñeiro-López, L.; Valverde-Muñoz, F.-J.; Trzop, E.; Muñoz, M.C.; Seredyuk, M.; Castells-Gil, J.; da Silva, I.; Martí-Gastaldo, C.; Collet, E.; Real, J.A. Guest induced reversible on-off switching of elastic frustration in a 3D spin crossover coordination polymer with room temperature hysteretic behaviour. *Chem. Sci.* **2021**, *12*, 1317–1326. [CrossRef]
48. Cingolani, A.; Galli, S.; Masciocchi, N.; Pandolfo, L.; Pettinari, C.; Sironi, A. Sorption-desorption behavior of bispirazolato-copper(II) 1D coordination polymers. *J. Am. Chem. Soc.* **2005**, *127*, 6144–6145. [CrossRef]
49. Qiu, L.-G.; Li, Z.-Q.; Wu, Y.; Wang, W.; Xu, T.; Jiang, X. Facile synthesis of nanocrystals of a microporous metal-organic framework by an ultrasonic method and selective sensing of orgonoamines. *Chem. Commun.* **2008**, 3642–3644. [CrossRef] [PubMed]
50. Bencini, A.; Casarin, M.; Forrer, D.; Franco, L.; Garau, F.; Masciocchi, N.; Pandolfo, L.; Pettinari, C.; Ruzzi, M.; Vittadini, A. Magnetic properties and vapochromic reversible guest-induced transformation in a bispyrazolato copper(II) polymer: An experimental and dispersion-corrected density functional theory study. *Inorg. Chem.* **2009**, *48*, 4044–4051. [CrossRef]
51. Zou, X.; Zhu, G.; Hewitt, I.J.; Sun, F.; Qiu, S. Synthesis of a metal-organic framework film by direct conversion technique for VOCs sensing. *Dalton Trans.* **2009**, 3009–3013. [CrossRef]
52. Chen, J.; Yi, F.-Y.; Yu, H.; Jiao, S.; Pang, G.; Sun, Z.-M. Fast response and highly selective sensing of amine vapors using a luminiscent coordination polymer. *Chem. Commun.* **2014**, *50*, 10506–10509. [CrossRef]
53. Mallick, A.; Garai, B.; Addicoat, M.A.; St. Petkov, P.; Heine, T.; Banerjee, R. Solid state organic amine detection in a photochromic porous metal-organic framework. *Chem. Sci.* **2015**, *6*, 1420–1425. [CrossRef]
54. Tan, B.; Chen, C.; Cai, L.-X.; Zhang, Y.-J.; Huang, X.-Y.; Zhang, J. Introduction of Lewis acidic and redox-active sites into a porous framework for ammonia capture with visual color response. *Inorg. Chem.* **2015**, *54*, 3456–3461. [CrossRef]
55. Chen, C.; Cai, L.-X.; Tan, B.; Zhang, Y.-J.; Yang, X.-D.; Zhang, J. Ammonia detection by using flexible Lewis acidic sites in luminiscent porous frameworks constructed from a bipyridinium derivative. *Chem. Commun.* **2015**, *51*, 8189–8192. [CrossRef] [PubMed]
56. Campbell, M.G.; Sheberla, D.; Liu, S.F.; Swager, T.M.; Dinca, M. Cu$_3$(hexaiminotriphenylene)$_2$: An electrically conductive 2D metal-organic framework for chemiresistive sensing. *Angew. Chem. Int. Ed.* **2015**, *54*, 4349–4352. [CrossRef] [PubMed]
57. Shen, X.; Yan, B. A novel fluorescence probe for sensing organic amine vapors from a Eu^{3+} β-diketonate functionalized bio-MOF-1 hybrid system. *J. Mater. Chem. C* **2015**, *3*, 7038–7044. [CrossRef]
58. Takahashi, A.; Tanaka, H.; Parajuli, D.; Nakamura, T.; Minami, K.; Sugiyama, Y.; Hakuta, Y.; Ohkoshi, S.; Kawamoto, T. Historical pigment exhibiting ammonia gas capture beyond standard adsorbents with adsorption sites of two kinds. *J. Am. Chem. Soc.* **2016**, *138*, 6376–6379. [CrossRef] [PubMed]
59. Zhao, S.-S.; Yang, J.; Liu, Y.-Y.; Ma, J.-F. Fluorescent aromatic tag-functionalized MOFs for highly selective sensing of metal ions and small organic molecules. *Inorg. Chem.* **2016**, *5*, 2261–2273. [CrossRef]
60. Yang, N.-N.; Sun, W.; Xi, F.-G.; Sui, Q.; Chen, L.-J.; Gao, E.-Q. Postsynthetic N-methylation making a metal-organic framework responsive to alkylamines. *Chem. Commun.* **2017**, *53*, 1747–1750. [CrossRef]
61. Świtlicka-Olszewska, A.; Machura, B.; Mroziński, J.; Kalińska, N.; Kruszynski, R.; Penkala, M. Effect of N-donor ancillary ligands on structural and magnetic properties of oxalate copper(II) complexes. *New J. Chem.* **2014**, *38*, 1611–1626. [CrossRef]
62. Calatayud, M.L.; Orts-Arroyo, M.; Julve, M.; Lloret, F.; Marino, N.; De Munno, G.; Ruiz-García, R.; Castro, I. Magneto-structural correlations in asymmetric oxalato-bridged dicopper(II) complexes with polymethyl-substituted pyrazole ligands. *J. Coord. Chem.* **2017**, *71*, 657–674. [CrossRef]
63. Castro, I.; Calatayud, M.L.; Orts-Arroyo, M.; Moliner, N.; Marino, N.; Lloret, F.; Ruiz-García, R.; De Munno, G.; Julve, M. Ferro- and Antiferromagnetic Interactions in Oxalato-Centered Inverse Hexanuclear and Chain Copper(II) Complexes with Pyrazole Derivatives. *Molecules* **2021**, *26*, 2792. [CrossRef]
64. Bruker. *SMART*; Data Collection Software (Version 4.050); Siemens Analytical Instruments Inc.: Madison, WI, USA, 1996.
65. Bruker. *SAINT*; (Version 7.68A); Bruker AXS Inc.: Madison, WI, USA, 2009.
66. Bruker. *SADABS*; (Version 2008/1); Bruker AXS Inc.: Madison, WI, USA, 2008.
67. Sheldrick, G.M. A short history of *SHELX*. *Acta Cryst.* **2008**, *A64*, 112–122. [CrossRef]
68. Sheldrick, G.M. Crystal structure refinement with *SHELXL*. *Acta Cryst.* **2015**, *C71*, 3–8.
69. *Diamond Software*; Version 4.6.4 (2021); Crystal Impact Kreuzherrenstr: Bonn, Germany, 2021.
70. Zeroka, D.; Jensen, J.O. Infrared spectra of some isotopomers of methylamine and the methylammonium ion: A theoretical study. *J. Mol. Struct.* **1998**, *425*, 181–192. [CrossRef]
71. Cavalca, L.; Villa, A.C.; Manfredotti, A.G.; Tomlinson, A.A.G. Crystal, molecular, and electronic structure of catena-μ-oxalato-ammine-copper(II). *J. Chem. Soc. Dalton Trans.* **1972**, *3*, 391–395. [CrossRef]

72. PubChem Database. *CID Numbers: 6329 (Methylamine), 674 (Dimethylamine), and 1146 (Trimethylamine)*; National Center for Biotechnology Information (NCBI): Bethesda, MD, USA.
73. Hall, J.W.; Marsh, W.E.; Weller, R.R.; Hatfield, W.E. Exchange coupling in the alternating-chain compounds *catena*-di-μ-chloro-bis(4-methylpyridine)copper(II), *catena*-di-μ-bromo-bis(*N*-methylimidazole)copper(II), *catena*-[hexadione bis(thiosemicarbazonato)]copper(II), and *catena*-[octanedione bis(thiosemicarbazonato)]copper(II). *Inorg. Chem.* **1981**, *20*, 1033–1037.
74. Girerd, J.J.; Kahn, O.; Verdaguer, M. Orbital reversal in (oxalato)copper(II) linear chains. *Inorg. Chem.* **1980**, *19*, 274–276. [CrossRef]
75. Julve, M.; Verdaguer, M.; Gleizes, A.; Philoche-Levisalles, M.; Kahn, O. Design of μ-oxalato copper(II) binuclear complexes exhibiting expected magnetic properties. *Inorg. Chem.* **1984**, *23*, 3808–3818. [CrossRef]
76. Journaux, Y.; Sletten, J.; Kahn, O. Tunable interactions in μ-oxamido copper(II) binuclear complexes. *Inorg. Chem.* **1985**, *24*, 4063–4069. [CrossRef]

 magnetochemistry

Article

In Quest of Molecular Materials for Quantum Cellular Automata: Exploration of the Double Exchange in the Two-Mode Vibronic Model of a Dimeric Mixed Valence Cell [†]

Boris Tsukerblat [1,*], Andrew Palii [2,*] and Sergey Aldoshin [2]

1 Department of Chemistry, Ben-Gurion University of the Negev, Beer-Sheva 84105, Israel
2 Laboratory of Molecular Magnetic Nanomaterials, Institute of Problems of Chemical Physics of Russian Academy of Sciences, 142432 Chernogolovka, Moscow Region, Russia; sma@icp.ac.ru
* Correspondence: tsuker@bgu.ac.il (B.T.); andrew.palii@uv.es (A.P.)
† Dedicated to the memory of Professor Peter Day.

Abstract: In this article, we apply the two-mode vibronic model to the study of the dimeric molecular mixed-valence cell for quantum cellular automata. As such, we consider a multielectron mixed valence binuclear $d^2 - d^1$–type cluster, in which the double exchange, as well as the Heisenberg-Dirac-Van Vleck exchange interactions are operative, and also the local ("breathing") and intercenter vibrational modes are taken into account. The calculations of spin-vibronic energy spectra and the "cell-cell"-response function are carried out using quantum-mechanical two-mode vibronic approach based on the numerical solution of the dynamic vibronic problem. The obtained results demonstrate a possibility of combining the function of molecular QCA with that of spin switching in one electronic device and are expected to be useful from the point of view of the rational design of such multifunctional molecular electronic devices.

Keywords: quantum cellular automata; molecular cell; mixed-valence; electron transfer; double exchange; magnetic exchange; dimeric mixed valence clusters

1. Introduction

This article is dedicated to the memory of Professor Peter Day with the question posed to him in 1998 ([1], see image below): "Molecular information processing: Will it happen?" This question and subsequent discussions in his inimitable manner was focused on the fundamental issues of the "design and manufacture artificial structures using molecules that will carry out" the function of storing and processing memory in living organisms. In his general arguments, Peter Day appealed to common problems of information (see highlights in the excerpt from the article by Peter Day published in *Proc. Royal Inst. Great Britain*) interconnected with the switching processes in a binary systems and discussed the fundamental limits of computing speed and power dissipation. These ideas presented in detail along with the discussion of the molecular aspects are in focus of the contemporary issues in the topic of Quantum Cellular Automata (QCA) and search for the new molecular materials for the nanoscale devices. In this regard, it is pertinent to note that the Robin and Day assignment [2] of mixed-valence compounds according to the degree of localization plays a guiding role in the search of the relevant molecules.

In accordance with the general ideas proposed in the pioneering study by Lent et al. [3], the electronic QCA devices are based on the square planar cells composed of quantum dots [3–5]. Two excess electrons captured by a square-planar four-dot cell provide a possibility to encode binary information (**0** and **1**) in the two antipodal (diagonal) distributions of the charges. To illustrate encoding and operating with binary information underlying the actions of electronic devices, a dimeric system can be used, as illustrated in Figure 1. The dimeric unit can be considered as a "half-cell" from which the "full-cell" (tetrameric unit) can be constructed. Figure 1 illustrates a dimeric cell in which the delocalized pair the mobile electron is evenly distributed between two sites and the two predominantly localized configurations corresponding to the binary **0** and **1**.

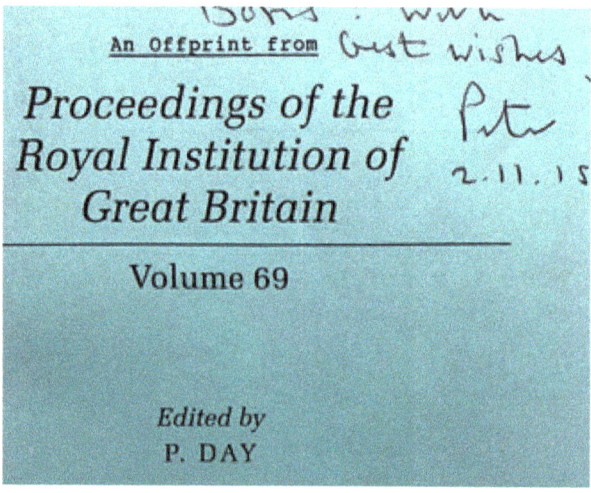

Excerpt from the article by Peter Day published in *Proc. Royal Inst. Great Britain*, v.69. pp. 85–106 (1998).

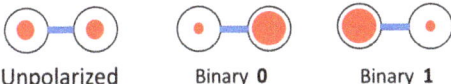

Figure 1. Two charge distributions in a two-dot cell or in a dimeric (mixed-valence) MV molecule with one mobile electron corresponding to the delocalized (unpolarized) configuration and localized configurations corresponding to binary and **1**. The red balls indicate the populated sites and their sizes symbolize the degree of localization of the mobile electrons.

The functional properties of devices are based on the concept of the action of the Coulomb forces that can control and transmit the binary information encoded in a cell. Let us consider the two dimeric cells 2 and 1 in a certain geometry (shown in Figure 2), one of which has a definite charge configuration (binary **1** in Figure 2), while the second one is unpolarized. Let us assume that the polarized state of the cell 2 can be induced and controlled so that this cell can be termed as the "driver cell". The electrostatic effect of the driver cell 2 with a given polarization affects the neighboring cell 1 forcing this cell to acquire polarization **0**. The polarization of the cell 1 obeys the action driver cell and in this way the driver cell can transmit the binary information to the surrounding cells. Thus the cell 1 can be referred to as the "working cell".

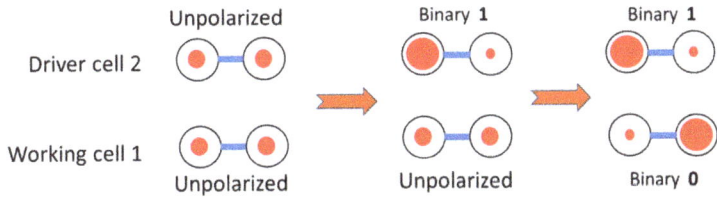

Figure 2. Scheme of the elementary process of the control of the binary information through the action of the dimeric driver cell 2 to the working cell 1. Left part shows the initial step of the information processing: unpolarized driver and unpolarized working cell. Then at the next step the driver gets polarization corresponding to the binary **1**. Finally, the polarized driver cell acts on the unpolarized working cell and gives rise to its polarization corresponding to binary **0**.

By combining such cells, one can obtain different QCA-based devices such as wires, majority logical gates, etc. QCA can be regarded as alternative to the traditional element base obtained with the aid of complementary metal–oxide–semiconductor (CMOS) technology for creating nanoscale devices capable of performing computations at very high switching rates. The advantages of the QCA devices as compared with CMOS ones are the smaller size of the devices and lower power consumption, which is a consequence of the current-less nature of the QCA devices.

As a further development of the concept of QCA based on quantum dots, a new fruitful idea of usage molecules as cells was proposed (see discussion in Reference [6] and a short overview in [7]). As natural candidates act as molecular cells, the dimeric MV molecules with one mobile excess electron and tetrameric MV molecules with two excess electrons have been proposed, as these bistable systems can encode the binary information [6]. The scope of this article does not allow to review a wealth of information on the results of molecular QCA, including the synthesis and study molecular systems, which are able to act as molecular cells. The reader can find it in references [8–25] and references thein.

In the systems considered until recently as tetrameric or dimeric cells (in both molecular and quantum dot-based implementations), the excess electrons were assumed to migrate over the diamagnetic centers (which can be alternatively referred to as "spinless cores"). The search for molecules that would be suitable for the design of molecular cells represents a problem whose complexity is caused by the requirements to such cells implied by their functional purposes, such as ability to encode the binary information and to easily control it by varying external field of the driver-cell. In terms of physical concepts, this can be formulated as the requirement of clearly pronounced property of bistability of the charge distribution in the working cell, which would ensure binary information is encoded. Moreover the high polarizability allows operations to be performed with the encoded information in the working cell by means of variable electric field created by the driver-cell.

To-date, the reported molecules possessing all properties required for the design of QCA are relatively scarce given that the search and synthesis of suitable molecules represents a very non-trivial task. At the same, this task lays the core of the design of the molecular QCA. In this regard, recently we have proposed [26], in order to expand the class of systems suitable as cells by including MV clusters, by which the excess electrons move over the network of localized spins (spin cores).

The presence of spin cores in such kind of systems (which here will be conventionally referred to as magnetic clusters) leads to the appearance of a specific kind of magnetic interaction, known as double exchange (DE). The DE is a spin-polarization mechanism resulting in the ferromagnetic spin alignment that occurs in MV clusters, containing mobile excess electrons, which produce polarization of the localized spins hat, to explain the ferromagnetic properties of some perovskites (see classical paper [27]. As far as the magnetic ions are involved, the Heisenberg-Dirac-Van Vleck (HDVV) exchange interaction between these ions is considered as well. As distinguishable from the traditional cells in which only charges are employed, the magnetic clusters considered here the spin degrees of freedom can be involved. Therefore, along with the QCA function proper, an additional useful functionality can be expected, such as spin switching in the working cell under the action of the electrostatic field induced by the driver-cell. Actually, the magnetic working cell has a ground state with a definite full spin that can be changed under the action of the purely electrostatic field of the magnetic driver cell. This phenomenon has been referred to as "spin switching" effect.

In the recent short communication [26], only a general idea of using magnetic MV clusters as cells for QCA devices and spin switchers has been proposed and theoretically supported in the framework of a simplified model, which takes into account only relevant spin-spin interactions and electron delocalization. At the same time, a number of topical issues, related to the theory of cells in which the DE involved has not been discussed, is provided in reference [26]. In particular, the previously developed two-mode vibronic model of an one-electron dimeric cell [28] should be generalized to the case of magnetic dimeric cells exhibiting DE and HDVV exchange forms. The model takes into account the interaction of an excess electron with both "breathing" local vibrations and the intercenter vibration.

In this article, we consider this problem for the case of the magnetic dimeric cells of the $d^2 - d^1$-type based on the transition metal ions (system in which the electron transfer occurs over the paramagnetic spin cores d^1). Although, the results born much wider frameworks of applicability. The aim of the present study is to develop the vibronic model for a free magnetic cell and the cell influenced by the driver cell. We attempted to reveal the conditions for spin switching under the action of the Coulomb field induced in the working cell by a neighboring driver-cell. On the basis of the developed model, we discuss both the spin switching effect and its influence on the cell-cell response function.

2. Magnetic Interactions in a $d^2 - d^1$-type Cell

We consider a $d^2 - d^1$-type MV dimer A-B (Figure 3), containing two equivalent paramagnetic centers playing a role of spin cores and an excess electron, migrating between these cores. We denote the spin of the core (d^1 ion) by S_0 ($S_0 = 1/2$ in the present case). It is assumed that we are dealing with the high-spin metal ions so that the spin of d^2 ion is $S_0 + 1/2 = 1$. These two spins are combined to give the total spin S of the dimer, which takes the values $1/2$ and $3/2$.

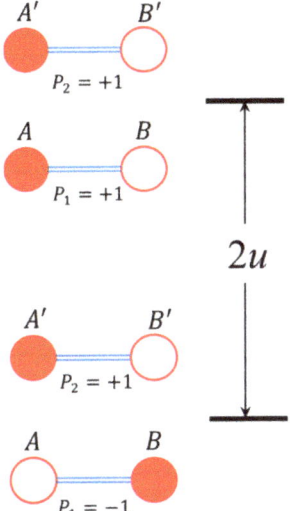

Figure 3. Mutual disposition of the driver-cell and the working cell, and the two possible electronic distributions in a pair of interacting dimeric cells shown to explain the physical meaning of the intercell Coulomb energy u. The sites belonging to the driver-cell and the working cell are primed and unprimed correspondingly, the site comprising (in a definite electronic distribution) the excess electron is shown as a red ball.

The electronic Hamiltonian of the dimer is represented as the following matrix defined for each value of the total spin:

$$\psi_A(S, M_S)\ \psi_B(S, M_S)\hat{H}(S) = \begin{pmatrix} -JS(S+1) + uP_2 & t_S \\ t_S & -JS(S+1) - uP_2 \end{pmatrix}. \quad (1)$$

In Equation (1) the basis $\psi_A(S, M_S)$ and $\psi_B(S, M_S)$ defining the 2 × 2-matrix includes the localized states (symbols A and B), which are characterized by the total spin S and quantum number of spin projection M_S. The off-diagonal matrix elements in Equation (1) are defined as follows:

$$t_S = \frac{t(S+1/2)}{2S_0 + 1} \quad (2)$$

The value t_S can be considered as the spin-dependent electron transfer parameter, or alternatively, DE parameter (deduced by Anderson and Hasegawa [27]), while t is the one-electron (bi-orbital) transfer parameter. Linear spin dependence of the transfer matrix element in Equation (2) is known as the main manifestation of the DE, and just this linear dependence predetermines ferromagnetic effect caused by this interaction. Diagonal matrix elements, include two types of contributions, namely, the exchange contribution of HDVV type, where J is the exchange parameter describing the interaction of centers with spins S_0 and $S_0 + 1/2$ (within each localized configuration of the dimer). In order to take into account the effect of the driver cell, we consider also the contribution describing the Coulomb interaction between the cells (terms (terms $\pm uP_2$), where P_2 is the polarization of the driver- cell $A' - B'$ (cell 2). The value $P_2 = +1$ corresponds to localization of the mobile charge on site A, while providing $P_2 = -1$ the charges localized on site B, as shown in Figure 3. Usually, in the theory of QCA, it is assumed that the driver-cell is a source of a Coulomb field acting on the working cell (cell 1), which causes its polarization. Finally, u is the characteristic energy of the Coulomb interaction between the cells. The physical meaning of this parameter is clear from Figure 3 showing the relative disposition of the working cell and the driver-cell, and the two possible electronic distributions in a pair of interacting cells. As observable, the energy $2u$ is the difference between the Coulomb repulsion energies of the excess electrons occupying neighboring and distant (energetically favorable) positions in the two interacting dimers.

3. Polarization of a Cell

The polarization of the driver-cell is determined by the following expression [3,4]:

$$P_2 = \frac{\rho_{A'} - \rho_{B'}}{\rho_{A'} + \rho_{B'}} \quad (3)$$

where $\rho_{A'}$ and $\rho_{B'}$ are the probabilities (electronic densities) of the two localizations of the excess electron ($\rho_{A'} + \rho_{B'} = 1$). The electronic densities on the centers have standard quantum-chemical definition through the eigenvectors of the system. It is assumed that polarization of the driver cell (in conformity with the definition of the driver cell) can be varied in a controllable manner from the value $P_2 = -1$ to value $P_2 = +1$, i.e., between the two fully polarized states. When the driver cell 2 is polarized, it induces polarization of the working cell 1. The latter polarization is also determined by Equation (3) in which the replacement $\rho_{A'} \to \rho_A, \rho_{B'} \to \rho_B$ is to be made.

While, the DE in dimers are known to always involve a ferromagnetic interaction [27], the HDVV exchange can, either be ferro- or antiferro-magnetic, depending on the physical conditions which determine the sign of the parameter J. We consider the most common situation when the HDVV exchange is antiferromagnetic ($J < 0$). In this case, the ground spin-state of the free cell (case of $P_2 = 0$) is determined by the competition between the ferromagnetic DE and antiferromagnetic HDVV exchange interactions. In most cases, DE dominates over the HDVV exchange and so the ground state proves to be ferromagnetic ($S_{gr} = 2S_0 + 1/2 = 3/2$). When $P_2 \neq 0$ the working cell is subjected to the action of an electrostatic field created by the driver-cell, which tends to localize the excess electron in the working cell. Since this field restricts the mobility of the excess electron it leads to a partial suppression of the ferromagnetic DE. As a result of such suppression, the antiferromagnetic HDVV exchange (that acts within localized configurations and hence is not affected by the field) can become the dominant interaction, which can lead to a stabilization of the spin-state with lower total spin value or, in other words, to cause a spin switching effect.

Effective control of the spin-state of the working cell by the Coulomb field of the driver-cell can be significantly complicated by the fact that, in strongly delocalized MV systems the ground ferromagnetic state is separated from the excited state, with a lower spin by a quite large energy gap and so spin switching is only possible is the Coulomb field is so strong that its effective energy exceeds this gap. However, the last condition is difficult to fulfill since the distances between the neighboring cells must be much longer than the distances between centers inside the cell, in order to prevent the electron transfer and the HDVV exchange between the cells. At the same time, the situation may be not so hopeless if we are dealing with the MV molecular systems, in which along with the electronic interactions, an essential role is played by the interaction of excess electrons with molecular vibrations of the cell. Therefore, we arrive at the conclusion regarding the crucial role of the vibronic coupling in MV molecules, in the context of discussion of the feasibility of spin switching effect.

4. Two-Mode Vibronic Model

In the vibronic model of a MV unit, we employ the interactions of the excess electron with the two types of vibrations that are taken into account. The first type is the active molecular vibration, which is composed of the totally symmetric "breathing" local modes spanning non-magnetic atoms of the redox sites. In inorganic metal clusters, the local vibrations are related to the displacements of the nearest ligand environments of the metal ions, while in organic compounds the redox sites have more complex structure and can involve several C-C bonds. In a particular case of the octahedrally coordinated metal sites in inorganic compounds, the local modes can be assigned to the full-symmetric vibrations (A_{1g}

symmetry). The vibronic interaction in MV compounds (particularly, in molecular cells) can be described in the framework of generally accepted Piepho-Kraus-Schatz (PKS) vibronic model [28–30]. Although this model is rather simplified, it successfully describes the key features of MV systems, such as the occurrence of a potential barrier between localized configurations. In particular, this model underlies the Robin and Day classification of MV compounds according to the degree of localization.

Within the conventional PKS vibronic model, the following symmetric and antisymmetric (with respect to inversion in the dimer resulting in the interchange $A \leftrightarrow B$) molecular coordinates can be composed of the local dimensionless coordinates q_A and q_B describing the "breathing" displacements [29,30]:

$$q_\pm = \frac{1}{\sqrt{2}}(q_A \pm q_B) \tag{4}$$

It can be shown that the totally symmetric (even) vibration q_+ can be excluded from the consideration since the corresponding contribution to the vibronic coupling is proportional to the unit matrix. From the physical point of view, this means that in course of this vibration, both sites are compressing and expanding in phase. This is obviously irrelevant to the charge transfer processes. On the contrary, the antisymmetric coordinate is interconnected with the vibration in course of which the expansion and compression of the sites occurs in the out-of-phase manner. The expanded site traps the mobile charge, while the compressed site tends to push it out. and thus, the PKS coupling is closely related to the electron transfer processes. The frequency of the vibration q_- will be denoted by ω and the parameter of vibronic coupling with this mode by v.

The second type of molecular vibrations involved in the model we use here is interrelated with the change of mutual disposition of the redox sites without changing their sizes. The corresponding part of the vibronic coupling describes the interaction of the excess electron with an intercenter vibration, which changes the distance between the sites [30]. Such kind of coupling arises from the modulation of the transfer parameter caused by the change of distance between the redox sites. To deduce this part of the vibronic coupling one should represent the transfer integral as a following series expansion:

$$t(R) = t(R_0) - \zeta (R - R_0) + \cdots \tag{5}$$

where $\zeta = -(\partial t/\partial R)_{R=R_0}$ is the parameter of vibronic interaction, and $t(R_0) \equiv t$ is the value of the transfer parameter evaluated at the equilibrium distance R_0 between the ions (below we will simply term it "transfer parameter"). The quantity $R - R_0$ plays a role of the vibrational coordinate associated with the inter-center vibration. We denote the frequency of this vibration as Ω and introduce the corresponding dimensionless normal coordinate as $Q = (R - R_0)/\sqrt{\hbar/M\Omega^2}$, where M is the effective mass. In contrast to local vibrations, which change the sizes of interacting mononuclear fragments, A and B as assumed in the PKS model, the inter-center vibration leaves the sizes of the coordination spheres unchanged. Therefore, we arrive at the two-mode vibronic problem, which takes into account the totally symmetric vibration Q with a frequency Ω and the antisymmetric (odd) vibration q_- with the frequency ω.

Then, the total Hamiltonian of a dimeric working cell subjected to action of the Coulomb field created by the adjacent driver-cell with polarization P_2 is obtained in the following matrix form (see Refs. [26,28]):

$$\hat{H} = \left[\frac{\hbar\omega}{2}\left(q^2 - \frac{\partial^2}{\partial q^2}\right) + \frac{\hbar\Omega}{2}\left(Q^2 - \frac{\partial^2}{\partial Q^2}\right) \right] \begin{pmatrix} 1 & 0 \\ 0 & 1 \end{pmatrix} + \begin{pmatrix} vq + uP_2 & t_S - \zeta_S Q \\ t_S - \zeta_S Q & -vq - uP_2 \end{pmatrix}. \tag{6}$$

In Equation (6) the following short notation is used:

$$\zeta_S = \frac{\zeta\left(S + \frac{1}{2}\right)}{2S_0 + 1} \tag{7}$$

The value ζ_S can be referred to as the spin-dependent coupling parameter with the intercenter vibration, which has the same spin dependence as the DE contribution in Equation (2). The matrices involved in Equation (6) are defined in the same bi-dimensional basis $\psi_A(S, M_S)$, $\psi_B(S, M_S)$ as the matrix of the electronic Hamiltonian, Equation (1). Equation (7) represents a block of the full Hamiltonian matrix with a definite set of spin quantum numbers S, M_S.

5. Dynamic Vibronic Problem

The commonly accepted tool in considering the energy pattern and electron localization in MV cluster is the adiabatic approximation, based on the assumption that the kinetic energy of the heavy ions can be neglected, or alternatively, that the nuclear motion is much slower than the electronic one. In this approach, the energy levels of the system are associated with the adiabatic potentials or potential curves. The applicability of the adiabatic approximation is invalid for the vibronic levels in the vicinity of the avoided crossing of the potential curves. This area is relevant to the process occurring when the localization in the working cell changes under the action of the driver cell. That is the reason why the subsequent analysis is based on the solution of the dynamic vibronic problem for full electron-vibrational Hamiltonian including kinetic energy of the ions. The importance of the non-adiabatic approach in the problem of mixed valency (for a free MV dimer) was realized long time ago (see reference [29] dealing with the quantum-mechanical evaluation of the profiles of the intervalence absorption) and applied to the study of a molecular cell for QCA in reference [28].

To solve the quantum-mechanical problem the matrix of the Hamiltonian, Equation (6), is to be presented in the basis composed of the products $\psi_A(SM)|nN\rangle$ and $\psi_B(SM_S)|nN\rangle$, where $|nN\rangle$ are the wave functions of the two-dimensional harmonic oscillator (first term in Equation (6)), n and N are vibrational quantum numbers related to the two types (PKS and inter-center) of vibrational modes under consideration. These functions are the eigen-functions of the unperturbed Hamiltonian that is the Hamiltonian from which the vibronic coupling is eliminated. To obtain a solution to the dynamic vibronic problem, diagonalization of this infinite matrix is required. The numerical solution in the truncated basis gives a set of spin-vibronic energy levels ε_k^S of the working cell and the corresponding spin-vibronic wave functions, which have the form of the following superpositions:

$$|k, S, M_S\rangle = \sum_{n,N} \left[c_{A,n,N}^{k,S} \psi_A(S, M_S)|n, N\rangle + c_{B,n,N}^{k,S} \psi_B(S, M_S)|n, N\rangle \right]. \qquad (8)$$

The truncation procedure restricts the size of the matrices to be diagonalized in such a way that it ensures a required accuracy (i.e., good convergence) in the evaluation of the low lying vibronic levels. Knowledge of the coefficients in the eigen-functions in Equation (8) means we can assess the electronic densities on the sites that are required for the evaluation of the polarization of the cell and consequently the cell-cell response function.

The dependences of the spin-vibronic wave functions on the polarization P_2 of the driver cell are then calculated for various sets of parameters and used to evaluate the key characteristics of the QCA cell (and QCA gate), known as the cell-cell response function, that is the dependence of the polarization P_1 of the working cell on the polarization P_2 of the driver-cell.

6. Spin-Vbronic Levels and Cell-Cell Response Function

Figure 4 shows the dependences of the spin-vibronic energy levels of the working cell represented by a MV dimer of $d^2 - d^1$–type on the polarization P_2 of the driver-cell calculated for fixed values of the parameters t, J and u and various ratios between the values of the vibronic parameters ζ and v. Figure 5 shows a family of the cell-cell response functions calculated with the same sets of parameters. To simplify the discussion the frequencies of the two modes are assumed to be equal ($\Omega = \omega$).

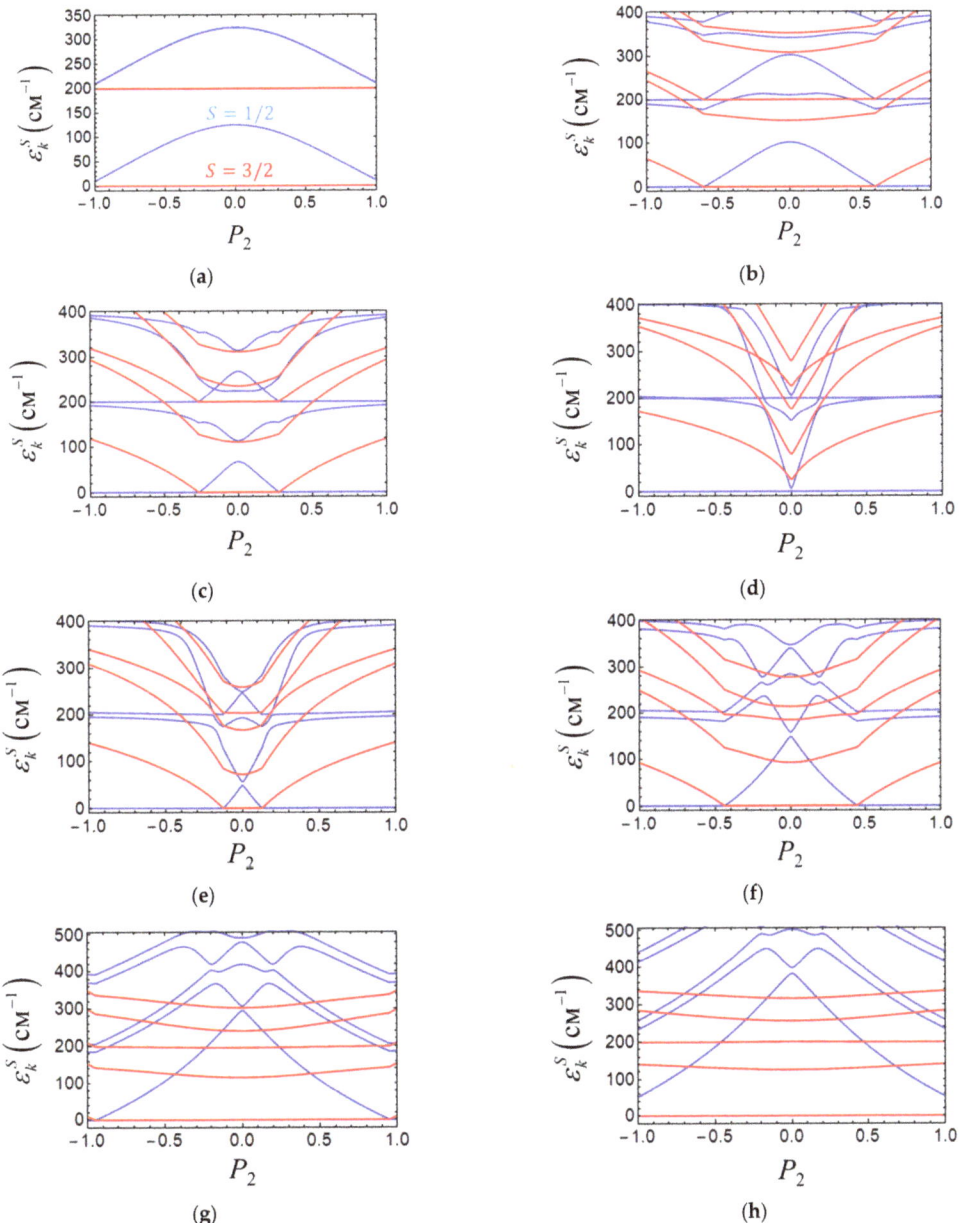

Figure 4. Spin-vibronic energy levels of the working cell $d^2 - d^1$ calculated as a function of polarization of the driver cell P_2 at $u = 600$ cm^{-1}, $\hbar\omega = \hbar\Omega = 200$ cm^{-1}, $t = 1000$ cm^{-1}, $J = -125$ cm^{-1} and the following sets of the vibronic coupling parameters: $v = 0$, $\zeta = 0$ (**a**); $v = 300$ cm^{-1}, $\zeta = 0$ (**b**); $v = 400$ cm^{-1}, $\zeta = 0$ (**c**); $v = 500$ cm^{-1}, $\zeta = 0$ (**d**); $v = 500$ cm^{-1}, $\zeta = 200$ cm^{-1} (**e**); $v = 500$ cm^{-1}, $\zeta = 300$ cm^{-1} (**f**); $v = 500$ cm^{-1}, $\zeta = 400$ cm^{-1} (**g**); $v = 500$ cm^{-1}, $\zeta = 450$ cm^{-1} (**h**). The ground spin-vibronic level is chosen as a reference point for the energy.

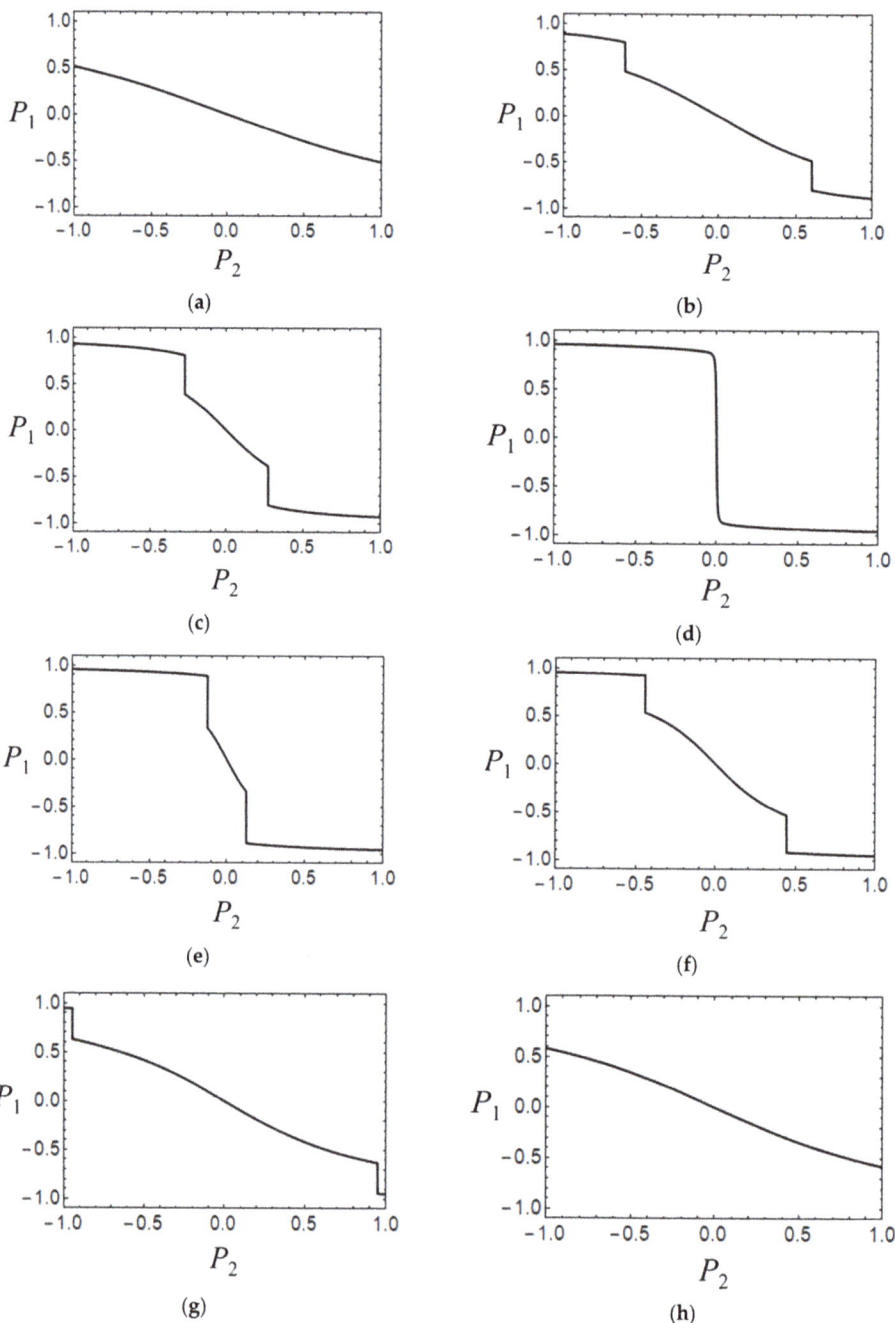

Figure 5. Cell-cell response functions evaluated for the $d^2 - d^1$-type cells at $u = 600$ cm^{-1}, $\hbar\omega = \hbar\Omega = 200$ cm^{-1}, $t = 1000$ cm^{-1}, $J = -125$ cm^{-1} and following sets of the vibronic coupling parameters: $v = 0$, $\zeta = 0$ (**a**); $v = 300$ cm^{-1}, $\zeta = 0$ (**b**); $v = 400$ cm^{-1}, $\zeta = 0$ (**c**); $v = 500$ cm^{-1}, $\zeta = 0$ (**d**); $v = 500$ cm^{-1}, $\zeta = 200$ cm^{-1}(**e**); $v = 500$ cm^{-1}, $\zeta = 300$ cm^{-1}(**f**); $v = 500$ cm^{-1}, $\zeta = 400$ cm^{-1}(**g**) and $v = 500$ cm^{-1}, $\zeta = 450$ cm^{-1}(**h**).

Figure 4a shows the limiting case of a negligibly weak coupling with both types of the vibrational modes ($v = \zeta = 0$). The selected sample values $t = 1000$ cm^{-1}, $J = -125$ cm^{-1} of the electronic parameters correspond to the indicated above-typical situation when the ferromagnetic contribution of DE significantly exceeds the antiferromagnetic contribution of the HDVV exchange. In this case, the ground state of an isolated cell $P_2 = 0$ has a maximum spin value $S = 3/2$ and it is separated from the first excited state with $S = 1/2$ by the energy gap of around 130 cm^{-1}. Figure 4a shows that the Coulomb field of the driver-cell tends to decrease the gap between these states due to suppression of the DE, but the ground state, in this case, remains at $S = 3/2$, even at the maximum polarization ($|P_2| = 1$) of the driver-cell.

The energy spectra in Figure 4b–d illustrate the influence of the PKS-type vibronic interaction on the field dependences of the energy levels. Since the PKS-type vibronic interaction tends to localize the excess electron, it weakens the ferromagnetic effect of the DE so that at a certain value of $|P_2|$ the spin switching $S = 3/2 \to S = 1/2$ occurs. It is notable that the larger the value of v, the lower the value of $|P_2|$, by which such spin switching occurs. This can be seen from the comparison of Figure 4b,c. Finally, for a sufficiently strong PKS-type vibronic coupling (the case shown in Figure 4d) the DE turns out to be almost completely suppressed and the ground spin-vibronic state is the low-spin one even provided that $P_2 = 0$.

Figure 4e–h demonstrate the effect of coupling of the excess electron with the inter-center vibration while the coupling with the PKS mode is assumed to be strong (the same as in the case shown in Figure 4d). As it follows from Figure 4e–h, the interaction with the inter-center vibration produces an effect that is in some sense opposite to the effect of the PKS interaction. Indeed, the PKS-type vibronic coupling tends to localize the electron at one of the redox sites, while the vibronic coupling with the intercenter mode promotes delocalization. In turn, an increase in the degree of delocalization leads to the enhancement of the ferromagnetic effect of the DE. As a result, the high-spin ground state of the working cell is restored, and at the same time the conditions required for the manifestation of spin switching become valid. Consequently, with an increase of the parameter ζ, the critical field (i.e., the value $|P_2|$) at which spin switching occurs also increases, which can be clearly seen through the comparison of the energy patterns, as shown in Figure 4e,f,g. Finally, in the case of a sufficiently strong interaction with the intercenter vibration, the self-trapping effect of the PKS-interaction proves to be fully compensated (Figure 4h), and consequently, the ground state remains the high-spin one regardless of the magnitude of $|P_2|$ exactly as in the case of a vanishingly weak vibronic coupling shown in Figure 4a.

The above effects of the two types of vibrations on the spin-vibronic energy levels of the cell manifest themselves also in the shapes of the cell-cell response function $P_1(P_2)$ (Figure 5). First, from Figure 5a,h, in the case of negligibly, we see weak vibronic PKS coupling. In the case of strong coupling with the inter-center vibration, the cell exhibits weak monotonic and almost linear response to the Coulomb field of the driver-cell. This feature of the cell-cell response function indicates that the cases mentioned so far are the least favorable ones from the point of view of functioning of the system, as both QCA cell and spin switcher.

In contrast, in the case of a strong PKS interaction and vanishingly weak interaction with the inter-center mode (case shown in Figure 5d), the cell-cell response function demonstrates a sharp nonlinear behavior, even at a very weak change in the polarization of the driver cell. This strongly non-linear behavior is indicative of the case for strong localization, which occurs when PKS coupling is strong. This case is the most favorable one for the functioning of the QCA devices.

Finally, in the intermediate cases (Figure 5b,c,e–g) the response function behaves non-monotonically demonstrating an abrupt change at critical values $|P_2|$ at which the spin switching in the cell occurs. This behavior is due to the fact that the states with lower values of the total spin are characterized by a higher polarizability as compared to the states for which the spin is higher. These cases are of practical interest from the point of view of the prospects in designing spin switching devices based on magnetic MV dimers.

7. Conclusions

This study was devoted to the problem of molecular implementation of QCA, a perspective technology with promising applications. Here, we attempted to proceed in studying the molecular cells as the central ingredient in the design of the QCA logical gates. We consider a cell represented by the multi-electron mixed valence binuclear $d^2 - d^1$-type cluster in which the double exchange, as well as the Heisenberg-Dirac-Van Vleck exchange interactions are operative. The dimeric unit can be considered also as a part of the bi-dimeric cell encoding binary information in the antipodal charge distributions.

Since the information is encoded in the charge distribution, an important issue we studied is interrelated with the vibronic coupling is known as a factor determining the charge localization in MV systems. We propose the two-mode vibronic model involving interactions with the local and intercenter modes and involves also the DE and HDVV exchange interactions have been applied to analyze the functional properties of MV dimer as the QCA cell.

It is demonstrated that the magnetic cell can encode the binary information and formulated the favorable conditions under which this functionality is efficient. An essentially new functionality closely interrelated with the spin degrees of freedom is the feature of spin switching in the working cell that was shown to occur under the electrostatic field of the driver cell. Therefore, it was shown that the magnetic cell can exhibit the property of multi-functionality being a reservoir for the binary information, and at the same time, act as a spin switcher.

The influence of both kind of vibrations on the dependences of the spin-vibronic levels of the working cell on the polarization of the driver-cell has been studied. The local "breathing" vibrations produce the trapping effect and increase the non-linearity of the cell-cell response function, while the inter-center vibrations tend to delocalize the system and therefore have destructive influence for the cell-response. Based on the developed vibronic model, the new features of the cell-cell response functions interrelated with the DE and HDVV exchange interactions in the magnetic molecular cell based on binuclear MV clusters of the $d^2 - d^1$-type have been revealed:

(1) in the case of the dominating DE, the character of the ground spin-state of a free and polarized cells has been shown to be strongly influenced by the interactions of the electronic subsystem with both types of vibrational modes included in the model. Therefore, in the case of vanishingly weak vibronic interaction, as well as in the case of strong coupling with the intercenter vibration, the ferromagnetic effect produced by the DE proves to be the dominating interaction;

(2) in this case of strong DE the $P_1(P_2)$ dependence (cell-cell response function) demonstrates weak and almost linear cell-cell response that is an unfavorable case for the QCA function. Moreover, the spin-switching is not possible in this case;

(3) in contrast, at strong vibronic PKS coupling, the ferromagnetic DE is largely suppressed, which leads to the stabilization of the state with a minimum total spin, along with the appearance of a strong nonlinear cell-cell response. This case is definitely favorable for the design of QCA-based devices.

(4) finally, when the contributions of both types of vibrations are comparable, the magnetic cell has been shown to exhibit properties of a spin-switcher, i.e., the electrostatic field of the driver change spin of the working cell along with the charge distribution. In relation to the QCA application, this feature manifests itself in the specific shape of the cell-cell response function exhibiting sharp steps. More precisely, the polarizability of the working cell is efficiently increased that is favorable for the QCA action.

The results, mentioned so far, create hope on their practical feasibility and relevance to the rational design of multifunctional molecular electronic devices that combine the function of the charge carriers of information in QCA with that of the spin-switchers.

Author Contributions: Conceptualization, B.T., A.P. and S.A.; methodology, A.P.; software, A.P.; writing—review and editing, B.T. and A.P.; All authors have read and agreed to the published version of the manuscript.

Funding: This research was funded by Ministry of Science and Higher Education of the Russian Federation, grant number AAAA-A19-119092390079-8.

Institutional Review Board Statement: Not applicable.

Informed Consent Statement: Not applicable.

Data Availability Statement: Not applicable.

Acknowledgments: A.P. and S.A. acknowledge support from the Ministry of Science and Higher Education of the Russian Federation (the state assignment no. AAAA-A19-119092390079-8).

Conflicts of Interest: The authors declare no conflict of interest.

Abbreviations

QCA	Quantum Cellular Automata
CMOS	Complimentary Metal-Oxide-Semiconductor Structure
MV	Mixed Valence
DE	Double Exchange
HDVV	Heisenberg-Dirac-Van Vleck (exchange, model)
PKS	Piepho-Kraus-Schatz (coupling, model)

References

1. Day, P. Molecular information processing: Will it happen? *Proc. R. Inst. Great Br.* **1998**, *69*, 85–106.
2. Robin, M.B.; Day, P. Mixed valence chemistry—A survey and classification. *Adv. Inorg. Chem. Radiochem.* **1967**, *10*, 247–422.
3. Lent, C.S.; Tougaw, P.D.; Porod, W.; Bernstein, G.H. Quantum Cellular Automata. *Nanotechnology* **1993**, *4*, 49–57. [CrossRef]
4. Lent, C.S.; Tougaw, P.; Porod, W. Bistable saturation in coupled quantum dots for quantum cellular automata. *Appl. Phys. Lett.* **1993**, *62*, 714–716. [CrossRef]
5. Lent, C.S.; Tougaw, P.D. Lines of interacting quantum-dot cells: A binary wire. *J. Appl. Phys.* **1993**, *74*, 6227–6233. [CrossRef]
6. Lent, C.S.; Isaksen, B.; Lieberman, M. Molecular Quantum-Dot Cellular Automata. *J. Am. Chem. Soc.* **2003**, *125*, 1056–1063. [CrossRef]
7. Tsukerblat, B.; Palii, A.; Aldoshin, S. Molecule Based Materials for Quantum Cellular Automata: A Short Overview and Challenging Problems. *Israel J. Chem.* **2020**, *60*, 527–543. [CrossRef]
8. Jiao, J.; Long, G.J.; Grandjean, F.; Beatty, A.M.; Fehlner, T.P. Building Blocks for the Molecular Expression of Quantum Cellular Automata. Isolation and Characterization of a Covalently Bonded Square Array of Two Ferrocenium and Two Ferrocene Complexes. *J. Am. Chem. Soc.* **2003**, *125*, 7522–7523. [CrossRef]
9. Li, Z.; Beatty, A.M.; Fehlner, T.P. Molecular QCA Cells. 1. Structure and Functionalization of an Unsymmetrical Dinuclear Mixed-Valence Complex for Surface Binding. *Inorg. Chem.* **2003**, *42*, 5707–5714. [CrossRef] [PubMed]
10. Li, Z.; Fehlner, T.P. Molecular QCA Cells. 2. Characterization of an Unsymmetrical Dinuclear Mixed-Valence Complex Bound to a Au Surface by an Organic Linker. *Inorg. Chem.* **2003**, *42*, 5715–5721. [CrossRef]
11. Qi, H.; Sharma, S.; Li, Z.; Snider, G.L.; Orlov, A.O.; Lent, C.S.; Fehlner, T.P. Molecular Quantum Cellular Automata Cells. Electric Field Driven Switching of a Silicon Surface Bound Array of Vertically Oriented Two-Dot Molecular Quantum Cellular Automata. *J. Am. Chem. Soc.* **2003**, *125*, 15250–15259. [CrossRef] [PubMed]
12. Braun-Sand, S.B.; Wiest, O. Theoretical Studies of Mixed Valence Transition Metal Complexes for Molecular Computing. *J. Phys. Chem. A* **2003**, *107*, 285–291. [CrossRef]
13. Braun-Sand, S.B.; Wiest, O. Biasing Mixed-Valence Transition Metal Complexes in Search of Bistable Complexes for Molecular Computing. *J. Phys. Chem. B* **2003**, *107*, 9624–9628. [CrossRef]
14. Qi, H.; Gupta, A.; Noll, B.C.; Snider, G.L.; Lu, Y.; Lent, C.; Fehlner, T.P. Dependence of Field Switched Ordered Arrays of Dinuclear Mixed-Valence Complexes on the Distance between the Redox Centers and the Size of the Counterions. *J. Am. Chem. Soc.* **2005**, *127*, 15218–15227. [CrossRef] [PubMed]
15. Jiao, J.; Long, G.J.; Rebbouh, L.; Grandjean, F.; Beatty, A.M.; Fehlner, T.P. Properties of a Mixed-Valence $(Fe^{II})_2(Fe^{III})_2$ Square Cell for Utilization in the Quantum Cellular Automata Paradigm for Molecular Electronics. *J. Am. Chem. Soc.* **2005**, *127*, 17819–17831. [CrossRef]
16. Lu, Y.; Lent, C.S. Theoretical Study of Molecular Quantum Dot Cellular Automata. *J. Comput. Electron.* **2005**, *4*, 115–118. [CrossRef]
17. Zhao, Y.; Guo, D.; Liu, Y.; He, C.; Duan, C. A Mixed-Valence $(Fe^{II})_2(Fe^{III})_2$ Square for Molecular Expression of Quantum Cellular Automata. *Chem. Commun.* **2008**, 5725–5727. [CrossRef]

18. Nemykin, V.N.; Rohde, G.T.; Barrett, C.D.; Hadt, R.G.; Bizzarri, C.; Galloni, P.; Floris, B.; Nowik, I.; Herber, R.H.; Marrani, A.G.; et al. Electron-Transfer Processes in Metal-Free Tetraferrocenylporphyrin. Understanding Internal Interactions to Access Mixed Valence States Potentially Useful for Quantum Cellular Automata. *J. Am. Chem. Soc.* **2009**, *131*, 14969–14978. [CrossRef]
19. Wang, X.; Yu, L.; Inakollu, V.S.S.; Pan, X.; Ma, J.; Yu, H. Molecular Quantum-Dot Cellular Automata Based on Diboryl Radical Anions. *J. Phys. Chem. C* **2018**, *122*, 2454–2460. [CrossRef]
20. Burgun, A.; Gendron, F.; Schauer, P.A.; Skelton, B.W.; Low, P.J.; Costuas, K.; Halet, J.-F.; Bruce, M.I.; Lapinte, C. Straightforward Access to Tetrametallic Complexes with a Square Array by Oxidative Dimerization of Organometallic Wires. *Organometallics* **2013**, *32*, 5015–5025. [CrossRef]
21. Schneider, B.; Demeshko, S.; Neudeck, S.; Dechert, S.; Meyer, F. Mixed-Spin [2 × 2] Fe$_4$ Grid Complex Optimized for Quantum Cellular Automata. *Inorg. Chem.* **2013**, *52*, 13230–13237. [CrossRef] [PubMed]
22. Christie, J.A.; Forrest, R.P.; Corcelli, S.A.; Wasio, N.A.; Quardokus, R.C.; Brown, R.; Kandel, S.A.; Lu, Y.; Lent, C.S.; Henderson, K.W. Synthesis of a Neutral Mixed-Valence Diferrocenyl Carborane for Molecular Quantum-Dot Cellular Automata Applications. *Angew. Chem. Int. Ed.* **2015**, *54*, 15448–15671. [CrossRef] [PubMed]
23. Makhoul, R.; Hamon, P.; Roisnel, T.; Hamon, J.-R.; Lapinte, C. A Tetrairon Dication Featuring Tetraethynylbenzene Bridging Ligand: A Molecular Prototype of Quantum Dot Cellular Automata. *Chem. Eur. J.* **2020**, *26*, 8368–8371. [CrossRef]
24. Rahimia, E.; Reimers, J.R. Molecular quantum cellular automata cell design trade-offs: Latching vs. power dissipation. *Phys. Chem. Chem. Phys.* **2018**, *20*, 17881–17888. [CrossRef] [PubMed]
25. Ardesi, Y.; Pulimeno, A.; Graziano, M.; Riente, F.; Piccinini, G. Effectiveness of Molecules for Quantum Cellular Automata as Computing Devices. *J. Low Power Electron. Appl.* **2018**, *8*, 24. [CrossRef]
26. Palii, A.; Clemente-Juan, J.M.; Rybakov, A.; Aldoshin, S.; Tsukerblat, B. Exploration of the double exchange in quantum cellular automata: Proposal for a new class of cells. *Chem. Comm.* **2020**, *56*, 10682–10685. [CrossRef] [PubMed]
27. Anderson, P.W.; Hasegawa, H. Consideration of Double Exchange. *Phys. Rev.* **1955**, *100*, 675–681. [CrossRef]
28. Palii, A.; Aldoshin, S.; Zilberg, S.; Tsukerblat, B. A parametric two-mode vibronic model of a dimeric mixed-valence cell for molecular quantum cellular automata and computational ab initio verification. *Phys. Chem. Chem. Phys.* **2020**, *22*, 25982–25999. [CrossRef] [PubMed]
29. Piepho, S.B.; Krausz, E.R.; Schatz, P.N. Vibronic coupling model for calculation of mixed valence absorption profiles. *J. Am. Chem. Soc.* **1978**, *100*, 2996–3005. [CrossRef]
30. Piepho, S.B. Vibronic coupling model for the calculation of mixed-valence line shapes: The interdependence of vibronic and MO effects. *J. Am. Chem. Soc.* **1988**, *110*, 6319–6326. [CrossRef]

Article

Magnetic and Structural Properties of Organic Radicals Based on Thienyl- and Furyl-Substituted Nitronyl Nitroxide

Tadashi Sugano [1,*], Stephen J. Blundell [2], William Hayes [2] and Hatsumi Mori [3]

[1] Department of Chemistry, Meiji Gakuin University, Kamikurata, Totsuka, Yokohama 244-8539, Japan
[2] Clarendon Laboratory, Department of Physics, Oxford University, Parks Road, Oxford OX1 3PU, UK; stephen.blundell@physics.ox.ac.uk (S.J.B.); bill.hayes@physics.ox.ac.uk (W.H.)
[3] The Institute for Solid State Physics, The University of Tokyo, Kashiwanoha, Kashiwa 277-8581, Japan; hmori@issp.u-tokyo.ac.jp
* Correspondence: sugano@law.meijigakuin.ac.jp

Citation: Sugano, T.; Blundell, S.J.; Hayes, W.; Mori, H. Magnetic and Structural Properties of Organic Radicals Based on Thienyl- and Furyl-Substituted Nitronyl Nitroxide. *Magnetochemistry* 2021, 7, 62. https://doi.org/10.3390/magnetochemistry7050062

Academic Editors: Lee Martin, Scott Turner, John Wallis, Hiroki Akutsu and Carlos J. Gómez García

Received: 31 March 2021
Accepted: 29 April 2021
Published: 6 May 2021

Publisher's Note: MDPI stays neutral with regard to jurisdictional claims in published maps and institutional affiliations.

Copyright: © 2021 by the authors. Licensee MDPI, Basel, Switzerland. This article is an open access article distributed under the terms and conditions of the Creative Commons Attribution (CC BY) license (https://creativecommons.org/licenses/by/4.0/).

Abstract: Magnetic properties of organic radicals based on thienyl- and furyl-substituted nitronyl nitroxide (NN) and iminonitroxide (IN) were investigated by measuring the temperature dependence of the magnetization. The magnetic behavior of 2-benzo[*b*]thienyl NN (2-BTHNN) is interpreted in terms of the two-magnetic-dimer model, in which one dimer exhibits ferromagnetic (FM) intermolecular interaction and the other dimer shows antiferromagnetic (AFM) interaction. The existence of two dimers in 2-BTHNN is supported by crystal structure analysis. The magnetic behaviors of 2-bithienyl NN, 4-(2′-thienyl)phenyl NN (2-THPNN), 2- and 3-furyl NN, 2-benzo[*b*]furyl NN, and 3-benzo[*b*]thienyl IN are also reported. The one-dimensional alternating AFM nature observed in 2-THPNN is consistent with its crystal structure.

Keywords: nitronyl nitroxide; iminonitroxide; magnetism; organic crystals

1. Introduction

Magnetism in neutral organic radicals based on nitronyl nitroxide (NN) (2-substituted 4,4,5,5-tetramethyl-4,5-dihydro-3-oxido-1*H*-imidazol-3-ium-2-yl-1-oxyl) and iminonitroxide (IN) (2-substituted 4,4,5,5-tetramethyl-4,5-dihydro-1*H* imidazol-2-yl-1-oxyl) has long been studied to find new magnetically interesting molecular crystals, since these radicals are usually stable in the solid state [1] and have been components of many molecule-based magnets [2–10].

Introducing sulfur atoms into the NN and IN derivatives would result in an increase in magnetic interactions between neighboring molecules in molecular crystals, since the sulfur atoms can make larger molecular orbital overlaps as observed in conducting organic materials [11]. We have, therefore, been preparing thienyl-substituted NN and IN, which include a sulfur atom, and investigating their magnetic properties [12–14]. We report here magnetic properties of three thienyl-substituted NN derivatives, 2-benzo[*b*]thienyl NN (2-BTHNN), 2-bithienyl NN (2-BiTHNN), and 4-(2′-thenyl)phenyl NN (2-THPNN) and two thienyl-substituted IN derivatives, 2- and 3-benzo[*b*]thienyl IN (2-BTHIN and 3-BTHIN). We also report here crystal structures of 2-BTHNN and 2-THPNN to discuss magneto-structural correlations in these radicals. To discuss the effects of sulfur substitution on magnetic interactions, magnetic properties of three furyl-substituted NN derivatives, 2- and 3-furyl NN (2-FNN and 3-FNN), and 2-benzo[*b*]furyl NN (2-BFNN), and a furyl-substituted IN derivative, 2-furyl IN (2-FIN), are also described, because the furyl ring is the oxygen analogue of the thienyl ring. The molecular structures of organic radicals reported here are listed in Figure 1.

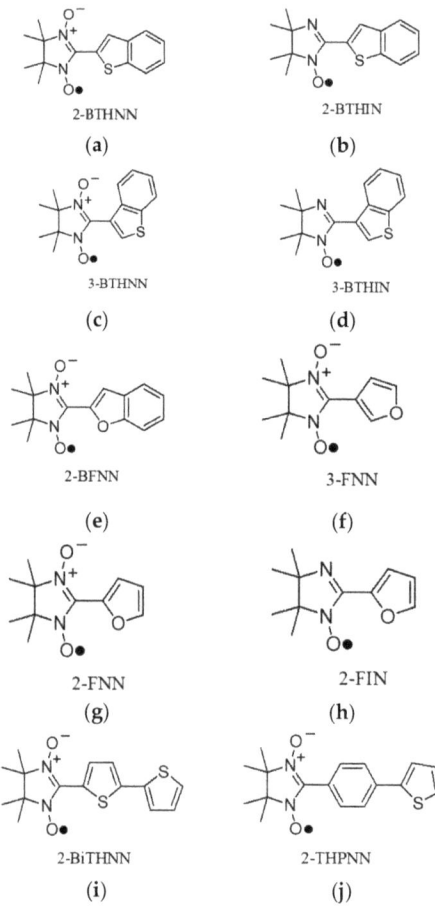

Figure 1. Molecular structures of organic radicals reported here. (**a**) 2-BTHNN, (**b**) 2-BTHIN, (**c**) 3-BTHNN, (**d**) 3-BTHIN, (**e**) 2-BFNN, (**f**) 3-FNN, (**g**) 2-FNN, (**h**) 2-FIN, (**i**) 2-BiTHNN, and (**j**) 2-THPNN (see the text for the abbreviated names).

This paper is a tribute to Professor Peter Day who gave us many suggestions and opportunities to carry out our studies of magnetochemistry.

2. Results and Discussion

2.1. Magnetic Properties

Figure 2 shows the temperature dependence of the product of paramagnetic susceptibility χ_p and temperature T of 2-BTHNN and 2-BTHIN. Upon lowering the temperature, the product, $\chi_p T$, of 2-BTHNN decreases monotonically from 0.374 emu·K·mol^{-1} at 300 K to 0.198 emu·K·mol^{-1} at around 10 K. Below about 10 K, however, $\chi_p T$ of 2-BTHNN increases slowly to 0.204 emu·K·mol^{-1} at 1.8 K, suggesting the existence of ferromagnetic (FM) interactions in 2-BTHNN.

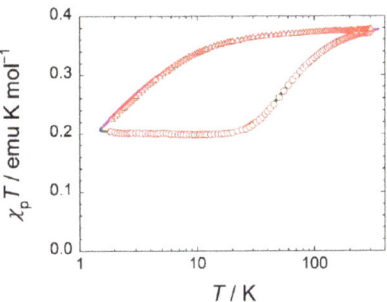

Figure 2. Temperature dependences of $\chi_p T$ of 2-BTHNN (open circles) and 2-BTHIN (open triangles). The solid line for 2-BTHNN indicates a fit using the two-magnetic-dimer model (see the text).

Since the temperature dependence of $\chi_p T$ of 2-BTHNN shown above appears not to be simple, we examined several models to fit the data and then found that the two-magnetic-dimer model, in which one molecular dimer (which we denote as the FM dimer hereafter) exhibits FM intermolecular interactions and the other dimer (which we denote as the AFM dimer) shows antiferromagnetic (AFM) intermolecular interactions, can explain the temperature dependence of $\chi_p T$ of 2-BTHNN, as represented by the solid line in Figure 2. At temperatures lower than about 10 K, the moderately strong AFM interaction operating in the AFM dimer, mentioned below, leads to an almost complete vanishing of the contribution of the AFM dimer to χ_p. This AFM interaction is interpreted in terms of the two-spin dimer model [15] with the exchange coupling constant $J/k = -55$ K and the Curie constant $C = 0.197$ emu·K·mol^{-1}. This magnitude of C is about a half of 0.376 emu·K·mol^{-1} for the uncorrelated $S = 1/2$ spins in 2-BTHNN

As a result, the temperature dependence of $\chi_p T$ of 2-BTHNN below about 10 K would come from only the contribution of the FM dimer. This contribution is modeled in terms of the Curie–Weiss law with the Weiss temperature $\theta = +0.06$ K and $C = 0.197$ emu·K·mol^{-1}. The positive Weiss temperature obtained here clearly indicates the existence of the FM intermolecular interactions in the FM dimer. In addition, we observed further evidence for the FM interactions by measuring the magnetization isotherms at low temperatures below 10 K. Upon lowering temperature, magnetization isotherms deviate from the $S = 1/2$ Brillouin function curve onto the $S = 1$ curve as shown in Figure 3. Since the Curie constant for the FM dimer is the same as that of the AFM dimer and it is just a half of the Curie constant for the uncorrelated $S = 1/2$ spins of 2-BTHNN, it is concluded that the half of the molecular spins exhibit the FM intermolecular interactions and the other half of spins show the AFM interactions in 2-BTHNN. This conclusion is supported further by analyzing the crystal structure of 2-BTHNN as mentioned below.

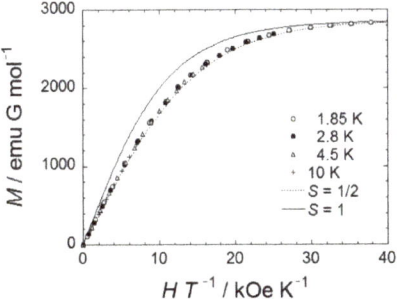

Figure 3. Magnetization isotherms of 2-BTHNN at 1.85 K (open circles), 2.8 K (closed circles), 4.5 K (open triangles), and 10 K (crosses).

In contrast, the temperature dependences of $\chi_p T$ of 2-BTHIN can be fitted simply to the Curie–Weiss law with $C = 0.378$ emu·K·mol^{-1} and $\theta = -1.23$ K. In this case, our results show that the elimination of an oxygen atom gives a drastic change of magnetic behavior.

Figure 4 shows the temperature dependences of $\chi_p T$ of 2-BiTHNN (open circles) and 2-THPNN (open triangles). These two radicals also have thienyl-including moieties that are longer than those of other radicals reported in this paper, as shown in Figure 1. These two radicals exhibit weak AFM intermolecular interactions, because the temperature dependences of $\chi_p T$ of 2-BiTHNN and 2-THPNN are interpreted in terms of the one-dimensional (1D) alternating Heisenberg model [16] with $J/k = -2.34$ K, alternating parameter $\alpha = 0.8$ and $C = 0.380$ emu·K·mol^{-1}, and with $J/k = -0.77$ K, $\alpha = 0.8$, and $C = 0.370$ emu·K·mol^{-1}, respectively, as represented by the solid lines in Figure 4. The origin of the alternating magnetic interactions in 2-THPNN is discussed below by referring to the crystal structure.

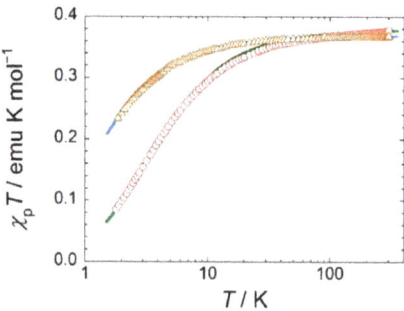

Figure 4. Temperature dependences of $\chi_p T$ of 2-BiTHNN (open circles) and 2-THPNN (open triangles). Solid lines represent theoretical fitting on the basis of the 1D alternating AFM Heisenberg model (see the text).

The magnetic properties of 4-(3′-thienyl)phenyl NN (3-THPNN), which is a structural isomer of 2-THPNN shown here, was previously reported by Coronado et al. about two decades ago [17]. The radical 3-THPNN shows weak AFM intermolecular interactions with $\theta = -1.5$ K and $C = 0.31$ emu·K·mol^{-1} similar to those found in 2-THPNN.

The temperature dependences of $\chi_p T$ of the three furyl-substituted nitronyl nitroxide radicals, 2-FNN (open circles), 3-FNN (open triangles), and 2-BFNN (open squares), are shown in Figure 5 together with those of furyl-substituted iminonitroxide radical, 2-FIN. The magnetic behaviors of 3-FNN, 2-BFNN, and 2-FIN are similar to each other, although the magnitude of the magnetic interaction is significantly different as mentioned below, while those of 2-FNN are quite different at temperatures lower than about 10 K. The values of $\chi_p T$ do not decrease steeply with lowering temperature but show a plateau between 4 and 10 K. Although this behavior appears to be reminiscent of that observed in 2-BTHNN, a similar kind of behavior is also characteristic of the four-spin linear tetramer model [18], because any upturn of $\chi_p T$ values at low temperatures is not observed. As represented by the solid line in Figure 5, we successfully reproduced the temperature dependence of $\chi_p T$ of 2-FNN in terms of the four-spin linear tetramer model with $J_1/k = -3.5$ K and $J_2/k = -13$ K, where J_1/k represents the intradimer interaction and J_2/k represents the interdimer interaction in the linear tetramer, having $S = 1/2$ spin on each molecule within the tetramer. The Curie constant used to fit the experimental data was $C = 0.340$ emu·K·mol^{-1}. This value is slightly lower than the value $C = 0.376$ emu·K·mol^{-1} that is expected for uncorrelated $S = 1/2$ spins. This difference comes from the contribution of impurity spins with $C_i = 0.038$ emu·K·mol^{-1} and $\theta_i = -0.1$ K used to obtain the best fit to the experimental data. The overall Curie constant is 0.378 emu·K·mol^{-1} and close to that expected for uncorrelated $S = 1/2$ spins. The origin of spin interactions within the tetramer is not clear, since we have no crystal information for 2-FNN at present.

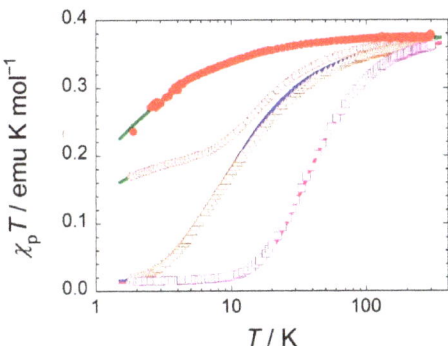

Figure 5. Temperature dependences of $\chi_p T$ of 2-FNN (open circles), 3-FNN (open triangles), 2-BFNN (open squares), and 2-FIN (closed circles). The solid lines represent fits using the four-spin linear tetramer model for 2-FNN, the 1D alternating AFM Heisenberg model for 3-FNN and 2-BFNN, and the Curie–Weiss law for 2-FIN (see the text).

The temperature dependences of $\chi_p T$ of 3-FNN and 2-BFNN can be fitted to the 1D alternating Heisenberg model [16] with $J/k = -7.5$ K, $\alpha = 0.6$, $C = 0.346$ emu·K·mol^{-1}, $C_i = 0.025$ emu·K·mol^{-1} and $\theta_i = -0.0$ K for 3-FNN, and with $J/k = -33$ K, $\alpha = 0.2$, $C = 0.369$ emu·K·mol^{-1}, $C_i = 0.022$ emu·K·mol^{-1}, $\theta_i = -1.0$ K for 2-BFNN. On the other hand, the temperature dependence of $\chi_p T$ of 2-FIN interpreted in terms of the Curie–Weiss law yields $C = 0.376$ emu·K·mol^{-1} and $\theta = -1.0$ K.

The sulfur analogues of 2- and 3-FNN, 2- and 3-THNN, show 1D alternating Heisenberg behavior with $J/k = -6.6$ K, $\alpha = 0.5$, $C = 0.359$ emu·K·mol^{-1}, $C_i = 0.009$ emu·K·mol^{-1}, $\theta_i = 0.0$ K for 2-THNN, and with $J/k = -5.3$ K, $\alpha = 0.6$, $C = 0.360$ emu·K·mol^{-1}, $C_i = 0.016$ emu·K·mol^{-1}, $\theta_i = 0.0$ K for 3-THNN [14]. The magnetic behavior of 2-FNN mentioned above is very different from that of 2-THNN. That is to say, 2-FNN shows the four-spin linear tetramer behavior and 2-THNN exhibits 1D alternating Heisenberg behavior. It is, therefore, difficult to compare magnetic interactions directly in both radicals. However, the magnetic behaviors of 3-FNN and 3-THNN are both interpreted in terms of the 1D alternating Heisenberg model with $J/k = -7.5$ K ($\alpha = 0.6$) and $J/k = -5.3$ K ($\alpha = 0.6$), respectively. This result suggests that the substitution of the oxygen atom in the furyl ring by the sulfur atom does not yield stronger magnetic interactions in this case.

The magnetic interactions in 2-BTHNN, which is the sulfur analogue of 2-BFNN, seem to become stronger. The magnetic behavior of 2-BFNN can also be explained by using the two-spin dimer model with $J/k = -35$ K, although the 1D alternating Heisenberg model with $J/k = -34$ K ($\alpha = 0.2$) gives slightly better fit as mentioned above. The value of $J/k = -35$ K is smaller than $J/k = -55$ K as observed in 2-BTHNN. This result indicates that the substitution of the oxygen atom in the furyl ring by the sulfur atom yields stronger magnetic interactions in this case.

The temperature dependences of $\chi_p T$ of 3-BTHIN is shown in Figure 6 together with that of 3-BTHNN [14]. The magnetic behavior of 3-BTHIN is reproduced in terms of the 1D AFM Heisenberg model with $J/k = -1.78$ K, $C = 0.348$ emu·K·mol^{-1}, $C_i = 0.032$ emu·K·mol^{-1} and $\theta_i = 0.0$ K, whereas that of 3-BTHNN is interpreted in terms of quasi-two-dimensional FM intermolecular interactions with $J/k = +0.16$ K within the layer and $J'/k = +0.02$ K for the interlayer and $C = 0.384$ emu·K·mol^{-1} [14]. Elimination of an oxygen atom from one of the NO groups of the nitronyl nitroxide moiety of 3-BTHNN results in a remarkable change in magnetic behavior due possibly to a change in molecular arrangements in the solid. To discuss further, it is indispensable to determine the crystal structures of 3-BTHIN.

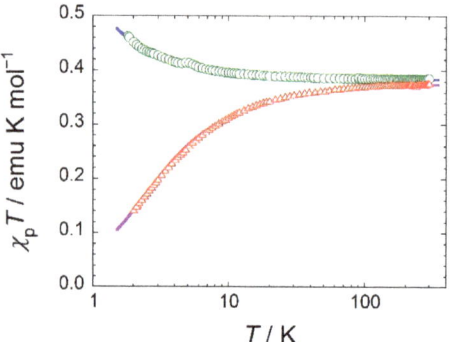

Figure 6. Temperature dependences of $\chi_p T$ of 3-BTHIN (open triangles) and 3-BTHNN (open circles). Solid lines represent theoretical fitting on the basis of the 1D AFM Heisenberg model for 3-BTHIN and the quasi-2D FM Heisenberg model for 3-BTHNN (see the text).

2.2. Crystal Structures

The radical 2-BTHNN crystallizes in the monoclinic space group C2/c. The crystallographic data of 2-BTHNN are listed in Table 1. Figure 7a shows an ORTEP view of the crystal structure along the b axis. The 2-BTHNN molecules form two different types of molecular dimers as denoted by A and B shown in Figure 7a. In dimer A, the 2-BTHNN molecules stack face-to-face and their molecular long axes make an angle of 73°, as shown in Figure 7b. The shortest intermolecular atomic distance between N and O atoms is 3.499(3) Å. In dimer B, the 2-BTHNN molecules stack face-to-face in a head-to-tail manner as shown in Figure 7c. The benzothienyl rings are close to each other to avoid steric hindrances due to bulky methyl groups on the nitronyl nitroxide moieties. The shortest intermolecular atomic distance is 3.267(5) Å between the C atom of the benzene ring and the C atom of the thiophene ring. Quite different molecular arrangements in these two dimers A and B mentioned above would yield very distinctive magnetic behaviors, i.e., FM and AFM interactions in 2-BTHNN. Although it is not easy to attribute the origin of FM and AFM interactions onto these different dimers, the nearly orthogonally arranged molecules in the dimer A seems to give the FM interactions, and the face-to-face stacking of benzothienyl groups appears to result in the AFM interactions in 2-BTHNN.

Table 1. Crystallographic data for the organic radicals 2-BTHNN and 2-THPNN.

	2-BTHNN	2-THPNN
Chemical formula	$C_{15}H_{17}N_2O_2S$	$C_{17}H_{19}N_2O_2S$
Formula weight	289.37	315.41
Crystal system	Monoclinic	Monoclinic
Space group	C2/c	$P2_1/n$
a (Å)	17.636 (6)	13.30 (7)
b (Å)	11.400 (4)	9.57 (4)
c (Å)	29.886 (9)	14.51 (7)
β (°)	95.442 (4)	117.28 (7)
V (Å^3)	5982 (3)	1642 (14)
Z value	16	4
Dcalc (Mg·m^{-3})	1.285	1.276
Reflections independent	6259	3631
R, R_w [$I > 2\sigma(I)$]	0.0808, 0.1002	0.0808, 0.0967

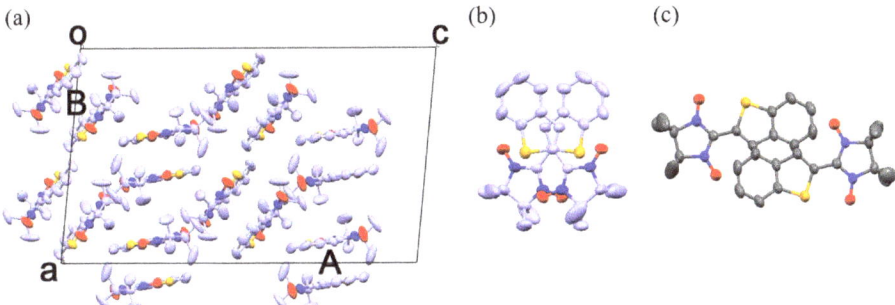

Figure 7. (**a**) An ORTEP view of the crystal structure of 2-BTHNN along the *b* axis. "A" and "B" denote the two kinds of molecular dimers. Molecular conformations of two neighboring 2-BTHNN in dimer A (**b**) and dimer B (**c**).

The radical 2-THPNN crystallizes in the monoclinic space group $P2_1/n$. The crystallographic data of 2-THPNN are also listed in Table 1. Figure 8 shows an ORTEP view of the crystal structure along the direction perpendicular to the molecular planes of the 2-THPNN within one of the molecular stacks. The 2-THPNN molecules stack side-by-side in a head-to-tail manner along the *a* axis. The molecular planes of the molecules belong to the neighboring stacks are arranged perpendicularly. In the molecular stacks, there are two types of atomic contacts between neighboring molecules. One type of atomic contact is formed between the O atom on the NO group and the two C atoms on the phenyl ring with the atomic distances of 3.404(2) Å and 3.449(3) Å. The other type of atomic contact is formed between the O atom and the C atom on the phenyl ring with the atomic distance of 3.309(3) Å and the C atom on the thienyl ring with the atomic distance of 3.462(3) Å. These two types of atomic contacts existing in the molecular stacks along the *a* axis probably result in the 1D alternating magnetic interactions observed in 2-THPNN.

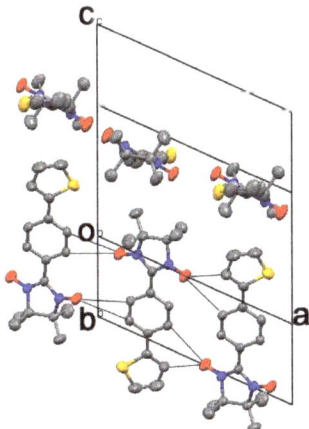

Figure 8. An ORTEP view of the crystal structure of 2-THPNN along the vertical direction to the molecular planes. The 2-THPNN molecules form side-by-side molecular stacks along the *a* axis in a head-to-tail manner.

3. Materials and Methods

The radicals were prepared according to the procedures reported in [1] and purified through column chromatography followed by a recrystallization. Commercially available (Aldrich) benzo[*b*]thiophene-2-carboxyaldehyde, benzo[*b*]thiophene-3-carboxyaldehyde, 2-benzofurancarboxyaldehyde, 2-furanaldehyde, 3-furanaldehyde, 2,2′-bithio-phene-5-

carboxyaldehyde, and 4-(2′-thienyl)benzaldehyde were used without further purification. N,N′-Dihydroxy-2,3-diamino-2,3-dimethylbutane was obtained according to the literature [1]. Other reagents and solvents were used as purchased.

2-BTHNN and 2-BTHIN. Benzo[b]thiophene-2-carboxyaldehyde (2.56 g, 15.8 mmol) and N,N′-dihydroxy-2,3-diamino-2,3-dimethylbutane (2.14 g, 14.4 mmol) were mixed in 15 mL of benzene at 40 °C. The reaction mixture was stirred for 24 h, after which the resulting white solid of 1,3-dihydroxy-2-(2-bebzo[b]thienyl)-4,4,5,5-tetramethylimidazolidine was filtered off and washed with 5 mL benzene twice and dried under vacuum. Yield: 98%. A solution of sodium periodate (2.42 g, 11.3 mmol) in 25 mL of water was added dropwise to a suspension of 1,3-dihydroxy-2-(2-bebzo[b]thienyl)-4,4,5,5-tetramethylimidazolidine (2.18 g, 7.47 mmol) in 100 mL dichloromethane at room temperature. The dark-green organic phase was separated and concentrated under vacuum. The crude product was separated and purified by column chromatography (eluent: ethyl acetate, alumina) to obtain 2-BTHNN (dark-green solid) and 2-BTHIN (red solid).

3-BTHNN and 3-BTHIN. A similar experimental procedure was used to obtain 3-BTHNN (dark-green/blue solid) and 3-BTHIN (red solid) by using benzo[b]thiophene-3-carboxyaldehyde.

2-BFNN. 2-Benzofurancarboxyaldehyde (2.58 g, 17.7 mmol) and N,N′-dihydroxy-2,3-diamino-2,3-dimethylbutane (2.39 g, 16.1 mmol) were mixed in 15 mL of benzene at 40 °C. The reaction mixture was stirred for 2 h, after which the resulting white solid of 1,3-dihydroxy-2-(2-bebzo[b]furyl)-4,4,5,5-tetramethylimidazolidine was filtered off and washed with 5 mL of benzene twice and dried under vacuum. Yield: 95%. A solution of sodium periodate (2.74 g, 12.8 mmol) in 25 mL of water was added dropwise to a suspension of 1,3-dihydroxy-2-(2-bebzo[b]furyl)-4,4,5,5-tetramethylimidazolidine (2.30 g, 8.33 mmol) in 100 mL of dichloromethane at room temperature. The dark-green organic phase was separated and concentrated under vacuum. The crude product was separated and purified by column chromatography (eluent: ethyl acetate, alumina) to obtain 2-BFNN (dark-green solid).

3-FNN. 3-Furanaldehyde (1.20 g, 12.4 mmol) and N,N′-dihydroxy-2,3-diamino-2,3-dimethylbutane (1.71 g, 11.5 mmol) were mixed in 10 mL of benzene at 40 °C. The reaction mixture was stirred for 24 h, after which the resulting light-brown solid of 1,3-dihydroxy-2-(3-furyl)-4,4,5,5-tetramethylimidazolidine was filtered off and washed with 5 mL of benzene twice and dried under vacuum. Yield: 75%. A solution of sodium periodate (2.80 g, 13.1 mmol) in 25 mL of water was added dropwise to a suspension of 1,3-dihydroxy-2-(3-furyl)-4,4,5,5-tetramethylimidazolidine (1.95 g, 8.63 mmol) in 100 mL of dichloromethane at room temperature. The dark-blue organic phase was separated and concentrated under vacuum. The crude product was separated and purified by column chromatography (eluent: ethyl acetate, alumina) to obtain 3-FNN (dark-blue solid). In this case, enough amount of 3-FIN was not obtained as a byproduct.

2-FNN and 2-FIN. 2-Furanaldehyde (1.93 g, 20.1 mmol) and N,N′-dihydroxy-2,3-diamino-2,3-imethylbutane (2.71 g, 18.3 mmol) were mixed in 10 mL of benzene at 40 °C. The reaction mixture was stirred for 20 h after which the resulting light-brown solid of 1,3-dihydroxy-2-(2-furyl)-4,4,5,5-tetramethylimidazolidine was filtered off and washed with 5 mL benzene twice and dried under vacuum. Yield: 74%. A solution of sodium periodate (2.14 g, 10.0 mmol) in 20 mL of water was added dropwise to a suspension of 1,3-dihydroxy-2-(2-furyl)-4,4,5,5-tetramethylimidazolidine (1.50 g, 6.64 mmol) in 100 mL of dichloromethane at room temperature. The dark-blue organic phase was separated and concentrated under vacuum. The crude product was separated and purified by column chromatography (eluent: ethyl acetate, alumina) to obtain 2-FNN (dark-blue solid) and 2-FIN (red solid).

2-BiTHNN. 2,2′-bithiophene-5-carboxyaldehyde (1.56 g, 7.98 mmol) and N,N′-dihydroxy-2,3-diamino-2,3-imethylbutane (1.10 g, 7.39 mmol) were mixed in 10 mL of benzene at 40 °C. The reaction mixture was stirred for 20 h, after which the resulting light-yellow/brown solid of 1,3-dihydroxy-2-(2′-bithienyl)-4,4,5,5-tetramethylimidazolidine was filtered off and washed with 3 mL of benzene five times and dried under vacuum. Yield: 63%. A solution of sodium periodate (1.50 g, 6.98 mmol) in 15 mL of water was added dropwise to a suspension of 1,3-dihydroxy-2-(2′-bithienyl)-4,4,5,5-tetramethylimidazolidine (1.50 g, 4.62 mmol) in 100 mL of dichloromethane at room temperature. The dark-green organic phase was separated and concentrated under vacuum. The crude product was separated and purified by column chromatography (eluent: ethyl acetate, alumina) to obtain 2-BiTHNN (dark-green solid).

2-THPNN. 4-(2′-thienyl)benzaldehyde (1.00 g, 5.32 mmol) and N,N'-dihydroxy-2,3-diamino-2,3-imethylbutane (0.726 g, 4.90 mmol) were mixed in 10 mL of benzene at 40 °C. The reaction mixture was stirred for 24 h, after which the resulting light-brown solid of 1,3-dihydroxy-2-[4-(2′-thienyl)phenyl]-4,4,5,5-tetramethylimidazolidine was filtered off and washed with 5 mL of benzene three times and dried under vacuum. Yield: 88%. A solution of sodium periodate (1.39 g, 6.51 mmol) in 15 mL of water was added dropwise to a suspension of 1,3-dihydroxy-2-[4-(2′-thienyl)phenyl]-4,4,5,5-tetramethylimidazolidine (1.38 g, 4.32 mmol) in 80 mL of dichloromethane at room temperature. The dark-green organic phase was separated and concentrated under vacuum. The crude product was separated and purified by column chromatography (eluent: ethyl acetate, alumina) to obtain 2-THPNN (dark-green solid).

Crystals suitable for X-ray diffraction studies were grown by slow evaporation from concentrated solutions of 2-BTHNN and 2-THPNN in toluene in the dark and cold room.

The magnetization isotherms up to 7 T and the magnetic susceptibility over the temperature range from 1.8 K to 300 K were measured using Quantum Design MPMSXL7 SQUID (superconducting quantum interference device) magnetometers. The contribution of the diamagnetism to the susceptibility was subtracted by extrapolating the temperature dependence of the susceptibility to high temperatures where the Curie–Weiss law is applicable.

X-ray diffraction intensities were recorded on a Rigaku AFC10 automatic four-circle diffractometer with graphite monochromated Mo-Kα (λ = 71.075 pm). Intensity data were corrected for Lorentz and polarization effects but not for absorption. The crystal structures were solved by the direct methods and the positions of hydrogen atoms were calculated. A full-matrix least-square refinement was carried out, in which non-hydrogen atoms were treated with anisotropic thermal parameters and those of hydrogen atoms were treated isotropic parameters. The X-ray crystallographic CIF files for 2-BTHNN and 2-TPHNN are available as CCDC2079877 and CCDC2079881, respectively.

4. Conclusions

We showed magneto-structural correlations in the radicals 2-BTHNN and 2-THPNN by considering the results of magnetic measurements and X-ray crystallographic analyses. The coexistence of the FM and AFM intermolecular interactions in 2-BTHNN arises from the formation of two different types of radical molecular dimers. The 1D alternating AFM intermolecular interactions in 2-THPNN come from the molecular arrangements of chain-like side-by-side and head-to-tail stacking. We discussed the atomic substitution effects on magnetism by comparing the magnetic behaviors of thienyl- and furyl-substituted nitronyl nitroxide. We also investigated the effects of O atom elimination from the NO group on magnetism by comparing the magnetic behaviors of nitronyl nitroxide and iminonitroxide having the same attached moieties.

Author Contributions: Writing—original draft preparation, T.S.; writing—review and editing, S.J.B.; H.M.; supervision, W.H. All authors have read and agreed to the published version of the manuscript.

Funding: This research received no external funding.

Conflicts of Interest: The authors declare no conflict of interest.

References

1. Ullman, E.F.; Osiecki, J.H.; Boocock, D.G.B.; Darcy, R. Studies of stable free radicals. X. Nitronyl nitroxide monoradicals and biradicals as possible small molecule spin labels. *J. Am. Chem. Soc.* **1972**, *94*, 7049–7059. [CrossRef]
2. Awaga, K.; Maruyama, Y. Ferromagnetic intermolecular interaction of the organic radical, 2-(4-nitrophenyl)-4,4,5,5-tetramethyl-4,5-dihydro-1*H*-imidazolyl-1-oxy-3-oxide. *Chem. Phys. Lett.* **1989**, *158*, 556–558. [CrossRef]
3. Kinoshita, M.; Turek, P.; Tamura, M.; Nozawa, K.; Shiomi, D.; Nakazawa, Y.; Ishikawa, M.; Takahashi, M.; Awaga, K.; Inabe, T.; et al. An organic radical ferromagnet. *Chem. Lett.* **1991**, *20*, 1225–1228. [CrossRef]
4. Awaga, K.; Inabe, T.; Maruyama, Y. Ferromagnetic intermolecular interaction and crystal structure of the *p*-pyridyl nitronyl nitroxiide radical. *Chem. Phys. Lett.* **1992**, *190*, 349–352. [CrossRef]
5. Nakazawa, Y.; Tamura, M.; Shirakawa, N.; Shiomi, D.; Takahashi, M.; Kinoshita, M.; Ishikawa, M. Low-temperature magnetic properties of the ferromagnetic organic radical, *p*-nitrophenyl nitronyl nitroxide. *Phys. Rev. B* **1992**, *46*, 8906–8914. [CrossRef] [PubMed]
6. Sugano, T.; Pratt, F.L.; Kurmoo, M.; Takeda, N.; Ishikawa, M.; Blundell, S.J.; Pattenden, P.A.; Valladares, R.M.; Hayes, W.; Day, P. Magnetic ordering in some organic molecular magnets. *Synth. Met.* **1995**, *71*, 1827–1828. [CrossRef]
7. Sugano, T.; Kurmoo, M.; Uekusa, H.; Ohashi, Y.; Day, P. Magneto-structural correlation in two isomeric series of nitronyl nitroxide molecular magnets: Intermolecular interactions relevant to ferromagnetic exchange in naphthl and quinolyl derivatives. *J. Solid State Chem.* **1999**, *145*, 427–442. [CrossRef]
8. Catala, L.; Feher, R.; Amabilino, D.B.; Wurst, K.; Veciana, J. Pyrazol-4-yl-substituted α-nitronyl and α-imino nitroxide radicals in solution and solid states. *Polyhedron* **2001**, *20*, 1563–1569. [CrossRef]
9. Sugano, T.; Blundell, S.J.; Lancaster, T.; Pratt, F.L.; Mori, H. Magnetic order in the purely organic quasi-one-dimensional ferromagnet 2-benzimidazolyl nitronyl nitroxide. *Phys. Rev. B* **2010**, *82*, 180401. [CrossRef]
10. Adriano, C.; Freitas, R.S.; Paduan-Filho, A.; Pagliuso, P.G.; Oliveira, N.F.; Lahti, P.M. Magnetic phase diagram of the organic antiferromagnet F4BImNN. *Polyhedron* **2017**, *136*, 2–4. [CrossRef]
11. Mori, T.; Kobayashi, A.; Sasaki, Y.; Kobayashi, H.; Saito, G.; Inokuchi, H. The intermolecular interaction of tetrathiafulvalene and bis(ethylenedithio)tetrathiafulvalene in organic metals. Calculation of overlaps and models of energy-band structures. *Bull. Chem. Soc. Jpn.* **1984**, *57*, 627–633. [CrossRef]
12. Sugano, T. Magnetic phase transition between ferromagnetic high-temperature phase and antiferromagnetic low-temperature phase in 5-carboxy-2-thienyl nitronyl nitroxide. *Chem. Lett.* **2001**, *29*, 32–33. [CrossRef]
13. Sugano, T. Magnetic phase transitions in organic radical crystals studied by electron spin resonance. *Synth. Met.* **2003**, *137*, 1167–1168. [CrossRef]
14. Sugano, T.; Blundel, S.J.; Hayes, W.; Day, P.; Mori, H. Magnetic and structural properties of monoradicals and diradicals based on thienyl-substituted nitronyl nitroxide. *Physica B* **2010**, *405*, S327–S330. [CrossRef]
15. Bleaney, B.; Bowers, K.D. Anomalous paramagnetism of copper acetate. *Proc. R. Soc. A* **1952**, *214*, 451–465. [CrossRef]
16. Bonner, J.C.; Fisher, M.E. Linear magnetic chains with anisotropic coupling. *Phys. Rev.* **1964**, *135*, A640–A658. [CrossRef]
17. Coronado, E.; Giménez-Saiz, C.; Nicolas, M.; Romero, F.M.; Rusanov, E.; Stoeckli-Evans, H. Synthesis, crystal structures and electronic properties of imidazoline nitroxide radicals bearing active groups in electropolymerisation. *New J. Chem.* **2003**, *27*, 490–497. [CrossRef]
18. Rubenacker, G.V.; Drumheller, J.E.; Emerson, K.; Willett, R.D. Magnetic susceptibility of $((CH)_3NH)_2Cu_4Br_{10}$, chains of stacked linear tetramers. *J. Magn. Magn. Mater.* **1986**, *54–57*, 1483–1484. [CrossRef]

Article

New Cyanido-Bridged Heterometallic 3d-4f 1D Coordination Polymers: Synthesis, Crystal Structures and Magnetic Properties

Diana Dragancea [1,2,*], Ghenadie Novitchi [3,*], Augustin M. Mădălan [1] and Marius Andruh [1,*]

1 Inorganic Chemistry Laboratory, Faculty of Chemistry, University of Bucharest, Str. Dumbrava Rosie nr. 23, 020464 Bucharest, Romania; augustin.madalan@chimie.unibuc.ro
2 Institute of Chemistry, Str. Academiei nr. 3, MD 2028 Chișinău, Moldova
3 CNRS, University Grenoble Alpes, LNCMI, F-38000 Grenoble, France
* Correspondence: ddragancea@gmail.com (D.D.); ghenadie.novitchi@lncmi.cnrs.fr (G.N.); marius.andruh@dnt.ro (M.A.)

Citation: Dragancea, D.; Novitchi, G.; Mădălan, A.M.; Andruh, M. New Cyanido-Bridged Heterometallic 3d-4f 1D Coordination Polymers: Synthesis, Crystal Structures and Magnetic Properties. *Magnetochemistry* 2021, 7, 57. https://doi.org/10.3390/magnetochemistry7050057

Academic Editors: Lee Martin, Scott Turner, John Wallis, Hiroki Akutsu and Carlos J. Gómez García

Received: 8 April 2021
Accepted: 26 April 2021
Published: 28 April 2021

Publisher's Note: MDPI stays neutral with regard to jurisdictional claims in published maps and institutional affiliations.

Copyright: © 2021 by the authors. Licensee MDPI, Basel, Switzerland. This article is an open access article distributed under the terms and conditions of the Creative Commons Attribution (CC BY) license (https://creativecommons.org/licenses/by/4.0/).

Abstract: Three new 1D cyanido-bridged 3d-4f coordination polymers, {[Gd(L)(H$_2$O)$_2$Fe(CN)$_6$]·H$_2$O}$_n$ (**1**$_{GdFe}$), {[Dy(L)(H$_2$O)$_2$Fe(CN)$_6$]·3H$_2$O}$_n$ (**2**$_{DyFe}$), and {[Dy(L)(H$_2$O)$_2$Co(CN)$_6$]·H$_2$O}$_n$ (**3**$_{DyCo}$), were assembled following the building-block approach (L = pentadentate *bis*-semicarbazone ligand resulting from the condensation reaction between 2,6-diacetyl-pyridine and semicarbazide). The crystal structures consist of crenel-like Ln^{III}-M^{III} alternate chains, with the Ln^{III} ions connected by the hexacyanido metalloligands through two *cis* cyanido groups. The magnetic properties of the three complexes have been investigated. Field-induced slow relaxation of the magnetization was observed for compounds **2**$_{DyFe}$ and **3**$_{DyCo}$. Compound **3**$_{DyCo}$ is a new example of chain of Single Ion Magnets.

Keywords: heterometallic complexes; cyanido-bridged complexes; coordination polymers; single molecule magnets; lanthanides

1. Introduction

The discovery of slow relaxation of the magnetization phenomena for discrete metal complexes (Single Molecule Magnets, SMMs) and 1D coordination polymers (Single Chain Magnets, SCMs) has stimulated the development of an intensive interdisciplinary research field. Beyond their relevance in fundamental Physics and Chemistry, spectacular applications in quantum computing and high-density information storage from these molecules are expected [1]. Although the field of SMMs was initially dominated by transition metal-based-systems, the focus of research shifted to lanthanides, which increase the energy barriers of SMMs [2,3]. The lanthanide ions (especially Tb^{III}, Dy^{III}, and Ho^{III}), bring large magnetic moments and high uniaxial magnetic anisotropy, which are essential prerequisites for the observation of slow relaxation of the magnetization. While SMMs can be mono- and oligonuclear (homo- and heteronuclear) complexes, most of the SCMs are constructed from two different spin carriers (e.g., 3d-3d', 3d-4f, 2p-3d, and 2p-4f) [4]. Homospin SCMs are rare and rather serendipitously obtained [5,6]. Examples of lanthanide-based Single-Chain Magnets are also limited, and most of them result from the association of lanthanide ions with nitronyl-nitroxide (paramagnetic) ligands [7,8]. Lanthanide ions can be linked to other paramagnetic metal ions through small bridging ligands, which facilitate the exchange interactions. The building-block approach, relying on the employment of metalloligands, represents an excellent strategy to generate heterometallic coordination compounds [9]. Anionic cyanido complexes are very popular in this respect. The self-assembly processes between [M(CN)$_6$]$^{3-}$ metalloligands and cationic Ln^{III} complexes led to a rich variety of structural architectures. The assembling complex cations are either solvated Ln^{III} species or heteroleptic complexes, containing chelating ligands and weakly coordinated anions or solvent molecules, which can be easily replaced by the cyanido bridge. The dimensionality of the resulting coordination polymers is dependent on the number of accessible positions at the lanthanide ions. For example, nitrogen donor blocking ligands attached to Ln^{III} ions, such as 1,10-phenanthroline, 2,2′-bipyridine, 2,2′:6′,2″—terpyridine, 2,4,6-tri(2-pyridyl)-1,3,5-triazine, favor the aggregation of 1D coordination polymers, employing [M(CN)$_6$]$^{3-}$ as metalloligands [10–18]. When the reactions between the lanthanide salts and the hexacyanido building block occur in dimethylformamide (DMF) or dimethyl sulfoxide (DMSO), depending on the experimental conditions, discrete species, 1D, 2D, or even 3D coordination polymers have been obtained [19–24]. By decreasing the number of cyanido groups within the metalloligand, the formation of low-dimensionality coordination polymers is favored [25]. Most of these 3d-4f cyanido-bridged complexes show interesting physical properties, mainly magnetic [26–28] and optical [29–31], and in some cases combined slow magnetic relaxation and light emission were revealed [32,33].

In this paper, we report on a new family of 1D coordination polymers, which are assembled from [LnL]$^{3+}$ and [M(CN)$_6$]$^{3-}$ ions (L = *bis*-semicarbazone ligand, M = Fe, Co). The pentadentate *bis*-semicarbazone ligand, L (Scheme 1), was previously used for the synthesis of both 3d-3d/3d-4d cyanido-bridged discrete [34] and polymeric structures [35–41].

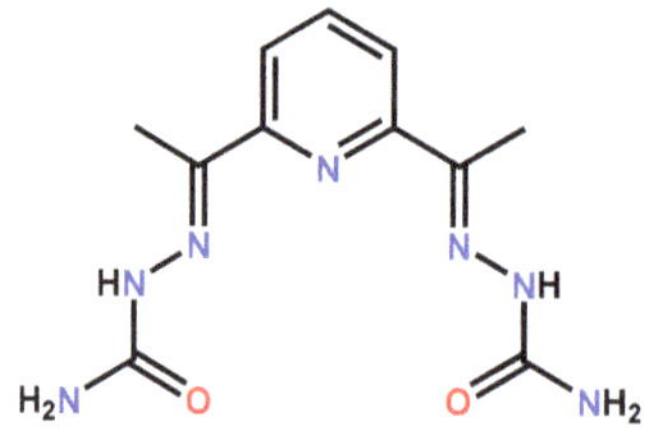

Scheme 1. The structure of *bis*-semicarbazone ligand, L.

The new compounds have been characterized by single-crystal X-ray diffraction, and their magnetic properties have been investigated.

2. Experimental Section

2.1. Materials and Physical Measurements

All reagents and solvents for synthesis were commercially purchased and used without any further purification. The *bis*-semicarbazone ligand L was synthesized according to the method reported in the literature [42].

IR spectra were recorded on a FTIR Bruker Tensor V-37 spectrophotometer (KBr pellets) in the range of 4000–400 cm^{-1}. Elemental analysis was performed on a EuroEA Elemental Analyzer.

Magnetic Studies: DC magnetic susceptibility data (2–300 K) were collected on powdered samples using a SQUID magnetometer (Quantum Design MPMS-XL), applying a magnetic field of 0.1 T. All data were corrected for the contribution of the sample holder and the diamagnetism of the samples estimated from Pascal's constants [43,44]. The field dependence of the magnetization (up to 5 T) was measured between 2.0 and 5.0 K. AC magnetic susceptibility was measured between 2 and 7 K with an oscillating field magnitude of H_{ac} = 3.0 Oe and frequency ranging between 1 and 1488 Hz in presence of a dc field up to H_{dc} = 4000 Oe. Fitting of the variable parameters and estimation of errors was performed with lsqcurvefit solver in MATLAB, and jacobian matrix was used to generate 95% confidence intervals on the fitted parameters. Typical examples of this analysis are presented in Figures S3–S6.

X-ray powder diffraction data were measured on a Proto AXRD benchtop using Cu-Kα radiation with a wavelength of 1.54059 Å in the range of 5–35° (2θ).

2.2. Single Crystal X-ray Crystallography

X-ray diffraction data were collected at 293 K on a Rigaku XtaLAB Synergy-S diffractometer operating with Mo-Kα (λ = 0.71073 Å) micro-focus sealed X-ray tube. The structures were solved by direct methods and refined by full-matrix least squares techniques based on F^2. The non-H atoms were refined with anisotropic displacement parameters. Calculations were performed using SHELX-2014 or SHELX-2018 crystallographic software package [45,46]. Supplementary X-ray crystallographic data in CIF format have been deposited with the CCDC with the following reference numbers: 2069217 (1_{GdFe}), 2069215 (2_{DyFe}), and 2069216 (3_{DyCo}). A summary of the crystallographic data and the structure refinement for crystals 1–3 are given in Table S1.

2.3. Synthesis of Complexes

The three compounds are synthesized following the same general procedure: $LnCl_3 \cdot 6H_2O$ (0.06 mmol) and L (0.06 mmol) in 10 mL H_2O were stirred at 80 °C for 30 min. The above-cooled solution was filtered and transferred to a 30 mL test tube. Additional 5 mL of water was layered over the aqueous solution of the mononuclear complexes, and finally, a 10 mL H_2O solution containing 0.06 mmol $K_3[Fe(CN)_6]$ or $K_3[Co(CN)_6]$ was then slowly layered on top. The whole set up was kept undisturbed and slow diffusion of these two solutions led, after 2 weeks, to single crystals. The reaction mixture was mechanically stirred and was filtered off through frit followed by drying under vacuum to obtain a polycrystalline solid. Single crystals required for the X-ray data collections were picked up from the crystalline mixtures prior to mechanical stirring.

{[Gd(L)(H$_2$O)$_2$Fe(CN)$_6$]·H$_2$O}$_n$, 1_{GdFe}: Orange crystalline solid, mass (yield): 22 mg (51 %). Selected IR data (KBr, cm^{-1}): 3457 (m), 3349 (m), 3189 (m), 2147 (mw), 2121 (vs), 1677 (vs), 1629 (m), 1614 (m), 1542 (s), 1461 (mw), 1367 (mw), 1307 (mw), 1266 (mw), 1196 (s), 1174 (mw), 1138 (mw), 1106 (mw), 1004 (w), 815 (mw), 769 (mw), 705 (mw), 656 (mw), 563 (mw), 485 (mw), 417 (mw). Elemental analysis. Calcd. for $C_{17}H_{21}N_{13}O_5FeGd$: C, 29.15; H, 3.02; N, 25.99%; found C, 29.09; H, 2.96; N, 26.01%.

{[Dy(L)(H$_2$O)$_2$Fe(CN)$_6$]·3H$_2$O}$_n$, 2_{DyFe}: Orange crystalline solid, mass (yield): 27 mg (60 %). Selected IR data (KBr, cm^{-1}): 3457 (m), 3348 (w), 3236 (mw), 1676 (m), 1631 (m), 1609 (m), 1542 (m), 1461 (m), 1371 (m), 1309 (s), 1267 (s), 1196 (s), 1136 (vs), 1105 (m), 816 (m), 769 (mw), 704 (m), 654 (mw), 570 (mw), 507 (mw), 487 (mw). Elemental analysis. Calcd. for $C_{17}H_{25}N_{13}O_7DyFe$: C, 27.53; H, 3.40; N, 24.55%; found C, 27.25; H, 3.37; N, 24.69%.

{[Dy(L)(H$_2$O)$_2$Co(CN)$_6$]·H$_2$O}$_n$, 3_{DyCo}: White crystalline solid, mass (yield): 22 mg (51 %). Selected IR data (KBr, cm^{-1}): 3458 (s), 3349 (s), 3238 (s), 3189 (mw), 2917 (mw), 2849 (w), 2362 (mw), 2165 (mw), 2148 (mw), 2132 (vs), 2091 (w), 1677 (vs), 1631 (m), 1614 (m), 1542 (vs), 1464 (m), 1368 (m), 1309 (mw), 1268 (m), 1199 (ms), 1174 (mw), 1138 (m), 1107 (m), 1005 (w), 816 (mw), 770 (mw), 708 (mw), 656 (mw), 556 (mw), 495 (m), 478 (m), 459 (mw), 422 (m). Elemental analysis. Calcd. for $C_{17}H_{21}N_{13}O_5CoDy$: C, 28.80; H, 2.99; N, 25.69%; found C, 28.64; H, 2.98; N, 25.45%.

3. Results and Discussion

3.1. Synthesis and Structures of the Complexes

The three compounds, {[Gd(L)(H$_2$O)$_2$Fe(CN)$_6$]·H$_2$O}$_n$, 1_{GdFe}, {[Dy(L)(H$_2$O)$_2$Fe(CN)$_6$]·3H$_2$O}$_n$, 2_{DyFe}, and {[Dy(L)(H$_2$O)$_2$Co(CN)$_6$]·H$_2$O}$_n$, 3_{DyCo}, have been obtained via slow diffusion of water solutions containing [Ln(L)(H$_2$O)$_4$]Cl$_3$-K$_3$[M(CN)$_6$] in 1:1 molar ratio. All complexes show C=O and C=N IR absorption peaks in the range of 1654–1657 cm^{-1}, which indicate the presence of the semicarbazone ligand. The split bands at 2200–2100 cm^{-1} are assigned to both the monodentate and bridging cyanido groups [47]. The crystalline phase purity of the samples was confirmed by the good agreement between the PXRD patterns and the ones simulated using single-crystal data (Figure S1). The FTIR spectra are displayed in Figure S2.

Complexes 1_{GdFe} and 3_{DyCo} are isostructural, and they crystallize in the orthorhombic space group *Pbca* with one crystallization water molecule/formula unit. Complex

2_{DyFe} crystallizes in the monoclinic system, space group $P2_1/c$, with three lattice water molecules/formula unit. In all complexes, the metal ions have similar coordination environments, and the topology of the heterometallic chains is identical.

Compounds 1_{GdFe} and 3_{DyCo} consist of heterometallic chains with alternating distributions of the 3d and 4f metal ions. Since the two compounds are isostructural, we will describe only the crystal structure of the compound 1_{GdFe}. The general appearance of the chains is crenel-like, due to the fact that the $[Fe(CN)_6]^{3-}$ metalloligand acts as a bridge trough two *cis* cyanido groups and the two neighboring connecting $[Fe(CN)_6]^{3-}$ moieties are placed on the same side of the organic ligand coordinated to the lanthanide ion (Figure 1). One of the two water molecules coordinated to the lanthanide ion is involved in intra-chain hydrogen interaction with a cyanido group from a $[Fe(CN)_6]^{3-}$ metalloligand coordinated to a neighboring lanthanide ion. The chains are running along the crystallographic *a* axis.

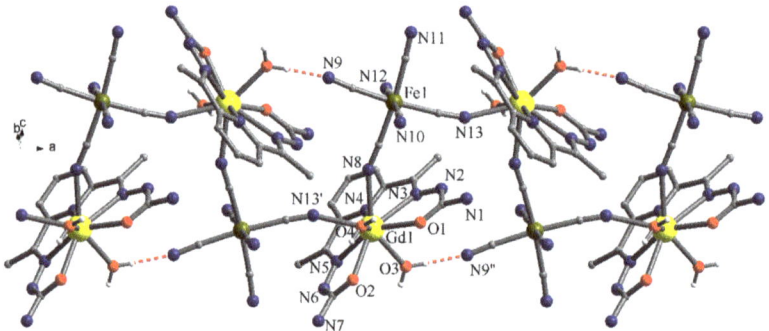

Figure 1. Perspective view of the 1D coordination polymer in compound 1_{GdFe} (symmetry codes: $' = -0.5 + x, 0.5 - y, 1 - z;$ $'' = 0.5 + x, 0.5 - y, 1 - z$).

The lanthanide ions are nine-coordinated by the pentadentate organic ligand (O1, O2, N3, N4, and N5), two nitrogen atoms arising from the cyanido bridges (N8, N13'; symmetry code: $' = -0.5 + x, 0.5 - y, 1 - z$), and two aqua ligands (O3, O4). The Ln-N bond lengths with the organic ligand are in the range of 2.584(6) − 2.598(6) Å for 1_{GdFe} and 2.562(2) − 2.574(2) for 3_{DyCo}, respectively, while for the cyanido groups are Gd1-N8 = 2.534(6), Gd1-N13' = 2.566(6), respectively, Dy1-N8 = 2.519(3) and Dy1-N13' = 2.544(3). The Ln-O bond lengths are slightly longer with the organic ligand than aqua ligands: Gd1-O1 = 2.423(5), Gd1-O2 = 2.406(5), Gd1-O3 = 2.330(5), Gd1-O4 = 2.375(5), Dy1-O1 = 2.398(2), Dy1-O2 = 2.382(2), Dy1-O3 = 2.308(2), Dy1-O4 = 2.3518(19) Å. The coordination geometry of the gadolinium ion can be described as spherical capped square antiprism, according to the calculations made with SHAPE software (Table S2) [48].

The O3 aqua ligands are further connected by hydrogen bonding to neighboring chains generating a 2D supramolecular architecture in the crystallographic *ac* plane (Figure 2). The distances for the hydrogen interactions are: (O3)H1O···N9'' = 1.85 and (O3)H2O···N11''' = 1.93 Å, while the corresponding angles are: O3-H1O···N9'' = 173.4 and O3-H2O···N11''' = 170.6° (symmetry codes: $'' = 0.5 + x, 0.5 - y, 1 - z;$ $''' = x, 0.5 - y, -0.5 + z$).

The extension of the supramolecular architecture to 3D is also mediated by hydrogen bond interactions involving the second aqua ligand, the crystallization water molecules and cyanido groups of the anionic metalloligand (Figure 3). Each O4 coordinated water molecule is involved as donor in hydrogen interactions with two crystallization water molecules. Each crystallization water molecule is acceptor for two hydrogen interactions with two coordinated water molecules from neighboring layers and donor for two cyan groups also from the two neighboring layers. The distances for these hydrogen interactions are: (O4)H4O···O5 = 1.95, (O4)H3O···O5^i = 1.92, (O5)H5O···N10 = 2.18 and (O5)H6O···N12^{ii} = 2.14 Å, while the corresponding angles are: O4-H4O···O5 = 166.7, O4-

H3O···O5^i = 161.8, O5-H5O···N10 = 158.7 and O5-H6O···N12^{ii} = 169.7° (symmetry codes: ^{i} = 1 − x, −y, 1 − z; ^{ii} = 1.5 − x, −0.5 + y, z).

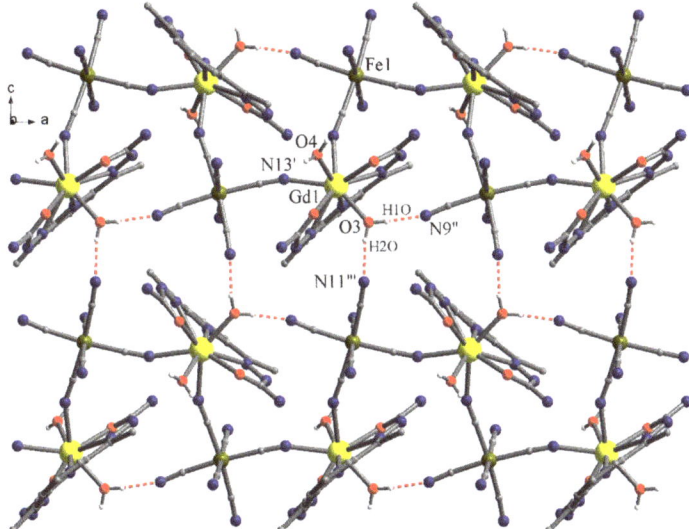

Figure 2. View along the crystallographic *b* axis of the packing diagram of compound **1**$_{GdFe}$ showing a supramolecular layer (symmetry codes: $'$ = −0.5 + x, 0.5 − y, 1 − z; $''$ = 0.5 + x, 0.5 − y, 1 − z; $'''$ = x, 0.5 − y, −0.5 + z).

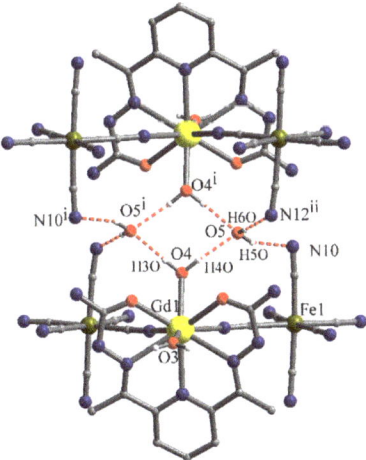

Figure 3. Detail of the packing diagram with the hydrogen interactions established between chains from neighboring supramolecular layers and crystallization water molecules in crystal **1**$_{GdFe}$ (symmetry codes: ^{i} = 1 − x, −y, 1 − z; ^{ii} = 1.5 − x, −0.5 + y, z).

Compound **2**$_{DyFe}$ consists also of heterometallic chains running in this case along the crystallographic *b* axis, and crystallization water molecules. The 1D chains are formed in a similar manner by connecting [Dy(L)(H$_2$O)$_2$]$^{3+}$ complex cations by the [Fe(CN)$_6$]$^{3−}$ metalloligands, which employ two *cis* cyanido groups for bridging (Figure 4). The Dy^{III} ion is nine-coordinated by the pentadentate ligand (O1, O2, N3, N4, and N5), two nitrogen atoms arising from the cyanido bridges (N8, N13), and two aqua ligands (O3, O4). The coordination geometry of the dysprosium ion is also spherical capped square antiprism (Table S2). The Dy-N and Dy-O bond lengths are in the range of 2.503(2)–2.553(2) and 2.3298(18)–2.3783(13) Å, respectively. The two Dy-N bond distances (nitrogen atoms arising from the bridging cyanido groups) are Dy1-N8 = 2.556(2) and Dy1-N13′ = 2.533(2) Å (symmetry code: ′ = 1 − x, −0.5 + y, 1.5 − z). The Fe^{III} ions show a slightly distorted octahedral geometry with Fe1-C bond lengths ranging from 1.929(3) to 1.957(3) Å. Each {Dy(L)(H$_2$O)$_2$} module links two {Fe(CN)$_6$} fragments in *cis* positions (the Dy···Fe···Dy angle is 95.78°), and each {Fe(CN)$_6$} metalloligand connects two Dy^{III} ions, resulting in a crenel-like chain structure.

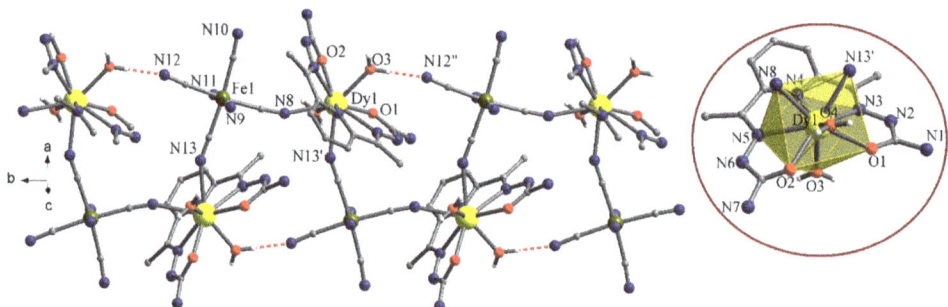

Figure 4. Perspective view of the 1D coordination polymer in compound **2**$_{DyFe}$. The inset shows a detail of the coordination environment of the Dy^{III} ion (symmetry codes: ′ = 1 − x, −0.5 + y, 1.5 − z; ″ = x, −1 + y, z).

The O3 aqua ligand is also involved in intra- and interchain hydrogen bonding generating an analogous 2D supramolecular architecture in the *ab* crystallographic plane (Figure 5). The distances for the hydrogen interactions are: (O3)H1O···N12″ = 1.88 and (O3)H2O···N10‴ = 2.05 Å, while the corresponding angles are: O3-H1O···N12″ = 172.4 and O3-H2O···N10‴ = 167.9° (symmetry codes: ″ = x, −1 + y, z; ‴ = 2 − x, −0.5 + y, 1.5 − z).

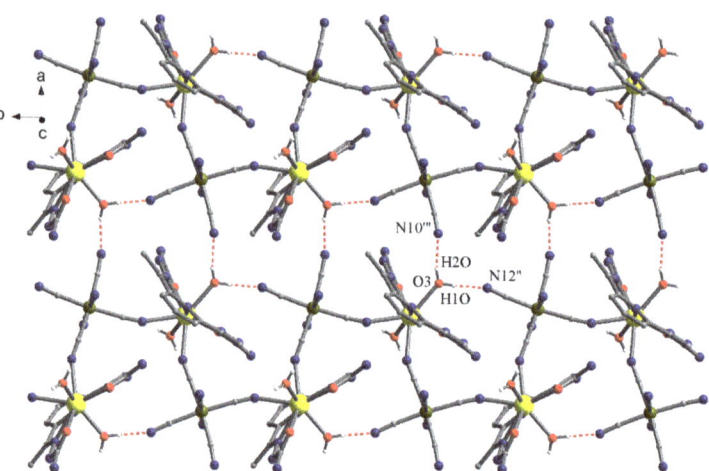

Figure 5. View along the crystallographic *c* axis of the packing diagram of compound **2**$_{DyFe}$ showing a supramolecular layer (symmetry codes: ″ = x, −1 + y, z; ‴ = 2 − x, −0.5 + y, 1.5 − z).

The main differences between crystals **1**$_{GdFe}$ and **2**$_{DyFe}$ appear in hydrogen interactions established between the supramolecular layers. Compound **2**$_{DyFe}$ has two more crystallization water molecules per unit comparing with the crystals $\mathbf{1}_{GdFe}$ and $\mathbf{3}_{DyCo}$. The O4 coordinated water molecule is also involved as donor in hydrogen interactions with two crystallization water molecules, O5 and O5^i, (Figure 6). Each of these crystallization water molecules is acceptor for two hydrogen interactions with two coordinated water molecules from neighboring layers and acts as donor for only one cyanido group (O5 is donor for N11^i atom). The other two crystallization water molecules are involved in hydrogen bonding with one NH$_2$ group and one cyanido group (O6), respectively, and two cyanido groups (O7). The distances for the hydrogen interactions are: (O4)H4O$\cdots$O5 = 1.91, (O4)H3O$\cdots$O5^i = 2.02, (O5)H5O$\cdots$N11^i = 2.28, (N7)H5N$\cdots$O6 = 2.16, (O6)H8O$\cdots$N12^{iii} = 2.28, (O7)H9O$\cdots$N9^{ii} = 2.15, and (O7)H10O$\cdots$N10^i = 2.20 Å, while the corresponding angles are: O4-H4O$\cdots$O5 = 174.4, O4-H3O$\cdots$O5^i = 171.4, O5-H5O$\cdots$N11^i = 146.7, N7-H5N$\cdots$O6 = 162.3, O6-H8O$\cdots$N12^{iii} = 166.1, O7-H9O$\cdots$N9^{ii} = 153.9, and O7-H10O$\cdots$N10^i = 148.8 ° (symmetry codes: ^{i} = 1 − x, 1 − y, 1 − z; ^{ii} = 1 − x, −0.5 + y, 1.5 − z; ^{iii} = x, 1.5 − y, −0.5 + z).

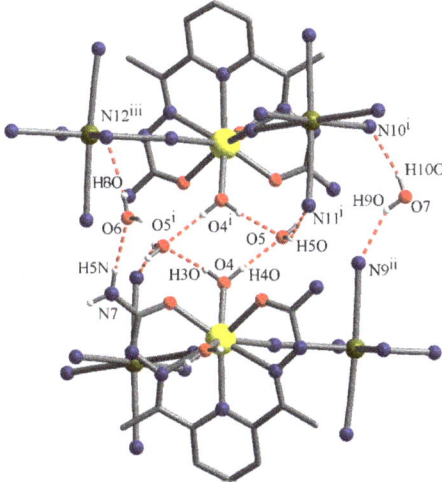

Figure 6. Detail of the packing diagram with the hydrogen interactions established between chains from neighboring supramolecular layers and crystallization water molecules in crystal $\mathbf{2}_{DyFe}$ (symmetry codes: ^{i} = 1 − x, 1 − y, 1 − z; ^{ii} = 1 − x, −0.5 + y, 1.5 − z; ^{iii} = x, 1.5 − y, −0.5 + z).

The shortest intramolecular Fe$\cdots$Gd distances in $\mathbf{1}_{GdFe}$ are 5.513 and 5.557 Å, while the intramolecular Dy$\cdots$Fe distances in $\mathbf{2}_{DyFe}$ are 5.542 and 5.494 Å. Selected bond distances and angles for compounds **1**–**3** are listed in Table S3.

3.2. Magnetic Properties of the Complexes

Static magnetic characterizations. The magnetic susceptibility data for compounds **1**–**3** were measured on polycrystalline samples in the temperature range of 2–300 K as shown in Figure 7, in the form of $\chi_M T$ vs. T curves. The observed $\chi_M T$ values at 300 K for $\mathbf{1}_{GdFe}$, $\mathbf{3}_{DyCo}$, and $\mathbf{2}_{DyFe}$ are of 8.265, 15.454, and 16.305 cm^3 mol^{-1}K, which are slightly higher than the expected values for a non-interacting spin system of one Gd^{III} (7.88 cm^3 mol^{-1} K, S = 7/2, $^8S_{7/5}$, g = 2.00), Dy^{III} (14.17 cm^3 mol^{-1} K, S = 5/2, $^6H_{15/2}$, g = 4/3) [49], and one low-spin $S = \frac{1}{2}$ Fe^{III} ion or one diamagnetic Co^{III} ion [44]. Upon cooling, the $\chi_M T$ values stay almost constant in the high temperature region, while at low temperatures, the $\chi_M T$ values show a rapid decrease and reach the values of 6.833, 12.722, and 11.210 cm^3 mol^{-1} K, respectively, at 2.0 K. In the case of $\mathbf{1}_{GdFe}$, the decrease in $\chi_M T$ with temperature may be associated with Fe^{III}-Gd^{III} antiferromagnetic interactions along the heterometallic alternating chain. The possible presence of intermolecular interactions can also contribute to this decrease. The expected ferrimagnetic behavior (i.e., the characteristic minimum on the

$\chi_M T$ vs. T curve) is not observed, probably due to the small magnitude of the exchange interactions along the chain. The evolution of the temperature dependence of the magnetic susceptibility for 3_{DyCo} is exclusively defined by the presence of strongly anisotropic Dy^{III} ions, which are isolated by diamagnetic low spin Co^{III} ions. The decrease in $\chi_M T$ with the temperature is due to the depopulation of M_J (Stark) sublevels of the Dy^{III} centers in 3_{DyCo} [50]. This effect is certainly present in the case of compound 2_{DyFe}. Additionally, a Fe^{III}-Dy^{III} magnetic coupling along the chain can be expected. In the case of 2_{DyFe}, the evolution of $\chi_M T$ shows a more important slope compared to compound 3_{DyCo} (Figure 7), with a lower value of susceptibility at 2.0 K (11.210 cm^3 mol^{-1} K). This indicates the presence of some antiferromagnetic impact, which contributes to the observed decreasing $\chi_M T$ values. For 2_{DyFe}, the existence of magnetic interactions similar to those in 1_{GdFe} also suggests the formation of a ferrimagnetic chain, which, associated with strong magnetic anisotropy, could lead to a Single Chain Magnet. Unfortunately, as in the case of 1_{GdFe}, the increase in $\chi_M T$ at low temperatures and the characteristic minimum were not detected for 2_{DyFe} (Figure 7). This behavior is probably due to the very small antiferromagnetic interactions along the chain. The magnetization measurements (Figure S3) support the presence of an important magnetic anisotropy in 2_{DyFe} and 3_{DyCo}.

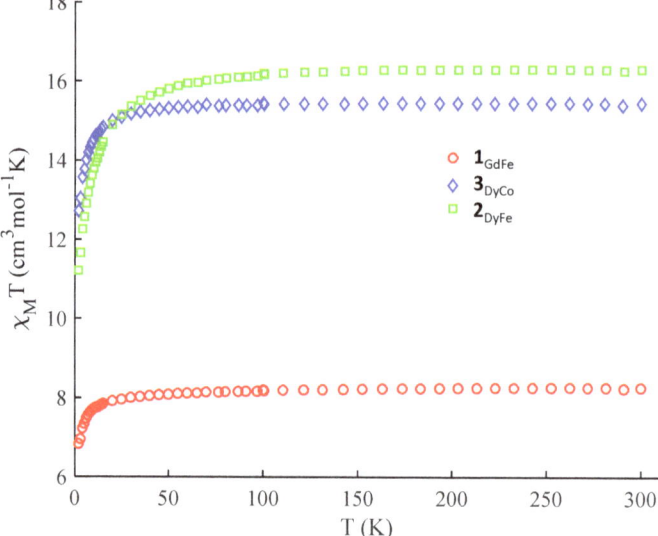

Figure 7. Temperature dependence of $\chi_M T$ vs. T for complexes 1_{GdFe}, 2_{DyFe}, and 3_{DyCo}.

Dynamic magnetic characterizations. Dynamic magnetic properties of the compounds 1_{GdFe}, 2_{DyFe}, and 3_{DyCo} were studied by measuring the temperature and field dependence ac (alternative current) magnetic susceptibility. Compound 1_{GdFe} does not have any manifestation of the out-of-phase component (χ''_{ac}) of the ac magnetic susceptibility at 2 K and zero dc (direct current) field. After applying small dc field (2000 Oe) no modification was observed in χ''_{ac} component of ac susceptibility.

For 3_{DyCo}, no signal was observed under zero dc field at 2.0 K, in χ''_{ac} component of ac susceptibility. After applying of small dc fields (up to 4000 Oe), a frequency dependent out-of-phase signal appears (Figure 8b) and has a rich evolution in function of the field. Such behavior is consistent with presence of strongly anisotropic paramagnetic centers Dy^{III} and indicates the presence of field-induced slow magnetic relaxation. The intensity of the out-of-phase signals gradually increases till about 2000 Oe, and then, it slightly decreases. To investigate the nature of slow magnetic relaxation, additional ac susceptibility data were collected under fixed *dc* field (2000 Oe) and stable temperatures between 2.0 and 5.0 K (with a 0.2 K increment)—Figure 8d–f. The temperature sweeping of the *ac* susceptibility shows the important evolution of the χ''_{ac} component and supports the presence of field-induced slow magnetic relaxation in 3_{DyCo}. Since the Co^{III} ion is diamagnetic, compound 3_{DyCo} can be described as being a chain of Single Ion Magnets.

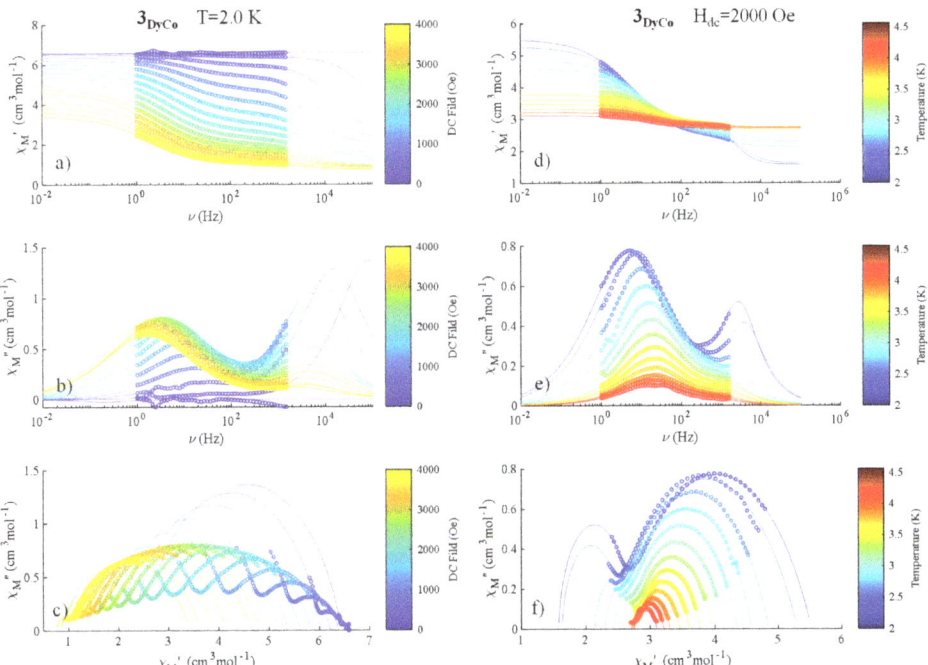

Figure 8. Field dependence (**left, a,b**) and temperature dependence (**right, d,e**) of *ac* susceptibility (H_{ac} = 3.0 Oe) and Cole–Cole plots, (**c,f**), for 3_{DyCo} at the indicated temperatures and fields. The solid lines represent the best fits according to the generalized Debye model for two relaxations processes (Equations (S1) and (S2)).

A similar strategy of measurements was used in the dynamic analysis of 2_{DyFe}. As in the case of 3_{DyCo}, at *T* = 2.0 K and zero *dc* field, no signal was detected in the out-of-phase component of the *ac* susceptibility. The signals appear when a small magnetic *dc* field was applied and has similar evolutions as in the case of 3_{DyCo} (Figure S8a–c). A *dc* field of 3000 Oe was used to perform the temperature sweeping measurements of the *ac* susceptibility in the case of 2_{DyFe} (Figure S8d–f).

For both compounds (2_{DyFe} and 3_{DyCo}), the visual analysis of the out-of-phase susceptibilities, as well of the χ''_{ac} vs. χ'_{ac} plots (Cole–Cole plots), suggests the presence of at least two distinct relaxation processes. In consequence, the *ac* susceptibility data (field sweeping and temperature sweeping measurements) for 3_{DyCo} and 2_{DyFe} were evaluated with generalized (extended) Debye equations combining two-relaxation processes [51–53]. The two relaxation times (τ_1, and τ_2,) and two distribution parameters (α_1, and α_2) occur along with two isothermal susceptibilities (χ_{T1} and χ_{T2}) and one common adiabatic susceptibility (χ_s) (see Equations (S1) and (S2)). The deconvolution of two relaxations process is presented in Figures S4–S7. Variable parameters derived from the best fits of the *ac* susceptibility are shown in Figures S9–S12.

In both compounds, the first relaxation process LF (Low Frequency) is well defined, while the second HF (High Frequency) process has large errors on the variable parameters. The distributions of relaxation times for the LF process are rather broad (α_1 = 0.3 ÷ 0.5). The extracted temperature and field dependence of relaxation times for 2_{DyFe} and 3_{DyCo} can be modulated based on four relaxation mechanisms according to the following equation [54–58]:

$$\tau^{-1}_{T/H}(T,H) = \frac{Q_1}{1+Q_2H^2} + \tau_0^{-1}\exp\left(-\frac{U_{eff}}{kT}\right) + AH^4T + CT^n. \quad (1)$$

The first term represents the Quantum Tunneling of Magnetization (QTM), the second term corresponds to Orbach, the third to Direct, and the last one to Raman process; moreover, H = applied dc magnetic field and T = temperature. In order to constrain the variable parameters and avoid the overparameterization problem, temperature and field dependence of relaxation times were fitted simultaneously [56] (vector of data: τ^{-1} in s^{-1}, T in Kelvin, and H in kOe). Only LF signals will be discussed below, as the second process (HF) is poorly defined. Different combinations of the four mechanisms of relaxation have been used in order to simulate the evolution of the relaxation times. For the LF signals, with both compounds, the contribution of Quantum Tunneling of Magnetization (QTM) is indispensable to simulate the relaxation data. The continuous decreasing trend of τ^{-1} vs. applied field (H) excludes the presence of significant contribution of Direct relaxation mechanism. The other contributions in relaxation times can be Raman and/or Orbach, which have the same increasing evolutions with temperature variation [57]. In this restricted range of temperature, it is difficult to separate these two components. In order to have some information regarding the manifestation of these mechanisms, the comparative fits on the temperature dependence of relaxation time for 2_{DyFe} and 3_{DyCo} have been done (Figures S13–S15). Both mechanisms can reproduce the time of evolution. The quality of the Orbach mechanism is slightly better. It should be mention here that the distribution parameters (α) have important impact on uncertainties relaxation time [59] and can also be an argument in favor of one or another mechanism. The analysis presented in Figures S13–S15 shows the similarity in uncertainties of relaxation times for both mechanisms. Based on this argument and a low temperature range (2–4 K) of relaxation data for 3_{DyCo} and 2_{DyFe}, as well the traditional representation of relaxation phenomena in SMM, our analysis of LF relaxation process is limited to two contributions: QTM and Orbach. The best fit for LF relaxation processes based on the two mechanisms has been obtained for the following sets of parameters:

3_{DyCo}: U_{eff}/k = 7.1 K; τ_0 = 7.5 × 10^{-5} s; Q_1 = 121 s^{-1}; Q_2 = 1.05 kOe^{-2}

2_{DyFe}: U_{eff}/k = 10.8 K; τ_0 = 5.9 × 10^{-4} s; Q_1 = 200 s^{-1}; Q_2 = 0.05 kOe^{-2}.

The obtained relaxation parameters are similar for compounds 2_{DyFe} and 3_{DyCo}. Due to the diamagnetic Co^{III} ions in 3_{DyCo}, the slow magnetic relaxation is solely associated with the anisotropic Dy^{III} ions. The existence of intrachain magnetic interaction in 2_{DyFe} does not change significantly the energy barrier of slow relaxation of the magnetization (see the temperature dependence in Figure 9), but affects more the field dependence, which becomes much more redistributed. This probably can be associated to redistribution/mixing the different energy levels in the 2_{DyFe} as a result of the small antiferromagnetic interaction along the chain and of intermolecular (interchain) interaction. The splitting of M_J (Stark) sublevels of the Dy^{III} centers under variation of the magnetic field also contributes to this redistribution. As in the case of static magnetic measurements, these competitive interactions cannot be quantified at the reported range of temperatures.

In a recent paper, Ma et al. report on a family of discrete, tetranuclear 3d-4f complexes assembled from a cationic lanthanide complexes and [M(CN)$_6$]$^{3-}$ metalloligands (M = Fe, Co), the ligand attached to the lanthanide(III) ions (Tb, Dy, and Ho) being also pentadentate [60]. The field-induced slow relaxation of the magnetization, with a low energy barrier (11.17 K), was observed only with the [Dy$_2$Co$_2$] derivative. The presence of the paramagnetic Fe^{III} ion does not improve the SMM behavior: for the [Dy$_2$Fe$_2$] derivative, the slow relaxation is not observed even by applying dc fields.

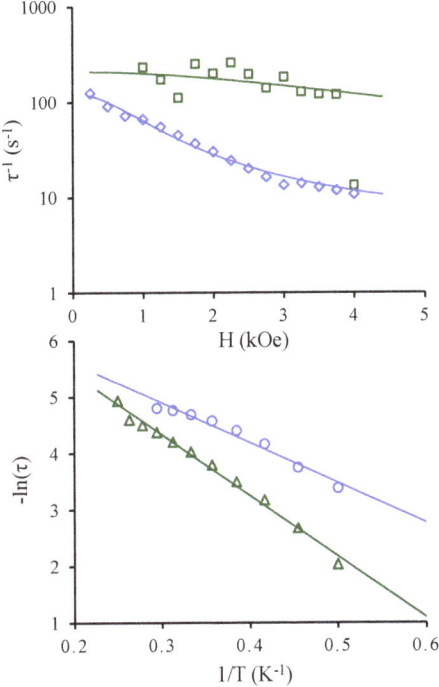

Figure 9. Field (**up**) and temperature (**down**) dependence of relaxation times for **3**$_{DyCo}$ and **2**$_{DyFe}$. The solid lines correspond to the fit using QTM and Orbach mechanisms for the magnetic relaxation.

4. Conclusions

In this paper, we have shown that the pentadentate *bis*-semicarbazone ligand, L, generates robust cationic Ln^{III} complexes, which are useful modules for constructing heterometallic coordination polymers. The metalloligands, [M(CN)$_6$]$^{3-}$, employ two *cis* cyanido groups as bridges against the Ln^{III} ions, resulting in a wave-like chain topology for the three compounds. The investigation of the magnetic properties reveals that the two Dy^{III}-containing coordination polymers exhibit slow relaxation of the magnetization, with rather low energy barriers. From the magnetic point of view, compound **3**$_{DyCo}$ behaves like a chain of Single Ion Magnets. These results open interesting perspectives for the synthesis of new cyanido-bridged 3d-4f complexes, using not only homoleptic but also heteroleptic cyanido tectons, as well as other types of metalloligands. Further work is in progress in our laboratory.

Supplementary Materials: The following are available online at https://www.mdpi.com/article/10.3390/magnetochemistry7050057/s1, Crystallographic data (single crystal and PXRD), bond distances and angles, infrared spectra; magnetic data; treatment of the ac magnetic data.

Author Contributions: Conceptualization, D.D. and M.A.; methodology, M.A., D.D. and G.N.; formal analysis, G.N. and A.M.M.; investigation, D.D., A.M.M. and G.N.; writing—original draft preparation, D.D. and M.A.; writing—review and editing, M.A. and G.N. All authors have read and agreed to the published version of the manuscript.

Funding: MAGMOLMET, grant ID: 867445.

Acknowledgments: D.D. is grateful to the European Union's Horizon 2020 research and innovation programme, for financial support (MAGMOLMET, grant ID: 867445).

Conflicts of Interest: The authors declare no conflict of interest.

References

1. Woodruff, D.N.; Winpenny, R.E.P.; Layfield, R.A. Lanthanide Single-Molecule Magnets. *Chem. Rev.* **2013**, *113*, 5110–5148. [CrossRef]
2. Goodwin, C.A.P.; Ortu, F.; Reta, D.; Chilton, N.F.; Mills, D.P. Molecular magnetic hysteresis at 60 Kelvin in dysprosocenium. *Nature* **2017**, *548*, 439–442. [CrossRef]
3. Guo, F.S.; Day, B.M.; Chen, Y.C.; Tong, M.L.; Mansikkamaki, A.; Layfield, R.A. Magnetic hysteresis up to 80 kelvin in a dysprosium metallocene single-molecule magnet. *Science* **2018**, *362*, 1400–1403. [CrossRef]
4. Dhers, S.; Feltham, H.L.C.; Brooke, S. A toolbox of building blocks, linkers and crystallisation methods used to generate single-chain magnets. *Coord. Chem. Rev.* **2015**, *296*, 24–44. [CrossRef]
5. Zheng, Y.-Z.; Lan, Y.; Wernsdorfer, W.; Anson, C.E.; Powell, A.K. Polymerisation of the Dysprosium Acetate Dimer Switches on Single-Chain Magnetism. *Chem. Eur. J.* **2009**, *15*, 12566–12570. [CrossRef] [PubMed]
6. Liu, Z.-Y.; Xia, Y.-F.; Jiao, J.; Yang, E.-C.; Zhao, X.-J. Two water-bridged cobalt(ii) chains with isomeric naphthoate spacers: From metamagnetic to single-chain magnetic behaviour. *Dalton Trans.* **2015**, *44*, 19927–19934. [CrossRef] [PubMed]
7. Bogani, L.; Sangregorio, C.; Sessoli, R.; Gatteschi, D. Molecular engineering for single-chain-magnet behavior in a one-dimensional dysprosium–nitronyl nitroxide compound. *Angew. Chem. Int. Ed.* **2005**, *44*, 5817–5821. [CrossRef] [PubMed]
8. Bernot, K.; Bogani, L.; Caneschi, A.; Gatteschi, D.; Sessoli, R. A family of rare-earth-based single chain magnets: Playing with anisotropy. *J. Am. Chem. Soc.* **2006**, *128*, 7947–7956. [CrossRef]
9. Pedersen, K.S.; Bendix, J.; Clérac, R. Single-molecule magnet engineering: Building-block approaches. *Chem. Commun.* **2014**, *50*, 4396–4415. [CrossRef]
10. Pal, S.; Dey, K.; Benmansour, S.; Gómez-García, C.J.; Nayek, H.P. Syntheses, structures and magnetic properties of cyano-bridged one-dimensional Ln3+–Fe3+ (Ln = La, Dy, Ho and Yb) coordination polymers. *New J. Chem.* **2019**, *43*, 6228–6233. [CrossRef]
11. Yu, D.-Y.; Li, L.; Zhou, H.; Yuan, A.-H.; Li, Y.-Z. Cyano-Bridged 4f-3d Assemblies with Achiral Helical Chains: Syntheses, Structures, and Magnetic Properties. *Eur. J. Inorg. Chem.* **2012**, 3394–3397. [CrossRef]
12. Estrader, M.; Ribas, J.; Tangoulis, V.; Solans, X.; Font-Bardía, M.; Maestro, M.; Diaz, C. Synthesis, Crystal Structure, and Magnetic Studies of One-Dimensional Cyano-Bridged Ln3+−Cr3+ Complexes with bpy as a Blocking Ligand. *Inorg. Chem.* **2006**, *45*, 8239–8250. [CrossRef] [PubMed]
13. Figuerola, A.; Ribas, J.; Casanova, D.; Maestro, M.; Alvarez, S.; Diaz, C. Magnetism of Cyano-Bridged Ln3+−M3+ Complexes. Part II: One-Dimensional Complexes (Ln3+ = Eu, Tb, Dy, Ho, Er, Tm; M3+ = Fe or Co) with bpy as Blocking Ligand. *Inorg. Chem.* **2005**, *44*, 6949–6958. [CrossRef] [PubMed]
14. Figuerola, A.; Diaz, C.; Ribas, J.; Tangoulis, V.; Sangregorio, C.; Gatteschi, D.; Maestro, M.; Mahía, J. Magnetism of Cyano-Bridged Hetero-One-Dimensional Ln3+−M3+ Complexes (Ln3+ = Sm, Gd, Yb; M3+ = FeLS, Co). *Inorg. Chem.* **2003**, *42*, 5274–5281. [CrossRef] [PubMed]
15. Petrosyants, S.P.; Ilyukhin, A.B.; Efimov, N.N.; Gavrikov, A.V.; Novotortsev, V.M. Self-assembly and SMM properties of lanthanide cyanocobaltate chain complexes with terpyridine as blocking ligand. *Inorg. Chim. Acta* **2018**, *482*, 813–820. [CrossRef]
16. Muddassir, M.; Song, X.-J.; Chen, Y.; Cao, F.; Weia, R.-M.; Song, Y. Ion-induced diversity in structure and magnetic properties of hexacyanometalate–lanthanide bimetallic assemblies. *CrystEngComm* **2013**, *15*, 10541–10549. [CrossRef]
17. Figuerola, A.; Ribas, J.; Solans, X.; Font-Bardía, M.; Maestro, M.; Diaz, C. One Dimensional 3d-4f Heterometallic Compounds: Synthesis, Structure and Magnetic Properties. *Eur. J. Inorg. Chem.* **2006**, 1846–1852. [CrossRef]
18. Zhao, H.; Lopez, N.; Prosvirin, A.; Chifotidesa, H.T.; Dunbar, K.R. Lanthanide–3d cyanometalate chains Ln(III)–M(III) (Ln = Pr, Nd, Sm, Eu, Gd, Tb; M = Fe) with the tridentate ligand2,4,6-tri(2-pyridyl)-1,3,5-triazine (tptz): Evidence of ferromagnetic interactions for the Sm(III)–M(III) compounds (M = Fe, Cr). *Dalton Trans.* **2007**, 878–888. [CrossRef]
19. Liu, J.; Knoeppel, D.W.; Liu, S.; Meyers, E.A.; Shore, S.G. Cyanide-Bridged Lanthanide(III)−Transition Metal Extended Arrays: Interconversion of One-Dimensional Arrays from Single-Strand (Type A) to Double-Strand (Type B) Structures. Complexes of a New Type of Single-Strand Array (Type C). *Inorg. Chem.* **2001**, *40*, 2842–2850. [CrossRef]
20. Kou, H.-Z.; Gao, S.; Sun, B.-W.; Zhang, J. Metamagnetism of the First Cyano-Bridged Two-Dimensional Brick-Wall-like 4f–3d Array. *Chem. Mater.* **2001**, *13*, 1431–1433. [CrossRef]
21. Wilson, D.C.; Liu, S.; Chen, X.; Meyers, E.A.; Bao, X.; Prosvirin, A.V.; Dunbar, K.R.; Hadad, C.M.; Shore, S.G. Water-Free Rare Earth-Prussian Blue Type Analogues: Synthesis, Structure, Computational Analysis, and Magnetic Data of {LnIII(DMF)6FeIII(CN)6}∞ (Ln = Rare Earths Excluding Pm). *Inorg. Chem.* **2009**, *48*, 5725–5735. [CrossRef]
22. Chen, W.-T.; Guo, G.-C.; Wang, M.-S.; Xu, G.; Cai, L.-Z.; Akitsu, T.; Akita-Tanaka, M.; Matsushita, A.; Huang, J.-S. Self-Assembly and Characterization of Cyano-Bridged Bimetallic [Ln–Fe] and [Ln−Co] Complexes (Ln = La, Pr, Nd and Sm). Nature of the Magnetic Interactions between the Ln3+ and Fe3+ Ions. *Inorg. Chem.* **2007**, *46*, 2105–2114. [CrossRef]
23. Chen, W.-T.; Wu, A.-Q.; Guo, G.-C.; Wang, M.-S.; Ca, L.-Z.; Huang, J.-S. Cyano-Bridged 2D Bimetallic 4f-3d Arrays with Monolayered Stair-Like, Brick-Wall-Like, or Bilayered Topologies–Rational Syntheses and Crystal Structures. *Eur. J. Inorg. Chem.* **2010**, 2826–2835. [CrossRef]
24. Xin, Y.; Wang, J.; Zychowicz, M.; Zakrzewski, J.J.; Nakabayashi, K.; Sieklucka, B.; Chorazy, S.; Ohkosh, S. Dehydration−Hydration Switching of Single-Molecule Magnet Behavior and Visible Photoluminescence in a Cyanido-Bridged DyIIICoIII Framework. *J. Am. Chem. Soc.* **2019**, *141*, 18211–18220. [CrossRef]
25. Visinescu, D.; Toma, L.M.; Fabelo, O.; Ruiz-Pérez, C.; Lloret, F.; Julve, M. Low-Dimensional 3d-4f Complexes Assembled by Low-Spin [FeIII(phen)(CN)4]− Anions. *Inorg. Chem.* **2013**, *52*, 1525–1537. [CrossRef] [PubMed]

26. Chorazy, S.; Rams, M.; Wyczesany, M.; Nakabayashi, K.; Ohkoshi, S.; Sieklucka, B. Antiferromagnetic exchange and long-range magnetic ordering in supramolecular networks constructed of hexacyanido-bridged LnIII(3-pyridone)–CrIII (Ln = Gd, Tb) chains. *CrystEngComm.* **2018**, *20*, 1271–1281. [CrossRef]
27. Xue, A.-Q.; Liu, Y.-Y.; Li, J.-X.; Zhang, Y.; Meng, Y.-S.; Zhu, W.-H.; Zhang, Y.-Q.; Sun, H.-L.; Wang, F.; Qiu, G.-X.; et al. The differential magnetic relaxation behaviours of slightly distorted triangular dodecahedral dysprosium analogues in a type of cyano-bridged 3d–4f zig-zag chain compounds. *Dalton Trans.* **2020**, *49*, 6867–6875. [CrossRef] [PubMed]
28. Petrosyants, S.P.; Ilyukhin, A.B.; Efimov, N.N.; Novotortsev, V.M. Cyano-Bridged d–f Ensembles of the Dysprosium Tetrapyridine Complexes with the Hexacyanoferrate Anion. *Russ. J. Coord. Chem.* **2018**, *44*, 660–666. [CrossRef]
29. Chorazy, S.; Zakrzewski, J.J.; Wang, J.; Ohkoshi, S.; Sieklucka, B. Incorporation of hexacyanidoferrate(III) ion in photoluminescent trimetallic Eu(3-pyridone)[Co1−xFex(CN)6] chains exhibiting tunable visible light absorption and emission properties. *CrystEngComm* **2018**, *20*, 5695–5706. [CrossRef]
30. Chorazy, S.; Kumar, K.; Nakabayashi, K.; Sieklucka, B.; Ohkoshi, S. Fine Tuning of Multicolored Photoluminescence in Crystalline Magnetic Materials Constructed of Trimetallic EuxTb1−x[Co(CN)6] Cyanido-Bridged Chains. *Inorg. Chem.* **2017**, *56*, 5239–5252. [CrossRef]
31. Chorazy, S.; Wyczesany, M.; Sieklucka, B. Lanthanide Photoluminescence in Heterometallic Polycyanidometallate-Based Coordination Networks. *Molecules* **2017**, *22*, 1902. [CrossRef]
32. Chorazy, S.; Rams, M.; Nakabayashi, K.; Sieklucka, B.; Ohkoshi, S. White Light Emissive DyIII Single-Molecule Magnets Sensitized by Diamagnetic [CoIII(CN)6]3− Linkers. *Chem. Eur. J.* **2016**, *22*, 7371–7375. [CrossRef]
33. Chorazy, S.; Rams, M.; Wang, J.; Sieklucka, B.; Ohkoshi, S. Octahedral Yb(III) complexes embedded in [CoIII(CN)6]-bridged coordination chains: Combining sensitized near-infrared fluorescence with slow magnetic relaxation. *Dalton Trans.* **2017**, *46*, 13668–13672. [CrossRef]
34. Qian, K.; Huang, X.-C.; Zhou, C.; You, X.-Z.; Wang, X.-Y.; Dunbar, K.R. A Single-Molecule Magnet Based on Heptacyanomolybdate with the Highest Energy Barrier for a Cyanide Compound. *J. Am. Chem. Soc.* **2013**, *135*, 13302–13305. [CrossRef]
35. Sasnovskaya, V.D.; Kopotkov, V.A.; Talantsev, A.D.; Morgunov, R.B.; Yagubskii, E.B.; Simonov, S.V.; Zorina, L.V.; Mironov, V.S. Synthesis, Structure, and Magnetic Properties of 1D {[MnIII(CN)6][MnII(dapsc)]}n Coordination Polymers: Origin of Unconventional Single-Chain Magnet Behavior. *Inorg. Chem.* **2017**, *56*, 8926–8943. [CrossRef]
36. Zorina, L.V.; Simonov, S.V.; Sasnovskaya, V.D.; Talantsev, A.D.; Morgunov, R.B.; Mironov, V.S.; Yagubskii, E.B. Slow Magnetic Relaxation, Antiferromagnetic Ordering, and Metamagnetism in MnII(H2dapsc)-FeIII(CN)6 Chain Complex with Highly Anisotropic Fe-CN-Mn Spin Coupling. *Chem. Eur. J.* **2019**, *25*, 14583–14597. [CrossRef]
37. Bar, A.K.; Gogoi, N.; Pichon, C.; Goli, V.M.L.D.P.; Thlijeni, M.; Duhayon, C.; Suaud, N.; Guihéry, N.; Barra, A.-L.; Ramasesha, S.; et al. Pentagonal Bipyramid FeII Complexes: Robust Ising-spin Units Towards Heteropolynuclear Nano-Magnets. *Chem. Eur. J.* **2017**, *23*, 4380–4396. [CrossRef] [PubMed]
38. Pichon, C.; Suaud, N.; Duhayon, C.; Guihéry, N.; Sutter, J.-P. Cyano-Bridged Fe(II)−Cr(III) Single-Chain Magnet Based on Pentagonal Bipyramid Units: On the Added Value of Aligned Axial Anisotropy. *J. Am. Chem. Soc.* **2018**, *140*, 7698–7704. [CrossRef]
39. Sommerer, S.O.; Westcott, B.L.; Cundari, T.R.; Krause, J.A. A structural and computational study of tetraaqua[2,6-diacetylpyridinebis-(semicarbazone)]-gadolinium(III) trinitrate. *Inorg. Chim. Acta* **1993**, *209*, 101–104. [CrossRef]
40. Gioia, M.; Crundwell, G.; Westcott, B.L. Tetraaqua[2,6-diacetylpyridine bis(semicarbazone)]samarium(III) trinitrate. *IUCrData* **2018**, *3*, x181454. [CrossRef]
41. Sasnovskaya, V.D.; Kopotkov, V.A.; Kazakova, A.V.; Talantsev, A.D.; Morgunov, R.B.; Simonov, S.V.; Zorina, L.V.; Mironov, V.S.; Yagubskii, E.B. Slow magnetic relaxation in mononuclear complexes of Tb, Dy, Ho and Er with the pentadentate (N3O2) Schiff-base dapsc ligand. *New J. Chem.* **2018**, *42*, 14883–14893. [CrossRef]
42. Palenik, G.J.; Wester, D.W.; Rychlewska, U.; Palenik, R.C. Pentagonal-Bipyramidal Complexes. Synthesis and Crystal Structures of Diaqua [2,6-diacetylpyridine bis(semicarbazone)]chromium(III) Hydroxide Dinitrate Hydrate and Dichloro[2,6-diacetylpyridine bis(semicarbazone)] iron(III) Chloride Dihydrate. *Inorg. Chem.* **1976**, *15*, 1814–1819. [CrossRef]
43. Pascal, P. Magnochemical studies. *Ann. Chim. Phys.* **1910**, *19*, 5–70.
44. Kahn, O. *Molecular Magnetism*; VCH Publishers: New York, NY, USA, 1993.
45. Sheldrick, G.M. SHELXT—Integrated space-group and crystal-structure determination. *Acta Cryst.* **2015**, *A71*, 3–8. [CrossRef]
46. Sheldrick, G.M. Crystal structure refinement with SHELXL. *Acta Cryst.* **2015**, *C71*, 3–8. [CrossRef]
47. Nakamoto, K. *Infrared and Raman Spectra of Inorganic and Coordination Compounds*, 4th ed.; John Wiley & Sons: Hoboken, NJ, USA, 1986; p. 245.
48. Llunell, M.; Casanova, D.; Cirera, J.; Alemany, P.; Alvarez, S. *SHAPE, Program. for the Stereochemical Analysis of Molecular Fragments by Means of Continuous Shape Measures and Associated Tools*, Version 2.1; University of Barcelona: Barcelona, Spain, 2013.
49. Benelli, C.; Gatteschi, D. Magnetism of Lanthanides in Molecular Materials with Transition-Metal Ions and Organic Radicals. *Chem. Rev.* **2002**, *102*, 2369–2388. [CrossRef]
50. Kahn, M.L.; Ballou, R.; Porcher, P.; Kahn, O.; Sutter, J.-P. Analytical Determination of the {Ln–Aminoxyl Radical} Exchange Interaction Taking into Account Both the Ligand-Field Effect and the Spin–Orbit Coupling of the Lanthanide Ion (Ln = DyIII and HoIII). *Chem. A Eur. J.* **2002**, *8*, 525–531. [CrossRef]
51. Guo, Y.-N.; Xu, G.-F.; Gamez, P.; Zhao, L.; Lin, S.-Y.; Deng, R.; Tang, J.; Zhang, H.-J. Two-Step Relaxation in a Linear Tetranuclear Dysprosium(III) Aggregate Showing Single-Molecule Magnet Behavior. *J. Am. Chem. Soc.* **2010**, *132*, 8538–8539. [CrossRef]

52. Grahl, M.; Kötzler, J.; Seßler, I. Correlation between Domain-Wall Dynamics and Spin-Spin Relaxation in Uniaxial Ferromagnets. *J. Magn. Magn. Mater.* **1990**, *90–91*, 187–188. [CrossRef]
53. Dolai, M.; Ali, M.; Titiš, J.; Boča, R. Cu(II)–Dy(III) and Co(III)–Dy(III) Based Single Molecule Magnets with Multiple Slow Magnetic Relaxation Processes in the Cu(II)–Dy(III) Complex. *Dalton Trans.* **2015**, *44*, 13242–13249. [CrossRef] [PubMed]
54. Lucaccini, E.; Sorace, L.; Perfetti, M.; Costes, J.-P.; Sessoli, R. Beyond the Anisotropy Barrier: Slow Relaxation of the Magnetization in Both Easy-Axis and Easy-Plane Ln(Trensal) Complexes. *Chem. Commun.* **2014**, *50*, 1648–1651. [CrossRef] [PubMed]
55. Zadrozny, J.M.; Atanasov, M.; Bryan, A.M.; Lin, C.Y.; Rekken, B.D.; Power, P.P.; Neese, F.; Long, J.R. Slow Magnetization Dynamics in a Series of Two-Coordinate Iron(II) Complexes. *Chem. Sci.* **2013**, *4*, 125–138. [CrossRef]
56. Feng, X.; Liu, J.L.; Pedersen, K.S.; Nehrkorn, J.; Schnegg, A.; Holldack, K.; Bendix, J.; Sigrist, M.; Mutka, H.; Samohvalov, D.; et al. Multifaceted Magnetization Dynamics in the Mononuclear Complex [ReIVCl4(CN)2]2−. *Chem. Commun.* **2016**, *52*, 12905–12908. [CrossRef] [PubMed]
57. Ding, Y.S.; Yu, K.X.; Reta, D.; Ortu, F.; Winpenny, R.E.P.; Zheng, Y.Z.; Chilton, N.F. Field- and Temperature-Dependent Quantum Tunnelling of the Magnetisation in a Large Barrier Single-Molecule Magnet. *Nat. Commun.* **2018**, *9*, 1–10. [CrossRef] [PubMed]
58. Liu, X.; Feng, X.; Meihaus, K.R.; Meng, X.; Zhang, Y.; Li, L.; Liu, J.; Pedersen, K.S.; Keller, L.; Shi, W.; et al. Coercive Fields Above 6 T in Two Cobalt(II)–Radical Chain Compounds. *Angew. Chem. Int.* **2020**, *59*, 10610–10618. [CrossRef] [PubMed]
59. Reta, D.; Chilton, N.F. Uncertainty estimates for magnetic relaxation times and magnetic relaxation parameters. *Phys. Chem. Chem. Phys.* **2019**, *21*, 23567–23575. [CrossRef]
60. Wang, R.; Wang, H.; Wang, J.; Bai, F.; Ma, Y.; Li, L.; Wang, Q.; Zhao, B.; Cheng, P. The different magnetic relaxation behaviors in [Fe(CN)6]3− or [Co(CN)6]3− bridged 3d–4f heterometallic compounds. *CrystEngComm* **2020**, *22*, 2998–3004. [CrossRef]

Article

New Radical Cation Salts Based on BDH-TTP Donor: Two Stable Molecular Metals with a Magnetic $[ReF_6]^{2-}$ Anion and a Semiconductor with a $[ReO_4]^-$ Anion

Nataliya D. Kushch [1,*], Gennady V. Shilov [1], Lev I. Buravov [1], Eduard B. Yagubskii [1,*], Vladimir N. Zverev [2,3], Enric Canadell [4] and Jun-ichi Yamada [5]

1. Institute of Problems of Chemical Physics RAS, 142432 Chernogolovka, Russia; shilg@icp.ac.ru (G.V.S.); buravov@icp.ac.ru (L.I.B.)
2. Institute of Solid State Physics, Russian Academy of Sciences, 142432 Chernogolovka, Russia; zverev@issp.ac.ru
3. Moscow Institute of Physics and Technology, 141700 Dolgoprudnyi, Russia
4. Institut de Ciència de Materials de Barcelona (ICMAB-CSIC), Campus de la U.A.B., E-08193 Bellaterra, Spain; canadell@icmab.es
5. Department of Material Science, Graduate School of Material Science, University of Hyogo, 3-2-1 Kyoto, Kamigori-choAko-gun, Hyogo 678-1297, Japan; yamada@sci.u-hyogo.ac.jp
* Correspondence: kushch@icp.ac.ru (N.D.K.); yagubski@gmail.com (E.B.Y.)

Citation: Kushch, N.D.; Shilov, G.V.; Buravov, L.I.; Yagubskii, E.B.; Zverev, V.N.; Canadell, E.; Yamada, J.-i. New Radical Cation Salts Based on BDH-TTP Donor: Two Stable Molecular Metals with a Magnetic $[ReF_6]^{2-}$ Anion and a Semiconductor with a $[ReO_4]^-$ Anion. Magnetochemistry 2021, 7, 54. https://doi.org/10.3390/magnetochemistry7040054

Academic Editors: Carlos J. Gómez García, Lee Martin, Scott Turner, John Wallis and Hiroki Akutsu

Received: 18 March 2021
Accepted: 15 April 2021
Published: 20 April 2021

Publisher's Note: MDPI stays neutral with regard to jurisdictional claims in published maps and institutional affiliations.

Copyright: © 2021 by the authors. Licensee MDPI, Basel, Switzerland. This article is an open access article distributed under the terms and conditions of the Creative Commons Attribution (CC BY) license (https://creativecommons.org/licenses/by/4.0/).

Abstract: Three radical cation salts of BDH-TTP with the paramagnetic $[ReF_6]^{2-}$ and diamagnetic $[ReO_4]^-$ anions have been synthesized: κ-(BDH-TTP)$_4$ReF$_6$ (**1**), κ-(BDH-TTP)$_4$ReF$_6$·4.8H$_2$O (**2**) and pseudo-κ''-(BDH-TTP)$_3$(ReO$_4$)$_2$ (**3**). The crystal and band structures, as well as the conducting properties of the salts, have been studied. The structures of the three salts are layered and characterized by alternating κ-(**1**, **2**) and κ''-(**3**) type organic radical cation layers with inorganic anion sheets. Similar to other κ-salts, the conducting layers in the crystals of **1** and **2** are formed by BDH-TTP dimers. The partial population of positions of Re atoms and disorder in the anionic layers of **1**–**3** are their distinctive features. Compounds **1** and **2** show the metallic character of conductivity down to low temperatures, while **3** is a semiconductor. The *ac* susceptibility of crystals **1** was investigated in order to test the possible slow relaxation of magnetization associated with the $[ReF_6]^{2-}$ anion.

Keywords: organic conductors; metal complex anions; molecular magnets; electrocrystallization; crystal and band structures; conductivity; magnetic properties

1. Introduction

Multifunctionality is one of the most attractive trends in the chemistry of modern materials. Among multifunctional materials, compounds combining electrical conductivity and magnetism in the same crystal lattice have been the object of intense study in recent decades [1–6]. This interest is associated with the search for synergy of these properties, which can lead to new phenomena and the tuning of one of the properties as a response to an external factor affecting the other. Research in this area is focused mainly on the family of quasi-two-dimensional (super)conductors based on the radical cation salts of bis(ethylenedithio)tetrathiafulvalene (BEDT-TTF, Scheme 1) and its derivatives with various paramagnetic metal complex anions. In such compounds, the electrical conductivity is associated with mobile electrons of the organic layers, and transition metal ions in the insulating anion layers are responsible for magnetism. Among the possible magnetic counterions for conducting radical cation salts, molecular nanomagnets—the so-called single molecular magnets (SMMs)—attract much attention. SMMs exhibit unique magnetic properties at liquid helium temperatures such as slow magnetization relaxation, blocking and quantum magnetization tunneling and, due to these properties, the SMMs are promising compounds [7–11]. The creation of multifunctional compounds combining conductivity and single molecule magnetism could open new directions in terms of further studying the fundamental properties and practical applications of molecular nanomagnets. In particular, the study of the relationship between relaxing local spins and conducting electrons in superparamagnetic/conducting hybrid materials is important for the development of quantum spintronic systems.

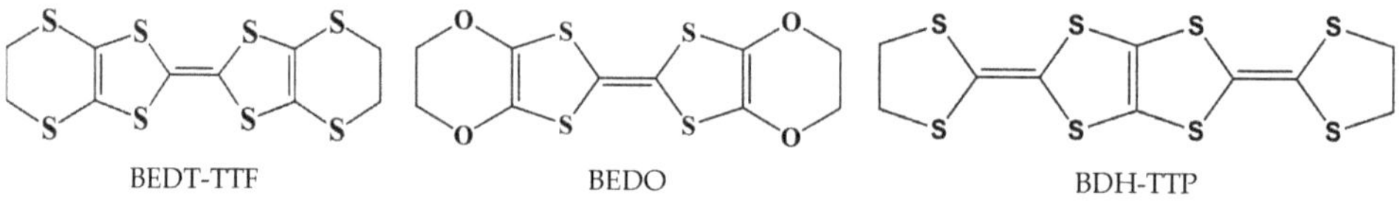

Scheme 1. π-donor molecules.

There are already several publications in the literature devoted to the design and synthesis of conducting radical anion salts based on the M(dmit)$_2$ complexes and tetracyanoquinodimethane containing cationic metal complexes as counterions, showing slow magnetic relaxation [6,12,13]. Recently, a radical cation salt of BEDT-TTF has also been synthesized with a single-molecule magnet as a counterion, namely, an anionic trifluoroacetate dysprosium complex [14]. However, in all these compounds, conductivity and single-molecule magnetism occur in different temperature ranges: the single-molecule magnetic properties appear at very low helium temperatures (usually below 20 K), while the conductivity is at best maintained up to the liquid nitrogen (77.2 K). In 2018, we used the oxygen analog of BEDT-TTF, bis(ethylenedioxo)tetrathiafulvalene (BEDO), Scheme 1, as an organic π-donor and investigated its electrochemical oxidation in the presence of the Re(IV) hexafluoride complex, [ReF$_6$]$^{2-}$, which in the composition of (PPh$_4$)$_2$[ReF$_6$]·2H$_2$O salt displays SMM properties below 5 K under application of a low permanent magnetic field [15]. As a result, the first conductive field-induced SMM was synthesized—(BEDO)$_4$[ReF$_6$]·6H$_2$O— in which conductivity and single-molecule magnetism coexist in the same temperature range [16]. More recently, another BEDO salt has been synthesized with a monomolecular magnet [Co(pdms)]$^{2-}$ as a counterion, which also shows the coexistence of conductivity and monomolecular magnetism up to 11 K [17].

In the present work, we investigated the electrochemical oxidation of 2,5-bis (1,3-dithiol-2-ylidene)-1,3,4,6-tetrathiapentalene donor, C$_{10}$S$_8$H$_8$, (BDH-TTP, Scheme 1) in the presence of (PPh$_4$)$_2$[ReF$_6$] 2H$_2$O, electrolyte. Our choice of the BDH-TTP donor, as in the case of BEDO, is determined by the fact that it forms radical cation salts with counterions of a different nature, size and shape, which retain metallic conductivity up to the temperature of liquid helium (4.2 K) [18,19].

Unlike BEDT-TTF and its derivatives, in the case of BDH-TTP and BEDO, the crystalline packing of the conducting layers is determined primarily by the nature of these donors themselves, and the counterions have almost no effect on the packing of the radical cation layers. Herein, we report the synthesis and the crystalline and band structures, as well as the transport properties under the normal and high pressure of new organic metals based on BDH-TTP, κ-(BDH-TTP)$_4$ReF$_6$ (**1**) and κ-(BDH-TTP)$_4$ReF$_6$·4.8H$_2$O (**2**), containing κ-type organic layers as well as the new semiconductor, pseudo-κ''-(BDH-TTP)$_3$(ReO$_4$)$_2$ (**3**). To probe the magnetization dynamics of **1**, the ac susceptibility was studied.

2. Results and Discussion
2.1. Synthesis

Electrochemical oxidation of the BDH-TTP donor has been studied in a medium chlorobenzene (CB) + 10% abs. ethanol, (CB) + (5 ÷ 15%) rectified (96%) ethanol containing 4% water and (CB) + 10% trifluoroethanol. Different types of alcohol have been used as additives to the main solvent CB. The (Ph$_4$P)$_2$[ReF$_6$]·2H$_2$O salt has been used as an electrolyte. Crystals of **1** grow when trifluoroethanol is used as an additive. In a mixture of CB with the addition of 96% ethanol, the crystals of **2** and **3** formed simultaneously at the anode while electrocrystallization was not observed in the presence of the absolute alcohol due to the inability to set the desired current. Thus, it can be stated that alcohol additives have a decisive influence on the formation of these BDH-TTP salts. Moreover, the formation of salt **3** indicates that, in the process of electrocrystallization, Re^{4+} is oxidized to Re^{7+}. The synthetic procedure for the preparation of salts **1**–**3** is given in the experimental section.

2.2. Crystal Structure

The prepared radical cation salts were characterized by the layered structures with alternating conductive radical cation layers and insulating anionic sheets formed with $[ReF_6]^{2-}$ octahedra in salt **1**, $[ReF_6]^{2-}$ octahedra and water molecules in **2** or $[ReO_4]^-$ tetrahedra in **3**. Crystallographic data and refined structural parameters for the crystals **1**–**3** are given in Table S1, Supplementary Materials.

2.2.1. Crystal Structure of the Salts κ-(BDH-TTP)$_4$ReF$_6$ (1) and κ-(BDH-TTP)$_4$ReF$_6$·4.8H$_2$O (2)

The compound **1** crystallizes in a monoclinic system ($P2_1/c$ space group). The asymmetric part of the crystal structure consists of one molecule of BDH-TTP and one anion in the general position with a population of 0.25 (see Figure S1, ESI). Thus, there is one anion and four radical cations per unit cell. Figure 1 shows the layered structure of the crystals of **1**, in which conductive and insulating layers alternate along the a-axis. The radical cationic layers in **1** are packed in a typical κ-type arrangement formed by BDH-TTP equivalent dimers (Figure 2a), which are located at an angle of 83.78° relative to each other.

The donor molecules are arranged into "face-to-face" dimers by a "ring-over-bond" type (Figure S2, Supplementary Materials) [19]. This configuration, in which two neighboring donor molecules are shifted in relation to each other along the long molecular axis by approximately a length of the C=C double bond (~1.4 Å), is typical for the BEDT-TTF and BDH-TTP κ-salts [18–20]. The terminal ethylene groups in **1** are disordered. The central C=C bond length of the radical cation, as well as two terminal C=C bond lengths in the TTP fragments, have almost the same values (1,353–1.357 Å), which correspond well to the charge +0.5 [18–21]. The anionic layer, formed by isolated anions with the incomplete population of Re (Figure 2b), is highly disordered. Since it has been impossible to accurately determine the position of fluoride ions in the octahedra, they have therefore been refined with restrictions on bond lengths and thermal parameters.

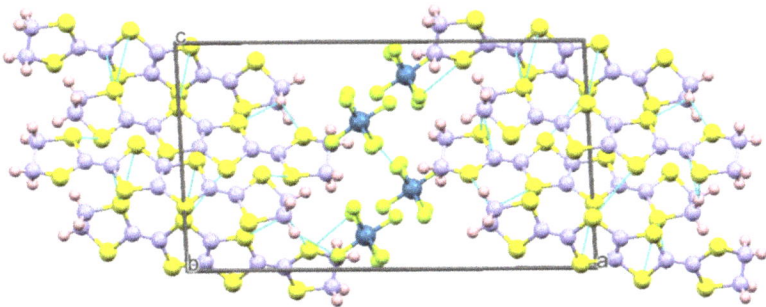

Figure 1. Crystal structure of the κ-(BDH-TTP)$_4$ReF$_6$ (**1**) salt projected on the ac-plane. Dashed lines show shortened intermolecular contacts between the BDH-TTP molecules.

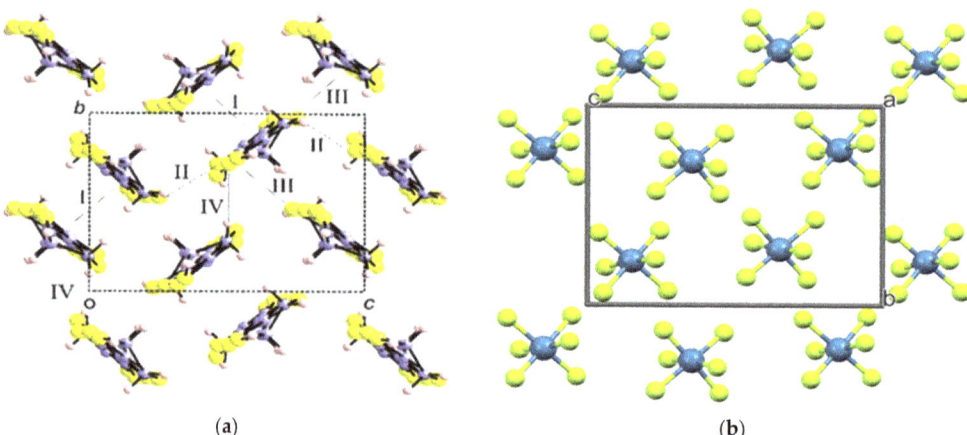

Figure 2. Projections of: (**a**) the radical cation layer, where the different intermolecular interactions are labeled and (**b**) the anionic layer along the *a* axis in the κ-(BDH-TTP)$_4$ReF$_6$ (**1**) salt.

Between the dimers there are four short S ... S contacts and one short S ... C contact, and the contacts involving hydrogen atoms (Table 1), while inside the dimers in the conducting layers, are only intermolecular contacts of the types S ... HC (S5 ... H9A) and C ... HC (C6 ... H9A).

Table 1. Short intermolecular contacts in the structure of the salt **1**.

Short Contact	Contact Length, Å	Symmetry Operation for the 1st Atom in Contact
S1 ... S2	3.584	x, y, z
S3 ... S1	3.531	x, y, z
S8 ... S3	3.549	x, y, z
S7 ... S5	3.516	x, y, z
S8 ... C5	3.434	x, y, z
S4 ... H10A	2.977	x, y, z
S6 ... H10A	2.942	x, y, z
S5 ... H9A	2.858	x, y, z
C6 ... H9A	2.822	x, y, z

Salt **2** crystallizes in a monoclinic system and, in contrast to **1**, was refined in the space group *C*2/*c*. The asymmetric part of the crystal structure comprises the [ReF$_6$]$^{2-}$ anion in a general position with a population of 0.25 disordered by the two-fold axis, a radical cation in a general position and three water molecules with a population of 0.72, 0.24, 0.24, respectively. Thus, there are two anions and eight radical cations per unit cell or, formally, one Re atom for four BDH-TTP molecules (see Figure S3, Supplementary Materials). Consequently, the composition of salt **2** differs from **1** by the presence of water molecules in its structure. The crystal structure of **2** is shown in Figure 3, in which conductive and insulating anionic layers alternate along the *a*- axis. The radical cation layers consist of BDH-TTP dimers built from two identical donor molecules with an average charge of +0.5 per molecule (κ-type packing) (Figure 4a). Terminal ethylene groups in the BDH-TTP radical cations are disordered.

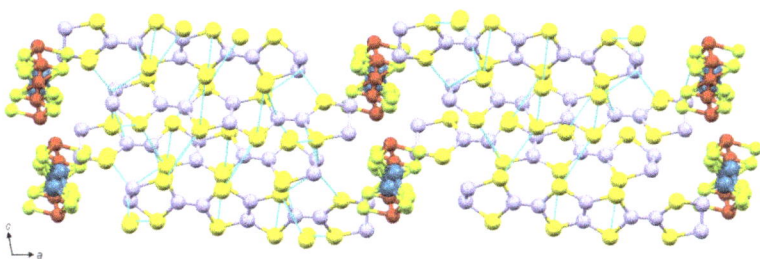

Figure 3. Crystal structure of the κ-(BDH-TTP)$_4$ReF$_6$·4.8H$_2$O (**2**) salt projected on the *ac*-plane. Dashed lines show shortened intermolecular contacts between the BDH-TTP molecules.

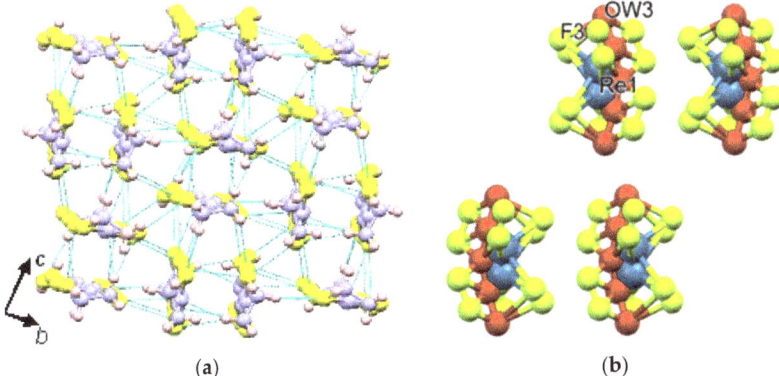

Figure 4. (**a**) Projection of the conducting layer on the *bc*-plane and (**b**) Optimal view of the anionic layer fragment in the κ-(BDH-TTP)$_4$ReF$_6$·4.8H$_2$O (**2**) salt.

As in salt **1**, the BDH-TTP dimers in **2** have a "ring-over-bond" configuration (Figure S4, Supplementary Materials). Inside the dimers, the central parts of the donor molecules are almost parallel. The dihedral angle between the neighboring dimers is 83.45°. As in salt **1**, this angle in **2** is smaller by ≈16° than those characteristic of classical BEDT-TTF and BETS κ-type salts [3,4,20].

The conducting layers are characterized by the presence of several short S...S contacts, some of which are shorter than the sum of the Van der Waals radii by 0.007–0.110 Å. In addition to these contacts there are also intermolecular contacts of the S...H-C and C...H-C types (Table 2). Between the dimers there are six short S...S- and one S...C contact and also intermolecular contacts S2...HC7 and S7...HC7. In the dimer, only intermolecular contacts S8...H8B and C6...H8B are present (Table 2).

The anionic layers of salt **2** consist of the isolated octahedral [ReF$_6$]$^{2-}$ anions and water molecules localized near them. Figure 4b shows a fragment of the anionic layer. The octahedral anions in **2** are more distorted, compared with those in **1**. As a result of significant disorder in the anionic layer, the atomic population of the [ReF$_6$]$^{2-}$ anion and water molecules have been refined with restrictions on bond lengths and thermal parameters and then fixed, as in salt **1**. As a consequence, it has not been possible to accurately analyze the different interactions leading to the actual crystal package in both salts.

2.2.2. Crystal Structure of the Pseudo-κ''-(BDH-TTP)$_3$(ReO$_4$)$_2$ (**3**) Salt

Radical cation salt **3** crystallizes in the triclinic system. The asymmetric unit includes the [ReO$_4$]$^-$ anion in the general position and two radical cations in the general position (Figure S5, Supplementary Materials). In the anion, oxygen atoms are disordered over two positions with occupancies 0.63 and 0.37. The radical cation and anion layers alternate along the c axis of the unit cell (Figure 5). The radical cations in the conducting layers form a two-dimensional network connected by several short contacts S...S and intermolecular contacts CH...S, Figure 6a. Values of short contacts and intermolecular CH...S contacts in the structure of **3** are listed in Table 3.

Table 2. Short intermolecular contacts in the structure of the salt **2**.

Short Contact	Contact Length, Å	Symmetry Operation for the 1st Atom in Contact
S3 ... S1	3.569	x, y, z
S2 ... S5	3.581	x, y, z
S4 ... S6	3.537	x, y, z
S4 ... S1	3.531	x, y, z
S3 ... S6	3.593	x, y, z
S5 ... S8	3.490	x, y, z
C5 ... S6	3.416	x, y, z
C7H ... S2	2.057	x, y, z
C7H ... S7	2.878	x, y, z
S8 ... H8B	2.957	x, y, z
C6 ... H8B	2.782	x, y, z

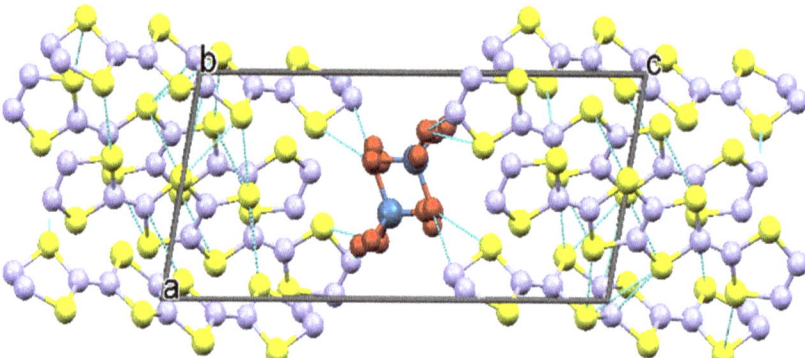

Figure 5. Projection of the crystal structure of the pseudo-κ"-(BDH-TTP)$_3$(ReO$_4$)$_2$ salt (**3**) on the ac plane; dashed lines show shortened intermolecular contacts between the BDH-TTP molecules.

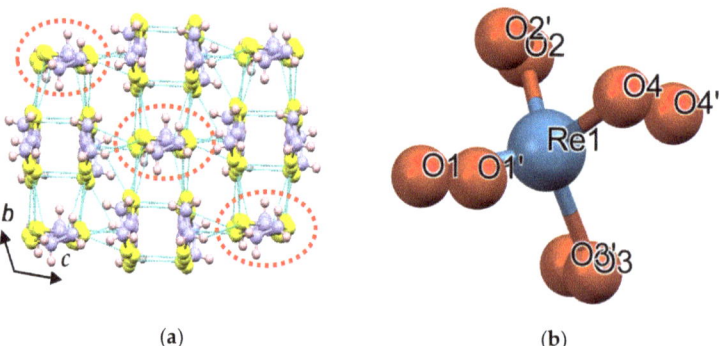

(a)   (b)

Figure 6. (**a**) Optimal view of the radical cation layers of the salt pseudo-κ"-(BDH-TTP)$_3$(ReO$_4$)$_2$ (**3**); The single molecules BDH-TTP are highlighted by circles; (**b**) The disordered anion [ReO$_4$]$^-$ in the structure of **3** with the numbering of the atoms in the tetrahedron.

Table 3. The short intermolecular contact values and hydrogen bonds in the structure of salt 3.

Short Cotact	Length, Å	Sym. Operation for the 1st Atom	Short Contact	Length, Å	Sym. Operation for the 1st Atom
O3 ... S10	3.045	x, y, z	S12 ... 3	3.498	x, y, z
O4 ... S5	3.045	x, y, z	S12 ... S2	3.260	x, y, z
O1 ... C5	3.126	x, y, z	S10 ... S4	3.579	x, y, z
O3 ... C14	3.160	x, y, z	C15 ... S3	3.466	x, y, z
O3 ... C6	3.204	x, y, z	S9 ... S6	3.482	x, y, z
O4 ... C10	3.160	x, y, z	S9 ... C7	3.296	x, y, z
O3 ... C6	3.204	x, y, z	S6 ... S2	3.423	x, y, z
O2 ... C6	3.059	x, y, z	S11 ... S7	3.244	x, y, z
O2 ... H6A	2.246	x, y, z	S11 ... S6	3.531	x, y, z
O1 ... H6A	2.707	x, y, z	C15 ... S7	3.383	x, y, z
O4 ... H14B	2.533	x, y, z	S9 ... S8	3.457	x, y, z
O2 ... H13A	2.496	x, y, z	S3 ... S7	3.466	x, y, z
O1 ... H5B	2.559	x, y, z	S10 ... H6B	2.943	−1 + x, 1 + y, z
O1 ... H9A	2.612	x, y, z	S4 ... H6B	2.873	x, y, z

The conducting layers in 3 are built from dimers and single donor molecules (so called pseudo-κ-type packing) [5,22]. In the dimers, the radical cations are located "face-to-face" with a longitudinal shift larger than the shifts in salts 1 and 2 (Figure S6, Supplementary Materials). The pseudo-κ'-type packing, where four single molecules surround one dimer, is described in BEDT-TTF salts [5,22]. Unlike the known BEDT-TTF salts with pseudo-κ'-type packing [5,22], we have observed four dimers surrounding a donor molecule (new pseudo-κ"-type packing) in 3 (Figure 6a). This type of packing of conducting organic layers has not been previously observed in radical cation salts. The dihedral angle between the single BDH-TTP molecule and dimer is 84.45°. The distances between the radical cations in dimer and between single molecules are equal to 3.289 Å and 5.374 Å, respectively. The distances along the *a* and *b* axes between the radical cations in the dimer and a single BDH-TTP molecule are 3.456 and 3.651 Å, respectively. The lengths of C=C bonds in the single radical cation and dimer are shown in Table 4. By the value of C=C bonds, one can estimate the charge state of the BDH-TTP molecule [18,21]. From the analysis of the values of these bonds, it can be assumed that the charge state of a single BDH-TTP molecule is close to 0, and that of dimer molecules is close to +2. However, the presence of short contacts between dimers and single ET molecules (Figure S7, Supplementary Materials) indicates the existence of some charge transfer between them. In order to clarify this situation, the electronic band structure of 3 was calculated (see the electronic structures section).

Table 4. The lengths of C=C bonds in the dimer and single radical cation of salt 3.

C=C Bond in the Dimer Radical Cation	Length of C=C Bond, Å	C=C Bond in the Single Radical Cation	Length of C=C Bond, Å
central C1=C2	1.374	central C15=C15	1.362
terminal C3=C4	1.358	terminal C11=C12	1.345
terminal C7=C8	1.365	terminal C11=C12	1.345

Anionic layers are formed by isolated anions $[ReO_4]^-$. Each Re atom has a tetrahedral environment and is bonded to four oxygen atoms (Figure 6b). Between the BDH-TTP molecules and $[ReO_4]^-$ tetrahedra there are short C...O and S...O contacts as well as intermolecular C-H ... O contacts (see Table 4). The Re-O bond lengths in the tetrahedra are within the interval 1.700–1.786 Å range, Table S2, Supplementary Materials

2.3. Electronic Structure of the BDH-TTP Salts

2.3.1. Electronic Structure of κ-(BDH-TTP)$_4$ReF$_6$ (1) and κ-(BDH-TTP)$_4$ReF$_6$·4.8H$_2$O (2)

The donor layers of **1** contain only one type of BDH-TTP dimer built from two identical donors and there are four different types of HOMO... HOMO (highest occupied molecular orbital) intermolecular interactions (see Section 2, Figure 2a): (i) the intra-dimer interaction (I); (ii) two interactions between donors in different dimers almost orthogonal (II and III); and (iii) one interaction between donors in different dimers forming a chain along the b-direction (IV). The strength of these interactions can be qualified from the so-called HOMO ... HOMO interaction energies [23]. Those calculated for the present salt are 0.3567 (I), 0.0826 (II) 0.1372 (III) and 0.2326 (IV) eV. These values implicate very similar inter-dimer interactions along the two main directions of the lattice. Thus, as far as the HOMO ... HOMO interactions are concerned, this salt must be very isotropic within the layers plane. This was also the case with other BDH-TTP κ-type salts such as κ-(BDH-TTP)$_2$X with X = [FeNO(CN)$_5$], PF$_6$, FeCl$_4$, and [Hg(SCN)$_4$].C$_6$H$_5$NO$_2$ [18,21,24,25]. For instance, the HOMO ... HOMO interaction energies calculated for the first of these salts are: 0.3613 (I), 0.0742 (II) 0.1476 (III) and 0.2420 (IV) eV, which are remarkably similar to those of κ-(BDH-TTP)$_4$ReF$_6$. Since the shape of the anions in these salts is quite different, it is clear that the inner structure of many BDH-TTP κ-salts is mostly determined by the intra-layer hydrogen bonding and S ... S interactions with only a minor influence of the anions, which tend to adapt to it. The calculated band structure of **1** is shown in Figure 7a. Since the repeat unit of the donor layer contains four donors, Figure 7a contains four HOMO-based bands. With one-quarter occupation of the anion positions, the HOMO bands must contain two holes so that the Fermi level cuts the two upper HOMO bands and the system should exhibit metallic behavior, in agreement with our transport measurements.

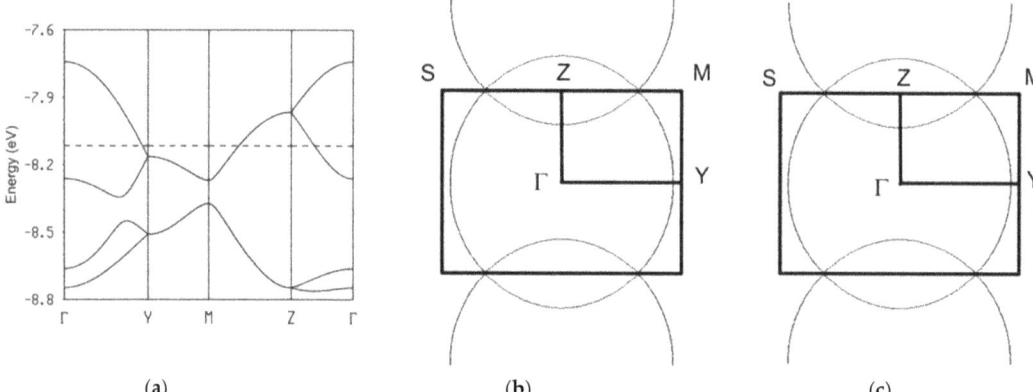

Figure 7. (**a**) Calculated band structure and (**b**) Fermi surface for the donor layers of κ-(BDH-TTP)$_4$ReF$_6$ (**1**); (**c**) Fermi surface calculated for the donor layers of κ-(BDH-TTP)$_4$ReF$_6$·4.8H$_2$O (**2**). In (**a**) the dashed line refers to the Fermi level. Γ = (0, 0), Y = (b*/2, 0), Z = (0, c*/2), M = (b*/2, c*/2) and S = (−b*/2, c*/2).

The calculated Fermi surface is shown in Figure 7b: it is made of a series of superposed cylinders with a practically circular cross section. The area of the large circles of Figure 7b amounts to 100% of the cross-sectional area of the Brillouin zone, whereas that of the closed circuit around Z amounts to 16.6%.

Despite the fact that the 3D crystal structure of κ-(BDH-TTP)$_4$ReF$_6$·4.8H$_2$O is different from that of κ-(BDH-TTP)$_4$[ReF$_6$], the donor layers are structurally very similar. In particular, they contain only one type of BDH-TTP dimer built from two identical donors. The calculated β$_{|HOMO-HOMO|}$ interactions energies for this salt are: 0.3536 (I), 0.0797(II) 0.1363 (III) and 0.2392 (IV) eV. These values are almost identical to those for κ-(BDH-TTP)$_4$ReF$_6$, thus confirming the small influence of the anion shape over the internal structure of the

BDH-TTP κ-type layers. This is in contrast to the situation for the κ- phases of BEDT-TTF and may be a useful guiding principle in the search for multifunctional materials. As expected from the strong similarity in HOMO ... HOMO interactions, the electronic structure of the two salts is completely equivalent. For instance, the Fermi surface calculated for the donor layers of κ-(BDH-TTP)$_4$ReF$_6$·4.8H$_2$O is shown in Figure 7c. The area of the large circles of Figure 7c amounts to 100% of the cross sectional area of the Brillouin zone, whereas that of the closed circuit around Z amounts to 17% (to be compared with 16.6% for κ-(BDH-TTP)$_4$ReF$_6$). Thus, our study suggests that both salts could exhibit Shubnikov-de Haas oscillations of the magnetoresistance with a frequency corresponding approximately to 17% of the cross-section area of the Brillouin zone. In short, despite the different crystal structures, κ-(BDH-TTP)$_4$ReF$_6$ and κ-(BDH-TTP)$_4$ReF$_6$·4.8H$_2$O are almost indistinguishable.

2.3.2. Electronic Structure of the Pseudo-κ"-(BDH-TTP)$_3$(ReO$_4$)$_2$ (3) Salt

The donor layers of this salt (see Figure 8) contain both dimeric units and single donors with an orthogonal orientation.

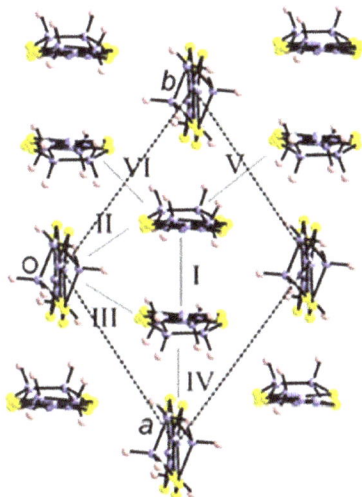

Figure 8. Donor layer of pseudo-κ"-(BDH-TTP)$_3$(ReO$_4$)$_2$ (3) where the different intermolecular interactions are labelled.

The dimeric units form an oblique lattice in the holes of which the single donors reside. The layer contains six different HOMO ... HOMO interactions of three different types: (i) an intradimer interaction (I), (ii) three different interactions between one donor of the dimer and the single donors (II-IV), and (iii) two different interactions between donors of the dimeric units (V-VI). The calculated HOMO ... HOMO interaction energies are: 0.7480 eV (I), 0.1190 eV (II), 0.1992 eV (III), 0.0585 eV (IV), 0.1260 eV (V) and 0.2431 eV (VI). The intra-dimer interaction is thus very strong and it will lead to bonding and antibonding combinations of the two HOMOs separated by a large energy gap. Because of the 3:2 stoichiometry, only one of the two levels will be filled since the HOMO of the single donor will be located within this energy gap. Thus, the single donor must be considered as neutral and the two donors of the dimer as positively charged despite the apparent similarity in C=C bond lengths. Note, however, that the C=C distances are shorter in the single donor (1.345/1.362/1.345 Å in the single donor versus 1.365/1.374/1.365 Å in the dimers), and this fact is consistent with the presence of both neutral and singly charge donor molecules.

The interactions between the two different types of molecule (interactions II to IV) are smaller but quite substantial because several S ... S contacts are relatively short.

Finally, the HOMO ... HOMO interactions between donors of the dimers (V-VI) are calculated to be very substantial, especially for the interactions approximately along the *a* direction. The calculated band structure is shown in Figure 9. As expected, the upper and lower bands are built from the antibonding and bonding combinations of the two HOMOs of the dimeric units. In between the two bands there is a third band based on the HOMO of the single molecule. This is the highest occupied band of the system which is considerably less dispersive. Note that the upper, empty band in Figure 9 exhibits a quite substantial energy dispersion of ~0.4 eV, although the dispersion along the *a*-direction is approximately twice as large as that along the *b*-direction, in agreement with the analysis of the different interactions.

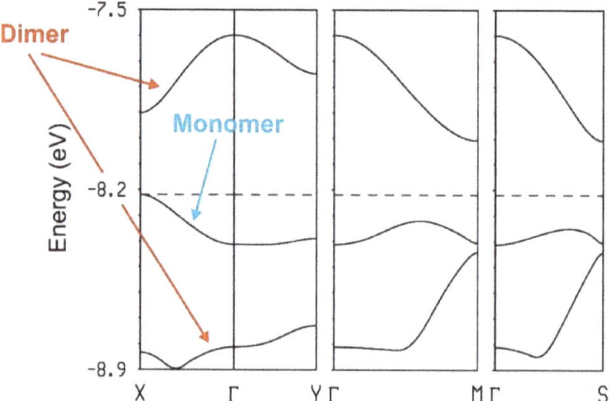

Figure 9. Calculated band structure for the donor layers of pseudo-κ''-(BDH-TTP)$_3$(ReO$_4$)$_2$ (**3**) crystals. The dashed line refers to highest occupied level and Γ = (0, 0), X = (a^*/2, 0), Y = (0, b^*/2), M = (a^*/2, b^*/2) and S = ($-a^*$/2, b^*/2).

Thus, this salt must be an anisotropic semiconductor with an indirect gap from X to M and better conductivity along a direction between *a* and *a* + *b*, since both holes and electrons have smaller effective masses along these directions. The calculated activation energy is 105 meV.

2.4. Conductivity and Magnetic Properties

2.4.1. Conducting Properties of the κ-(BDH-TTP)$_4$ReF$_6$ (**1**) and κ-(BDH-TTP)$_4$ReF$_6$·4.8H$_2$O (**2**) Crystals

Crystals of **1** and **2** were first characterized by resistance measurements at ambient pressure and pressure up to 10 kbar. At room temperature, the ambient pressure conductivity of crystals of **1** and **2** is 3–5 and 8–10 Ohm^{-1} cm^{-1}, respectively. Figure 10a,b shows the temperature dependences of the resistivity of the crystal of **1** measured both in the conducting plane of the BDH-TTP radical cation layers and in the perpendicular direction. As one can see from Figure 10, in both directions at ambient pressure and, under the pressure, the sample shows a metallic behavior of the resistivity with a significant drop at 300–50 °C and a weak growth of the resistivity at lower temperatures, probably due to the localization of carriers resulting from the disorder in the crystal structure of the salt.

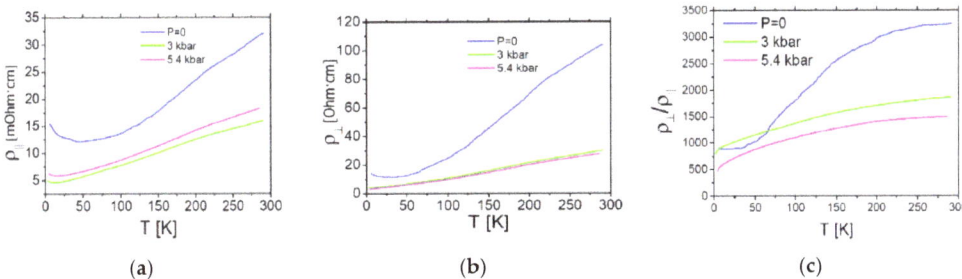

Figure 10. (a) Temperature dependence of the resistivity of the κ-(BDH-TTP)$_4$ReF$_6$ sample (**1**) measured in the conducting plane and (**b**) perpendicular to it at different pressures; (**c**) The temperature dependence of the resistivity anisotropy of (**1**).

Under external pressures of 3 and 5 kbar, the crystal's resistivity also exhibits a metallic behavior but no transition toward a superconducting state was observed. Interestingly, the growth of the conductivity with decreasing temperature is stronger in the direction perpendicular to the conducting layers. The dependence of the resistivity anisotropy with temperature and pressure is presented in Figure 10c. The room-temperature anisotropy of the resistivity at ambient pressure is approximately 3300. When the temperature drops to 4.2 K, it falls about 3.5 times. When external pressure is applied (in the range from 3 to 5.4 kbar), the anisotropy decreases but depends weakly on the pressure, while a decrease in the temperature from 300 to 4.2 K contributes to its decrease by about 2.3 times.

The crystal of **2** shows a metallic behavior in the 300–4.2 K temperature range at ambient pressure as well as under 1 kbar along the conducting plane (see Figure 11). At ambient pressure, the sample resistance is significantly reduced with the temperature. Under the pressure P = 1 kbar a monotonous and considerable drop in the resistance is observed in the 300–50 K temperature range. However, with a further decrease in the temperature, the drop of the resistance flattens. The metallic character of the conductivity of salts **1** and **2** correlates well with the calculations of their electronic structures (see Section 2.3).

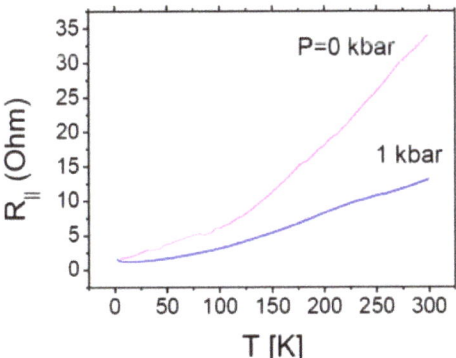

Figure 11. The temperature and pressure dependencies of the resistance measured in the conducting *bc* plane at P = 0 kbar and 1 kbar of the κ-(BDH-TTP)$_4$ReF$_6$·4.8H$_2$O (**2**) crystals.

2.4.2. Conductivity of the Pseudo-κ″-(BDH-TTP)$_3$(ReO$_4$)$_2$ (3) Crystals

The room temperature conductivity of crystals of **3** measured in the plane of the conducting layers is ≈1–1.5 Ohm^{-1} cm^{-1}. With the temperature decrease (Figure 12), the sample shows an exponential dependence of the resistivity with an activation energy equal to 0.11 eV, which correlates well with the value of the activation energy obtained from calculations of the electronic band structure of this salt. It should be noted that the sample conductivity value is quite high in comparison with that of typical organic semiconductors.

Looking at the band structure of **3** (Figure 9), it is clear that both the top of the middle band (occupied and centered on the single molecules) and, especially, the bottom of the upper band (empty and centered on the dimers), exhibit a significant energy variation, that is, a relatively low effective mass. Thus, both the electron and hole carriers have small effective masses due to the strong intermolecular interactions and are likely responsible for the high conductivity even if the salt exhibits an activated conductivity.

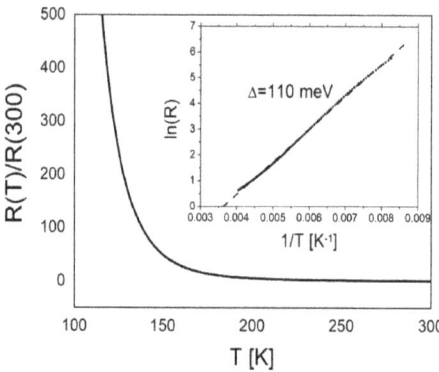

Figure 12. The temperature and pressure dependence of the resistance of the pseudo-κ″-(BDH-TTP)$_3$(ReO$_4$)$_2$ (**3**) salt measured in the conducting *ab* plane.

2.4.3. *ac*- Magnetic Properties of the κ-(BDH-TTP)$_4$[ReF$_6$] Salt (**1**)

In order to test whether the anion of [ReF$_6$]$^{2-}$ as a counterion in the radical cation salt κ-(BDH-TTP)$_4$[ReF$_6$] retains the properties of single ion magnet (SIM), we have investigated the ac magnetic susceptibility of **1** at 2 K under zero applied *dc* field and different *dc* fields (0–0.6 Tesla). Within the frequency range 10–10,000 Hz, compound **1** does not show a signal for out-of-phase magnetic susceptibility (χ'') under zero applied *dc* field. The maximum of χ'' is not observed on either the field dependence of χ'' at a frequency of 100 Hz or on the frequency dependence of χ'' in a field of 0.4 Tesla; the χ'' signal hovers around 0 (Figure S8, Supplementary Materials). The presence of maxima in the frequency or field dependences of χ'' indicates a slow relaxation of the magnetization, which is one of the distinctive features of SMMs [7–11]. Thus, the study of the *ac* susceptibility of salt **1** shows that this compound is not an SMM, unlike the parent complex (PPh$_4$)$_2$[ReF$_6$]·2H$_2$O [15] and the organic metal (BEDO)$_4$[ReF$_6$]·6H$_2$O [16]. The loss of SIM properties by the anion [ReF$_6$]$^{2-}$ in the composition of salt **1** is possibly associated with the presence of greater disorder in the structure of the anionic layers in crystals of **1**.

3. Materials and Methods

All solvents were obtained, chemically clean or extremely clean, from Merk-Sigma-Aldrich and have been used without further purification. The donor BDH-TTP was synthesized according to the method described in [21] and was recrystallized from CS$_2$ (carbon disulfide). The electrolyte (PPh$_4$)$_2$[ReF$_6$]·2H$_2$O was prepared according to the procedure described in [15].

3.1. Synthesis of the Salts

3.1.1. Synthesis of the Crystals κ-(BDH-TTP)$_4$ReF$_6$ (**1**)

The rhombus-shaped crystals of **1** were prepared under argon atmosphere by electrochemical oxidation of BDH-TTP (C$_1$ = 8 mg, 0.021·10^{-3} mmol) in the presence of the supporting electrolyte (PPh$_4$)$_2$[ReF$_6$]·2H$_2$O (C$_2$ = 21 mg, 0.028 mmol) in a chlorobenzene (CB, 18 mL)–trifluoroethanol (2 mL) mixture. The reaction was performed at constant current, I = 1.25 µA, at 25 K. The synthesis was carried out in an H-shaped two-electrode glass cell with cathodic and anodic chambers separated by a porous glass membrane. The electrodes were 1 mm diameter platinum wires, electrochemically purified in a 0.1 N sulfuric acid solution. The crystals of **1** formed on the anode within 7–10 days. Electron-probe X-ray microanalysis (EPMA) showed a ratio of S: Re: F atoms to be $\cong$ 33.1:1:6.2.

3.1.2. Synthesis of the Crystals κ-(BDH-TTP)$_4$ReF$_6$·4.8H$_2$O (**2**) and Pseudo-κ''-(BDH-TTP)$_3$(ReO$_4$)$_2$ (**3**)

The rhombic crystals of **2** and plate-like crystals of **3** were prepared under argon atmosphere by electrochemical oxidation of BDH-TTP. Unlike the synthesis procedure for crystals **1**, a mixture of chlorobenzene (CB, 18 mL)–96% alcohol (2 mL) was used as the reaction medium. The crystals of **2** and **3** were formed simultaneously on the anode within three weeks.

A preliminary analysis of the crystal compositions determined by electron-probe X-ray microanalysis (EPMA) showed a ratio of S:Re:F and S:Re:O atoms $\cong$ 33.9:1:6.7 and $\cong$ 23.5:1:3.7 for rhombic and plate-like crystals, respectively. The final composition of crystals of **2** and **3** was established from complete X-ray diffraction analysis.

3.2. Electron-Probe X-ray Microanalysis

Preliminary composition of the salts was determined with the electron-probe X-ray microanalysis (EPMA) on a JEOL JSM-5800L scanning electron microscope (SEM) at 100-fold magnification and 20 keV electron beam density. The depth of beam penetration into the sample was 1–3 μm.

3.3. Single Crystal X-ray Analysis

X-ray diffraction analyses of the salts **1–3** were carried out on a CCD Agilent XCalibur diffractometer with an EOS detector (Agilent Technologies UK Ltd., Yarnton, Oxfordshire, England). Data collection, determination and refinement of unit cell parameters were carried out using the CrysAlis PRO program suite [26]. X-ray diffraction data at 150(1) K or 100(1) K for the salts **1–2** and **3**, respectively, were collected using MoKα (λ = 0.71073 Å) radiation.

The structure (**1**) was solved by the direct methods. The positions and thermal parameters of non-hydrogen atoms were refined isotropically and then anisotropically by the full-matrix least-squares method. At the first stages, the atomic population of the [ReF$_6$]$^{2-}$ anion was refined; at latter stages it was fixed. The positions of the hydrogen atoms were calculated geometrically.

The structure (**2**) was solved with the direct methods. The positions and thermal parameters of non-hydrogen atoms were refined isotropically and then anisotropically by the full-matrix least-squares method. The anion was found to be disordered over two positions; the refinement was carried out, imposing restrictions on bond lengths and thermal parameters. In difference syntheses of the electron density near the Re positions, peaks of the electron density were revealed; they were taken as disordered water atoms. Taking these peaks into account made it possible to improve the refinement of the structure by about 3%. In the first stages, the population of the anion and water molecules was refined and during the final refinement it was fixed. The positions of the hydrogen atoms were calculated geometrically.

The structure of **3** was solved by the direct methods. The positions and thermal parameters of non-hydrogen atoms were refined isotropically and then anisotropically by the full-matrix least-squares method. Oxygen atoms in the tetrahedron, disordered over two positions, were refined with restrictions imposed on bond lengths, position populations and thermal parameters.

The X-ray crystal structures data have been deposited with the Cambridge Crystallographic Data Center, with reference codes CCDC 2069204–2069206. Selected crystallographic parameters and the data collection and refinement statistics are given in Table S1. All calculations were performed with the SHELX-97 program package [27].

3.4. Band Structure Analysis

The tight-binding band structure calculations were of the extended Hückel type [28]. A modified Wolfsberg-Helmholtz formula was used to calculate the non-diagonal $H_{\mu\nu}$ values [29]. All valence electrons were taken into account in the calculations and the basis set consisted of Slater-type orbitals of double-ζ quality for C 2s and 2p, S 3s and 3p, and of single-ζ quality for H. The ionization potentials, contraction coefficients and exponents were taken from previous work [30]. The possible role of the disorder of the terminal ethylene groups was checked by carrying out calculations with different combinations of the major and minor configurations. As it is generally found, there was no effect on the results because the HOMO does not exhibit a noticeable contribution of these terminal ethylene groups.

3.5. Conducting Properties

Resistivity measurements were carried out using a four-probe technique and a lock-in amplifier at 20 Hz alternating current. The samples were thin plates with a characteristic lateral size of about 0.4–1 mm and the thickness was in the range 15–50 µm. The surface of the plate was oriented along the conducting layers, that is, parallel to the (*bc*) plane. Depending on the size and the shape of the samples, we could measure either in-plane resistance or resistances in both in-plane and out-of-plane directions. The first case concerns the samples of crystals of **2** and **3**, which were thin and comparatively long plates, so we could make four contacts along one of the sample's surfaces and measure the in-plane resistance. As for the samples of crystals of **1**, we could make two contacts, attached to each of two opposite sample surfaces and measured both in-plane $R_{||}$ and out-of-plane $R_{\perp}$ resistances and calculated the resistivity using the modified Montgomery method [31]. The contacts were made using the conducting graphite paste. The measurements in the 1.3–300 K temperature range were carried out in a ^{4}He cryostat with a variable temperature insert. To create an external pressure of up to 5.4 kbar, the samples were subjected to quasi-hydrostatic pressure using a Cu-Be clamp-cell with silicone oil as a pressurized medium and a manganin sensor for pressure control.

3.6. Magnetic Properties

Alternating-current (*ac*) magnetic susceptibility measurements were performed on the polycrystalline sample of salt **1** using a PPMS magnetometer (Quantum Design) under an *ac* field of 3 Oe. The data were collected in a zero *dc* field and different applied *dc* fields.

4. Conclusions

Three new radical cation salts: κ-(BDH-TTP)$_4$ReF$_6$ (**1**), κ-(BDH-TTP)$_4$ReF$_6$·4.8H$_2$O (**2**) and pseudo-κ″-(BDH-TTP)$_3$[ReO$_4$]$_2$ (**3**) have been obtained with original electrocrystallization methods using various mixtures of solvents and electrolyte (PPh)$_2$[ReF$_6$]·2H$_2$O that is field-induced SIM. The addition of rectified (96%) ethanol and trifluoroethanol to chlorobenzene as the main solvent is critical for the formation of these salts. Crystals of salt **1** are formed if trifluoroethanol is used, while the addition of 96% ethanol leads to partial oxidation of paramagnetic Re(IV) to diamagnetic Re(VII) during the electrocrystallization, resulting in the formation of salts **2** and **3**.

The radical cation salts are characterized by the alternation of layers of organic radical cations with layers composed of isolated anions [ReF$_6$]$^{2-}$, [ReF$_6$]$^{2-}$·4.8H$_2$O and [ReO$_4$]$^{-}$ for salts **1**, **2** and **3**, respectively. In compounds **1** and **2**, the radical cation layers have the κ-type of molecular packing formed by BDH-TTP equivalent dimers, while in **3**, the organic layers consist of radical cation dimers (BDH-TTP)$_2^{2+}$ and single molecules BDH-TTP (a new type of pseudo-κ″-packing). A distinctive feature of these salts is the occurrence of disorder in the anionic layers and incomplete population of the Re anions in the crystalline lattice.

The study of the temperature dependences of the conductivity for salts **1–3** shows that **1** and **2** are stable molecular metals, and **3** is a semiconductor with a conductivity activation energy of 110 meV. Calculations of the electronic band structures of the salts correlate well with the data on conductivity measurements. The Fermi surface of **1** and **2** contains closed circuits with areas of 100% and approximately 17% of the cross section of the Brillouin zone. The band structure of **3** clearly shows that this salt should be a semiconductor with high enough conductivity and that in the salt crystal structure the single donors must be considered as neutral molecules and the dimers as $(BDH-TTP)_2^{2+}$.

Analysis of the *ac* magnetic susceptibility of salt **1** shows that this compound does not exhibits the properties of SMMs, which could be expected from the presence of the $[ReF_6]^{2-}$ anion in the composition of salt. Perhaps this is due to the presence of significant disorder in the structure of the anionic layers in crystals **1**.

Supplementary Materials: The following are available online at https://www.mdpi.com/article/10.3390/magnetochemistry7040054/s1, Figure S1: Asymmetric unit and the designations of the radical cation and anion in the crystals κ-(BDH-TTP)$_4$ReF$_6$ salt (**1**); Figure S2. Mutual arrangement of the radical cations in the dimer and the designations of central C=C bond in the molecules κ-(BDH-TTP)$_4$ReF$_6$ salt (**1**); Figure S3. Asymmetric unit and the designations of the radical cation and anion in the crystals κ-(BDH-TTP)$_4$ReF$_6$·4.8H$_2$O (**2**); Figure S4. Mutual arrangement of the radical cations in the dimer and the designations of central C=C bond in the molecules κ-(BDH-TTP)$_4$ReF$_6$·4.8H$_2$O salt (**2**); Figure S5. Asymmetric unit and the designations of the radical cations and anion in the crystals pseudo-κ''-(BDH-TTP)$_3$(ReO$_4$)$_2$ (**3**); Figure S6. Mutual arrangement of the radical cations in the dimer and the designations of central C=C bond in the molecules of the pseudo-κ''-(BDH-TTP)$_3$([ReO$_4$)$_2$ salt (**3**). Dashed lines show shortened contacts designated as S2 . . . S6 and S3 . . . S7 between the radical cation in the dimer; Figure S7. The short contacts between the radical cation inside the dimer and between the dimer and single BDH-TTP molecules; Figure S8. Frequency dependences of the out-of-phase (χ'') ac susceptibility for κ -(BDH-TTP)$_4$ReF$_6$ salt (**1**) at temperature of 2.0 K in a dc field of 0.4 Tesla; Table S1. Crystallographic data and refined structural parameters for the crystals **1–3**; Table S2. Bond length of Re=O in a tetrahedron $[ReO_4]^-$.

Author Contributions: Data curation, N.D.K., G.V.S., L.I.B., V.N.Z., E.C. and J.-i.Y.; Formal analysis, N.D.K., G.V.S., L.I.B., V.N.Z., E.C. and J.-i.Y.; Investigation, G.V.S., L.I.B., E.B.Y., V.N.Z., E.C. and J.-i.Y.; Project administration, E.B.Y.; Writing—original draft, N.D.K., G.V.S., L.I.B. and E.B.Y.; Writing—review & editing, G.V.S., L.I.B. and E.B.Y. All authors have read and agreed to the published version of the manuscript.

Funding: This research was funded by the Ministry of Science and Higher Education of the Russian Federation (Grant No. 075-15-2020-779). Work in Spain was supported by MICIU (through the Severo Ochoa FUNFUTURE (CEX2019-000917-S) Excellence Centre distinction and Grant PGC 2018-096955-B-C44), and by Generalitat de Catalunya (2017SGR1506).

Acknowledgments: The authors would like to thank O.V. Maximova for *ac* magnetic measurements.

Conflicts of Interest: The authors declare no conflict of interest. The funders had no role in the design of the study; in the collection, analyses, or interpretation of data; in the writing of the manuscript, or in the decision to publish the results.

References

1. Ouahab, L. *Multifunctional Molecular Materials*, 1st ed.; Pan Stanford Publishing Pte. Ltd.: Singapore, 2013.
2. Coronado, E.; Day, P. Magnetic Molecular Conductors. *Chem. Rev.* **2004**, *104*, 5419–5449. [CrossRef] [PubMed]
3. Kobayashi, H.; Cui, H.; Kobayashi, A. Organic Metals and Superconductors Based on BETS (BETS = Bis(ethylenedithio)-tetraselenafulvalene. *Chem. Rev.* **2004**, *104*, 5265–5288. [CrossRef] [PubMed]
4. Kushch, N.D.; Yagubskii, E.B.; Kartsovnik, M.V.; Buravov, L.I.; Dubrovskii, A.D.; Chekhlov, A.N.; Biberacher, W. π-donor BETS Based Bifunctional Superconductor with Polymeric Dicyanamidomanganate(II) Anion Layer: κ-(BETS)$_2$Mn[N(CN)$_2$]$_3$. *J. Am. Chem. Soc.* **2008**, *130*, 7238–7240. [CrossRef] [PubMed]
5. Prokhorova, T.G.; Yagubskii, E.B. Organic Conductors and Superconductors Based on Bis(ethylenedithio)tetrathia-fulvalene Radical Cation Salts with Supramolecular Tris(oxalato)metallate Anions. *Russ. Chem. Rev.* **2017**, *86*, 164–180. [CrossRef]
6. Cosquer, G.; Shen, Y.; Almeida, M.; Yamashita, M. Conducting Single-molecule Magnet Materials. *Dalton Trans.* **2018**, *47*, 7616–7627. [CrossRef]
7. Gatteschi, D.; Sessoli, R.; Villain, J. *Molecular Nanomagnets, Mesoscopic Physics and Nanotechnology*; Oxford University Press: Oxford, UK, 2006.

8. Milios, C.; Winpenny, R.E.P. Cluster-Based Single-Molecule Magnets. In *Structure and Bonding*; Gao, S., Ed.; Springer: Berlin/Heildelberg, Germany, 2015; Volume 164, pp. 1–109.
9. Bogani, L.; Wernsdorfer, W. Molecular Spintronics Using Single-molecule Magnets. *Nature Mater.* **2008**, *7*, 179–186. [CrossRef]
10. Aromi, G.; Aguilà, D.; Gamez, P.; Luis, F.; Roubeau, O. Molecules as Prototypes for Spin-Based CNOT and SWAP Quantum Gates. *Chem. Soc. Rev.* **2012**, *41*, 537–546.
11. Bartolomé, J.; Luis, F.; Fernandez, J. *Molecular Magnets–Physics and Applications*; Springer: Berlin/Heidelberg, Germany, 2014.
12. Kubo, K.; Hiraga, H.; Miyasaka, H.; Yamashita, M. Multifunctional Single-molecule Magnets with Electrical Conductivity. In *Multifunctional Molecular Materials*; Ouahab, L., Ed.; Pan Stanford Publishing Pte. Ltd.: Singapore, 2013; pp. 61–103.
13. Zhang, X.; Xie, H.; Ballesteros-Rivas, M.; Woods, T.J.; Dunbar, K.R. Conducting Molecular Nanomagnet of Dy(III) with Partially Charged TCNQ Radicals. *Chem. Eur. J.* **2017**, *23*, 7448–7452. [CrossRef]
14. Shen, Y.; Cosquer, G.B.; Breedlove, K.; Yamashita, M. Hybrid Molecular Compound Exhibiting Slow Magnetic Relaxation and Electrical Conductivity. *Magnetochemistry* **2016**, *2*, 44. [CrossRef]
15. Pedersen, K.S.; Sigrist, M.; Sorensen, M.A.; Barra, A.-L.; Weyhermuller, T.; Piligkos, S.; Thuesen, C.A.; Vinum, M.G.; Mutka, H.; Weihe, H.; et al. [ReF$_6$]$^{2-}$: A Robust Module for the Design of Molecule-Based Magnetic Materials. *Angew. Chem. Int. Ed.* **2014**, *53*, 1351–1354. [CrossRef]
16. Kushch, N.D.; Buravov, L.I.; Kushch, P.P.; Shilov, G.V.; Yamochi, H.; Ishikawa, M.; Otsuka, A.; Shakin, A.; Maximova, O.V.; Volkova, O.S.; et al. The Multifunctional Compound Combining Conductivity and Single-Molecule Magnetism in the Same Temperature Range. *Inorg. Chem.* **2018**, *57*, 2386–2389. [CrossRef]
17. Shen, Y.; Ito, H.; Zhang, H.; Yamochi, H.; Katagiri, S.; Yoshina, S.K.; Otsuka, A.; Ishikawa, M.; Cosquer, G.; Uchida, K.; et al. Simultaneous Manifestations of Metallic Conductivity and Single-Molecule Magnetism in a Layered Molecule-based Compound. *Chem. Sci.* **2020**, *11*, 11154–11161. [CrossRef]
18. Yamada, J.-I.; Akutsu, H.; Nishikawa, H.; Kikuchi, K. New Trends in the Synthesis of π-Electron Donors for Molecular Conductors and Superconductors. *Chem. Rev.* **2004**, *104*, 5057–5083. [CrossRef] [PubMed]
19. Bardin, A.; Akutsu, H.; Yamada, J.-I. New Family of Six Stable Metals with a Nearly Isotropic Triangular Lattice of Organic Radical Cations and Diluted Paramagnetic System of Anions: κ(κ$_\perp$)-(BDH-TTP)$_4$MX$_4$·Solv, Where M = CoII, MnII.; X = Cl, Br and Solv = (H$_2$O)$_5$, (CH$_2$X). *Cryst. Growth Des.* **2016**, *16*, 1228–1246. [CrossRef]
20. Williams, J.M.; Ferraro, R.J.; Thorn, R.J.; Carlson, K.D.; Geiser, U.; Wang, H.H.; Kini, A.M.; Whangbo, M.-H. *Organic Superconductors (Including Fullerenes)*; Prentice Hall, Englewood Cliffs: New Jersey, NJ, USA, 1992.
21. Yamada, J.; Watanabe, M.; Anzai, H.; Nishikawa, H.; Ikemoto, I.; Kikuchi, K. BDH-TTP as a Structural Isomer of BEDT-TTF, and Its Two-Dimensional Hexafluorophosphate Salt. *Angew. Chem. Int. Ed.* **1999**, *38*, 810–813. [CrossRef]
22. Kurmoo, M.; Graham, A.W.; Day, P.; Coles, S.J.; Hursthouse, M.B.; Caulfield, J.L.; Singleton, J.; Pratt, F.L.; Hayes, W.; Ducasse, L.; et al. Superconducting and Semiconducting Magnetic Charge Transfer Salts: (BEDT-TTF)$_4$AFe(C$_2$O$_4$)$_3$·C$_6$H$_5$CN (A = H$_2$O, K, NH$_4$). *J. Am. Chem. Soc.* **1995**, *49*, 12209–12217. [CrossRef]
23. Whangbo, M.H.; Williams, J.M.; Leung, P.C.W.; Beno, M.A.; Emge, T.J.; Wang, H.H. Role of the Intermolecular Interactions in the Two-Dimensional Ambient-Pressure Organic Superconductors β-(ET)$_2$I$_3$ and β-(ET)$_2$IBr$_2$. *Inorg. Chem.* **1985**, *24*, 3500–3502. [CrossRef]
24. Shevyakova, I.; Buravov, L.; Tkacheva, V.; Zorina, L.; Khasanov, S.; Simonov, S.; Yamada, J.; Canadell, E.; Shibaeva, R.; Yagubskii, E. New Organic Metals Based on BDH-TTP Radical Cation Salts with the Photochromic Nitroprusside Anion [FeNO(CN)$_5$]$^{2-}$. *Adv. Funct. Mater.* **2004**, *14*, 660–668. [CrossRef]
25. Kushch, N.D.; Kazakova, A.V.; Buravov, L.I.; Yagubskii, E.B.; Simonov, S.V.; Zorina, L.V.; Khasanov, S.S.; Shibaeva, R.P.; Canadell, E.; Son, H.; et al. The First BDH-TTP Radical Cation Salts with Mercuric Counterions, κ-(BDH-TTP)$_4$[Hg(SCN)$_4$]C$_6$H$_5$NO$_2$ and κ-(BDH-TTP)$_6$[Hg(SCN)$_3$][Hg(SCN)$_4$]. *Synth. Met.* **2005**, *155*, 588–594. [CrossRef]
26. Agilent. *CrysAlis PRO Version171.35.19*; Agilent Technologies UK Ltd.: Yarnton, Oxfordshire, UK, 2011.
27. Sheldrick, G.M. *SHELXL-97. Program for Crystal Structure Refinement*; University of Göttingen: Göttingen, Germany, 1997.
28. Whangbo, M.H.; Hoffmann, R. The Band Structure of the Tetracyanoplatinate Chain. *J. Am. Chem. Soc.* **1978**, *100*, 6093–6098. [CrossRef]
29. Ammeter, J.H.; Buergi, H.B.; Thibeault, J.C.; Hoffmann, R. Counterintuitive Orbital Mixing in Semiempirical and ab initio Molecular Orbital Calculations. *J. Am. Chem. Soc.* **1978**, *100*, 3686–3692. [CrossRef]
30. Penicaud, A.; Boubekeur, K.; Batail, P.; Canadell, E.; Auban-Senzier, P.; Jerome, D. Hydrogen-bond Tuning of Macroscopic Transport Properties from the Neutral Molecular Component Site along the Series of Metallic Organic-inorganic Solvates (BEDT-TTF)$_4$Re$_6$Se$_5$C$_{19}$·[G], [Gt = DMF, THF, dioxane]. *J. Am. Chem. Soc.* **1993**, *115*, 4101–4112. [CrossRef]
31. Buravov, L.I. Calculation of Resistance Anisotropy with Allowance for the Ends of the Sample with the Help of a Conformal Transformation. *Sov. Phys. Tech. Phys.* **1989**, *34*, 464–469.

MDPI
St. Alban-Anlage 66
4052 Basel
Switzerland
Tel. +41 61 683 77 34
Fax +41 61 302 89 18
www.mdpi.com

Magnetochemistry Editorial Office
E-mail: magnetochemistry@mdpi.com
www.mdpi.com/journal/magnetochemistry

www.ingramcontent.com/pod-product-compliance
Lightning Source LLC
LaVergne TN
LVHW070211100526
838202LV00015B/2034